Birdsong, Speech, and Language

BIRDSONG, SPEECH, AND LANGUAGE

Exploring the Evolution of Mind and Brain

edited by Johan J. Bolhuis and Martin Everaert

foreword by Robert C. Berwick and Noam Chomsky

The MIT Press
Cambridge, Massachusetts
London, England

MIT Press books may be purchased at special quantity discounts for business or sales promotional use. For information, please email special_sales@mitpress.mit.edu or write to Special Sales Department, The MIT Press, 55 Hayward Street, Cambridge, MA 02142.

This book was set in Syntax and Times Roman by Toppan Best-set Premedia Limited, Hong Kong. Printed and bound in the United States of America.

Library of Congress Cataloging-in-Publication Data

Birdsong, speech, and language: exploring the evolution of mind and brain / edited by Johan J. Bolhuis and Martin Everaert; foreword by Robert C. Berwick and Noam Chomsky.
    p. cm.
Includes bibliographical references and index.
ISBN 978-0-262-01860-9 (alk. paper)
1. Language acquisition.   2. Birds—Vocalization.   3. Birdsongs.   4. Speech acts (Linguistics)   5. Human evolution.   6. Cognitive neuroscience.   7. Neurolinguistics.   I. Bolhuis, Johan J.   II. Everaert, Martin.
P118.B56 2013
401'.93—dc23
2012028611

10 9 8 7 6 5 4 3 2 1

Contents

Foreword: A Bird's-Eye View of Human Language and Evolution

Robert C. Berwick and Noam Chomsky

Scholars have long been captivated by the parallels between birdsong and human speech and language. Over two thousand years ago, Aristotle had already observed in his *Historia Animalium* (about 350 BCE) that some songbirds, like children, acquire sophisticated, patterned vocalizations, "articulated voice," in part from listening to adult "tutors" but also in part via prior predisposition: "Some of the small birds do not utter the same voice as their parents when they sing, if they are reared away from home and hear other birds singing. A nightingale has already been observed teaching its chick, suggesting that [birdsong] . . . is receptive to training" (*Hist. Anim.* 1970, 504a35–504b3; 536b, 14–20). Here Aristotle uses the Greek word *dialektos* to refer to song variation, paralleling human speech, and even anticipates recent work on how the songs of isolated juvenile vocal learning birds might "drift" from those of their parents over successive generations. Given two millennia of progress from neuroscience to genomics, we might expect that our insights regarding the parallels between birdsong and human language have advanced since Aristotle's day. But how much have we learned? That is the aim of this book: What can birdsong tell us today about the biology of human speech and language?

From an evolutionary standpoint, birds are particularly well placed to probe certain biolinguistic questions. The last common ancestor of birds and mammals (the clade *Amniotes*) lived about 310–330 million years ago, so 600 million years of evolutionary time in all separates humans from *Aves*, 300 million years from this common ancestor to humans, plus 300 million years from this ancestor to birds. This gulf of more than half a billion years provides an opportunity to resolve certain vexing questions about the adaptive significance of particular biological traits, because given such a large gap of evolutionary time, analogous "solutions" are more likely to have arisen as a result of independent, convergent evolution, rather than by shared descent from a common ancestor—the classic example being the independent development of wings in bats and birds (Stearns & Hoekstra, 2005). Since the last common ancestor of birds and bats did not have wings, we can more readily conclude that these distinct "solutions" arose independently as adaptive solutions to the same common functional problem of flying. Paradoxically, if two species are extremely closely related—humans and chimpanzees—it can be much more challenging to sort out which traits are due to shared ancestry (i.e., homology) and which are true functional adaptations. It is thus crucial to explore in depth the extent to which the many parallels between human speech and birdsong, ranging from vocal learning, to vocal imitation and vocal production, to analogous brain regions and neural pathways in both songbirds and humans, might best be thought of as the result of converging mechanisms. From this vantage point, on balance it would seem that birdsong is most comparable to the mechanisms of human speech, not language in the broad sense, with both solving the common problem of "externalizing" some internal representation as a set of serially ordered motor commands to distinct vocal "output machines."

On the other hand, one should not be too hasty in dismissing the possibility of shared ancestry and the insights it might provide into language. For example, though bird wings and bat wings may have arisen independently, both feathers and hair share keratin genes derived from some common ancestor of both, and so the "solution" to flying remains a more nuanced interplay between shared ancestry and common descent (Eckharta et al., 2008). Indeed, since the rise of the "evo-devo" revolution, over the past several decades biologists have grown to appreciate that there has been a surprising amount of conservation across species in the tree of life, sometimes revealed only by a deeper look at shared traits at the cellular and molecular levels, including regulatory and ontogenetic effects, sometimes called "deep homology." On this account, it would be no surprise to find much common ground between birdsong and human speech, even down to the level of corresponding brain regions. If this commonality turns out to be correct, it would also be a favorable state of affairs since it would reinforce the possibility of using songbirds as animal models of language, especially speech in certain respects. Perhaps the most famous current example of such a case centers on the gene encoding forkhead-box protein P2 (FoxP2), a highly conserved DNA regulatory factor, which apparently plays a role in guiding normal neuronal development involving both vocal learning and production in humans and songbirds (Fisher & Scharff, 2009; Vernes et al., 2011). How far one can drive this genomic work upward into neuronal assemblies—ultimately, the dissection of the underlying circuitry responsible for vocal production—remains to be seen, but the current "state of play" in this area is covered by several chapters that follow.

In any case, the bridge between birdsong research and speech and language dovetails extremely well with recent developments in certain strands of current linguistic thinking, which aim to identify the assumed species-specific biological substrate for language, so reducing to a minimum any language-specific cognitive traits (Berwick & Chomsky, 2011). This stands in sharp contrast to the earliest attempts at developing explicit rule systems that even began to approach descriptive adequacy in terms of accounting for the properties of human language. The complexity of such rule systems poses a seemingly insurmountable biolinguistic puzzle, because it requires that one assumes substantial, language-particular machinery without any clear path as to how this highly specific cognitive capability might have arisen. Now however, according to some linguists, one can strip away all this complexity, arriving at a system that requires much less in the way of language-particular rules. This system contains just a single operation that combines hierarchical structure into larger representations, along with a storehouse of conceptual "atoms," roughly corresponding to individual words, along with two interface systems, one an external "input-output" system mapping internal representations to speech or manual signs, and the second an internal mapping between these internal representations and the cognitive systems of thought (Hornstein, 2009). If this reduction is on the right track, and some of the chapters in this book address this very point, it would go a long way to resolving what some have called "Darwin's problem"—the biolinguistic question as to the origin of language. Such a "bare-bones" linguistic account would also accord with the view that the capacity for language apparently emerged relatively late and rapidly in evolutionary terms and has not changed substantially since then. Biologically, this points to a common evolutionary scenario: most of the substrate for language must have already been in place, and what we see in the case of language is evolutionary opportunism—the assembly of already-existing abilities into a novel phenotype. For example, during the past few years alone, at least two "input-output" system abilities long thought to be the sole province of humans have been claimed to be attested in other vocal-learning animals: (1) perception of synthetic "auditory caricatures" of spoken words in chimpanzees (Heimbauer, Beran, & Owren, 2011), and (2) rhythmic entrainment to music in birds (Patel, Iversen, Bregman, & Schulz, 2009). To be sure, these abilities focus only on acoustic input, and we do not yet know what role, if any, these abilities play in human speech and language; nothing comes close to human language in other animal species. But by understanding the scope of what other animals can do, the continued exploration of birdsong as pursued in this book can only boost our understanding of how the interface between language and the external world evolved and works, thus improving our focus on that part of language that remains uniquely human.

References

Aristotle. (1970). *Historia Animalium* (A. Peck, Trans.). Cambridge, MA: Loeb Classical Library, Harvard University Press.

Berwick, R., & Chomsky, N. (2011). The biolinguistic program: The current state of its development. In A. Di Sciullo & C. Boeckx (Eds.), *The biolinguistic enterprise* (pp. 19–41). Oxford: Oxford University Press.

Preface

The idea for this book arose out of a workshop and public symposium, "Birdsong, Speech and Language: Converging Mechanisms," which we organized at Utrecht jointly with Peter Hagoort of the Max Planck Institute for Psycholinguistics (Nijmegen) in April 2007. This symposium was the first of its kind, and set the stage for fruitful discussions between birdsong researchers and linguists in the context of behavioral biology that have continued ever since. Reflections on the parallels and differences between birdsong and language are being published with some regularity, and birdsong features prominently at the biannual Evolution of Language conferences (Evolang).

The behavioral parallels between birdsong learning and speech acquisition had been known for some time. Darwin indicated he had noticed the similarity when he quoted the 18th-century vice president of the Royal Society, Daines Barrington, to the effect that "[birdsongs] are no more innate than language is in man" and the first singing attempts "may be compared to the imperfect endeavour in a child to babble" (*The Descent of Man*, 1871, p. 55). After Darwin, the study of birdsong was taken up in earnest in the middle of the 20th century, first by William Thorpe at Cambridge, then in the United States after it was spread there via Thorpe's pupil, Peter Marler. Through the work of Marler and his scientific progeny it became clear that, apart from a transitional "babbling" phase, there are other fascinating parallels between song learning in songbirds and speech acquisition in human infants. In addition, Marler noted that birdsong can have quite a complex structure, which he termed "phonological syntax."

In linguistics the groundbreaking work of Eric Lennenberg in his *Biological Foundations of Language* (1967) led to interest in the development of language, not only in the individual but also in the species. More recently this has culminated in a renewed interest in the biological foundations of human language, explored in the "biolinguistic program" of generative grammar.

We are grateful to the Netherlands Organization for Scientific Research (NWO-ALW) for the generous support that enabled us to organize the 2007 conference, and to Utrecht University for hosting the conference. We also want to thank the symposium participants for their willingness to engage in scientific discussions with colleagues from very different research fields. We are grateful to our referees for critically evaluating the draft chapters, particularly to Robert Berwick, who read almost all the chapters and gave invaluable advice. We are very pleased that he and Noam Chomsky wrote such an excellent foreword to the book. Thanks also to a series of editors at The MIT Press, particularly to James DeWolf and Marc Lowenthal for their continued help and support during the preparation of the manuscript. Finally, we wish to thank the book's contributors for their fine chapters, which we hope will provide the impetus to further research on birdsong, speech, and language.

J.J.B.
M.B.H.E.
Utrecht, February 2012

INTRODUCTION

1   The Design Principles of Natural Language

Martin Everaert and Riny Huybregts

This book explores the cognitive and neural similarities between birdsong and language, and between birdsong learning and the acquisition of speech and language. This chapter is meant to give nonlinguists some information on what linguistics is about, and on how different perspectives on linguistics shape the debate on the relationship between human language and animal communication systems, or animal languages, if one wants to use that notion. As a preview, we also highlight some core notions in other chapters.

The Generative Enterprise

Linguistics investigates the systems underlying language and speech. Linguists seek to develop an understanding of the rules and laws that govern the structure and use of particular languages. They search for the general laws and principles governing all natural languages—that is, the nature of human language in its many guises and the way it fulfills our communicative needs. In other words, linguistics aims at a deeper understanding of the constraints and forces that shape human languages.

Traditionally, linguistics was firmly rooted in the research traditions of the humanities. In the last decades of the 20th century, the study of language moved away from its philological roots, positioning itself more and more as a cognitive science. In this approach, language is an interesting phenomenon because it represents a structured and accessible product of the human mind, a window into human cognitive abilities: language studied as a means of understanding the nature of the mind that produces it. Linguistics thus became part of cognitive science, the study of the higher-level mental functions, including memory, reasoning, visual perception/recognition, and so on. Underlying the cognitive sciences is the conception of human beings—individually as well as collectively—as information processing systems. This conception has fundamentally changed the outlook of linguistics by establishing close connections both with the formal sciences, leading to computational linguistics, and with psychology, resulting in the joint venture of psycholinguistics. The recent efforts to link linguistics to biology and cognitive neuroscience—producing the field of bio-linguistics—hold considerable promise for a further dimension in the understanding of the cognitive faculties involved in language.

Linguists study all aspects of language: from the generation of speech sounds and their acoustic properties, to the role of language in social cohesion, to how language gets processed by the brain and interpreted. To do so, we need a thorough understanding of the architecture of the language system and in-depth knowledge of its main components. There are many approaches to the study of language. We think that Wikipedia (http://en.wikipedia.org/wiki/Language) gives a fair assessment of the different approaches to the study of language that one can distinguish. First, language is the study of the mental faculty that allows humans to learn languages (language acquisition) and produce and understand utterances (psycholinguistics). Second, language is a formal system of signs governed by grammatical rules of combination to communicate meaning. From this perspective language is studied as an independent system (grammatical architecture), describing and showing the interaction between the various parts of that system (phonology/phonetics, morphology/lexicon, syntax, semantics/pragmatics). Third, language is a system of communication that enables humans to cooperate, focusing on the social functions of language (sociolinguistics). Language is primarily taken as a cultural phenomenon, as a tool for social interaction.

Generative grammar is a linguistic tradition concerned with human language taken as a "mental organ," a language faculty, and therefore naturally assumes that the process of language acquisition falls within biologically determined cognitive capacities. As such it studies language from the standpoint of the first and second definitions, simultaneously, above. This language faculty is also called "Universal Grammar"—the system of principles, conditions, and rules that are elements or properties of all human languages, not merely by accident, but by biological necessity. This language faculty is at work in the acquisition of the knowledge of language. Triggered by appropriate and continuing experience, this innate endowment creates a specific grammar, the grammar of L, put to use for the interaction with others, in speaking and understanding language. Generative grammar is interested in the "logical problem" of language acquisition—how children move from a state of having no knowledge of a language to a state having full knowledge, given limited input, and in a relatively short time.[1] In his influential work, *Lectures on Government and Binding*, Noam Chomsky (1981, pp. 224, 231) formulates it as follows: "One proposal, which I think is basically correct, is that this innate endowment consists of a system of principles, each with certain possibilities of parametric variation, and that acquisition of knowledge of grammar with all it entails is, in part, a matter of setting the parameters one way or another on the basis of presented experience."

This "generative enterprise" (Chomsky, Huybregts, and Van Riemsdijk, 1982) has been a revolution of goals and interests as much as a revolution in method or content (formal description of hierarchically structured strings of morphemes or phonemes, subject to the empirical test of acceptability judgments). The formal foundations of generative grammar have always played a major role in theoretical developments. For many linguists not working within the generative grammar tradition this is precisely the problem. Generative grammar focuses on analytic statements (about structural properties of language) that go beyond "what you see/hear."

The methods used in the study of language are diverse and can be roughly divided into three groups: behavioral, computational, and, more recently, neurophysiological. It depends on the subdiscipline as to which methods are preferred. Any evidence will do, whether acquired by experimentation, by the use of corpora, or by assessing linguistic structures with our own judgments.[2] As long as one can show that the data are relevant to one's hypothesis, no methodology has any a priori preferential status.

Generative grammar is first and foremost linked to Chomsky, whose ideas on the study of language shaped the field. Chomsky's ideas were made more widely accessible for the first time in his 1956 paper "Three Models for the Description of Language." This paper was followed by his 1957 book *Syntactic Structures*, which had an impact on a far broader audience. But note that, a half century later, it is important to realize that generative grammar, as a research program, encompasses many different, competing, frameworks, such as Lexical Functional Grammar (Bresnan, 2001), Head Phrase Structure Grammar (Pollard & Sag, 1994), the Minimalist Program (Chomsky, 1995), Optimality Theory (Prince & Smolensky, 2004), and even Relational Grammar (Perlmutter, 1983). Proponents of these frameworks (by and large) address similar empirical and theoretical problems and publish in the same journals. These theories are very different in their styles of analysis, but they share the common goal of finding a contentful theory of the limits of natural language.

Any generative linguistic theory attempts to describe the tacit knowledge of a native speaker through a system of rules that characterizes a *discrete infinity* of *hierarchically structured* expressions: a set of ordered pairs (π,λ) over an infinite range, in which π is a phonological expression ("sound") and λ a semantic representation ("meaning"). The study of the acquisition of language must be solidly grounded in this study of the internal structure of the grammar. Their common goals are much more important than what divides them, although that might sometimes be difficult to see or accept for adherents of these specific frameworks. Linguists in these frameworks all take linguistics as a science that should provide a precise (explicit and formal) model of a cognitively embedded computational system of human language, and make clear how it is acquired. Even formal semanticists who do not adhere to some of the cognitive underpinnings of generative grammar, and therefore would not call themselves generative grammarians, would have no problem in accepting the goal of linguistics as just described.³ Perhaps it is, therefore, better to say that there is a substantial group of linguists who could be called formal linguists. Divisions are never crystal clear, but opposed to formal linguists, one could position functional linguists (Halliday, 1985, and Van Valin & LaPolla, 1997, among others). For functional linguistics the conceptual problem of language acquisition is not necessarily an issue. Neither are they committed to the idea that language has domain-specific properties that have a genetic basis.

Design Principles

Languages look quite different from each other on the surface, exhibiting a tremendous amount of variation, manifesting itself in many aspects of the structure of languages. The question, however, is whether "languages could differ from each other without limit and in unpredictable ways" (Joos 1957, p. 96). This quote is often misinterpreted but it reflects an attitude to linguistic theorizing that takes language diversity to be its most remarkable property, and rejects the hypothesis that there are universal properties of language as a "myth." In this view "languages differ so fundamentally from one another at every level of description (sound, grammar, lexicon, meaning) that it is very hard to find any single structural property they share" (Evans & Levinson, 2009, p. 429). Some linguists appear to endorse this statement,⁴ but not generative grammarians. In the generative research tradition it is hypothesized that certain principles of the language faculty are innate, universal properties of languages, which impose restrictions on diversity. Some of these universals are supposed to apply to all languages uniformly, while other universals are formulated in such a way that there is room for variation, so-called parameters. These parameters can take one of two or more values depending on the structure of the specific language being learned (see Smith & Law, chapter 6, this volume). For formal linguists, notwithstanding the enormous variety one can observe, languages show a remarkable degree of similarity, which takes the form of a set of common mechanisms and principles. Together these universal properties of language define the basic layout of the language system, the design principles of language. Such design principles should not be confused with the notion "universals" used in crosslinguistic work (like in typology). Universals like "all languages have verbs" are simply robust generalizations, and they might be true for all languages, but that does not necessarily mean that it follows from what Chomsky called Universal Grammar. For Chomsky, studying Universal Grammar is about finding the design features of the language capacity, a biological system, through studying the grammars of specific languages. Berwick and Chomsky (2011) discuss features such as "digital infinity and recursive generative capacity," "displacement," "locality constraints," and so on. We will focus on two of them here.

Structure Sensitivity

Rules in language are, at any level, sensitive to the structure of language. Take a sentence as in (1a), with its (very simplified) structure in (1b):

(1) a. The man you were looking at has left.
    b. [$_{S1}$ [$_{NP}$ the man [$_{S2}$ you **were** looking at]] **has** left]

The analysis in (1b) reflects that the noun phrase (NP) *the man who you were looking at* is a substring of the sentence (S) *the man who you were looking at has left*. Furthermore, we have boldfaced the finite verbs in this sentence. If we create a yes-no question on the basis of this sentence, what would it look like? The rule is that you create a yes-no question in English by moving the finite auxiliary verb, if there is one, to the first position of the sentence:

(2) a. You were looking at the man.
    b. Q: Were you looking at the man? A: No, I wasn't.
(3) a. The man has left.
    b. Q: Has the man left? A: Yes, he has.

What would the yes-no question associated with (1) look like, (4a) or (4b)?

(4) a. *Were the man you looking at has left?
    b. Has the man you were looking at left?

Of course it is (4b). Example (4a) is structurally so deviant that, when uttered, it would take some time to understand what could have been meant (which is reflected by linguists putting an asterisk in front of the sentence). What (4b) shows is that the finite verb *were*, which is the nearest to the target position (in linear distance), must remain inert, while the finite verb *has*, which is the farthest removed from the target position, can move to the target position:

(5) a. [$_{S1}$ **Were** [$_{NP}$ the man [$_{S2}$ you _ looking at]] **has** left]?
    b. [$_{S1}$ **Has** [$_{NP}$ the man [$_{S2}$ you **were** looking at]] _ left]?

Why? The finite verb *were* is contained in A = {S1, S2, NP}; the finite verb *has* is contained in B = {S1}. Since B is properly included in A, it means that *has* is selected for movement. Since *has* must be selected for movement, this means that structural distance, and not linear distance, is computationally significant, with structural distance measured in terms of least number of phrasal containments ("closest to the root").

The example in (6) exemplifies a similar point. The interpretation of (6a) is as in (6b), and not as in (6c):

(6) a. The brother of the butcher looked at himself.
    b. '*The brother of the butcher* looked at *the brother of the butcher.*'
    c. 'The brother of *the butcher* looked at *the butcher.*'

Again, in terms of linear order the noun phrase *the butcher* is the closest antecedent to which the reflexive *himself* could refer, but that is not a possible interpretation. Why not? Because, again, linear order is not relevant here, but hierarchical
structure is. Without going into technical details, one could say that the reflexive
needs a prominent antecedent, "higher in the tree," NP_1, but not NP_2 contained
in NP_1:[5]

(7)

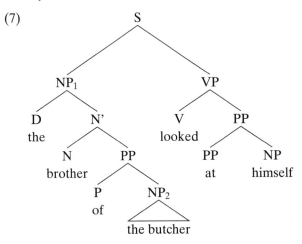

Dependency

Another possible design principle is the fact that natural language exhibits "dependencies." Words in an utterance are connected. The example in (1a), repeated here,
illustrated that point well:

(1a) The man you were looking at has left.

This simple example illustrates that natural language abounds with "dependencies"
and that these dependencies are often nonlocal, affecting nonadjacent words/
phrases:

(8) a. *The man(you−at)−has*: The subject and the verb agree; if the subject was
        The men (you−at): The verb would have to change to *have*.
     b. *you were*: The subject and the verb agree; if the subject was *I* the verb has
        to change to *was*.
     c. *look−at*: The choice of the preposition is dependent on the choice of the
        verb; some verbs do not select prepositions (like *hate*), but if a verb selects
        a prepositional phrase, the choice of preposition is limited: *John looked
        at/*to Mary*.

   d. *were—looking*: The form of the verb, ending in *-ing* (a progressive), is dependent on the use of the verb *to be*.

   e. *has—left*: The form of the verb, a participle, is dependent on the use of the verb *to have*.

   f. *the man—at*: *The man* is interpreted as the object of *talk to*, the noun phrase that normally follows the preposition to, but not in (1a) because it is a relative clause with a "gap."

The example in (9) highlights other types of dependencies, specified in (10):

(9) *John* did **not** talk about *himself* **at all.**

(10) a. *John—himself*: The form of reflexive is dependent on its antecedent, in this case *himself*. If the subject had been *Mary*, the reflexive would have to be *herself*.

   b. The phrase *at all* needs to be licensed by a negative element, a property of so-called negative polarity items: **John did talk about himself **at all**.

It might very well be that such dependencies "define" human language when compared to animal language, and, perhaps more important, set natural language apart from human communication systems other than language. Instead of assuming that language should be understood as a system of "communication," and should be understood solely in those terms, it is at least a hypothesis worth investigating.

Biolinguistics: Formal Language Theory and Its Relevance for Birdsong

In formal language theory, "language" is taken as a set of strings, strings being sequences of symbols generated by a finite set of production rules working on an agreed-on (finite) set of terminal and nonterminal symbols. There is long-standing interest in the formal complexity of such grammars, the so-called Chomsky Hierarchy. Depending on the type of rules allowed in the grammar, grammars can be defined as in (11) (Wall, 1972, p. 212):

(11) a. finite-state: $A \rightarrow xB$ or $A \rightarrow x$
   b. context-free: $A \rightarrow \omega$, where $\omega \neq$ the null string
   c. context-sensitive: $\varphi A \psi \rightarrow \varphi \omega \psi$, where φ and ψ, but not ω, may be the null string
   d. unrestricted rewriting system: no restriction (Turing machine)
(A, B nonterminals; x terminals; φ, ω, ψ sequences of nonterminals and terminals)

   In early work, Chomsky discusses the properties of natural languages in terms of this formal language theory. The question was what kind of rules were required to accurately describe natural languages. In Chomsky (1956), "Three Models for the Description of Language," the following is stated:

(12) a. Finite state systems are inadequate in their weak generative capacity, and a fortiori strongly inadequate to properly characterize natural language
  b. Even if simple phrase structure systems are weakly inadequate they still fail massively in their strong generative capacity
  c. Transformational generative systems are strongly (and weakly) adequate to characterize natural language

Demonstration of the correctness of (12b) provided sufficient reason for starting to work on "transformational" models that could account for structural properties of language such as phrase-structural dependencies. But generative systems have succeeded in simplifying and unifying their computational systems by doing away with complexities and stipulations.[6]

In (12) Chomsky not only refers to the strength of the possible grammar type to be used for the description of natural language, but he also mentions the "weak" and "strong" generative capacity of these grammars. Unfortunately, this is an aspect of formal grammar theory, noted from the very beginning, that has often been completely neglected, and has led to misinterpretation of many results in birdsong research, among others. We will, therefore, briefly discuss this issue.

Weak and Strong Generative Capacity

What is the difference between weak and strong generative capacity? In weak generative capacity, what counts is whether a grammar will generate correct strings of terminal symbols (words in natural language); strong generative capacity adds the requirement that the right hierarchical structure is accounted for. And this latter point is of essence for the study of natural language, as we have noted. Let us illustrate the point.

Suppose we have the "language" consisting of the strings in (13):

(13) ab, aabb, aaabbb, aaaabbbb, . . .

This language could be generated by the grammar in (14), creating (15).

(14) S → Z b
  Z → a S
  S → ab

However, the same language can also be generated by the grammar in (16), creating a structure as in (17):

(15)

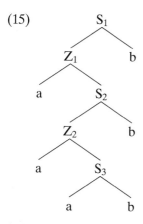

(16) S → a Z
Z → S b
S → ab

(17)

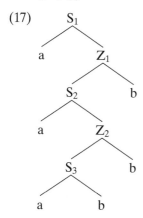

Observe that the terminal string of *a*'s and *b*'s in (15) and (17) is the same, but the way the *a*'s and *b*'s are clustered is crucially different. If we put it in "labeled brackets," an equivalent notation to trees, we can see the difference (the subscripts indicate constituency at the same hierarchical level):

(15′) $[_{S1} [_{Z1}$ a $[_{S2} [_{Z2}$ a $[_{S3}$ ab $]_{S3}]_{Z2}$ b $]_{S2}]_{Z1}$ b $]_{S1}$

(17′) $[_{S1}$ a $[_{Z1} [_{S2}$ a $[_{Z2} [_{S3}$ ab $]_{S3}$ b $]_{Z2}]_{S2}$ b $]_{Z1}]_{S1}$

So, in the grammar (14) *ab* is a unit, a constituent, like *aabb, aaabbb*, or *aab, aaabb*. But not *abb, aabbb*. In (16), likewise, *ab, aabb*, and *aaabbb* are constituents, and also *abb, aabbb*. But not *aab, aaab*. So the two grammars are weakly equivalent—generating *ab, aabb, aaabbb*, etc.—but not strongly equivalent, and that is what is important.

Let us also illustrate this with some linguistic examples. If we observe "the string of words" in (18), what we want is a grammar that gives us the phrase structure in (19), not the one in (20):[7]

(18) the friend of the neighbor of the boy

(19) [NP3 the friend of [NP2 the neighbor of [NP1 the boy]]]

(20) [NP3 [NP2 [NP1the friend]] of the neighbor] of the boy]

The "right-branching" structure of (19) reflects the semantics of the phrase in the sense that, for instance, (18) is about "the friend of some neighbor" (in particular "the boy's neighbor") and not about "some friend of the boy" (in particular "the neighbor's friend"), a reading which would be consistent with the "left-branching" structure of (20). Clearly, the structure of (20) reflects neither the syntax nor the semantics of the phrase (18).[8]

In the following example both the sound level and the meaning level are dependent on hierarchical structure. Observe the compounds in (21a,b) and their respective structures in (22a,b):

(21) a. kítchen towel racks   meaning: 'racks for kitchen towels'
      b. kitchen tówel racks   meaning: 'towel racks for kitchens'

(22) a. [[kítchen towel] racks]
      b. [kitchen [tówel racks]]

These structures provide the right interpretation for these compounds, both at the *sensorimotor* interface (i.e., different prosodic prominence patterns) and at the *semantic* interface (i.e., the different meanings as a result of applying a semantic function compositionally to morphological structure). For example, the stress patterns follow from applying the simple structure-dependent recursive rule (23) to the different hierarchically structured expressions of (22):

(23) A in compound C, [C A B], is prosodically prominent unless B is a compound.

It is clear that even at this simple level of word morphology, finite-state systems do not suffice to properly characterize syntactic patterns that are required for phonetic and semantic interpretation. A *Bayesian* statistical analysis may (more or less successfully) approximate an unanalyzed corpus of compounds that incorporates the compound strings in (21), but crucially, they would also (more or less successfully) approximate a corpus of compounds that would contradict the recursive application of (23), and instead consists of compounds that are all prosodically prominent on the first or the last word. Apparently, as even these simple cases of word formation demonstrate, a simple notion of linearity (concatenation) is rejected by the language system in favor of a more complex notion of structure (hierarchy), an observation that, in our mind, must be explained in terms of intrinsic and domain-specific properties of a biolinguistic system.

In conclusion, we take human language as a biological system that is internal to an individual and that intensionally specifies an infinite set of structured expressions. It is a unique system with properties that are uniquely human and uniquely linguistic. It will not do, therefore, to weakly characterize its externalized strings in any of a number of ways. Rather its structure must be strongly characterized by explanatory principles and rules of the underlying biological system.[9]

Formal Grammars and Birdsong

Since the very beginnings of the scientific study of language it has been made clear that properties of *competence* systems (cognitive systems) must be sharply distinguished from limitations of *performance* systems (systems of use), particularly when these are domain-neutral—that is, not specific to language, or even cross-specific as may plausibly be the case for constraints on memory (Miller & Chomsky, 1963; Chomsky, 1965). Since these limitations seem neither language-specific nor species-specific, the study of language should properly abstract away from them in an effort to get further insight into domain-specific properties of language design. This is precisely what should also be done in the study of song-learning capacities of songbirds. However, Bloomfield, Gentner, and Margoliash (2011) characterize the generative study of language as "descriptive" and "analytical" in contrast with the "empirical" nature of their own research on the computational abilities of birds. What they mean is that generative linguistics describes distributional data and extends observed patterns of analysis to empirically unattested or unrealizable behavioral data, leading to a kind of "Platonist" model of computation. But linguistic data are simply behavioral data that can in principle be presented as behavioral experiments. Furthermore, linguistic principles can easily be understood to be predictive and quite open to experimental verification. Contrary to what they assert, generative grammar's natural place is therefore within "the general program of experimental cognitive neuroscience" (Marantz, 2005). Generative linguistics is "biolinguistics" and deals, for example, with properties of the genetic endowment of a human biological system for language (UG).

Their conclusion to the effect that "the conjunction of these two hypotheses (infinite recursion and the distinction between competence and performance), both necessary to explain the observed human behavior, is awkward" (p. 947) simply amounts to a failure to recognize the relevance of these distinctions in standard scientific methods. The capacities for recursive nesting are in fact biolinguistic, not "Platonic," but apparently are both species-specific (or "uniquely human") and domain-specific (or "uniquely linguistic"), while constraints on output capacity are widely shared with animals and are therefore both cross-specific and domain-general, thus providing principled reasons for abstracting away from them in the study of human language. In fact, factoring out the independent effects of memory constraints may very well serve to explain why songbirds do not seem to have "human" problems with processing self-embedding structures. The reason is that no self-embedding is involved in birdsong, hence no memory effects that are typical of nested embeddings. What is involved in this perceptual discrimination task is a cross-species capacity for computing small numerosities (numerical quantity) and serial order (numerical ranking). See Nieder (2005) for review and discussion of animal numerical competence.

Self-embedding, however, is the "fundamental property that takes a system outside of the generative capacity of a finite device" (Miller and Chomsky, 1963, p. 470) and is an automatic consequence of nesting of dependencies given a finite lexicon. For human language self-embedding with its associated compositional semantics results from phrase structure recursion (an aspect that relates to a cognitive or "intensional" property of language) and is constrained, independently and irrelevantly for the purposes of studying computational capacity, by language-external conditions on memory organization (an aspect that relates to a behavioral or "extensional" property of language). In a recent study Abe and Watanabe (2011) argue that Bengalese finches have a capacity for hierarchical structure that enables them to recognize nested dependencies. In fact, they show a capacity for correctly recognizing syllable strings of a strictly finite length. They do so by computing small numbers of syllables (three or less), s_1-s_2-s_3, and identifying the associates of these syllables in reversed order, $\hat{s}_3$-$\hat{s}_2$-$\hat{s}_1$, in the syllable string s_1-s_2-s_3-c-$\hat{s}_3$-$\hat{s}_2$-$\hat{s}_1$. But their achievements whatever their nature need not really surprise us. We have known about the remarkable "counting" and "ranking" abilities of birds ever since Koehler's (1941, 1951) studies on animal numerical competence and the work by Straub and others on serial learning of pigeons in the late 1970s, recently reviewed by Nieder (2005). But these abilities have nothing to do with a capacity for recursion or phrase structure grammar. The finches simply generalize from a strictly finite set S_0 of syllable strings to which they have been habituated (i.e., 16 simple strings without "embedding" and 36 strings with one level of "embedding") to an equally strictly finite set of syllable strings S_f (with a nesting depth of 2 or less and a cardinality of 160). The ability to recognize "context-free" patterns in finite sets cannot be used to infer any capacity for phrase structure recursion. We are dealing here with strictly finite sets that can simply be listed without use for recursive operations in the first place. In fact no recursion can be demonstrated in this case or in other widely publicized cases of recursive birdsong.

Abe and Watanabe's argument fails (as does the one by Gentner, Fenn, Margoliash, & Nusbaum, 2006, for precisely the same reasons) and, in fact, cannot be saved since its resurrection will only be self-defeating. Their Bengalese finches can select from four syllabic flavors of A (each uniquely paired with a specific syllabic flavor of F). Further, C also has four syllabic flavors, which, however, can be freely selected. Finally, each of the three selected flavors of A in A_X-A_Y-A_Z-C-F_Z-F_Y-F_X is meant to be distinct from the others. It follows that this "avian language" is strictly finite—that is, 160 strings all in all. Without loss of generality we can simplify matters by abstracting away from the effects of C. The total language will then amount to precisely 1!(4-over-1) + 2!(4-over-2) + 3!(4-over-3) = 40 strings that conform to the A^nF^n syllabic format, where each jth syllabic A is uniquely "linked" with its $(n - (j - 1))^{th}$ syllabic associate F. Inspection of the grammar they provide for this "self-embedding language" (their Figure 3a) teaches us that it has no (in)direct recursion and that self-embedding (their "center-embedding") is absent from the language. Contrary to their intention, the rules they propose, namely, $S \Rightarrow AP \wedge BP$, $AP \Rightarrow A$, $BP \Rightarrow CP \wedge F$, $CP \Rightarrow C$, or $CP \Rightarrow AP' \wedge BP'$ (with $AP \neq AP'$ and $BP \neq BP'$), exclude self-embedding, a fundamental property of phrase structure recursion. This is a consequence of the stipulative condition of distinctness they impose on the sequence of A's (and of F's) in the syllable strings. Self-embedding is excluded by stipulation, and recursion does not apply. The language can simply be finitely listed. So the argument as presented collapses.[10]

But someone could argue that the effects of phrase structure recursion could be guaranteed by equating AP' and BP' (in their rule $CP \Rightarrow AP' \wedge BP'$) with AP and BP (in their rule $S \Rightarrow AP \wedge BP$). But this will not do. First of all, if this had been their intention, their formal grammar should have included rule $CP \Rightarrow AP \wedge BP$ to formalize the self-embedding recursive property of CP. Second, any such reformulation would be self-defeating. It would violate the distinctness condition they have stipulated for the selection of A's in the A^nCF^n strings to which the finches were habituated. A different language would therefore have been generated. On the other hand, the assumption of unbounded recursion together with the stipulation that each nested dependency be distinct from any other will give rise to a *nonrecursively enumerable* language—that is, an uncountably infinite language whose grammar is no longer finitely representable and therefore cannot be part of any biological system whatsoever.

Nevertheless, the grammar perhaps equivocates between imposing strict distinctness and allowing nondistinctness. In that case, let us do away with the stipulative distinctness condition and reformulate the grammar in an effort to resurrect Abe and Watanabe's argument. Substitution of rule $CP \Rightarrow AP \wedge BP$ for their rule $CP \Rightarrow AP' \wedge BP'$ suffices to generate a recursive language that is essentially one version of the well-known context-free "mirror-image" language xcx^{-1}. Apart from the effects of the now abandoned stipulation, this language conforms to the A-C-F sequences of the original language. However, there is another important question, which has to do with choice of "phrases" and their "labeling." These aspects are pertinent to the structure of human language but are arbitrary choices in artificial language contexts, highly misleading, and often unnecessary unless for specific purposes. It is permissible, therefore, to do away with "phrases" like *AP*, *BP*, and *CP* since their constituency and labeling play no role in the experiment, and in fact cannot do so. Compositional semantics, displacement effects, and labeling requirements are properties that relate to the structure of a natural language (its "intension" or its strong generative capacity), fundamental and unique properties of a biological system but irrelevant for artificial languages. We think that their unavoidable absence in artificial birdsong will turn out to be an important reason for why these experiments, which aim to demonstrate human computational properties in birds, will invariably fail to be successes.

To be concrete, stripping away everything that is not essential for the argument, let us reformulate the relevant rules as $S \Rightarrow C$ and $S \Rightarrow A_X \wedge S \wedge B_X$, the latter an abbreviation of four rewrite rules that account for the "lexical" dependencies between choices of A, namely $A_1 = \alpha$, $A_2 = \beta$, $A_3 = \gamma$, $A_4 = \delta$, and of B, namely $B_1 = \alpha'$, $B_2 = \beta'$, $B_3 = \gamma'$, $B_4 = \delta'$. As before, C allows a free choice out of four options. We now have a version of "Artificial Bengalese Finch" (*ABF*), which is a properly context-free language and is in the spirit of the original argument. If Bengalese finches, or for that matter starlings, are now tested for their discriminating capacities for *ABF*, then, on the assumption that cross-specificity for working memory constraints is the norm for mammalian and avian brains, there will be essentially two logically possible outcomes.[11]

The first possibility is that the birds show good discriminative capacity (say up to a total string length of five plus or minus two syllables). Since human performance is very poor for self-embedding in natural language, and memory constraints hold across species, the reasonable conclusion to draw must be that what the birds are doing is not processing self-embedding structures but rather applying their numerical competence (numerosity and serial order). Perruchet and Rey (2005) arrive at a conclusion consistent with this finding. If songbirds really apply a counting strategy, it is to be expected that humans detached from a natural language context will tap the same resources when their capacities for discriminating "nested dependencies" in artificial languages are tested experimentally. Furthermore, humans will in fact be expected do so with comparable success, an expectation that is found to be true with some qualifications (Müller, Bahlmann, & Friederici, 2010; Lai & Poletiek, 2011). A "nested dependency depth" of two or three is the upper bound for good performance. This is also consistent with recent experiments on the limits of human working memory, which downsize Miller's (1956) "magical number seven, plus or minus two" to a relative working memory capacity that is more likely to be two or three, accuracy declining sharply with four or more (Cowan, 1998). Note that there would be no contradiction between the failure of humans to process self-embedding in natural language (a system of discrete infinity) and their far better results in processing "nested dependencies" in artificial strings (finite sets of discrete elements). These are incomparable phenomena, involving different biological systems, each with their own explanations.

The second possibility is that the birds show poor discriminative power. In this case, nothing follows. There is no way of knowing what is going on. It is impossible to demonstrate the presence of a human phrase-structural competence that is substantially incapacitated by more general cross-specific performance limitations, a highly unlikely eventuality. A more parsimonious conclusion would be to say that phrase structure competence is unnecessary and redundant, and what is involved is just limitations on working memory, a result that could be more plausibly argued to be closer to the biological truth. If true, this would also be a welcome result since a constrained working memory is a bare necessity anyway and may possibly be sufficient to explain the outcome of the behavioral experiment. We may expect strictly finite sets of strings X-C-Y (with X longer than 3 or 4 and the syllables of Y reversely and uniquely paired with the syllables of X) to be outside the learning range of these songbirds because their short-term memory may provide inadequate pushdown storage support.

The "language capacity" these birds were tested for could therefore be simply characterized by a *one-sided linear* system (24) incorporating some kind of counting device (which can be thought of as simulating a bounded working memory). Only strings of syllables $A^n B^n$ with n smaller than some fixed arbitrary integer i (here 3) will be accepted by (24).[12]

Elimination of diacritic subscripts of the terminal symbols gives the European starling language (Gentner et al., 2006). The finite Bengalese finch language, with cardinality 40, discussed by Abe & Watanabe, strictly conforms to (24) if a_j, b_j are understood to stand for specifically linked choices of A,F but its full grammar will need 120 rules (abstracting from choices of c) to fully characterize its 40 distinct syllable strings. In contrast, the infinite language L_2 of (25) needs only 13 rules, the rules (25a–c) being abbreviations of 4 rules each ($i = 1,2,3$, or 4), one for every a,b pair. The comparison nicely illustrates a well-known result: to assume a language is infinite is to make a simplifying move.

(24) *One-sided linear grammar*
  $L_1 = a^n c b^n$ for $n \leq 3$
  a. $\Sigma^0 \Rightarrow a_1 \Sigma^1$
  b. $\Sigma^1 \Rightarrow a_2 \Sigma^2$
     $\Sigma^1 \Rightarrow c\, S^1$
  c. $\Sigma^2 \Rightarrow a_3 \Sigma^3$
     $\Sigma^2 \Rightarrow c\, S^2$

(25) *Two-sided linear grammar*
  $L_2 = a^n c b^n$ for $n \geq 0$
  a. $S \Rightarrow a_i \Sigma^i$
  b. $\Sigma^i \Rightarrow C\, b_i$
  c. $\Sigma^i \Rightarrow S\, b_i$
  d. $C \Rightarrow c$

    d. $\Sigma^3 \Rightarrow c \; S^3$
    e. $S^3 \Rightarrow b_3 \; S^2$
    f. $S^2 \Rightarrow b_2 \; S^1$
    g. $S^1 \Rightarrow b_1$

The conclusion is unavoidable that these arguments do not succeed in demonstrating what they intend to show. This lack of success results from a deeper failure to understand what Bloomfield et al. (2011) incorrectly calls the "awkward distinction" between cognitive systems of computation ("competence") and use systems of communication ("performance"). The title of their article "What Birds Have to Say about Language" should rather be reformulated as a question receiving the reasonable answer "Nothing much thus far." If songbirds share any computational ability for language with humans (perhaps they do), this is not likely to be a capacity for phrase structure systems (see Box 1.1). In fact, as has been argued several times in the past and more recently (e.g., Chomsky, 1986, 2010), a capacity for recursive language in birds yielding systems of digital infinity would pose the very hard problem of "unused capacity," a property unknown in the biological world except as an "exaptive" by-product of some other evolutionary development. But whereas arithmetical and maybe musical competences could be argued to be offshoots and actually simplifications of the language faculty to other capacities, no such argument can be offered for the recursive capacity for birdsong. On the other hand, humans and songbirds may share a computational ability to solve the interface problem of externalization for rhythmic structure. This is the idea proposed and discussed in Berwick (2011). Here Berwick, adopting a comparative-biological approach, explores the similarities in metrical phonology and merge-based syntax in a promising effort to relate birdsong to human merge-based rhythm.

    How then could the behavioral experiment demonstrate an "animal model" for human language when what it does is simply confirming the use of *WMC* in a task of auditory perception? The conclusion must be that birdsong does not show phrase structure recursion characteristic of human language. However, this result does not exclude the possibility of finite-state recursion for natural birdsong, whose syntax is probably not more powerful than *k-reversible regular languages* (Berwick, Okanoya, Beckers, & Bolhuis, 2011).

Evolution of Language

An overwhelming number of scientists calling themselves biologists do not work on evolutionary biology. A similar situation holds for linguists. Not many linguists are interested in the evolution of language, and, therefore, not many linguists address the issue in their work. At the same time it is evident that there is a very active community working on language evolution. Why? The reason is that the majority

Box 1.1
Summary of the argument

A. Artificial language

1. *Strong generative capacity* (*SGC*) is nonarbitrary and specific for natural language. In contrast, *SGC* is arbitrary and stipulative for artificial languages.

2. The need for *context-free phrase structure grammar* (or its strong generative properties) cannot possibly be demonstrated for finite sets.

3. In the behavioral experiment both the habituation set and the test set of syllable strings are *strictly finite*.

4. Therefore, hierarchical phrase structure is nondemonstrable for familiar or novel strings.

B. Behavioral experiment

1. The experiment is a behavioral experiment testing *auditory discrimination*.

2. Auditory discrimination involves several cognitive systems, including *working memory capacity* (*WMC*).

3. Limitations of *WMC* are *domain-general* (independent of language) and *cross-specific*.

4. Birds perform excellently in tasks of visual and auditory discrimination: the relative limits of their *WMC* for *numerosity* and *serial-order* tasks is something like 3 or 4.

C. Conclusion

1. Birds generalize from one *finite* set to an equally restricted *finite* set within these *WMC* boundaries.

2. Therefore, the discrimination experiment does not show anything beyond the limitations of *WMC* and by the logic of mathematical linguistics fails to establish hierarchical structure.

---

of contributors to Evolution of Language conferences[13] or books on this topic are nonlinguists.[14] Why would that be the case? We believe that Chomsky's ideas on this represent the point of view of many (even though they would strongly disagree on many other points with him), but surely not all (e.g., Jackendoff 2011). Chomsky addresses the question of why one would want to work on language evolution, and comes to a negative conclusion on the basis of considerations like the following. There are many much simpler questions that are scarcely investigated, such as the evolution of communication in the hundreds of species of bees, because they are regarded as much too hard. Another reason is that very little is known about the evolution of cognition generally, and it is quite possible that nothing much will ever be known, at least by any methods that are understood at all today (Lewontin 1998).

Despite his reservations, Chomsky himself does publish on this topic. One of the most important publications in this area is the well-known Hauser, Chomsky, and Fitch (2002) paper. This paper is primarily a methodological exposé on how one could study the evolution of language. Addressing the question of the evolution of language is not immediately relevant for linguistic theorizing as such, but it could be relevant for a discussion of the conceptual foundations of linguistic theorizing. However, such relevance can only be achieved if publications in this area adhere to a certain methodological rigor that is well explained in Botha (2006). He points out that empirical work on language evolution has to overcome the obstacle of evidential paucity: the lack of *direct* evidence about the forces, processes, events, and other factors that might or might not have been involved in the first emergence and subsequent development of language in humans. Often evidence outside language is taken as the starting point of inferences about factors in the evolution of language (or speech). So, in order to avoid speculation one has to be explicit about the components of the inferences that one draws about the evolution of language (or speech), including the assumptions by which these inferences are supported. For these inferences to be sound, Botha argues that the evidence from which they are drawn has to be shown to actually bear on the factors in language evolution at issue. This requires an elucidation of the ways the "other" phenomena are believed to be interlinked with them.

It is often postulated that Noam Chomsky takes the position that there could be no evolutionary explanation of language. Chomsky does not take that position, but his view on this matter, the "emergence" view (see below), does not always find great support in the community of linguists working on the evolution of language.[15] In a reaction to Hauser, Chomsky, and Fitch (2002), Ray Jackendoff and Steven Pinker (2005) reject emergence and have articulated an opposing view. The difference is between the assumption that language (in the UG sense of the word) could have arisen through a sequence of small changes, as Jackendoff and Pinker, among others, have advocated, or through a radical phenotypic change, as Chomsky argues for. In the incremental scenario, "individuals endowed with a more highly developed language faculty had some adaptive advantage over those with a less highly developed one, so natural selection played a role in each step" (Jackendoff , 2011, p. 588). The crucial question in such a line of reasoning is what the nature of that adaptive advantage is. The bottom line is that this incremental view can only be upheld if there is a communicative advantage to having language. Chomsky has always rejected the position that the essential properties of language are based on its communicative function. That is not to say that language is not used for communicative purposes; it only disputes the assumption that the "design principles" of languages need to be understood in those terms. This is one of the dividing lines between formal and functional linguists. Functional linguists almost by definition assume that language is communication, and that many of its features can be derived from that feature. For formal linguists the picture is not straightforward; the "regular" choices would include either a position such as taken by Chomsky or a position as is advocated by Jackendoff.

Contrary to Jackendoff's position, Chomsky and his coauthor on these issues, Robert Berwick, argue that the evolution of language may involve "emergence," the appearance of a qualitatively different phenomenon at a specific stage of organizational complexity (Chomsky 1968). It is a position on the evolution of language that, for instance, the paleoanthropologist Ian Tattersall (2007, p.58) adheres to: "Language is almost certainly a truly emergent quality, built upon what went before but entirely unpredicted by it."

Tattersall's line of argumentation is the following (based on Tattersall 2011).[16] He takes symbolic mode of reasoning as crucially human, and for him "language," "imagination," and "creativity" are the ultimate symbolic activity. Tattersall observes, for instance, a lack of creativity in tool making in *Homo sapiens'* predecessors (*Homo ergaster, Homo heidelbergensis*). Over a long period of time tools become more sophisticated, but only very slowly, and behavioral innovations do not tend to be associated with new kinds of humans.[17] The same holds for *Homo neanderthalensis*. They made tools monotonously, in the sense that there was a *sameness* to them over the whole vast expanse of time they lived in and the space these hominids inhabited. Even the early anatomical *Homo sapiens,* as far as it is possible to ascertain, actually behaved very much as the Neanderthals did. And it is not until significantly later, around 100,000 years ago, that we find evidence of the first stirrings of the symbolic spirit that we associate with *Homo sapiens* today. For Tattersall "symbolic activity" in itself is not enough to assume that there is language, but "a restless appetite for change" (for instance, in tool making) is what might indicate the presence of cognitive capacities that one can associate with the presence of language. It is the change in the tempo of innovation in general that is noteworthy, and that might reflect a fundamental transformation of how hominins did business in the world. Tattersall concludes that we cannot look to classical natural selection as an explanation of this new phenomenon. The reason is that natural selection is an opportunistic process, and not one that can conjure advantageous novelties into existence. One is, then, obliged to look to emergence, whereby a chance coincidence of acquisitions led to an entirely new level of complexity in function. It seems pretty clear that the brain of the immediate ancestor of modern humans had evolved to a point where a small neural change was sufficient to create a structure with an entirely new potential (Chomsky 2010).

Hauser, Chomsky and Fitch (2002) have been widely cited because of the hypothesis that the property of recursion would be a feature of language that would set it aside from other nonhuman communication systems. Unfortunately, a very important methodological point of the paper, a plea for "a comparative evolutionary approach to language," did not get the attention it deserved. Such an enterprise has to start with a well-defined idea of what "language" is. Hauser, Chomsky, and Fitch (2002, p. 1570) claim that one should make a distinction between a faculty of language in the broad sense (FLB), and a faculty of language in the narrow sense (FLN): "FLB includes sensory-motor, conceptual-intentional, and other possible systems (which we leave open); FLN includes the core grammatical computations that we suggest are limited to recursion." On this view, language is an efficient computational system linking sensorimotor systems of perception and production ("sound") to the cognitive interfaces of conception and intention ("meaning"). Various aspects of these interface systems might have been derived from traits of mammalian vocal and perceptual systems that proved useful for language but did not evolve "for" language. Likewise, categorical perception, prototypical magnet effect, and vocal imitation occur elsewhere in the animal world and may have served as preadaptive precursors to human sensorimotor systems. Similarly, at least primates (and maybe also corvids) may exhibit some traits of a primitive "theory of mind" (e.g., attribution of beliefs to other minds). Nevertheless, animal communication systems lack defining properties of human language, in particular recursive combination of (atomic) concepts and compositional meaning. Therefore, Hauser, Chomsky, and Fitch argue that questions of development and implementation of mechanisms should be carefully distinguished from questions of their functions and evolutionary origin. Though "language genes" do not exist, some genes (e.g., FoxP2) are involved in the neurogenesis of language and speech and show signs of recent positive selection in the human lineage, highlighting the recent emergence in the species. It is precisely for this reason that Hauser, Chomsky, and Fitch argue that it would be a worthwhile enterprise to engage in comparative studies (in biology), looking for evidence of aspects of both FLN and FLB in other species, and, more important, also outside the domain of communication. For theorizing in the domain of language evolution this might be the most fruitful route to take.

Notes

1. We are not talking about the "truth" of this statement, although we believe it is true, but simply mention it as a defining feature of a group of linguists. Whether or not the input is "limited" is a matter of fierce debate (Berwick, Pietroski, Yankama, & Chomsky, 2011).

2. The use of intuitive judgments has long been the hallmark of theoretical research in syntax, in semantics, and to a lesser degree in morphology and phonology. We think that that no longer holds true for much grammatical work. If we look at the work done in major linguistics departments, we see syntacticians, semanticists, and phonologists working together with psycholinguists, language acquisitionists, and sociolinguists, using experimental work to support

their theoretical claims, but also making use of intuitive judgments (sometimes supported with magnitude estimation experiments or corpora searches). (See Marantz, 2005).

3. See Van Riemsdijk (1984) for an attempt to define this common core.

4. Pullum and Scholz (2009) adhere to Evans and Levinson's position, but are, simultaneously, able to accept that not all conceivable differences between languages will be attested.

5. S = Sentence, D = Determiner, NP = Noun Phrase, N′ = part of a Noun Phrase, N = Noun, PP = Prepositional Phrase, P = Preposition, VP = Verb Phrase, V = Verb.

6. In particular the residue of phrase structure grammar, such as X'-system, and the residue of transformations, such as Move α, are unified under Merge.

7. NP = Noun Phrase, PP = Prepositional Phrase, D = Determiner, N = Noun, P = Preposition.

8. There are, of course, also syntactic arguments. In English the "possessor-possessee" relation is manifested either as a genitive (*John's* book) or as a prepositional phrase (*the book of John*). Likewise *the friend of the neighbor of the boy* is related to *the friend of the boy's neighbor* (based on a structure like (19)), but cannot be related to *the neighbor's friend of the boy* (based on a structure like (20)).

9. From this biolinguistic point of view, Pullum's (2011) criticism that Chomsky's revolution in linguistics was based on a failed argument in one of the founding publications, "Three Models for the Description of Language" (Chomsky, 1956), is rather beside the point. The criticism is highly misleading and, more importantly, completely irrelevant. It is completely irrelevant because what is at issue is not the weak generative capacity but the strong generative capacity of grammars. It is also misleading because it invites the inference that present-day understanding of language is built on a foundation that is empirically and conceptually flawed. But the criticism is not even true. Although it is essentially without consequence, human languages are easily shown to be beyond the weak generative capacity of finite-state systems (the first model for language discussed in Chomsky, 1956). The infinite sublanguage *rats^n smell^n* can be used to prove this simple theorem. In fact, it only suffices to rephrase the original argument in these terms, a point already suggested in Chomsky (1956, 1957). Surely, nothing dramatic is at stake here. What is at stake, however, is a better insight into the strong generative capacity of language. In "Three Models" it is argued, first, that even phrase structure grammars (the second model for language discussed there) are inadequate in this respect, and, second, that transformational versions of generative grammars (the third model for language) seem to succeed in characterizing the structure of language adequately. Therefore, even if phrase structure grammar suffices for weak generation, this model for language must be rejected for a more principled reason, namely, inadequacy in its strong generative capacity. See Huybregts (1984) for demonstrating that context-free phrase structure grammars are also insufficient even to weakly generate natural language.

10. For related discussion and further relevant criticism see Beckers, Bolhuis, Okanoya, and Berwick (2012). We are indebted to Noam Chomsky (p.c.) for pointing out the different properties of self-embedding (which humans are not good at) and center-embedding (which humans are much better at). Self-embedding seems to involve constraints on the organization of memory that go beyond mere memory limitations. In particular "calling in a parsing subroutine while executing it is particularly difficult" for a finite perceptual device. See Miller and Chomsky (1963, p. 470ff.) and Chomsky (1965, p. 14) for relevant discussion.

11. Without affecting the argument or its conclusions we could have simplified ABF to language $L = \{xx^{-1} \mid x$ is in$\{a, b\}^*\}$, and its grammar $G(L)$ to a set of rules (i) $S \Rightarrow a \wedge (S) \wedge a$ and (ii) $S \Rightarrow b \wedge (S) \wedge b$.

12. More generally, a finite state transducer incorporating some finitely bounded finite-state system will be sufficient to explain all of these experiments, including the birdsong experiments of Gentner et al. (2006) and Abe and Watanabe (2011), which are published in quite respectable scientific journals such as *Nature* or *Nature Neuroscience*, and are highly publicized in the professional as well as in the popular literature.

13. The International Conference on the Evolution of Language (Evolang) has been organized every two years since 1996.

14. Including the book that one of the authors, Martin Everaert, is presently editing: Rudie Botha and Martin Everaert (in press), *The Evolutionary Emergence of Human Language*.

15. According to Chomsky, after the emergence of concepts and merge there may have been no evolution of language at all: the rest is just historical language change.

16. Much of what is said in the text is directly derived from a written version of this talk.

17. Tattersall observes that, in contrast, brain enlargement forced dietary changes, but we see no material indication of this.

References

Abe, K., & Watanabe, D. (2011). Songbirds possess the spontaneous ability to discriminate syntactic rules. *Nature Neuroscience, 14*, 1067–1074.

Beckers, G. J. L., Bolhuis, J. J., Okanoya, K., & Berwick, R. C. (2012). Birdsong neurolinguistics: Songbird context-free grammar claim is premature. *NeuroReport, 23*, 139–145.

Berwick, R. C. (2011). All you need is merge: Biology, computation, and language from the bottom up. In A. M. di Sciullo & C. Boeckx (Eds.), *The biolinguistic enterprise: New perspectives on the evolution and nature of the human language faculty* (pp. 461–491). Oxford: Oxford University Press.

Berwick, R. C., & Chomsky, N. (2011). The biolinguistic program: The current state of its development. In A. M. di Sciullo & C. Boeckx (Eds.), *The biolinguistic enterprise: New perspectives on the evolution and nature of the human language faculty* (pp. 19–41). Oxford: Oxford University Press.

Berwick, R. C., Okanoya, K., Beckers, G. J. L., & Bolhuis, J. J. (2011). Songs to syntax: The linguistics of birdsong. *Trends in Cognitive Sciences, 15*, 113–121.

Berwick, R. C., Pietroski, P., Yankama, B., & Chomsky, N. (2011). Poverty of the stimulus revisited. *Cognitive Science, 35*, 1207–1242.

Bloomfield, T. C., Gentner, T. Q., & Margoliash, D. (2011). What birds have to say about language. *Nature Neuroscience, 14*, 947–948.

Botha, Rudie, and Martin Everaert. (in press). *The evolutionary emergence of human language: evidence and inference*. Oxford: Oxford University Press.

Bresnan, J. (2001). *Lexical-Functional Syntax*. Malden, MA: Blackwell.

Chomsky, N. (1956). Three models for the description of language. *IRE Transactions on Information Theory, 2,* 113–124.

Chomsky, N. (1957). *Syntactic structures.* The Hague: Mouton.

Chomsky, N. (1965). *Aspects of the theory of syntax.* Cambridge, MA: MIT Press.

Chomsky, N. (1968). *Language and mind.* New York: Harcourt, Brace & World.

Chomsky, N. (1981). *Lectures on government and binding.* Dordrecht: Foris.

Chomsky, N. (1986). *Language and problems of knowledge: The Managua lectures.* Cambridge, MA: MIT Press.

Chomsky, N. (1995). *The minimalist program.* Cambridge, MA: MIT Press.

Chomsky, N. (2010). Some simple evo-devo theses: How true might they be for language? In R. K. Larson, V. Déprez, & H. Yamakido (Eds.), *The evolution of human language: Biolinguistic perspectives* (pp. 45–62). Cambridge: Cambridge University Press.

Chomsky, N. (2011). Language and other cognitive systems: What is special about language? *Language Learning and Development, 7,* 263–278.

Chomsky, N., Huybregts, M. A. C., & Van Riemsdijk, H. C. (1982). *Noam Chomsky on the generative enterprise.* Berlin: Mouton de Gruyter.

Cowan, N. (1998). Visual and auditory working memory capacity. *Trends in Cognitive Sciences, 2,* 77–78.

Evans, N., & Levinson, S. C. (2009). The myth of language universals: Language diversity and its importance for cognitive science. *Behavioral and Brain Sciences, 32,* 429–492.

Gentner, T. Q., Fenn, K. M., Margoliash, D., & Nusbaum, H. C. (2006). Recursive syntactic pattern learning by songbirds. *Nature, 440,* 1204–1207.

Halliday, M. A. K. (1985). *An introduction to functional grammar.* London: Arnold.

Hauser, M., Chomsky, N., & Fitch, W. T. (2002, November 22). The faculty of language: What is it, who has it, and how did it evolve? *Science, 298,* 1569–1579.

Huybregts, M. A. C. (1984). The weak inadequacy of context-free phrase structure grammars. In G. de Haan, M. Trommelen, & W. Zonneveld (Eds.), *Van Periferie naar Kern* (pp. 81–99). Dordrecht: Foris.

Jackendoff, R. (2011). What is the human language faculty? Two views. *Language, 83,* 586–624.

Jackendoff, R., & Pinker, S. (2005). The nature of the language faculty and its implications for the evolution of language. *Cognition, 97,* 211–225.

Joos, M. ed. (1957). *Readings in linguistics.* Chicago: University of Chicago Press.

Koehler, O. (1941). Vom Erlernen unbenannter Anzahlen bei Vögeln. *Naturwissenschaften, 29,* 201–218.

Koehler, O. (1951). The ability of birds to "count." *Bulletin of Animal Behavior, 9,* 41–45.

Lai, J., & Poletiek, F. H. (2011). The impact of adjacent-dependencies and staged-input on the learnability of center-embedded hierarchical structure. *Cognition, 118,* 265–273.

Lewontin, R. (1998). The evolution of cognition: Questions we will never answer. In D. N. Osherson, D. Scarborough, & S. Sternberg (Eds.), *Methods, models, and conceptual issues: An invitation to cognitive science* (4th ed., Vol. 4, pp. 108–132). Cambridge, MA: MIT Press.

Marantz, A. (2005). Generative linguistics within the cognitive neuroscience of language. *Linguistic Review*, *22*, 429–445.

Miller, G. (1956). The magical number seven, plus or minus two: Some limits on our capacity for processing information. *Psychological Review*, *63*, 81–97.

Miller, G., & Chomsky, N. (1963). Finitary models of language users. In R. D. Luce, R. Bush, & E. Galanter (Eds.), *Handbook of mathematical psychology* (Vol. 2, pp. 419–492). New York: Wiley.

Müller, J. L., Bahlmann, J., & Friederici, A. D. (2010). Learnability of embedded syntactic structure depends on prosodic cues. *Cognitive Science*, *34*, 338–349.

Nieder, A. (2005). Counting on neurons: The neurobiology of numerical competence. *Nature Reviews: Neuroscience*, *6*, 177–190.

Perlmutter, D. M. (1983). *Studies in Relational Grammar*. Chicago: University of Chicago Press.

Perruchet, P., & Rey, A. (2005). Does the mastery of center-embedded linguistic structures distinguish humans from non-humans? *Psychonomic Bulletin & Review*, *12*, 307–313.

Pollard, C., & Sag, I. (1994). *Head-driven phrase structure grammar*. Chicago: University of Chicago Press.

Prince, A. S., & Smolensky, P. (2004). *Optimality theory: Constraint interaction in generative grammar*. Malden, MA: Blackwell.

Pullum, G. K. (2011). On the mathematics of *Syntactic structures*. *Journal of Logic, Language, and Information*, *20*, 277–296.

Pullum, G. K., & Scholz, B. C. (2009). For universals (but not for finite-state learning) visit the zoo. *Behavioral and Brain Sciences*, *32*, 466–467.

Riemsdijk, H. C. van. (1984). Introductory remarks. In W. de Geest & Y. Putseys (Eds.), *Sentential complementation*. Dordrecht: Foris.

Tattersall, I. (2007). Comments on "The evolution of human speech: Its anatomical and neural bases" by Philip Lieberman. *Current Anthropology*, *48*, 57–58.

Tattersall, I. (2011). A context for the emergence of UG and language. Paper presented at the conference "The Past and Future of Universal Grammar," University of Durham, December 15–18, 2011.

Van Valin, R. D., & LaPolla, R. J. (1997). *Syntax: Structure, meaning and function*. Cambridge: Cambridge University Press.

Wall, R. (1972). *Introduction to mathematical linguistics*. Englewood Cliffs, NJ: Prentice-Hall.

2 Evolution, Memory, and the Nature of Syntactic Representation

Gary F. Marcus

Natural language expressions are communicated in sequences of articulations, either sounds or gestures. Where sequence is known to matter, language clearly demands *some* sort of sequence-sensitive structure. *The dog bit the man* is not equivalent to *The man bit the dog*, and any system that represented sentences as unordered lists of words simply could not capture that fact.

A standard (though not universal) assumption in most theories of linguistics and psycholinguistics is that phrase structure is represented as a tree (Partee, ter Meulen, & Wall, 1990), as in Figure 2.1.

Trees are widespread, and their formal properties are well understood (Knuth, 1973). They play a central role not only in linguistics, but also in disciplines such as mathematics, computer science, taxonomy, and even genealogy. For example, computer programmers use trees to represent the nested directory (or folder) structures in which files are stored. In linguistics, tree-theoretic concepts, such as *path* and *distance*, often seem to provide important purchase on *syntactic* phenomena (Kayne, 1984; Pesetsky, 1982).

It is an open question, however, whether human beings can actually use trees as a data structure for representing sentences in a stable and reliable fashion. Although it is clear that educated humans can understand at an abstract level what trees are, implement them in computers, and use them as external representational devices, the evidence for their existence qua mental representation has been remarkably indirect, and there has not thus far been any fully satisfying account for how trees as such could be neurally realized.

The goal of this chapter is to suggest an alternative: in the course of ordinary linguistic processing, mental representations of sentences may be incomplete, inconsistent, or partial—realized via a system of overlapping but incompletely bound subtrees or *treelets* (Fodor, 1998; Fodor & Sakas, 2004; Kroch & Joshi, 1985; Lewis & Vasishth, 2005; Marcus, 2001, 2008) that are unified only in transitory and imperfect ways, sharply limited by the evolved nature of biological memory.

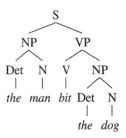

Figure 2.1
A simple phrase structure tree for *The man bit the dog*

Trees: Theoretical Objects and Psychological Objections

There is nothing difficult in principle about building a physical device that can represent trees; computers do so routinely. Each node in the tree is assigned a specific location in memory, with fields that identify both its content (say "NP" or "John") and pointers that lead to the memory location of the other immediately connected nodes, such as a given node's daughters (see Box 2.1). Such a solution fits naturally with the basic structure of computer memory, in which all elements are stored and retrieved based on their exact locations in memory.

In formal terms, a tree (as the term is used in linguistics) is a directed, acyclic graph (Partee et al., 1990), though see Chomsky (2001, 2008) for a minimalist conception. Graphs, in this sense, are objects consisting of nodes and the edges connecting them. A directed graph is one in which the edges have directionality—for example, from ancestor to descendant, or root to leaves. To say a graph is acyclic is simply to say that there are no loops. Figure 2.1, for example, meets these criteria, in that the essential elements are nodes and edges, there is a directionality from root (S) to leaves (*the*, *dog*, etc.), and no node loops back to an ancestor. Finally, syntactic trees are standardly taken to be singly rooted: there is a single distinguished node, like the S node in Figure 2.1, which no other category dominates. Given this sort of representation, one can transparently infer hierarchical relationships between elements, simply by traversing the tree using a straightforward algorithm (see Box 2.1).

This question about trees can thus be represented more precisely: Are human beings equipped to represent directed acyclic graphs of arbitrary size or complexity? Do the mental representations of sentences that human beings use ensure reliable traversability? There are at least three reasons why, even in advance of any linguistic investigation, one might be suspicious.

First, the capacity to represent a directed acyclic graph of arbitrary complexity is enormously demanding in terms of "binding" resources—that is, in terms of

Box 2.1
How computers implement syntactic trees

As Figure 2.2 illustrates, syntactic trees can be represented graphically as a set of labeled nodes—syntactic categories like noun, verb, sentence, etc., as well as terminal elements (individual words)—and edges that link each category with its constituents (Partee, ter Meulen, & Wall, 1990). In the illustration, a sentence (*S*), *The man bit the dog,* is represented as consisting of an NP subject and a VP predicate (Figure 2.2), the NP subject can be represented as consisting of a noun and a determiner, and so forth.

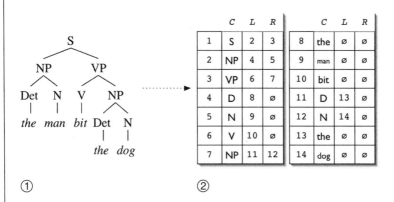

① ②

In an architecture with discrete, addressable memory locations, trees can be encoded by assigning each node to a particular memory location. For example, Node [S] might be assigned to location 1, the upper-left Node [NP] location 2, Node [VP] to location 3, and so on. Each memory location, a series of bytes indexed by address, carries three bits of information: the node's identity (e.g., NP, *dog*), a pointer to the location containing the left branch of the tree, and a pointer containing the right branch of the tree, as illustrated in Figure 2.2. (The *C* field encodes the category, *L* points to the location of the left branch, *R* to location of the right edge.)

Although the specifics often vary, the core organization is the same: a set of atomic memory locations, linked together by a pointer system obeying a fixed convention. Within this family of representations, structure-sensitive processing is readily realized with algorithms that "climb" the tree by moving along its edges and visiting its nodes (Knuth, 1973).

requiring a speaker or listener to maintain a large inventory of connections between specific elements in working memory. For example, by rough count, the tree in Figure 2.1 demands the stable encoding of at least a dozen bindings, on the reasonable assumption that each connection between a node and its daughters (e.g., S & NP) requires at least one distinct binding (perhaps as many as 42, if the table were translated into an array, such as in Box 2.1).

Although numbers of between 12 and 42 (more in more complex sentences) might at first blush seem feasible, they far exceed the amount of short-term information-binding bandwidth seen in other domains of cognition (Treisman & Gelade, 1980). George Miller (1956) famously put the number of elements a person could remember at 7 ± 2, and more recent work yields estimates closer to 4 or even fewer (Cowan, 2001; McElree, 2001). Similarly low limits seem to hold in the domain of visual object tracking (Pylyshyn & Storm, 1988). Although it is certainly possible that language affords a far greater degree of binding than in other domains (see Kintsch, Patel, & Ericsson, 1999), the general facts about human memory capacity clearly raise questions.

Second, there has thus far been no satisfying account of how neurally-realized networks would represent trees of arbitrary complexity. Three broad (overlapping) classes of solutions have been proposed: localizt, connectivity models (e.g., Selman & Hirst, 1985; Van der Velde & De Kamps, 2006); distributed, holistic models (e.g., Plate, 1994; Pollack, 1990; Smolensky, 1990); and temporal synchrony models (e.g., Henderson, 1994; Shastri & Ajjanagadde, 1993). Both connectivity and holistic models require either an exponentially increasing number of network units or arbitrary numerical precision (Plate, 1994). (For example, if sentences are mapped onto numeric values between 0 and 1, each additional element requires greater precision on the part of the mapping; a similar scale-up is required in the number of nodes used in Smolensky's tensor calculus.) Meanwhile, mechanisms that rely on temporal synchrony are limited to a single level of hierarchy or abstraction (Van der Velde & De Kamps, 2006), and no more than a handful of bindings—approximately seven, according to estimates derived from the properties of the most commonly entertained neural substrate for binding, gamma oscillations (Lisman & Idiart, 1995). In sum, there is no convincing account for how complete, unified trees of arbitrary size could plausibly be implemented in the brain.

Third, the fundamental technique by which computers represent trees—mapping those trees onto a lookup table stored in a location-addressable memory—seems not to be available to the human brain, which appears to organize memory by cues rather than specific locations, even for relatively recent information (Jonides, et al., 2008; McElree, 2006). Rather, biological memory—in species from spiders (Skow & Jakob, 2006) to snails (Haney & Lukowiak, 2001), rats (Carr, 1917), and humans (Smith, 2006)—is thought to be "content-addressable," meaning that memories are retrieved by content or context, rather than location. Given the short evolutionary history of language (Marcus, 2008), and the fundamentally conservative nature of evolution (Darwin, 1859; Jacob, 1977), context-addressable memory may be the only memory substrate that is available to the neural systems that support language. Although content addressability affords rapid retrieval, by itself it does not suffice to yield tree-geometric traversability. With content addressability, one can retrieve elements from memory based on their properties (e.g., animate, nominative, plural, etc.), but (absent location-addressable memory) not their location, affording some sort of approximate reconstruction (as discussed below), but not with the degree of reliability and precision that veridically represented trees would demand.

In Lieu of Trees

Absent discrete, location-addressable memory, how might human speakers and listeners represent syntactic structure at all? One possibility is that something that is little more than a notational variant, such as labeled bracketings, stand in. A more radical suggestion is that speakers may rely on a workaround or a "kluge" (Clark, 1987; Marcus, 2008): a clumsy patchwork of incomplete linguistic fragments stored in context-dependent memory that emulate—roughly—tree structure, but with a limited amount of precision and accuracy. Each treelet in itself would have the formal properties of a tree—directed, acyclic, and graph-theoretic in form—but be of limited size, some as small as a single hierarchical combination of mother and daughter nodes, none larger than could be encoded by independently motivated considerations about memory chunking. Crucially, treelets would (except perhaps in the case of memorized idioms or routine expressions) be smaller than full sentences, and *collections* of treelets themselves could not be fully unified, such that traversability and internal consistency could not be guaranteed. Rather, sets of treelets would potentially be underspecified (Sanford & Sturt, 2002) such that the bindings between them would often need to be reconstructed (or deduced) rather than directly read off of a unified internal representation (see Marcus, Hindle, & Fleck, 1983).

To put this somewhat differently, one would expect to have reliable access to hierarchical relations within a treelet (itself a proper directed acyclic graph), but not *between* treelets. Hierarchical relations between treelets would have to be *reconstructed*, rather than read directly from the representation, with a reliability that could be compromised by factors such as memory interference and the number of competing candidates. Box 2.2 works through two examples in comprehension. One consequence is that each sentence might require a large number of rather small elements to be distinctly represented; comprehension would be impaired to the extent that individual elements are confusable.

Box 2.2
Treelets illustrated

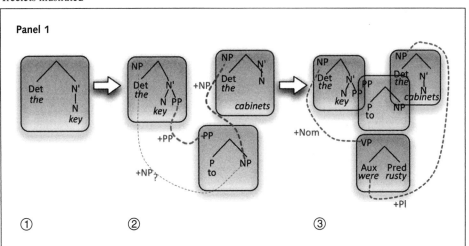

This series of treelet sets reflects the unfolding comprehension of the sentence *The key to the cabinets were rusty*, an ungrammatical string often perceived as grammatical (Pearlmutter et al., 1999; Wagers et al., 2009). **1**: A single treelet encompasses the structural description of the phrase *the key*. **2**: A set of treelets corresponds to the complex subject. Dashed lines indicate how the comprehender would associatively reconstruct relations between treelets. A spurious weak association is possible between the PP and the NP that contains it, but also unlikely given the left-to-right construction of the phrase. **3**: The presence of a VP with conflicting cues for identifying the subject leads comprehenders to reconstruct conflicting relations. For example, one treelet (*the key*) is selected for predication, but the embedded treelet (*the cabinets*) erroneously licenses agreement.

This proposal builds on earlier suggestions that subtrees might play an important explanatory role in linguistic representation, as in Tree-Adjoining Grammar (Kroch & Joshi, 1985), in the process of language acquisition, where the term *treelet* was also employed (Fodor, 1998; Fodor & Sakas, 2004); some trace the idea to Chomsky's (1957) proposal of generalized transformations. Furthermore, it relies closely on recent psycholinguistic research that sentence processing is vulnerable to retrieval-based interference (Van Dyke & Lewis, 2003; Van Dyke & McElree, 2006). Correspondingly it is particularly closely related to recent models of sentence processing that posit spreading activation or competition at the chunk/fragment level (Lewis & Vasishth, 2005; Stevenson, 1994). It moves beyond these earlier works in the suggestion that these small bits of structure are not only relevant as theoretical primitives, but in fact *the largest denominations of structure that can be stably encoded as trees.*

Box 2.2
(continued)

Panel 2

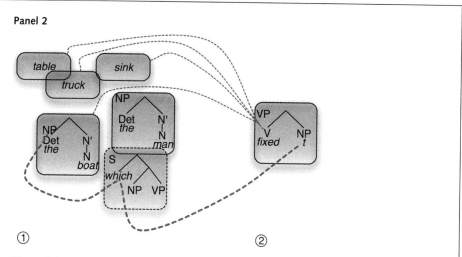

Figure 2.4

1 The set of treelets encoded during the comprehension of a sentence beginning *The boat which the man . . .*; **2** . . . *fixed*: the VP's syntactic cues point to the correct constituent (illustrated by the heavy dashed line), but semantic properties of the head can lead erroneously to the extrasyntactic fragments (illustrated by the light dashed line). Van Dyke and McElree thus find increased comprehension difficulty for **2**, relative to a control verb (. . . *sailed*) in which all cues converge on a single constituent.

One question raised by the treelet framework is whether only syntactic fragments are liable to be used in comprehension. If syntactic operations were limited to operations on a unified tree, one would not expect this to be the case. But Van Dyke and McElree (2006) have argued that even extrasyntactic items in memory, like the words in a memorized list, can interfere. They claim this occurs with filler-gap dependency completion; a way that could be captured is illustrated here.

There are many open questions, including, most crucially, the size, scope, and shape of individual treelets, and the nature of the mechanisms that mediate their bindings. See Box 2.4 for discussion.

Meanwhile, in some ways this proposal might seem to resemble the spirit of some connectionist proposals, such as Elman (1990) and McClelland (St. John & McClelland, 1990), that imply that syntactic structure might not be explicitly represented, but it differs sharply from those proposals in detail. According to the view developed here, treelets themselves are fundamentally symbolic representations, not the sort of system that "eliminative connectionists" have proposed that lack symbolic representations altogether, which lead to the limits I discussed in Marcus (2001).

If sentences were internally represented by means of loosely affiliated treelets rather than fully connected, well-defined trees (or equivalent labeled bracketings), everyday performance might be fine—to the extent that simple sentences tend to contain relatively few overlapping elements and to the extent that dependency construction is kept effectively local (see Box 2.3). However, as the confusability of elements increases, one might expect at least two empirical reflections in the (psycho) linguistic system: cases in which the parser can only make structurally imprecise reference to syntactic context—because a single set of cues might pick out more than one treelet—and cases in which local structural information otherwise predominates in parsing decisions.

Evidence That Human Representations of Linguistic Structure Are Incomplete, Transitory, and Occasionally Inconsistent

Consider first one of the oldest observations in modern linguistics: center embeddings, like (1), are nearly impossible to understand (Miller & Chomsky, 1963; Miller & Isard, 1964).

(1) The badger <u>that the dachshund</u> <u>the farmer bought</u> <u>harassed</u> left the forest.

Indeed, undergraduates are at chance at distinguishing the grammatical (2a) from the ungrammatical (2b), despite the omission (in 2b) of a whole verb phrase (Frazier, 1985; Gibson & Thomas, 1999).

(2) a. The ancient manuscript that the graduate student who the new card catalog had confused a great deal was studying in the library was missing a page.

    b. The ancient manuscript that the graduate student who the new card catalog had confused a great deal was missing a page.

What is surprising about center embeddings is not that they tax online processing (Abney & Johnson, 1991; Cowper, 1976; Miller & Chomsky, 1963; Miller & Isard, 1964; Stabler, 1994) but that they seem to confound even deliberate attempts at repair. If sentence structure could be unambiguously represented and fully connected, the recalcitrance of center embeddings "in the steady state" is puzzling. If

Box 2.3:
Why sentence processing is generally reliable and accurate

If the fundamental substrate for linguistic representation is as impoverished and vulnerable as I have suggested, why is it that sentence processing on the whole so often seems rapid, effortless, and relatively error-free? (See Boland, Tanenhaus, Garnsey, & Carlson, 1995; Kazanina, Lau, Lieberman, Phillips, & Yoshida, 2007; Phillips, 2006; Stowe, 1986; Sturt, 2003; Traxler & Pickering, 1996.) The apparent discrepancy between the relative reliability of everyday sentence processing and the weaknesses of the underlying substrate can be traced to four distinct factors.

First, morphological and semantic features can often be used as rough proxies for structural relations. For example, many of the relational properties of constituents, like that of being a subject, can be recovered from item features, like case, even in the absence of a capacity to reliably traverse a complete tree structure. To the extent that the vast majority of ordinary speech consists of simple sentences with relatively little embedding (Karlsson, 2007), such features suffice to point unambiguously; only in more complicated scenarios is breakdown a problem (Van Dyke & Lewis, 2003).

Second, the widespread existence of locality constraints in grammar may assist (or be a reflex of) an architecture in which global structure is only weakly representable. Thus, for example, on the treelet view, recovering the relative configuration of two elements should be more reliable when they are hierarchically local to one another. The success of comprehenders at resolving reflexive anaphora, whose antecedents are typically highly local, compared with difficulty in resolving pronominal anaphora, whose antecedents are antilocal (Badecker & Straub, 2002), seems consistent with this assumption, though more research is required.

Third, predictive processing strategies may enable the comprehender to maintain globally consistent hypotheses throughout the parse of a sentence by lessening the need to probe the syntactic context (Wagers, 2011). For example, it is well established that *wh*-dependencies are parsed actively: comprehenders successively formulate hypotheses about which position a *wh*-phrase has been displaced from before encountering any evidence that a constituent is missing from that position (Frazier & D'Arcais, 1989; Stowe, 1986). The basic syntactic properties of the displaced phrase thus need not be recovered and are effectively available to the parser until the dependency is completed.

Finally, a great deal of ordinary language processing appears to rapidly integrate extrasyntactic supports, guided by real-world knowledge and the representation of extralinguistic context (Sedivy, Tanenhaus, Chambers, & Carlson, 1999; Tanenhaus, Spivey-Knowlton, Eberhard, & Sedivy, 1995). By relying on a mixture of real-world knowledge, local cues, and well-adapted strategies, our linguistic system is able to compensate for the shortcomings of its representational architecture and achieve a degree of sentence processing that is in most instances, "good enough" (Ferreira et al., 2002), albeit not fully on a par with what might have been expected had human beings actually had recourse to a system for mentally representing trees of arbitrary size and complexity.

the representation is mediated by context-dependent memory fragments that resist a stable unification, confusion is expected to persist *to the extent that constituents are structurally confusable*. And indeed, when structures of different types are center embedded, the difficulty in resolving the embedded dependencies is ameliorated (Chomsky, 1965; DeRoeck et al., 1982; Lewis, 1996). Any account that relied on purely tree-geometric notions would have difficulty capturing properties such as increasing difficulty with self-similarity, structural forgetting, and a persistent resistance to repair.

Similarly, in sentences like (3), the ease with which clefted objects are bound to the embedded verbs (both underlined) is heavily influenced by the confusability of extraneous information (Gordon, Hendrick, & Johnson, 2001; Gordon, Hendrick, Johnson, & Lee, 2006).

(3) a. It was the dancer that the fireman liked.
    b. It was Tony that Joey liked.

In laboratory experiments, participants who are asked (beforehand) to memorize a list of nouns (*actor*, *baker*, *carpenter*) show greater impairment on (3a) than (3b), while those who memorize a list of names (*Andy*, *Barney*, *Charlie*) are more impaired on (3b) than (3a). These results, like those with center embedding, are surprising if processing could rely on a well-organized tree representation: the structural address of the clefted object is unambiguous. However, they fit naturally if access to syntactic structure is stored in fragmentary and confusable fashion. Further consolidating this view is evidence from response-signal techniques, which have shown that the time it takes to access discontinuous constituents (like clefted objects) in online comprehension does not depend on how far up the tree those constituents are located (McElree, Foraker, & Dyer, 2003).

Agreement processing provides another example where the properties of irrelevant constituents can lead the processing system astray. For example, in sentences like (4), an NP intervening between subject head noun and verb often leads to the incorrect production of verb agreement or failure to notice agreement errors in comprehension (compare with 5) (Bock & Miller, 1991; Pearlmutter, Garnsey, & Bock, 1999).

(4) ? The key to the cabinets are on the table.

(5) * The key are on the table.

What is especially striking about this phenomenon, frequent in speech and writing, is that one could easily imagine a system that would not be liable to these errors. With fully connected trees, it would be possible to refer unambiguously to the subject head noun to check or copy agreement. Treelets offer a natural account of why agreement attraction errors are prevalent: on the treelet view, there is no stable or permanent "path" between verb and its subject. Component representations have to be coordinated in a framework that is inherently transitory and vulnerable to confusion (Wagers, Lau, & Philips, 2009).

On a longer time scale, there is evidence that downstream memory for structural information is also highly vulnerable to confusion. Memory for the exact form of sentences is notably limited and highly labile (absent rehearsal or external supports like musical rhythm). In a widely cited study, Sachs (1967) showed that when just a few sentences intervened between a target sentence and a memory test, correct recognition for the target's exact form dropped dramatically, though memory for semantic contents declined relatively little. More recently, Lombardi and Potter (1992) demonstrated that recall for the most recently parsed sentence can be disrupted by intrusions of words with similar semantic content but different syntactic projections.

Ferreira and her colleagues have demonstrated that comprehenders often derive interpretations that are locally consistent but globally incompatible. For example, individuals who read (6) often come to the conclusion both that Anna bathed the baby, and that the baby slept (Christianson, Hollingworth, Halliwell, & Ferreira, 2001).

(6) (a) While Anna bathed the baby slept.
    (b) [[While Anna bathed][the baby slept.]]

A globally consistent parse of this sentence (bracketed in (b)) would require that the person who was bathed be Anna herself, but subjects often wrongly conclude that Anna bathed the baby. Errors such as these suggest that information that should be clear at a global level sometimes gets lost in the presence of locally well-formed fragments.

This conflict between local and global may also help to explain the existence of "syntactic illusions" like (7), which people initially perceive as grammatically acceptable, even though it lacks a coherent interpretation (Montalbetti, 1984).

(7) More people have <u>been to Russia</u> than I have [ ]

Similarly, consider (8a).

(8) a. The player <u>tossed the frisbee</u> fell on the grass.
    b. The coach smiled at the player <u>tossed the frisbee</u>.

Comprehenders initially analyze the phrase *tossed the frisbee* as the main-clause predicate, only to find out later it modified the subject as a reduced relative clause (the classic "garden-path" effect (Bever, 1970). In (8b), the reduced relative analysis is the only globally supportable option, because the NP is now in an oblique object position. Yet Tabor and colleagues have provided data that difficulty in accessing the reduced relative analysis persists even in (8b), and have argued that the difficulty stems from the "local coherence" of the string following *the player* (Gibson, 2006; Tabor, Galantucci, & Richardson, 2004).

Any one of these phenomena might, in itself, be explained in a variety of ways, but taken together, they suggest that the human linguistic system may muddle by without trees, instead relying on an incomplete, partial, and occasionally inconsistent collection of subtrees that are context dependent, cue-driven, and highly vulnerable to interference.

Discussion

Tree structures provide a powerful and useful means of organizing complex relationships, and anyone designing a brain from scratch would be well advised to incorporate them. But evolution does not always evolve the most useful solution for a given problem—only the fittest among available options. Instead, the logic of evolution is such that new systems are often built on spare parts (Jacob, 1977; Marcus, 2008), in ways that sometimes lead to suboptimal solutions. If this conjecture is correct, proper tree structures qua human representational format never evolved; instead, speakers and listeners are forced to rely on incompletely bound sets of context-dependent, interference-vulnerable treelets as an alternative.

This possibility could help make sense of a number of puzzles in linguistics, such as why human performance so often deviates—in very specific ways—from idealized competence. The fact that human listeners are "lazy decoders" (Stolz, 1967) who frequently settle for "good enough" (Ferreira, Bailey, & Ferraro, 2002) parses rather than parses that are strict or complete, may stem from a fundamental limitation in the underlying representational capacity.

At the same time, this underlying representational limit may also cast some light on linguistic *competence*. For example, many linguistic frameworks require rules to operate over local domains, like the principles of "minimality" that require certain syntactic dependencies to be established with the structurally closest element of the right type (Chomsky, 1995; Rizzi, 1990), which have no correlate in formal languages (such as predicate calculus or C++). Such constraints may arise not from any a priori requirement on how a linguistic system would necessarily be constructed, but simply (on the treelet-based view) as a strategy or accommodation to the limits imposed by the potentially deleterious consequences of fragment-based confusability. Along those lines, several recent proposals in theoretical syntax also eschew or minimize the role of a global syntactic structure—for example, Combinatory Categorial Grammar (Steedman, 2001), the Multiple Spell-Out hypothesis in minimalism (Uriagereka, 1999) (or, similarly, the Phonological Merge proposal (Frank, 2010)).

A treelet approach can both make sense of why language might be structured in this way: chaining together locally specified dependencies may be the most reliable

Box 2.4
Directions for future research

What is the size or domain of a treelet? Answers to this question are likely to come from several sources:

• From neurocomputational models, in which depth of embedding and self-similarity play a controlling role in determining how much information can be reliably encoded in one representation.

• From syntactic theories, in which natural local domains can be defined. In particular, there seems to be close kinship with the elementary trees of Tree-Adjoining Grammar (Kroch & Joshi, 1985).

• From psycholinguistic experimentation that carefully assesses what properties of constituents, and what structural domains, are capable of inducing interference in online sentence comprehension or production.

How do alternate conceptions of phrase structure interact with the nature of human memory? Of particular interest are sets-of-strings based implementations (Chomsky, 1955/1975; Lasnik & Kupin, 1977), in which many important structural relationships can be inferred but are not explicitly encoded; and representations that underspecify structure (Marcus et al., 1983).

The strategy of "chunking" information, which the treelet approach embodies, has long been appreciated in cognitive psychology (Anderson, 2005) and plays an important role in a recent model of syntactic parsing (Lewis & Vasishth, 2005). What are its implications for the formulation of syntactic theories?

way of building unbounded dependencies out of context-dependent fragments. In short, a shift from trees to a clumsier but more psychologically plausible system of treelets—in conjunction with experimental work that aims to link treelets with specific syntactic analyses (see Box 2.4)—might give linguists and psycholinguists not only new tools with which to understand linguistic performance but also a new a perspective on why language has the character it does.

Acknowledgments

This chapter was prepared with the support of NIH HD04873; I also thank Matt Wagers for extensive and tremendously helpful discussion.

References

Abney, S., & Johnson, M. (1991). Memory requirements and local ambiguities for parsing strategies. *Journal of Psycholinguistic Research, 20*, 233–250.

Anderson, J. R. (2005). Human symbol manipulation within an integrated cognitive architecture. *Cognitive Science, 29*, 313–341.

Badecker, W., & Straub, K. (2002). The processing role of structural constraints on the interpretation of pronouns and anaphors. *Learning & Memory (Cold Spring Harbor, NY)*, *28*, 748–769.

Bever, T. G. (1970). The cognitive basis for linguistic structures. In J. R. Hayes (Ed.), *Cognition and the development of language* (pp. 279–352). New York: Wiley.

Bock, J. K., & Miller, C. A. (1991). Broken agreement. *Cognitive Psychology*, *23*, 45–93.

Boland, J. E., Tanenhaus, M. K., Garnsey, S. M., & Carlson, G. N. (1995). Verb argument structure in parsing and interpretation: Evidence from wh-questions. *Journal of Memory and Language*, *34*, 774–806.

Carr, H. (1917). Maze studies with the white rat. *Journal of Animal Behavior*, *7*(4), 259–275.

Chomsky, N. A. (1955/1975). *The logical structure of linguistic theory*. Chicago: University of Chicago Press.

Chomsky, N. A. (1957). *Syntactic structures*. The Hague: Mouton.

Chomsky, N. A. (1965). *Aspects of the theory of syntax*. Cambridge, MA: MIT Press.

Chomsky, N. A. (1995). *The Minimalist Program*. Cambridge, MA: MIT Press.

Chomsky, N. A. (2001). Derivation by phase. In Michael Kenstowicz (Ed.), *Ken Hale: A life in language* (pp. 1–52). Cambridge, MA: MIT Press.

Chomsky, N. A. (2008). On phases. In Robert Freidin, C. P. Otero, & Maria Luisa Zubizarreta (Eds.), *Foundational issues in linguistic theory: Essays in honor of Jean-Roger Vergnaud* (pp. 133–166). Cambridge, MA: MIT Press.

Christianson, K., Hollingworth, A., Halliwell, J. F., & Ferreira, F. (2001). Thematic roles assigned along the garden path linger. *Cognitive Psychology*, *42*(4), 368–407.

Clark, A. (1987). The kludge in the machine. *Mind & Language*, *2*(4), 277–300.

Cowan, N. (2001). The magical number 4 in short-term memory: A reconsideration of mental storage capacity. *Behavioral and Brain Sciences*, *24*(1), 87–114, discussion 114–185.

Cowper, E. A. (1976). *Constraints on sentence complexity: A model for syntactic processing*. Providence, RI: Brown University.

Darwin, C. (1859). *On the origin of species*. London: Murray.

DeRoeck, A., Johnson, R., King, M., Rosner, M., Sampson, G., & Varile, N. (1982). A myth about centre-embedding. *Lingua*, *58*, 327–340.

Elman, J. L. (1990). Finding structure in time. *Cognitive Science*, *14*, 179–211.

Ferreira, F., Bailey, K. G. D., & Ferraro, V. (2002). Good-enough representations in language comprehension. *Current Directions in Psychological Science*, *11*(1), 11–15.

Fodor, J. D. (1998). Unambiguous triggers. *Linguistic Inquiry*, *29*(1), 1–36.

Fodor, J. D., & Sakas, W. G. (2004). *Evaluating models of parameter setting*. Paper presented at the 28th Annual Boston University Conference on Language Development.

Frank, R. (2010). Lexicalized syntax and phonological merge. In S. Bangalore & A. Joshi (Eds.), *Supertagging: Complexity of lexical descriptions and its relevance to natural language processing* (pp. 373–4066). Cambridge, MA: MIT Press.

Frazier, L. (1985). Syntactic complexity. In D. Dowty, L. Karttunen, & A. Zwicky (Eds.), *Natural language processing: Psychological, computational and theoretical perspectives* (129–189). Cambridge: Cambridge University Press.

Frazier, L., & D'Arcais, G. B. F. (1989). Filler driven parsing: A study of gap filling in Dutch. *Journal of Memory and Language, 28,* 331–344.

Gibson, E. (2006). The interaction of top-down and bottom-up statistics in the resolution of syntactic category ambiguity. *Journal of Memory and Language, 54,* 363–388.

Gibson, E., & Thomas, J. (1999). Memory limitations and structural forgetting: The perception of complex ungrammatical sentences as grammatical. *Language and Cognitive Processes, 14*(3), 225–248.

Gordon, P. C., Hendrick, R., & Johnson, M. (2001). Memory interference during language processing. *Journal of Experimental Psychology: Learning, Memory, and Cognition, 27*(6), 1411–1423.

Gordon, P. C., Hendrick, R., Johnson, M., & Lee, Y. (2006). Similarity-based interference during language comprehension: Evidence from eye tracking during reading. *Journal of Experimental Psychology: Learning, Memory, and Cognition, 32*(6), 1304–1321.

Haney, J., & Lukowiak, K. (2001). Context learning and the effect of context on memory retrieval in Lymnaea. *Learning & Memory (Cold Spring Harbor, NY), 8*(1), 35–43.

Henderson, J. B. (1994). Connectionist syntactic parsing using temporal variable binding. *Journal of Psycholinguistic Research, 23,* 353–379.

Jacob, F. (1977). Evolution and tinkering. *Science, 196,* 1161–1166.

Jonides, J., Lewis, R. L., Nee, D. E., Lustig, C. A., Berman, M. G., & Moore, K. S. (2008). The mind and brain of short-term memory. *Annual Review of Psychology, 59,* 193–224.

Karlsson, F. (2007). Constraints on multiple center-embedding of clauses. *Journal of Linguistics, 43,* 365–392.

Kayne, R. (1984). *Connectedness and binary branching.* Dordrecht: Foris.

Kazanina, N., Lau, E., Lieberman, M., Phillips, C., & Yoshida, M. (2007). The effect of syntactic constraints on the processing of backwards anaphora. *Journal of Memory and Language, 56,* 384–409.

Kintsch, W., Patel, V., & Ericsson, K. A. (1999). The role of long-term working memory in text comprehension. *Psychologia, 42,* 186–198.

Knuth, D. E. (1973). *The art of computer programming* (2nd ed.). Reading, MA: Addison-Wesley.

Kroch, A., & Joshi, A. (1985). *The linguistic relevance of Tree Adjoining Grammar* (Report No. MS-CIS-85-16). Philadelphia: University of Pennsylvania Department of Computer and Information Science.

Lasnik, H., & Kupin, J. J. (1977). A restrictive theory of transformational grammar. *Theoretical Linguistics, 4,* 173–196.

Lewis, R. L. (1996). Interference in short-term memory: The magical number two (or three) in sentence processing. *Journal of Psycholinguistic Research, 25,* 193–215.

Lewis, R. L., & Vasishth, S. (2005). An activation-based model of sentence processing as skilled memory retrieval. *Cognitive Science, 29,* 375–419.

Lisman, J. E., & Idiart, M. A. (1995). Storage of 7 +/– 2 short-term memories in oscillatory subcycles. *Science*, *267*(5203), 1512–1515.

Lombardi, L., & Potter, M. C. (1992). The regeneration of syntax in short-term memory. *Journal of Memory and Language*, *31*, 713–733.

Marcus, G. F. (2001). *The algebraic mind: Integrating connectionism and cognitive science.* Cambridge, MA: MIT Press.

Marcus, G. F. (2008). *Kluge: The haphazard construction of the human mind.* Boston: Houghton Mifflin.

Marcus, M. D., Hindle, D., & Fleck, M. (1983). D-theory: Talking about talking about trees. *Association for Computational Linguistics*, *21*, 129–136.

McElree, B. (2001). Working memory and focal attention. *Journal of Experimental Psychology: Learning, Memory, and Cognition*, *27*(3), 817–835.

McElree, B. (2006). Accessing recent events. In B. H. Ross (Ed.), *The psychology of learning and motivation: Advances in research and theory* (155–200). New York: Academic Press.

McElree, B., Foraker, S., & Dyer, L. (2003). Memory structures that subserve sentence comprehension. *Journal of Memory and Language*, *48*, 67–91.

Miller, G. A. (1956). The magical number seven, plus or minus two: Some limits on our capacity for processing information. *Psychological Review*, *63*, 81–97.

Miller, G. A., & Chomsky, N. (1963). Finitary models of language users. In R. D. Luce, R. R. Bush, & E. Galanter (Eds.), *Handbook of mathematical psychology* (Vol. 2, pp. 419–492). New York: Wiley.

Miller, G. A., & Isard, S. (1964). Free recall of self-embedded English sentences. *Information and Control*, *7*, 292–303.

Montalbetti, M. (1984). *After binding.* Unpublished doctoral dissertation, MIT.

Partee, B. H., ter Meulen, A. G., & Wall, R. (1990). *Mathematical methods in linguistics.* Dordrecht: Kluwer.

Pearlmutter, N. J., Garnsey, S. M., & Bock, K. (1999). Agreement processes in sentence comprehension. *Journal of Memory and Language*, *41*, 427–456.

Pesetsky, D. (1982). *Paths and categories.* Cambridge, MA: MIT Press.

Phillips, C. (2006). The real-time status of island constraints. *Language*, *82*, 795–823.

Plate, T. (1994). *Distributed representations and nested compositional structure.* Unpublished doctoral dissertation, University of Toronto.

Pollack, J. B. (1990). Recursive distributed representations. *Artificial Intelligence*, *46*, 77–105.

Pylyshyn, Z. W., & Storm, R. W. (1988). Tracking multiple independent targets: Evidence for a parallel tracking system. *Spatial Vision*, *3*, 179–197.

Rizzi, L. (1990). *Relativized minimality.* Cambridge, MA: MIT Press.

Sachs, J. (1967). Recognition memory for syntactic and semantic aspects of connected discourse. *Perception & Psychophysics*, *2*, 437–442.

Sanford, A., & Sturt, P. (2002). Depth of processing in language comprehension: not noticing the evidence. *Trends in Cognitive Sciences*, *6*(9), 382.

Sedivy, J. C., Tanenhaus, M. K., Chambers, C. G., & Carlson, G. N. (1999). Achieving incremental semantic interpretation through contextual representation. *Cognition*, *71*(2), 109–147.

Selman, B., & Hirst, G. (1985). *A rule-based connectionist parsing system.* Paper presented at the Proceedings of the Seventh Annual Cognitive Science Society, 212–221, Irvine, CA.

Shastri, L., & Ajjanagadde, V. (1993). From simple associations to systematic reasoning: A connectionist representation of rules, variables, and dynamic bindings using temporal synchrony. *Behavioral and Brain Sciences*, *16*, 417–494.

Skow, C. D., & Jakob, E. M. (2006). Jumping spiders attend to context during learned avoidance of aposematic prey. *Behavioral Ecology*, *17*(1), 34–40.

Smith, S. M. (2006). Context and human memory. In H. L. Roediger, Y. Dudai, & S. Fitzpatrick (Eds.), *Science of memory: Concepts* (111–114). New York: Oxford University Press.

Smolensky, P. (1990). Tensor product variable binding and the representation of symbolic structures in connectionist systems. *Artificial Intelligence*, *46*, 159–216.

St. John, M., & McClelland, J. L. (1990). Learning and applying contextual constraints in sentence comprehension. *Artificial Intelligence*, *46*, 217–257.

Stabler, E. P. (1994). The finite connectivity of linguistic structures. In C. Clifton, L. Frazier & K. Rayner (Eds.), Perspectives in sentence processing (pp. 303–336). Hillsdale, NJ: Erlbaum.

Steedman, M. (2001). *The syntactic process* (Vol. 25). Cambridge, MA: MIT Press.

Stevenson, S. (1994). Competition and recency in a hybrid network model of syntactic disambiguation. *Journal of Psycholinguistic Research*, *23*, 295–322.

Stolz, W. (1967). A study of the ability to decode grammatically novel sentences. *Journal of Verbal Learning and Verbal Behavior*, *6*(6), 143–163 .

Stowe, L. A. (1986). Evidence for on-line gap-location. *Language and Cognitive Processes*, *1*, 227–245.

Sturt, P. (2003). The time-course of the application of binding constraints in reference resolution. *Language and Cognitive Processes*, *18*, 867–873.

Tabor, W., Galantucci, B., & Richardson, D. (2004). Effects of merely local syntactic coherence on sentence processing. *Journal of Memory and Language*, *50*(4), 355–370.

Tanenhaus, M. K., Spivey-Knowlton, M. J., Eberhard, K. M., & Sedivy, J. C. (1995). Integration of visual and linguistic information in spoken language comprehension. *Science*, *268*(5217), 1632–1634.

Traxler, M. J., & Pickering, M. J. (1996). Plausibility and the processing of unbounded dependencies: An eye-tracking study. *Journal of Memory and Language*, *35*, 454–475.

Treisman, A. M., & Gelade, G. (1980). A feature-integration theory of attention. *Cognitive Psychology*, *12*(1), 97–136.

Uriagereka, J. (1999). Multiple spell-out. In S. D. Epstein & N. Hornstein (Eds.), *Working minimalism* (pp. 251–282). Cambridge, MA: MIT Press.

Van der Velde, F., & de Kamps, M. (2006). Neural blackboard architectures of combinatorial structures in cognition. *Behavioral and Brain Sciences*, *29*, 37–70.

Van Dyke, J. A., & Lewis, R. L. (2003). Distinguishing effects of structure and decay on attachment and repair: A cue-based parsing account of recovery from misanalyzed ambiguities. *Journal of Memory and Language*, *49*(3), 285–316.

Van Dyke, J. A., & McElree, B. (2006). Retrieval interference in sentence comprehension. *Journal of Memory and Language*, *55*, 157–166.

Wagers, M. W. (2011). *The structure of memory meets memory for structure in linguistic cognition*. Ann Arbor, MI: Proquest.

Wagers, M. W., Lau, E. F., & Philips, C. (2009). Agreement attraction in comprehension: Representations and processes. *Journal of Memory and Language*, *61*, 206–237.

3 Convergence and Deep Homology in the Evolution of Spoken Language

W. Tecumseh Fitch and Daniel Mietchen

Introduction: Homology and Analogy

The concept of homology—a biologically essential form of "sameness"—is one of the most powerful and useful ideas in evolutionary biology, and yet one of the most vexed. Darwin's early opponent, the great comparative anatomist Richard Owen, coined the term. For Owen, the skeletal homology between different forms of vertebrate limb (seal flippers, bird wings, and human hands) revealed an underlying "type" or Platonic order underlying biological reality. But Darwin recognized that such morphological homologies reflected a far more concrete underlying order: evolutionary descent with modification. In the post-Darwinian world, homology quickly and irrevocably transmuted its core meaning to the modern interpretation: structures or traits in different species that, though potentially different in form or function, descend from an ancestral structure or trait that was present in a common ancestor (Lankester, 1870; Mivart, 1870). Although Owen's rather different concept of "serial homology" (the "sameness" inherent in the repeated appendages of an arthropod, or the ribs or fingers of a vertebrate) persists, it has taken a backseat relative to this modern, Darwinian, notion of traits inherited from a common ancestral trait (what Owen termed *special homology*). Unadorned, the term *homology* today denotes a character shared by two taxa by virtue of inheritance from a common ancestor, regardless of current form or function. Homologies are typically used by systematists to construct taxonomies, and in phylogenetic analysis to reconstruct ancestral traits.

Owen also recognized a second form of structural similarity, an accidental and inessential sameness he referred to as "analogy." For Owen, analogies were similar traits that did not reflect the form of the underlying type. In Darwinian terms, such inessential similarities reveal independent evolutionary histories, rather than descent from a common trait. Today, analogies (traits arising by convergent evolution) and other forms of homoplasy (characters shared by two taxa, not present in their last common ancestor) are also interpreted in an evolutionary framework. Thus, for example, the superficial similarity between insect and bird wings does not stem from any winglike structure in the common ancestor of birds and insects, but rather reflects the exigencies of flight and the aerodynamic requirements it imposes. However, in Owen's philosophy analogies were uninteresting distractions from the search for underlying Baupläne, and even today a certain disdain for "analogies," particularly in systematics, has persisted. Nonetheless, comparative biologists recognize that examples of convergent or parallel evolution often provide valuable insights into natural selection. Analogies may result from the functional constraints of the problem space shaping a structure, and can play a crucial role in formulating and more importantly *testing* adaptive hypotheses (Harvey & Pagel, 1991; Tinbergen, 1963). They are thus crucial components of the comparative method, as we will elaborate below.

Genetic Homology

More recently, homology received a major boost in importance as molecular biological techniques matured to the point of providing protein and gene sequences in multiple organisms (Fitch, 1970, 2000). It rapidly became clear that both protein (amino acid) sequences and gene (base-pair) sequences were often evolutionarily conserved, allowing quantitative comparisons across widely divergent species. Because chunks of DNA evolve via descent with modification, a directly analogous Darwinian concept of homology could be applied to sequence data: homologous genes are those derived from the same gene in a common ancestor. Because of the great density of information in even a short gene sequence, and the precision with which such sequences can be determined, it is often easy to recognize genetic homologs by comparing sequence data. Furthermore, the redundancy of the genetic code means that certain base-pair substitutions have no effect on the protein "phenotype," and such synonymous substitutions provide a ready test for functional convergence in proteins ("analogy") versus true genetic homology. These virtues make ribonucleic and protein sequences perfect for testing phylogenetic hypotheses (Fitch & Margoliash, 1967). Thus, the concept of genetic homology was clarified and codified early, and continues to play a central role in modern molecular genetics and genomics (Fitch, 2000; Wagner, 2007).

Nonetheless, genetic interpretations of homology are not without complexity. For example, there are two common ways two or more copies of the "same" gene may come about. The first, directly analogous to the traditional concept, are duplicate genes that arise with speciation, when two species diverge from one previously interbreeding population of an ancestral species. In such genes, termed *orthologs*, the genealogy of the gene often maps directly onto that of the species possessing it. However, genes are often duplicated *within* a single lineage as well (Ohno, 1970), leading to another common possibility, where different genes within a single organism originated via gene duplication from a common ancestral gene. Such genes, termed *paralogs*, are akin to serial homologs, and the tree describing their genealogy clearly will not map isomorphically onto the phylogenetic trees of species possessing them. This adds complexity to the interpretation of gene families among different species, especially if independent duplication events occurred in two separate lineages. Nonetheless, practically speaking, genetic homology remains better defined, and far easier to measure, than its conceptual predecessor, morphological homology (Wagner, 2007).

The nature of the *relationship* between genetic and phenotypic homology remains far less clear. One might expect (or hope) that genetic homology directly reflects morphological homology, and vice versa. Unfortunately, it has become increasingly evident that the links between traditional morphological or behavioral homology, on the one hand, and genetic homology on the other, are complex and indirect, giving rise to a continuum of possible homologs, from genes to developmental processes through to structure (Hall, 2007). Two extreme possibilities are well attested:

1. Structures that are clearly homologous at the morphological level in two different organisms are not necessarily controlled by homologous genes. Some classic cases include segmentation genes in drosophila versus grasshoppers (see French, 2001; Nagy, 1994), or the role of Ubx in insect wing development (Weatherbee et al., 1999).
2. Structures that are clearly *not* homologous in the traditional sense, but represent analogs resulting from convergent evolution in different lineages, may nonetheless be governed by genes that are homologs, a situation dubbed "deep homology" by Shubin, Tabin, and Carroll (1997).

Despite their close conceptual relationship, "homology" at genetic, developmental, and phenotypic levels is not the same thing. The disjunctions above mean that we cannot expect simple relationships among phenotypic and genotypic levels, and must abandon the hope of recognizing morphological homology via its genetic basis alone (e.g., Dickinson, 1995; Gehring, 1998). Instead, the developmental/genetic bases for homologous traits must be sought at a more abstract level (e.g., Gilbert and Bolker's "process homology," Davidson's "kernel" gene regulatory networks, or Wagner's "character identity networks" (Davidson, 2006; Gilbert & Bolker, 2001; Wagner, 2007).

Our purpose here is not to further discuss this difficult issue, but rather to explore its consequences for comparative biologists, and particularly for comparisons between birds and mammals. For, we will argue, the discovery and increasing recognition of "deep homology" as an important phenomenon in evolution (cf., Shubin et al., 2009; Fitch, 2011; Scharff & Petri, 2011) provides a new impetus for neural and behavioral comparisons among highly divergent organisms.

Indeed, to the extent that deep homology is a common phenomenon, it provides a powerful new theoretical grounding, and practical justification, for studies of the mechanisms underlying convergently evolved traits. This prominently includes traits like vocal learning in humans and birds. This comparison, recognized as convergence, played a central role for Charles Darwin in his development of a model of language evolution (see Fitch, chapter 24, this volume). Unfortunately, the relevance to human evolution of convergently evolved traits like birdsong remains insufficiently appreciated today, particularly by those in the human-oriented sciences (cognitive psychology, anthropology, and linguistics).

Where deep homology involves humans, detailed molecular and experimental investigation of convergent traits in "model organisms" offers insights into their genetic, developmental, and ultimately computational basis in humans, at a far deeper and richer level than previously expected. Given that the main alternative is laborious genetic engineering of model organisms like mice and fruit flies, which do not naturally exhibit the trait of interest, and can hardly be expected to do so under the influence of a genetic manipulation, a comparative approach that exploits "experiments in nature" may provide the simplest empirical route to mechanistic insights for unusual human traits that is currently available. We will now try to justify this assertion by reviewing some recent results.

Deep Homology: History Repeats Itself

The term *deep homology* designates a situation in which traits in two widely separated species are generated by one or more genes or genetic networks that are homologous (Gilbert, 2003). Deep homology can exist even when the traits are analogous (or more generally, homoplastic) at the organismic and phenotypic level (Shubin, Tabin, & Carroll, 1997). That such a situation might be common was surprising at first, and was one of the first truly novel insights of evolutionary developmental biology ("evo-devo"). The first, now classic, example discovered was the role of *Hox* genes in limb development in insects and land vertebrates (Shubin et al., 1997). Appendages in these two groups are not homologous: early chordates present in the fossil record did not possess limbs at all. Thus limbs postdate the common ancestor of vertebrates and arthropods. Nonetheless, the complex genetic regulatory network that underlies the development of appendages is strikingly similar, with both the regulatory genes themselves and the interactions between them showing repeated and nonrandom identities, including a host of genes homologous at the sequence level. Indeed, even wing development in insects and birds, that classic example of convergent evolution, is controlled by a set of homologous genes, and unfolds in a comparable developmental sequence (Gilbert, 2003). Thus, genetic and developmental homologies may extend far deeper down into the tree of life than the phenotypic structures that they generate.

These early findings galvanized the field of evo-devo, and have led to an increasing use of comparative genetics to understand the evolutionary developmental bases for a diversity of morphological traits from branchial arches to image-forming eyes (Carroll, Grenier, & Weatherbee, 2005). The core insight is that the remarkable diversity of animal forms results from different deployments of a widely shared "developmental toolkit," much of it probably inherited from early metazoans or single-celled eukaryotes. Precisely because this toolkit is so widely shared, we can confidently expect studies of organisms as distantly related as yeast, sponges, octopuses, fruit flies, parrots, and mice to all inform one another in numerous ways, at a detailed mechanistic level (Carroll, 2005, 2006).

Another early example of deep homology, again occurring in insects and vertebrates, is the conserved role for the paired-box gene *Pax6* in eye development. The evolution of image-forming "camera" eyes is a classic case of convergent evolution, already exploited by (Darwin, 1859) to demonstrate the role of natural selection in shaping what he called "organs of extreme perfection." Further, detailed explorations of the variety of eyes in the animal world suggested that eyes of various types have evolved convergently at least 40 times (Land & Fernald, 1992; Von Salvini-Plawen & Mayr, 1977). Nonetheless, a second early evo-devo success story was the discovery that *Pax6* plays a similar role in eye specification in flies and mice, and indeed that defects in this same gene lead to congenital aniridia in humans. Further research has demonstrated that this developmental role, like that of *Hox* genes, is also widely conserved in animals, including such marvels of convergent evolution as the camera eye in squid and other cephalopods (Tomarev et al., 1997). Thus, again, a classic case of convergent evolution turned out to have a homologous genetic basis. This led to early suggestions that eyes of far greater complexity than previously imagined may have been present in early metazoans (i.e., that the genetic homology truly indicated a morphological homology). But the situation as now understood is more complex than this, and gives a clear indication of both the promise and the dangers of deep homology (Simpson & Price, 2002; Van Heyningen & Williamson, 2002). First, *Pax6* plays many roles in development, and is not limited to being a "master control gene for eyes." Indeed, recent evidence suggests that it may play an important role in regulating neurogenesis during cortical development (Mo & Zecevic, 2008). Second, a diversity of other, nonhomologous genes play important roles in eye development in arthropods, mollusks, and vertebrates, supporting the traditional interpretation that these eyes are morphologically convergent analogs (Gilbert & Bolker, 2001). In some ways, then, *Pax6* is at the other extreme from the *Hox* networks involved in appendage development: it is a relatively isolated example of genetic homology in an otherwise diverse and homoplastic set of genetic regulatory networks. Although there are other genetic homologs involved in eyes, most notably the opsin genes involved in photoreception, these appear to be fully homologous, inherited as such from the common ancestor of all eye-bearing animals.

FoxP2 as an Example of Deep Homology for a Convergent Behavioral Trait

The examples of *Hox, Ubx,* and *Pax* genes discussed above are all involved in developmental genetic control of morphological traits. But a central insight in modern ethology is that behavioral traits are just as amenable to evolutionary and phylogenetic analysis as morphology (Tinbergen, 1963). Thus, with equal validity, genetically influenced behavioral characteristics, such as vocalizations or components of courtship displays, can be compared among different species and classified as homologous or convergent (Lorenz, 1953; Tinbergen, 1952). We can therefore ask whether examples of deep homology are present in the behavioral realm as well. The answer, we will argue, is yes: *FoxP2* provides a well-studied example of a homologous gene that plays a direct causal role in vocal control and coordination, and in particular in vocal learning, in birds and humans (cf. Fitch, 2011; Scharff & Petri, 2011; Fisher, chapter 21, this volume). Because vocal learning has evolved convergently in these two clades, this is an example of deep homology at the neural/behavioral level.

The forkhead-box gene *FoxP2* is a highly conserved transcription factor found in all tetrapods. Like most transcription factors, it is pleiotropically expressed in many different tissues including the lungs, esophagus, and brain (Shu et al., 2007), but its role in vocal coordination has been a major focus of interest and research in the last decade (see Fisher, chapter 21, this volume). This role was initially uncovered, fortuitously, in the course of detailed studies of developmental speech impairment in a large British family (Vargha-Khadem & Passingham, 1990), which led first to isolation of the chromosomal region (Fisher, Vargha-Khadem, Watkins, Monaco, & Pembrey, 1998) and finally of the specific gene involved, *FoxP2* (referred to as *FOXP2* in humans, *Foxp2* in mice, and *FoxP2* in all other vertebrate species) (Lai, Fisher, Hurst, Vargha-Khadem, & Monaco, 2001). *FoxP2* shows a broad expression pattern in the developing mammalian brain consistent with a central role in motor control and coordination (including cerebral cortex, basal ganglia, and cerebellum) as well as auditory-motor coordination (e.g., the inferior colliculus), consistent with the motoric deficits observed in humans bearing a *FOXP2* mutant allele (Lai, Gerrelli, Monaco, Fisher, & Copp, 2003). These correlative findings are reinforced by genetic engineering experiments in mice that are knocked out for the gene, which show deficits in motor behavior, including the frequency (but not the fine structure) of vocal output (Shu et al., 2005).

Birds also possess a *FoxP2* ortholog, and its expression pattern in the avian brain shows a concordance with that in mammals, particularly in the striatum (part of the basal ganglia) (Haesler et al., 2004). Unlike mice or most other mammals, songbirds exhibit complex vocal learning—the capacity to produce novel vocalizations closely matched acoustically to vocalizations that they hear, either during formative developmental stages ("close-end learners") or throughout the lifespan ("open-ended learners"). This capacity for vocal learning has clearly evolved convergently in humans and songbirds: many bird orders, and most mammals, including all nonhuman primates, lack complex vocal learning, making it extremely unlikely that the last common ancestor of birds and mammals, a stem amniote, possessed this ability. Nonetheless, recent experimental work from Constance Scharff's laboratory in Berlin clearly demonstrates a role for *FoxP2* and other forkhead-box genes in songbird vocal control and vocal learning. First, correlational studies of *FoxP2* expression in closed-ended learners like finches showed strong expression in the striatal vocal Area X during the sensitive period for song learning. This was not the case in other control regions uninvolved in vocal learning. Second, seasonal open-ended learners like adult canaries exhibit *FoxP2* up-regulation during the annual period of song instability (Haesler et al., 2004). Third, in a technological tour de force, RNA interference via a viral vector was used to experimentally down-regulate *FoxP2* in developing zebra finches, during the sensitive song-learning period. When *FoxP2* was thus manipulated in Area X, the young finches showed less accurate, more variable song learning than normal birds, or birds with *FoxP2* manipulations to nonsong regions (Haesler et al., 2007). Together, these papers demonstrate a causal role for *FoxP2* in vocal learning in zebra finches, and thus reveal the *FoxP2* gene as a first example of deep homology for a convergent behavioral trait—a trait of central interest in the evolution of birdsong, speech, and language.

Other recent work on echolocating bats also indicates some vocal role for *FoxP2* in this large and highly-vocal group (Li, Wang, Rossiter, Jones, & Zhang, 2007). Compared to other mammalian clades, bats have a highly variable *FoxP2* gene sequence, and the ratio of nonsynonymous to synonymous base-pair changes, together with previous work on vocal function, strongly suggests accelerated evolution of this gene in echolocating bats. Interestingly, different changes have occurred in different bat lineages, especially those adopting different styles of echolocation (e.g., constant-frequency versus frequency-modulated signals), and at least one of these duplicates an unusual protein variant found in humans but not chimpanzees or other primates. The specific role of *FoxP2* in bat vocalization remains to be determined, and it is not apparently tied to vocal learning (although our knowledge of vocal learning in bats remains limited), but we can expect rapid progress on this front. Interestingly, Li et al. (2007) also sequenced *FoxP2* in echolocating cetaceans, and found no evidence for its differential evolution in this clade. Given that echolocating cetaceans have evolved a novel phonatory system, derived from nasal structures around the blowhole, and use this rather than the larynx to produce their echolocation clicks (Dormer, 1979; Norris, 1969), this apparent difference is not unexpected. However, the bird syrinx is also a novel production system (Fitch & Hauser, 2002; Suthers, 1999), and differences in the production of song may lead to differences in genetic developmental networks as well.

In summary, an important role for *FoxP2* in the neural systems underlying vocal learning has clearly been demonstrated in birds and humans, and some more general role in producing complex, coordinated vocalizations is suggested by the bat data. Because vocal learning has evolved convergently in all three lineages, and the *FoxP2* gene is clearly a highly conserved ortholog, this provides a beautiful neural/behavioral example of deep homology for a convergent trait. This discovery opens the door to experimental investigations of the specific role of *FoxP2* expression in vocal learning, and of the many other genes with which it interacts (see Fisher, chapter 21, this volume). Because many brain regions in birds are now understood to be directly homologous to mammalian brain regions (Avian Brain Nomenclature Consortium, 2005; Jarvis, 2004; Striedter, 2004), this approach promises rich new insights into the genetic basis of vocal learning, and many of the results are likely to have direct applicability to human speech and language learning. We conclude that this example of deep homology provides a powerful rationale for expanded comparative research using birdsong as a model system for human speech (Doupe & Kuhl, 1999; Doupe, Perkel, Reiner, & Stern, 2005; Scharff & Haesler, 2005). Even if the *FoxP2* example turned out to be unique, it would be momentous. However, we will now argue that this is more likely to be a common finding in the future, as our understanding of the genetic bases for neural circuits is further expanded.

Deep Homology as an Expected Correlate of Epigenesis

The highly conservative nature and deep homology of *Hox* genes in segmentation and limb development were considered surprising, contradicting the expectations of previous developmental or evolutionary biologists. Nonetheless, as awareness of these phenomena has permeated evo-devo circles, genetic conservatism and deep homology have come to seem less unexpected, or even predictable, for three reasons. First, comparative genomics has now demonstrated that complex organisms have far fewer genes than initially expected (down from initial estimates of 100,000 genes, early in the Human Genome Project, to a current good estimate of roughly 25,000 genes (International Human Genome Sequencing Consortium, 2004). Most genes are highly conserved in both sequence structure (with most human genes having a clear mouse ortholog) and structural/developmental function.

Second, there is little correlation between the total number of genes in a species and any obvious measure of complexity: mice and humans have comparable numbers of genes, while amphibians or plants like rice have more than either (40,000–60,000:

rice.plantbiology.msu.edu). Some fish have more, and some less, genes than humans, with no obvious correlation with "complexity" (Vandepoele, De Vos, Taylor, Meyer, & Van de Peer, 2004). These facts have necessitated a wholesale revision of ideas about how complex phenotypes and evolutionary novelties arise.

Third, and most importantly, the growing awareness of the importance of cellular behavior as a central explanatory factor in epigenesis leads directly to an expectation of deep homology. All metazoans begin their life as a single diploid cell, the zygote, and from this arise the trillions of cells that make up an adult vertebrate body. This process involves changes in the daughter cells over time, which are responses to the local environment in which each individual cell finds itself. Processes of cell-to-cell signaling, interactions between cells and their noncellular environment (e.g., the extracellular matrix), and physical forces experienced by individual cells all play crucial roles in cell type specification; each individual cell plays out its own life in close interaction with its local (within-organism) environment (Gordon, 2006; Keller, Davidson, & Shook, 2003; Kirschner & Gerhart, 2005). Many of these processes are shared among all metazoans, for the simple reason that they evolved in the single-celled organisms that were the common ancestors of all extant metazoans. Such processes as stabilization of the cytoskeleton by external molecules, or robust and appropriate responses to molecular gradients, evolved long before metazoans, but continue to play a central role in metazoan development today. Given that about half of the entire 4 billion year evolution of any extant life form occurred in the 2 billion initial years when all life was single-celled, we should not be surprised to find that half of our genes are involved in determining cellular behavior, rather than that of larger or smaller units of organization (Mouse Genome Sequencing Consortium, 2002, Figures 17 and 18). In fact, though, the influence of cellular behavior is probably much higher than this suggests, since all subsequent evolution "leveraged" these earlier evolved responses, which had already been strongly selected for robustness, survival, and self-replication in the face of contingent extracellular events (Kitano, 2004).

Epigenesis, the sensitivity of different groups of cells and tissues to their biochemical context in the developing organism, can be seen more and more to build on robust behaviors of individual cells. Whole complex organisms "self-organize" by virtue of these behaviors, which channel development, and thus evolution, in particular ways. These again lead to robustness of whole-organism development: the embryo flexibly deals with missing or additional limbs, and the brain flexibly adjusts to unusual informational inputs, in ways that are the direct causal end product, with elaboration, of cellular behavior. Thus, we argue, we should not be at all surprised to find deep homologies of the sort described above for birds and humans who are, in the big picture, very closely related metazoans, and who follow mostly the same developmental rules using very similar genomes.

Toward the Future: High Expectations for Bird-Human Comparisons?

We have suggested above that, if the study of bacteria, flies, and roundworms can provide profitable insights into human biology, we should not be surprised that work on birds can do the same. But this may seem a rather vague hope, even if true. Here we would like to suggest that bird-mammal comparisons provide a concrete set of possibilities to test quite specific hypotheses about the evolution of mechanisms underlying specific traits like vocal learning.

There is a long history of using bird-human comparisons to test specific functional and physiological hypotheses. A nice example is provided by endothermy, the capacity to maintain and regulate body temperature independent of the environment. The ancestral tetrapods and amniotes were nonendothermic, as are most fish, amphibians, and reptiles alive today. Birds and mammals independently evolved endothermy, despite its "wastefulness" in terms of energy usage, and this underlies many aspects of behavior in both clades (particularly the evolution of complex, energy-hungry brains). Because endothermy relies on basic metabolic processes that are widely shared among all organisms, we might expect some similar constraints to have acted in both clades.

Basal metabolic rate (BMR) differs greatly in absolute terms in organisms of different sizes, but only covers a narrow range in relative terms (Makarieva et al., 2008), presumably restricted by the physics of heat dissipation. Allometric hypotheses, developed using "mouse-to-elephant" correlations, were initially developed for mammals and were later tested in birds. These data were used to test and support the "membrane pacemaker" hypothesis, which suggests that the detailed composition of cellular membranes governs basic metabolic processes (Hulbert, 2007). Again, very similar cellular processes appear to underlie a convergently evolved trait, and this example illustrates the value of similar bird-mammal comparisons to actually test evolutionary hypotheses, using convergence as an "experiment in nature."

What sort of similar discoveries may we hope for in the future, specifically related to birdsong and human language? For example, how can the potentially huge list of genes involved in some aspect of acoustic communication be narrowed down to candidate genes specifically involved in vocal learning? Several approaches have been used to address such issues. First, in vocal learners, genes can be identified by comparing gene expression profiles before, during, and after singing and other behaviors. Gene expression in zebra finches has been studied at several time points within 3 hours after singing (Wada et al., 2006), yielding a list of 33 genes (out of 41 tested) whose expression pattern differed from that in the nonsinging controls. Repeating such analyses with more genes, at different stages of vocal learning, as well as in other oscine and nonoscine vocal learners could help to pinpoint further details.

Second, while 80% of the mouse genome (and probably a very similar percentage of the human genome) is expressed in the brain (Lein et al., 2007), it seems unlikely that any gene would be specific to any behavioral trait; rather the timing and location of gene expression is crucial (Lasky-Su et al., 2008). Given that *FoxP2* is already known to be involved in vocal learning, its targets (and those of related forkhead genes like *FoxP1*) can be screened, which is likely to turn up further candidates. Such analyses identified sets of 175 genes serving as targets of human *FOXP2* in the basal ganglia, 144 in the inferior frontal cortex and 192 in the lung of postmortem human fetuses (Spiteri et al., 2007), and 303 in human neuronal cell cultures (Vernes et al., 2007).

In a third line of evidence, intraspecific variation in vocal learning can be genetically profiled. This could be achieved by forward genetics in model organisms, or by screening relevant subpopulations. For example, adult songbirds may vary widely in the complexity of their songs, and this variation often appears to have some genetic basis (e.g., Catchpole, 1980); alternatively, song complexity may be affected by early life events, which in turn influence gene expression and nervous development in ways that can be experimentally explored (Buchanan, Spencer, Goldsmith & Catchpole, 2003; Spencer, Buchanan, Goldsmith, & Catchpole, 2003). Intraspecific variation in clinically normal humans may also prove to be a fruitful topic. For example, in healthy humans, tonal language has been associated with *MCPH1* (also known as microcephalin) and *MCPH5* (also known as *ASPM*) across multiple populations (Dediu & Ladd, 2007; Ladd, 2008). Finally, human patients with speech and language disorders should provide many more candidate genes involved in these capacities, as for *FOXP2*. Dyslexia appears particularly promising at present: it is associated with mutations in at least four genes—*DYX1C1*, *KIAA0319*, *DCDC2*, and *ROBO1*— all known or suggested to be involved in neuronal migration (Fisher & DeFries, 2002; Fisher & Francks, 2006).

Fourth, given that many of the brain development genes considered so far have homologs across vertebrates (and often beyond), studies of their rates of evolution in closely related lines leading to taxa with or without vocal learning might yield further insights into the underlying genetic mechanisms. For *FoxP2*, evidence of accelerated evolution is clear in echolocating bats (Li et al., 2007) but ambivalent in humans (Arbiza, Dopazo, & Dopazo, 2006; Zhang, Webb, & Podlaha, 2002), possibly because more chimp than human genes experienced positive selection (Bakewell, Shi, & Zhang, 2007). However, studies comparing homologous genes in humans and other primates identified several hundred genes positively selected in humans, including the above-mentioned *MCPH5*. The function of most of the genes that stood out in these studies is often not well understood. However, a subset of them seem to merit further study. For example, the sushi-repeat gene *SRPX2*, heavily expressed in neurons, was found to be implicated in orofacial dyspraxia and speech impairment (Roll et al., 2006). The Abelson-helper integration gene *AHI1* was found to be altered in about 10% of patients with Joubert syndrome (Parisi, Doherty, Chance, & Glass, 2007), most of whom show breathing abnormalities and speech disorders necessitating speech therapy (Hodgkins et al., 2004; Joubert, Eisenring, Robb, & Andermann, 1969; Steinlin, Schmid, Landau, & Boltshauser, 1997). Mouse and zebra finch orthologs of *AHI1* show neural expression patterns similar to humans (Doering et al., 2008), suggesting that these genetic model organisms can help uncover the basic developmental functions of these genes.

Fifth, and finally, the spatiotemporal expression patterns of any strong candidate gene can be manipulated, in animals, by means of conditional knockdown studies, as detailed above for the case of *FoxP2* (Haesler et al., 2007). Existing hypotheses concerning vocal learning can be tested experimentally using such techniques. For example, the hypothesis that increased vocal control entails direct connections from the motor cortex to the brainstem motor neurons controlling the vocal organs (Deacon, 1997; Jürgens, 2002; Kuypers, 1958) can be examined by manipulating candidate gene expression during development, and tracing the neural connectivity that results. In humans, similar lines of evidence could be produced by investigating the correlation between genetic profiles and phenotypes observable noninvasively by structural and functional neuroimaging (Gregório et al., 2009; Liégeois et al., 2003; Poretti et al., 2007). Such methods can also be applied in songbirds (Boumans et al., 2008; Poirier et al., 2008). Finally, single-gene imaging, already standard in invasive studies, is on the horizon for noninvasive techniques (Gade et al., 2008; Liu et al., 2007; Weissleder et al., 2000). Noninvasive imaging methods will allow far more temporal information to be extracted from developing organisms than current techniques (see Lee et al., 2007). Such experimental tools are maturing rapidly, and we need to update our conceptual framework(s) accordingly.

Conclusion

We have argued here that deep homology exists not only in the morphological domain but also in behavioral/neural comparisons, using the example of *FoxP2* as a "proof of principle." More importantly, we have argued that we can expect many more examples of such "deep homologies" to be uncovered by future research, and have sketched a theoretical argument to support this contention. Although these are still early days, we think that such considerations provide a powerful justification for further exploration of the manifest parallels between human spoken language and learned birdsong (and other examples of avian vocal learning, like those of parrots). This is particularly relevant in cases where traits central to human speech, like vocal learning, have no equivalent in the normal mammalian model species for genetics, development, and neuroscience (like mice, cats, or monkeys). Deep homology provides a rationale for employing a wide range of model systems. Regarding vocal learning, multiple bird species (not just the ubiquitous zebra finch) will be valuable in the lab. In addition, it will be desirable to complement typical lab species with studies in species that stand out in less lab-specific characteristics—be these brain size (e.g., lyrebirds; Sol & Price, 2008), or vocal labeling of conspecifics (spectacled parrotlets, *Forpus conspicillatus*; cf. Wanker, Sugama, & Prinage, 2005). Comparisons with closely related species that differ in the respective trait may prove crucial in identifying candidate genes: so there is a role for naturalists and fieldworkers here, not just laboratory workers.

Despite a historical resistance from some psychologists, anthropologists, and linguists to data from species other than our closest relatives, the primates, we see recent developments in evo-devo as providing powerful justification for a much broader application of the comparative method, in the pluralistic tradition best exemplified by Darwin himself.

References

Arbiza, L., Dopazo, J., & Dopazo, H. (2006). Positive selection, relaxation, and acceleration in the evolution of the human and chimp genome. *PLoS Computational Biology*, *2*, e38.

Avian Brain Nomenclature Consortium. (2005). Avian brains and a new understanding of vertebrate brain evolution. *Nature Reviews: Neuroscience*, *6*, 151–159.

Bakewell, M. A., Shi, P., & Zhang, J. (2007). More genes underwent positive selection in chimpanzee evolution than in human evolution. *Proceedings of the National Academy of Sciences of the United States of America*, *104*, 7489–7494.

Boumans, T., Gobes, S. M. H., Poirier, C., Theunissen, F. E., Vandersmissen, L., Pintjens, W., et al. (2008). Functional MRI of auditory responses in the zebra finch forebrain reveals a hierarchical organisation based on signal strength but not selectivity. *PLoS ONE*, *3*, e3184.

Buchanan, K. L., Spencer, K. A., Goldsmith, A. R., & Catchpole, C. K. (2003). Song as an honest signal of past developmental stress in the European starling (Sturnus vulgaris). *Proceedings of the Royal Society B: Biological Sciences*, *270*, 1149–1156.

Carroll, S. B. (2005). *Endless forms most beautiful*. New York: Norton.

Carroll, S. B. (2006). *The making of the fittest: DNA and the ultimate forensic record of evolution*. New York: Norton.

Carroll, S. B., Grenier, J. K., & Weatherbee, S. D. (2005). *From DNA to diversity: Molecular genetics and the evolution of animal design* (2nd ed.). Malden, MA: Blackwell Science.

Catchpole, C. K. (1980). Sexual selection and the evolution of complex songs among warblers of the genus *Acrocephalus*. *Behaviour*, *74*, 149–166.

Darwin, C. (1859). *On the origin of species*. London: John Murray.

Davidson, E. H. (2006). *The regulatory genome: Gene regulatory networks*. New York: Academic Press.

Deacon, T. W. (1997). *The symbolic species: The co-evolution of language and the brain*. New York: Norton.

Dediu, D., & Ladd, D. R. (2007). Linguistic tone is related to the population frequency of the adaptive haplogroups of two brain size genes, *ASPM* and *Microcephalin*. *Proceedings of the National Academy of Sciences of the United States of America, 104*, 10944–10949.

Dickinson, W. J. (1995). Molecules and morphology: Where's the homology? *Trends in Genetics, 11*, 119–121.

Doering, J. E., Kane, K., Hsiao, Y.-C., Yao, C., Shi, B., Slowik, A. D., et al. (2008). Species differences in the expression of Ahi1, a protein implicated in the neurodevelopmental disorder Joubert syndrome, with preferential accumulation to stigmoid bodies. *Journal of Comparative Neurology, 511*, 238–256.

Dormer, K. J. (1979). Mechanisms of sound production and air recycling in delphinids: Cineradiographic evidence. *Journal of the Acoustical Society of America, 65*, 229–239.

Doupe, A. J., & Kuhl, P. K. (1999). Birdsong and human speech: Common themes and mechanisms. *Annual Review of Neuroscience, 22*, 567–631.

Doupe, A. J., Perkel, D. J., Reiner, A., & Stern, E. A. (2005). Birdbrains could teach basal ganglia research a new song. *Trends in Neurosciences, 28*, 353–363.

Fisher, S. E., & DeFries, J. C. (2002). Developmental dyslexia: Genetic dissection of a complex cognitive trait. *Nature Reviews: Neuroscience, 3*, 767–780.

Fisher, S. E., & Francks, C. (2006). Genes, cognition and dyslexia: Learning to read the genome. *Trends in Cognitive Sciences, 10*, 250–257.

Fisher, S. E., Vargha-Khadem, F., Watkins, K. E., Monaco, A. P., & Pembrey, M. E. (1998). Localisation of a gene implicated in a severe speech and language disorder. *Nature Genetics, 18*, 168–170.

Fitch, W. M. (1970). Distinguishing homologous from analogous proteins. *Systematic Zoology, 19*, 99–113.

Fitch, W. M. (2000). Homology: A personal view on some of the problems. *Trends in Genetics, 16*, 227–231.

Fitch, W. M., & Margoliash, E. (1967). Construction of phylogenetic trees. *Science, 155*, 279–284.

Fitch, W. T., (2009). The Biology & Evolution of Language: "Deep Homology" and the Evolution of Innovation. In M. S. Gazzaniga (Ed.), *The Cognitive Neurosciences IV* (pp. 873–883) Cambridge, MA: MIT Press.

Fitch, W. T., & Hauser, M. D. (2002). Unpacking "honesty": Vertebrate vocal production and the evolution of acoustic signals. In A. M. Simmons, R. F. Fay, & A. N. Popper (Eds.), *Acoustic communication* (Vol. 16, pp. 65–137). New York: Springer.

French, V. (2001). Insect segmentation: Genes, stripes and segments in "Hoppers." *Current Biology, 11*, R910–R913.

Gade, T. P. F., Koutcher, J. A., Spees, W. M., Beattie, B. J., Ponomarev, V., Doubrovin, M., et al. (2008). Imaging transgene activity in vivo. *Cancer Research, 68*, 2878–2884.

Gehring, W. J. (1998). *Master control genes in development and evolution: The homeobox story*. New Haven, CT: Yale University Press.

Gilbert, S. F. (2003). *Developmental biology* (7th ed.). Sunderland, MA: Sinauer Associates.

Gilbert, S. F., & Bolker, J. A. (2001). Homologies of process and modular elements of embryonic construction. *Journal of Experimental Zoology, Part B: Molecular and Developmental Evolution, 291*, 1–12.

Gordon, R. (2006). Mechanics in embryogenesis and embryonics: Prime mover or epiphenomenon? *International Journal of Developmental Biology, 50*, 245–253.

Gregório, S. P., Sallet, P. C., Do, K.-A., Lin, E., Gattaz, W. F., & Dias-Neto, E. (2009). Polymorphisms in genes involved in neurodevelopment may be associated with altered brain morphology in schizophrenia: Preliminary evidence. *Psychiatry Research, 165*, 1–9.

Haesler, S., Rochefort, C., Geogi, B., Licznerski, P., Osten, P., & Scharff, C. (2007). Incomplete and inaccurate vocal imitation after knockdown of FoxP2 in songbird basal ganglia nucleus Area X. *PLoS Biology, 5*, e321.

Haesler, S., Wada, K., Nshdejan, A., Morrisey, E. E., Lints, T., Jarvis, E. D., et al. (2004). FoxP2 expression in avian vocal learners and non-learners. *Journal of Neuroscience, 24*, 3164–3175.

Hall, B. K. (2007). Homoplasy and homology: Dichotomy or continuum? *Journal of Human Evolution, 52*(5), 473–479.

Harvey, P. H., & Pagel, M. D. (1991). *The comparative method in evolutionary biology.* Oxford: Oxford University Press.

Hodgkins, P. R., Harris, C. M., Shawkat, F. S., Thompson, D. A., Chong, K., Timms, C., et al. (2004). Joubert syndrome: Long-term follow-up. *Developmental Medicine and Child Neurology, 46*, 694–699.

Hulbert, A. J. (2007). Membrane fatty acids as pacemakers of animal metabolism. *Lipids, 42*, 811–819.

International Human Genome Sequencing Consortium. (2004). Finishing the euchromatic sequence of the human genome. *Nature, 431*, 931–945.

Jarvis, E. D. (2004). Learned birdsong and the neurobiology of human language. *Annals of the New York Academy of Sciences, 1016*, 749–777.

Joubert, M., Eisenring, J. J., Robb, J. P., & Andermann, F. (1969). Familial agenesis of the cerebellar vermis: A syndrome of episodic hyperpnea, abnormal eye movements, ataxia, and retardation. *Neurology, 19*, 813–825.

Jürgens, U. (2002). Neural pathways underlying vocal control. *Neuroscience and Biobehavioral Reviews, 26*, 235–258.

Keller, R., Davidson, L. A., & Shook, D. R. (2003). How we are shaped: The biomechanics of gastrulation. *Differentiation, 71*, 171–205.

Kirschner, M. W., & Gerhart, J. C. (2005). *The plausibility of life: Resolving Darwin's dilemma.* New Haven, CT: Yale University Press.

Kitano, H. (2004). Biological robustness. *Nature Reviews: Genetics, 5*, 826–837.

Kuypers, H. G. J. M. (1958). Corticobulbar connections to the pons and lower brainstem in man: An anatomical study. *Brain, 81*, 364–388.

Ladd, D. R. (2008). Languages and genes: Reflections on biolinguistics and the nature-nurture question. *Biolinguistics, 2*, 114–126.

Lai, C. S. L., Fisher, S. E., Hurst, J. A., Vargha-Khadem, F., & Monaco, A. P. (2001). A forkhead-domain gene is mutated in a severe speech and language disorder. *Nature*, *413*, 519–523.

Lai, C. S. L., Gerrelli, D., Monaco, A. P., Fisher, S. E., & Copp, A. J. (2003). FOXP2 expression during brain development coincides with adult sites of pathology in a severe speech and language disorder. *Brain*, *126*, 2455–2462.

Land, M. F., & Fernald, R. D. (1992). The evolution of eyes. *Annual Review of Neuroscience*, *15*, 1–29.

Lankester, E. (1870). On the use of the term homology in modern zoology, and the distinction between homogenetic and homoplastic agreements. *Annals & Magazine of Natural History*, *6*, 34–43.

Lasky-Su, J., Lyon, H. N., Emilsson, V., Heid, I. M., Molony, C., Raby, B. A., et al. (2008). On the replication of genetic associations: Timing can be everything! *American Journal of Human Genetics*, *82*, 849–858.

Lee, S.-C., Mietchen, D., Cho, J.-H., Kim, Y.-S., Kim, C., Hong, K. S., et al. (2007). *In vivo* magnetic resonance microscopy of differentiation in Xenopus laevis embryos from the first cleavage onwards. *Differentiation*, *75*, 84–92.

Lein, E. S., Hawrylycz, M. J., Ao, N., Ayres, M., Bensinger, A., Bernard, A., et al. (2007). Genome-wide atlas of gene expression in the adult mouse brain. *Nature*, *445*, 168–176.

Li, G., Wang, J., Rossiter, S. J., Jones, G., & Zhang, S. (2007). Accelerated FoxP2 evolution in echolocating bats. *PLoS ONE*, *2*, e900.

Liégeois, F., Baldeweg, T., Connelly, A., Gadian, D. G., Mishkin, M., & Vargha-Khadem, F. (2003). Language fMRI abnormalities associated with FOXP2 gene mutation. *Nature Neuroscience*, *6*, 1230–1237.

Liu, C. H., Kim, Y. R., Ren, J. Q., Eichler, F., Rosen, B. R., & Liu, P. K. (2007). Imaging cerebral gene transcripts in live animals. *Journal of Neuroscience*, *27*(3), 713–722.

Lorenz, K. (1953). *Comparative studies on the behaviour of the Anatinae*. London: Read Books.

Makarieva, A. M., Gorshkov, V. G., Li, B.-L., Chown, S. L., Reich, P. B., & Gavrilov, V. M. (2008). Mean mass-specific metabolic rates are strikingly similar across life's major domains: Evidence for life's metabolic optimum. *Proceedings of the National Academy of Sciences of the United States of America*, *105*, 16994–16999.

Mivart, S. G. (1870). On the use of the term "homology." *Annals & Magazine of Natural History*, *6*, 113–121.

Mo, Z., & Zecevic, N. (2008). Is Pax6 critical for neurogenesis in the human fetal brain? *Cerebral Cortex*, *18*, 1455–1465.

Mouse Genome Sequencing Consortium. (2002). Initial sequencing and comparative analysis of the mouse genome. *Nature*, *420*, 520–562.

Nagy, L. M. (1994). Insect segmentation: A glance posterior. *Current Biology*, *4*, 811–814.

Norris, K. S. (1969). The echolocation of marine mammals. In H. T. Andersen (Ed.), *The biology of marine mammals* (pp. 391–423). New York: Academic Press.

Ohno, S. (1970). *Evolution by gene duplication*. Heidelberg: Springer.

Parisi, M. A., Doherty, D., Chance, P. F., & Glass, I. A. (2007). Joubert syndrome (and related disorders) (OMIM 213300). *European Journal of Human Genetics, 15,* 511–521.

Poirier, C., Vellema, M., Verhoye, M., Meir, V. V., Wild, J. M., Balthazart, J., et al. (2008). A three-dimensional MRI atlas of the zebra finch brain in stereotaxic coordinates. *NeuroImage, 41,* 1–6.

Poretti, A., Boltshauser, E., Loenneker, T., Valente, E. M., Brancati, F., Il'yasov, K., et al. (2007). Diffusion tensor imaging in Joubert syndrome. *American Journal of Neuroradiology, 28,* 1929–1933.

Roll, P., Rudolf, G., Pereira, S., Royer, B., Scheffer, I. E., Massacrier, A., et al. (2006). SRPX2 mutations in disorders of language cortex and cognition. *Human Molecular Genetics, 15,* 1195–1207.

Scharff, C., & Haesler, S. (2005). An evolutionary perspective on FoxP2: Strictly for the birds? *Current Opinion in Neurobiology, 15,* 694–703.

Scharff, C., & Petri, J. (2011). Evo-devo, deep homology and FoxP2: Implications for the evolution of speech and language. *Philosophical Transactions of the Royal Society B, 366,* 2124–2140.

Shu, W., Cho, J. Y., Jiang, Y., Zhang, M., Weisz, D., Elder, G. A., et al. (2005). Altered ultrasonic vocalization in mice with a disruption in the *Foxp2* gene. *Proceedings of the National Academy of Sciences of the United States of America, 102,* 9643–9648.

Shu, W., Lu, M. M., Zhang, Y., Tucker, P. W., Zhou, D., & Morrisey, E. E. (2007). Foxp2 and Foxp1 cooperatively regulate lung and esophagus development. *Development, 134,* 1991–2000.

Shubin, N., Tabin, C., & Carroll, S. (1997). Fossils, genes and the evolution of animal limbs. *Nature, 388,* 639–648.

Shubin, N., Tabin, C., & Carroll, S. (2009). Deep homology and the origins of evolutionary novelty. *Nature, 457,* 818–823.

Simpson, T. I., & Price, D. J. (2002). Pax6: A pleiotropic player in development. *BioEssays, 24,* 1041–1051.

Sol, D., & Price, T. D. (2008). Brain size and the diversification of body size in birds. *American Naturalist, 172,* 170–177.

Spencer, K. A., Buchanan, K. L., Goldsmith, A. R., & Catchpole, C. K. (2003). Song as an honest signal of developmental stress in the zebra finch (Taeniopygia guttata). *Hormones and Behavior, 44,* 132–139.

Spiteri, E., Konopka, G., Coppola, G., Bomar, J., Oldham, M., Ou, J., et al. (2007). Identification of the transcriptional targets of FOXP2, a gene linked to speech and language, in developing human brain. *American Journal of Human Genetics, 81,* 1144–1157.

Steinlin, M., Schmid, M., Landau, K., & Boltshauser, E. (1997). Follow-up in children with Joubert syndrome. *Neuropediatrics, 28,* 204–211.

Striedter, G. F. (2004). *Principles of brain evolution.* Sunderland, MA: Sinauer.

Suthers, R. A. (1999). The motor basis of vocal performance in songbirds. In M. D. Hauser & M. Konishi (Eds.), *The design of animal communication* (pp. 37–62). Cambridge, MA: MIT Press/Bradford Books.

Tinbergen, N. (1952). Derived activities: Their causation, biological significance, origin and emancipation during evolution. *Quarterly Review of Biology, 27*, 1–32.

Tinbergen, N. (1963). On aims and methods of ethology. *Zeitschrift für Tierpsychologie, 20*, 410–433.

Tomarev, S. I., Callaerts, P., Kos, L., Zinovieva, R., Halder, G., Gehring, W., et al. (1997). Squid Pax-6 and eye development. *Proceedings of the National Academy of Sciences of the United States of America, 94*, 2421–2426.

Van Heyningen, V., & Williamson, K. A. (2002). PAX6 in sensory development. *Human Molecular Genetics, 11*, 1161–1167.

Vandepoele, K., De Vos, W., Taylor, J. S., Meyer, A., & Van de Peer, Y. (2004). Major events in the genome evolution of vertebrates: Paranome age and size differ considerably between ray-finned fishes and land vertebrates. *Proceedings of the National Academy of Sciences of the United States of America, 101*, 1638–1643.

Vargha-Khadem, F., & Passingham, R. (1990). Speech and language deficits. *Nature, 346*, 226.

Vernes, S. C., Spiteri, E., Nicod, J., Groszer, M., Taylor, J. M., Davies, K. E., et al. (2007). High-throughput analysis of promoter occupancy reveals direct neural targets of FOXP2, a gene mutated in speech and language disorders. *American Journal of Human Genetics, 81*, 1232–1250.

Von Salvini-Plawen, L., & Mayr, E. (1977). On the evolution of photoreceptors and eyes. In M. K. Hecht, W. C. Sterre, & B. Wallace (Eds.), *Evolutionary biology* (Vol. 10, pp. 207–263). London: Plenum Press.

Wada, K., Howard, J. T., McConnell, P., Whitney, O., Lints, T., Rivas, M., et al. (2006). A molecular neuroethological approach for identifying and characterizing a cascade of behaviorally regulated genes. *Proceedings of the National Academy of Sciences of the United States of America, 103*, 15212–15217.

Wagner, G. P. (2007). The developmental genetics of homology. *Nature Reviews: Genetics, 8*, 473–479.

Wanker, R., Sugama, Y., & Prinage, S. (2005). Vocal labelling of family members in spectacled parrotlets, *Forpus conspicillatus. Animal Behaviour, 70*, 111–118.

Weatherbee, S. D., Nijhout, H. F., Grunert, L. W., Halder, G., Galant, R., Selegue, J., et al. (1999). Ultrabithorax function in butterfly wings and the evolution of insect wing patterns. *Current Biology, 9*, 109–115.

Weissleder, R., Moore, A., Mahmood, U., Bhorade, R., Benveniste, H., Chiocca, E. A., et al. (2000). In vivo magnetic resonance imaging of transgene expression. *Nature Medicine, 6*, 351–355.

Zhang, J., Webb, D. M., & Podlaha, O. (2002). Accelerated protein evolution and origins of human-specific features: Foxp2 as an example. *Genetics, 162*, 1825–1835.

4 Evolution of Brain Pathways for Vocal Learning in Birds and Humans

Erich D. Jarvis

What Is Vocal Learning?

Vocal learning is the ability to modify the acoustic and/or syntactic structure of sounds produced, including imitation and improvisation. It is distinct from auditory learning, which is the ability to make associations with sounds heard. Vocal learning depends on auditory learning, but auditory learning does not depend on vocal learning. Most, if not all, vertebrates are capable of auditory learning, but few are capable of vocal learning. The latter has been found to date in at least five distantly related groups of mammals (humans, bats, cetaceans such as dolphins and whales, pinnipeds like seals and sea lions, and elephants) and three distantly related groups of birds (parrots, hummingbirds, and songbirds) (Figure 4.1A,B) (Marler, 1970; Caldwell & Caldwell, 1972; Ralls, Fiorelli, & Gish, 1985; Baptista & Schuchmann, 1990; Pepperberg, 1994; Poole, Tyack, Stoeger-Horwath, & Watwood , 2005; Foote et al., 2006; Sanvito, Galimberti, & Miller, 2007); reviewed in Nottebohm, 1972; Janik & Slater, 1997; Jarvis, 2004; Bolhuis, Okanoya, & Scharff, 2010). For some species, such as most parrots, some songbirds (e.g., crows, starlings, mynahs, and lyrebirds), and harbor seals, vocal mimicry of human speech has been found (Ralls et al., 1985; MacKay, 2001). The most famous nonhuman speech mimic is the African grey parrot Alex (Pepperberg, 1999). Some studies have claimed evidence of vocal learning in nonhuman primates, including chimpanzees (Masataka & Fujita, 1989; Marshall, Wrangham, & Arcadi, 1999; Crockford, Herbinger, Vigilant, & Boesch, 2004). However, these nonhuman primate findings have been disputed, because either others have not been able to replicate them or when vocal changes have been found, they are subtle, representing call convergence of small changes in pitch of existing innate syllables, or imitation using sounds made by the lips but not by the larynx (Owren, Dieter, Seyfarth, & Cheney, 1993; Egnor & Hauser, 2004).

An example helps in understanding the distinction between vocal imitation learning found in humans and songbirds and auditory learning found in most species. A dog can learn the meaning of the human words *sit* (in English), *sientese* (in Spanish),

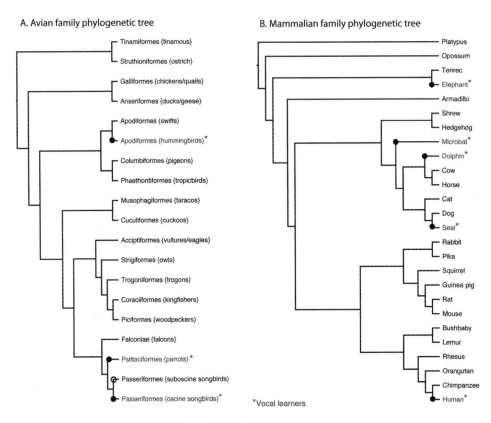

A. Avian family phylogenetic tree

- Tinamiformes {tinamous}
- Struthioniformes {ostrich}
- Galliformes {chickens/quails}
- Anseriformes {ducks/geese}
- Apodiformes {swifts}
- Apodiformes {hummingbirds}*
- Columbiformes {pigeons}
- Phaethontiformes {tropicbirds}
- Musophagiformes {taracos}
- Cuculiformes {cuckoos}
- Acciptiformes {vultures/eagles}
- Strigiformes {owls}
- Trogoniformes {trogons}
- Coraciiformes {kingfishers}
- Piciformes {woodpeckers}
- Falconiae {falcons}
- Psittaciformes {parrots}*
- Passeriformes {suboscine songbirds}
- Passeriformes {oscine songbirds}*

*Vocal learners

B. Mammalian family phylogenetic tree

- Platypus
- Opossum
- Tenrec
- Elephant*
- Armadillo
- Shrew
- Hedgehog
- Microbat*
- Dolphin*
- Cow
- Horse
- Cat
- Dog
- Seal*
- Rabbit
- Pika
- Squirrel
- Guinea pig
- Rat
- Mouse
- Bushbaby
- Lemur
- Rhesus
- Orangutan
- Chimpanzee
- Human*

Figure 4.1

Family trees of living avian and mammalian orders. (A) Tree of major bird orders based on DNA sequence of 19 genes (Hackett et al., 2008). The Latin name of each order is given along with examples of common species. The Passeriforme (songbird) order is divided into two separate groups, oscine songbirds (vocal learners) and suboscine songbirds (non–vocal learners) orders. (B) Tree of mammalian orders based on DNA sequence of 18 genes (Murphy et al., 2001), updated with additional genomic and fossil data (Springer, Murphy, Eizirik, & O'Brien, 2003; Murphy, Pringle, Crider, Springer, & Miller, 2007; Spaulding, O'Leary, & Gatesy, 2009). The relationships among bats, dolphins, and carnivores (cat, dog, and seal) vary among studies. Closed and open circles show the minimal ancestral nodes where vocal learning could have evolved as independent gains or losses, respectively. Independent losses would have at least required one common vocal-learning ancestor. The trees are not meant to present the final dogma of mammalian and avian evolution, because there are some significant differences of among studies and scientists.

or *osuwali* (in Japanese) or of a sentence like *Come here boy!* Dogs are not born with this knowledge of human words or syntax. They acquire it through auditory learning. However, a dog cannot imitate the sounds *sit, sientese,* or *osuwali.* Humans, parrots, and some songbirds can. This is vocal learning in the more advanced sense, and is considered mimicry when imitation occurs of other species and environmental sounds or simply imitation of species-specific vocalizations. A limited form of vocal modification may be present in many species, such as learning to change pitch (Egnor & Hauser, 2004), but this still needs further experimental study.

I believe that vocal learning is the trait most critical for spoken language, because of all the behavioral traits necessary for spoken language; vocal learning appears to be necessary and sufficient for nonhuman species to imitate and produce rudimentary speech, even with some limited meaning (Pepperberg, 1999; Hauser, Chomsky, & Fitch, 2002). The ability to engage in auditory learning and even cognition of auditory signals is not unique. Such cognitive auditory traits include the ability to understand rudimentary human speech by nonhuman animals and to form semantic meanings of sounds heard, as in the alarm calls that identify specific predators (Seyfarth, Cheney, & Marler, 1980; Kuhl, 1986). In addition, although non–vocal learners cannot perform de novo imitation of sounds, some can produce learned communication signals in another modality, as with the ability of chimpanzees to produce limited sign language (Gardner & Gardner, 1971; Rivas, 2005).

Most vocal learners only imitate the sounds of their own species (i.e., are nonmimics), and not all species have the same vocal-learning capacity. Humans are the most prolific vocal learners, because they learn to produce a seemingly infinite number of combinations of learned vocalizations. Some parrot species, such as the African grey, are prolific, learning to reciprocally communicate with humans (Pepperberg, 1999). Others, such as mockingbirds, brown thrashers, and some parrots mimic other species and environmental sounds, producing hundreds or thousands of warble-song combinations (Kroodsma, 1982; Derrickson, 1987). Less prolific species, such as canaries, learn 20–70 song types; still others are more limited, such as the zebra finch songbird and aphantochroa hummingbird, producing one distinct song type (Nottebohm, Nottebohm, & Crane, 1986; Catchpole & Slater, 1995; Farabaugh & Dooling, 1996; Ferreira, Smulders, Sameshima, Mello, & Jarvis, 2006).

Each of the vocal-learning avian and mammalian groups has close non-vocal-learning relatives (Figure 4.1). Thus, it has been argued that vocal learning evolved independently of a common ancestor in the three bird groups and in the three to five mammalian groups (Nottebohm, 1972; Janik & Slater, 1997; Jarvis, 2004). According to a view of avian phylogeny using the DNA sequence of 19 genes, parrots were placed as a sister group to songbirds (Hackett et al., 2008); this view is supported by DNA sequence deletion and insertion data (Suh et al., 2011), but alternative phylogenies have also been proposed since (Pratt et al., 2009; Nabholz,

Kunstner, Wang, Jarvis, & Ellegren, 2011). One possible interpretation based on the Hackett et al. (2008) and Suh et al. (2011) phylogeny (Figure 4.1B) is that that vocal learning evolved twice among birds, in hummingbirds and in the ancestor of parrots and songbirds, and was then lost in suboscine songbirds. Regardless of which evolutionary scenario is correct, both have led to the question of whether there is something special about the brains of these animals that can imitate sounds. The answer is yes.

Consensus Brain Systems of Vocal Learners

Of the species studied to date, only vocal learners (songbirds, parrots, hummingbirds, and humans) have been found to contain regions in their cerebrums (i.e., telencephalon) that control vocal behavior (Figure 4.2) (Jurgens, 1995; Jarvis et al., 2000). Vocal-control brain regions have not yet been investigated in cetaceans, bats, elephants, or pinnipeds. Non–vocal learners, including chickens and cats, only have midbrain, medulla, and thalamic (in cats) regions that control innate vocalizations (Wild, 1994a; Farley, 1997). Nonhuman primates have a laryngeal premotor cortex that is connected to other forebrain areas and has an indirect connection to brainstem vocal motor neurons (Figure 4.3), but this region is not required to produce species-specific vocalizations (Kirzinger & Jurgens, 1982; Jurgens, 2002; Simonyan & Jurgens, 2005). By comparing these vocal brain regions of different vocal-learning and non-vocal-learning species, it has been possible to generate a consensus vocal-learning pathway (Jarvis, 2004).

All three vocal-learning bird groups have seven comparable cerebral song nuclei: four posterior forebrain nuclei and three anterior forebrain nuclei (Figure 4.2A–C; abbreviations in Table 4.1; comparative anatomy in Table 4.2) (Jarvis et al., 2000). These brain nuclei have been given different names in each bird group because of the possibility that each evolved their song nuclei independently of a common ancestor (Striedter, 1994; Jarvis et al., 2000). Despite the possible independent origins, there are significant parallels in connectivity in all three bird groups (Figure 4.3). The posterior nuclei form a pathway that projects from a nidopallial song nucleus (HVC, NLC, VLN) to an arcopallial song nucleus (RA, AAC dorsal part, VA), to the nuclei in the midbrain (DM) and medulla (12th vocal motor neurons) (Figure 4.3C,E,F; black arrows) (Striedter, 1994; Durand, Heaton, Amateau, & Brauth, 1997; Vates, Vicario, & Nottebohm, 1997; Gahr, 2000). The 12th motor neurons project to the muscles of the syrinx, the avian vocal organ. DM and 12th are present in non-vocal-learning birds and control production of innate vocalizations, but no projections to them from the arcopallium occur (Wild, 1994a; Wild, Li, & Eagleton, 1997). Nuclei higher up in the posterior pathway have not been as well studied. However, songbird Av was recently shown to be highly interconnected with

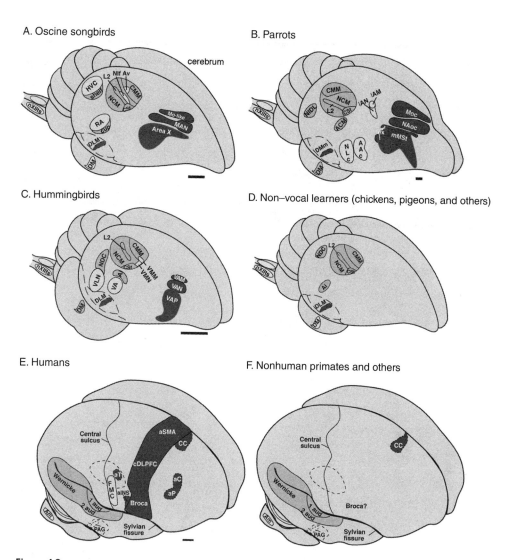

Figure 4.2
Proposed comparable vocal and auditory brain areas among birds and primates. (A) songbird, (B) parrot, (C) hummingbird, (D) chicken, (E) human, and (F) chimp. White regions, proposed posterior vocal pathway regions. Dark gray regions, proposed anterior vocal pathways. Light gray regions, auditory pathway regions. Basal ganglia, thalamic, and midbrain (for the human brain) regions are drawn with dashed-line boundaries to indicate that they are deeper in the brain relative to the anatomical structures above them. The anatomical boundaries drawn for the proposed human brain regions involved in vocal and auditory processing should be interpreted conservatively and for heuristic purposes only. Human brain lesions and brain imaging studies do not allow one to determine functional anatomical boundaries with high resolution. Scale bar: ~7mm. Abbreviations are in Table 4.1. Figure based on Jarvis (2004).

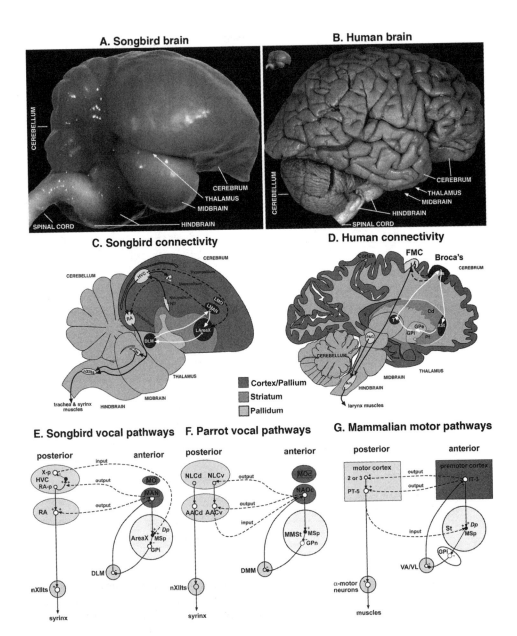

Figure 4.3
Comparative brain relationships, connectivity, and cell types among vocal learners and non-learners. (A) Zebra finch brain. (B) Human brain, with zebra finch brain to scale in upper-left corner. (C) Drawing through finch brain showing song system connectivity. (D) Drawing through human brain showing proposed vocal-pathway connectivity. (E) Songbird detailed cellular connectivity. (F) Parrot detailed cellular connectivity. (G) Mammalian motor pathway

HVC and NIf nuclei (Akutagawa & Konishi, 2010). The anterior forebrain nuclei (connectivity examined only in songbirds and parrots) form a closed loop, connecting a pallial song nucleus (MAN, NAO) to a striatal song nucleus (Area X, MMSt), to a dorsal thalamic nucleus (aDLM, DMM), back to the pallial song nucleus (MAN, NAO; Figure 4.3C,E,F; white arrows) (Durand et al., 1997; Vates et al., 1997). The anterior song nucleus of DLM (aDLM) is a renaming of songbird DLM, because only an anterior nucleus within DLM is active during singing and is not found in non-vocal-learner species (pigeon and chicken) (Wada, Sakaguchi, Jarvis, & Hagiwara, 2004; Horita et al., 2012). The parrot pallial MO song nucleus also projects to the striatal song nucleus (MMSt) (Durand et al., 1997). Connectivity of the songbird MO analog has not yet been determined.

The major differences among vocal-learning birds are in the connections between the posterior and anterior vocal pathways (Jarvis & Mello, 2000). In songbirds, the posterior pathway sends input to the anterior pathway via HVC to Area X; the anterior pathway sends output to the posterior pathway via lateral MAN (LMAN) to RA and medial MAN (mMAN) to HVC (Figure 4.3C,E,F; dashed lines) (Foster & Bottjer, 2001). In contrast, in parrots, the posterior pathway sends input into the anterior pathway via ventral AAC (AACv, parallel to songbird RA) to NAO (parallel to songbird MAN) and MO; the anterior pathway sends output to the posterior pathway via NAO to NLC (parallel to songbird HVC) and AAC (Figure 4.3F) (Durand et al., 1997).

With respect to vocal-learning mammals, ethical and practical issues prevent long-distance tract-tracing connectivity experiments in humans. These experiments are difficult to do with large-brained animals like cetaceans and elephants, and no forebrain vocal-pathway studies have been performed on bats that I am aware of. Thus, detailed tract-tracing connectivity of vocal-learning pathways is not known for any mammal. However, studies have been performed on other brain pathways of non-vocal-learning mammals. In this regard, the avian posterior song pathways are

---

connectivity. Dashed lines: connections between anterior and posterior pathways; inputs and outputs are labeled relative to anterior pathways. Output from songbird MAN to HVC and RA are not from the same neurons; medial MAN neurons project to HVC, lateral MAN neurons project to RA. **O**: excitatory neurons; •: inhibitory neurons; +: excitatory glutamate neurotransmitter release; –: inhibitory GABA release. MSp: medium spiny neuron. GPn: globus pallidus-like neuron in songbird Area X and parrot MMSt. Only the direct pathway through the mammalian basal ganglia (St to GPi) is shown because this is the one most similar to Area X connectivity (MSp to GPn) (Reiner et al., 2004) X-p: X-projecting neuron of HVC. RA-p: RA-projecting neuron of HVC. PT-5: pyramidal tract neuron of motor cortex layer 5. IT-3: intratelencephalic projecting neuron of layer 3. Human brain image reproduced, with permission, courtesy of John W. Sundsten, Digital Anatomist Project. Figure based on Jarvis (2004); Jarvis et al. (2005).

Table 4.1
Abbreviations used in this chapter

AAC—central nucleus of the anterior arcopallium	MG—medial geniculate
AACd—central nucleus of the anterior arcopallium, dorsal part	MLd—mesencephalic lateral dorsal nucleus
AACv—central nucleus of the anterior arcopallium, ventral part	MMSt—magnocellular nucleus of the anterior striatum
AI—intermediate arcopallium	MOc—oval nucleus of the mesopallium complex
ACM—caudal medial arcopallium	NAOc—oval nucleus of the anterior nidopallium complex
aDLPFC—anterior dorsal lateral prefrontal cortex	NCM—caudal medial nidopallium
aCC—anterior cingulate cortex	NDC—caudal dorsal nidopallium
aCd—anterior caudate	NIDL—intermediate dorsal lateral nidopallium
aINS—anterior insula cortex	NIf—interfacial nucleus of the nidopallium
Am—nucleus ambiguous	NLC—central nucleus of the lateral nidopallium
aP—anterior putamen	nXIIts—tracheosyringeal subdivision of the hypoglossal nucleus
aT—anterior thalamus	Ov—nucleus oviodalis
aST—anterior striatum	PAG—periaqueductal gray
Area X—Area X of the striatum	preSMA—presupplementary motor area
Av—avalanch	Pt—putatmen
Cd—caudate	St—striatum
CM—caudal mesopallium	RA—robust nucleus of the arcopallium
CSt—caudal striatum	Uva—nucleus uvaeformis
DLM—medial nucleus of dorsolateral thalamus	VA—vocal nucleus of the arcopallium
DM—dorsal medial nucleus of the midbrain	VA/VL—ventral anterior/ventral lateral nuclei of the mammalian thalamus
DMM—magnocellular nucleus of the dorsomedial thalamus	VAM—vocal nucleus of the anterior mesopallium
FMC—face motor cortex	VAN—vocal nucleus of the anterior nidopallium
HVC—(a letter-based name)	VAS—vocal nucleus of the anterior striatum
IC—inferior colliculus	VLN—vocal nucleus of the lateral nidopallium
L2—field L2	VMM—vocal nucleus of the medial mesopallium
MAN—magnocellular nucleus of anterior nidopallium	VMN—vocal nucleus of the medial nidopallium

Table 4.2
Comparable brain areas of vocal learners

Modality	Vocal			Auditory		
Species	Song	Parrot	Hummingbird	Human	Birds	Human
Subdivision						
Nidopallium	HVC	NLc	VLN	FMC – 2,3	L2,	1o aud – 4
	NIf	LAN	VMN	FMC – 2,3	L1, L3,	1o aud – 2,3
	MAN	NAO	VAN	Broca -2,3	NCM	2o aud – 2,3
Mesopallium	Av	LAM	VMM	FMC - ?	CM	2o aud - ?
	MO	MO	VAM	Broca - ?		
Arcopallium	RA	AAc	VA	FMC - 5	AI	2o aud - 5
Striatum	Area X	MMSt	VASt	Cd head	CSt	CSt
Thalamus	aDLM	DMM	aDLM	VL	Ov	MG
	Uva					
Midbrain	DM	DM	DM	PAG	MLd	IC

similar in connectivity to mammalian motor corticospinal pathways (Figure 4.3G). Specifically, the projecting neurons of songbird RA, parrot dorsal AAC, and hummingbird VA are similar to pyramidal tract (PT) neurons of lower layer 5 of the mammalian motor cortex (Matsumura & Kubota, 1979; Glickstein, May, & Mercier, 1985; Karten & Shimizu, 1989; Keizer & Kuypers, 1989; Reiner, Jiao, Del Mar, Laverghetta, & Lei, 2003). These neurons send long axonal projections out of the cortex through pyramidal tracts to synapse onto brainstem and spinal cord premotor or α-motor neurons that control muscles. The projection neurons of songbird HVC and parrot NLC are similar to layer 2 and 3 neurons of the mammalian cortex that send intrapallial projections to mammalian cortex layer 5 (Figure 4.3E,G) (Aroniadou & Keller, 1993; Capaday, Devanne, Bertrand, & Lavoie, 1998). Mammalian parallels to songbird NIf and Av are less clear.

The direct projection of songbird RA to the 12th vocal motor neurons is similar to the only connection physically determined in human cerebral vocal areas, the face motor cortex (FMC) projection directly to the brainstem vocal motor neurons called nucleus ambiguous (Am) (Figure 4.3D) (Kuypers, 1958a; Iwatsubo, Kuzuhara, Kanemitsu, Shimada, & Toyokura, 1990). Am projects to the muscles of the larynx, the mammalian vocal organ (Zhang, Bandler, & Davis, 1995; Jurgens, 1998), and is thus the mammalian parallel of avian 12th tracheosyringeal motor neurons. The FMC connection to Am in humans was determined using silver staining of degenerated axons in postmortem brains of patients that had had vascular strokes to the FMC (Kuypers, 1958a, 1958b; Iwatsubo et al., 1990). Similar lesions in macaque monkeys and chimpanzees (Kuypers, 1958b) as well as tract-tracing experiments in various monkey species confirmed the absence of a direct projection from the nonhuman primate motor cortex region to Am (Jurgens, 2002; Simonyan & Jurgens, 2003). Thus, the direct projection from the RA analog in birds and the face motor cortex in humans to brainstem vocal motor neurons that control the syrinx and larynx, respectively, has been argued to be a fundamental change that led to the evolution of learned song in birds and spoken language in humans (Kuypers, 1958a, 1958b; Iwatsubo et al., 1990; Wild, 1994a; Wild et al., 1997; Jurgens, 2002; Simonyan & Jurgens, 2003; Jarvis, 2004; Fitch, Huber, & Bugnyar, 2010; Fischer & Hammerschmidt, 2011; Simonyan & Horwitz 2011). I caution, though, that a sparse projection has been found in mice in our studies indicating that this connection may not be dichotomous among mammals but more continuous, potentially correlating with the level of vocal plasticity (Arriaga, Zhou, & Jarvis 2012).

The avian anterior vocal pathways are similar in connectivity to mammalian cortical-basal ganglia-thalamic-cortical loops (Figure 4.3E,G) (Bottjer & Johnson, 1997; Durand et al., 1997; Jarvis, Scharff, Grossman, Ramos, & Nottebohm, 1998; Perkel & Farries, 2000). Specifically, the projection neurons of songbird MAN and parrot NAO (Vates & Nottebohm, 1995; Durand et al., 1997; Foster, Mehta, & Bottjer, 1997) are similar to intratelencephalic (IT) neurons of layer 3 and upper layer 5 of the mammalian premotor cortex, which send two collateral projections: one to medium spiny neurons of the striatum ventral to it and the other to other cortical regions, including the motor cortex (Figure 4.3G) (Avendano, Isla, & Rausell, 1992; Reiner et al., 2003). In contrast to the situation in mammals, the spiny striatal neurons in songbird Area X project to pallidal-like cells within Area X and to a separate structure consisting only of pallidal cells (Perkel & Farries, 2000; Carrillo & Doupe, 2004; Reiner, Laverghetta, Meade, Cuthbertson, & Bottjer, 2004). This striatal-pallidal cell intermingling may be a general trait of the anterior avian striatum (Farries, Meitzen, & Perkel, 2005). The Area X pallidal cell types appear to be of two types: one type that like the mammalian internal globus pallidus (GPi) projects to the dorsal thalamus (aDLM), and another type like the mammalian external globus pallidus (GPe) that projects to other pallidal cells within Area X to the ventral pallidum proper (Figure 4.3E,F) (Gale & Perkel, 2010; Goldberg, Adler, Bergman, & Fee, 2010). From there, the connections are more similar to mammals again, where in both GPi-like cells project to the dorsal thalamus, which in turn projects back to layer 3 neurons in mammals or the LMAN analog in birds, closing parallel loops (Figure 4.3E,F) (Jacobson & Trojanowski, 1975; Alexander, DeLong, & Strick, 1986; Luo, Ding, & Perkel, 2001; Gale & Perkel, 2010).

Because connections between the posterior and anterior vocal pathways differ between songbirds and parrots, comparisons between them and mammals will also differ. In mammals, the PT-layer 5 neurons of the motor cortex have axon collaterals that project into the striatum, as well as the medulla and spinal cord (Figure 4.3G) (Alexander & Crutcher, 1990; Reiner et al., 2003). This is different from the songbird, where a specific cell type of HVC, called X-projecting neurons, projects to the striatum separately from the neurons of RA of the arcopallium that project to the medulla (Figure 4.3E). This is also different from the parrot, where AAC of the arcopallium has two anatomically separate neuron populations, AACd that projects to the 12th motor neurons and AACv that projects to the anterior pallial song nuclei NAO and MO (Figure 4.3F) (Durand et al., 1997). The outputs of neurons of the mammalian anterior pathways are proposed to be the collaterals of the IT layer 3 and IT upper layer 5 neurons that project to other cortical regions (Figure 4.3G) (Reiner et al., 2003; Jarvis, 2004).

This comparative analysis suggests that there are gross similarities between the connectivity of the specialized consensus bird-brain nuclei for learned vocalizing with nonvocal motor pathways (a posterior-like pathway) and cortical-basal-ganglia-thalamic-cortical loops (an anterior-like pathway) of mammals (Figure 4.3A-C). Differences between birds and mammals appear to be in the details, particularly with nuclear organization of the avian pallium versus layered organization in mammals, with pallidal cell types, and connectivity between posterior and anterior pathways.

Brain Lesion and Disorders of Vocal Learners

There are some gross similarities in behavioral deficits following lesions in specific brain areas of vocal-learning birds (experimentally placed) and of humans (due to stroke or trauma) that cannot be found in non-vocal-learning birds or mammals. For the posterior pathway, lesions in songbird HVC and RA cause deficits that are most similar to those found after damage to the human face motor cortex, more than anywhere else in the avian or human brain, this being muteness for song and speech—that is, learned vocalizations (Figure 4.4A) (Valenstein, 1975; Nottebohm, Stokes, & Leonard, 1976; Jurgens, Kirzinger, & Von Cramon, 1982; Simpson & Vicario, 1990; Jurgens, 1995). HVC lesions in zebra finches lead to adult animals producing subsong, babbling-like, juvenile vocalizations (Aronov, Andalman, & Fee, 2008) and lesions in RA lead to the complete inability to produce subsong or adult learned song (Nottebohm et al., 1976; Simpson & Vicario, 1990; Aronov et al., 2008). In parrots, lesions in the HVC analog (NLC) cause deficits in producing the correct acoustic structure of learned vocalizations, particularly for learned speech (Lavenex, 2000). Specific lesions in the analogous layer 3 and 5 cell types in the human face motor cortex have not been found nor are likely to occur, making a cell-type comparison difficult. There is also a dominant hemisphere for such an effect, this being the left side in canaries and humans and the right side in zebra finches (Nottebohm,

A. Lesions to the posterior song pathway in adult canaries

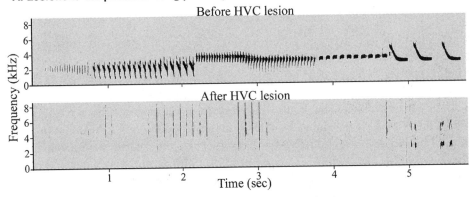

Before HVC lesion

After HVC lesion

Frequency (kHz)

Time (sec)

B. Lesions of the anterior song pathway in juvenile zebra finches

Juvenile ℓAreaX ℓMAN mMAN

Before

After

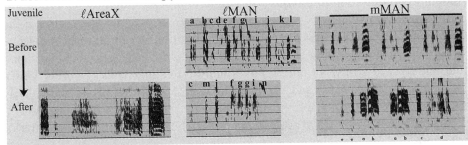

C. Lesion to the anterior song pathway in adult zebra finches

Adult ℓAreaX ℓMAN mMAN

Before

After

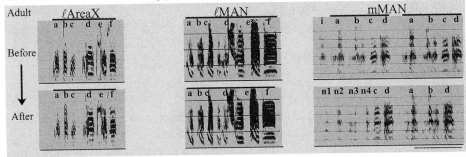

D. Lesions to Area X and stuttering

Other Adult Area X (Bengalese finch)

Before

After

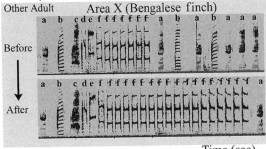

E. Lesions to LMAN and deaf

Deaf Deaf+ℓMAN

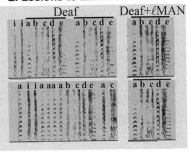

-Time (sec)-

1977; Williams, Crane, Hale, Esposito, & Nottebohm, 1992). When the lesion is uni-lateral, especially during juvenile development, both birds and human patients often recover some learned vocal behavior because the opposite hemisphere appears to take over some functions (Nottebohm, 1977; Rey, Dellatolas, Bancaud, & Talairach, 1988; Hertz-Pannier et al., 2002). If the lesions are bilateral, there is permanent loss of the ability to produce learned vocalizations. Innate sounds, such as contact and alarm calls in birds, or crying, screaming, and groaning in humans, can still be pro-duced. One difference relative to the human face motor cortex is that lesions in songbird NIf or parrot LAN of the posterior pathway do not prevent production of learned vocalizations or cause dysarthic-like vocalizations, but lead to produc-tion of more varied syntax or impaired vocal imitation (Hosino & Okanoya, 2000; Plummer & Striedter, 2002). In non-vocal learning avian species, lesions in the avian arcopallium where RA would be expected to be located, or in the face motor cortex in nonhuman primates, does not affect their ability to produce normal vocalizations (Kuypers, 1958b; Jurgens et al., 1982; Kirzinger & Jurgens, 1982; Lowndes & Davies, 1995).

For the anterior song pathway, lesions in songbird MAN cause deficits that are most similar to those found after damage to anterior parts of the human anterior premotor cortex (e.g., anterior insula, Broca's area, DLPFC, presupplementary motor area) more than anywhere else in the brain, this being disruption of imita-tion and/or inducing sequencing problems, but not preventing the ability to produce song (Figure 4.4B) or speech sounds (Nielsen & Jacobs, 1951; Barris, Schuman, & Schuman, 1953; Rubens, 1975; Valenstein, 1975; Mohr, 1976; Jonas, 1981; Nottebohm et al., 1990; Scharff & Nottebohm, 1991; Benson & Ardila, 1996; Dronkers, 1996; Foster & Bottjer, 2001). Specifically, lesions in songbird LMAN (Bottjer, Miesner, & Arnold, 1984; Scharff & Nottebohm, 1991; Kao, Doupe, & Brainard, 2005) and in the human insula and Broca's area (Mohr, 1976; Benson & Ardila, 1996; Dronkers,

◀ Figure 4.4

Behavioral deficits following forebrain lesions in song pathways in songbirds. Shown are sound spectrograms (i.e., sonographs) of songs bird produced before and after lesion of specific song nuclei. (A) Song of an adult male canary before and after lesion in HVC result-ing only in faint peeping sounds (Nottebohm et al., 1976). (B) Song of three juvenile zebra finches before and after lesion of lateral Area X (lArea X), lateral MAN (lMAN), and medial MAN (mMAN) (Scharff & Nottebohm, 1991; Foster & Bottjer, 2001). The birds could still sing, but song learning was disrupted. (B) Song of three adult zebra finches before and after lesion of lArea X, lMAN, and medial mMAN (Scharff & Nottebohm, 1991; Foster & Bottjer, 2001). No major changes are observed. Recordings before the Area X lesion were not reported in the example shown. (D) Song of an adult Bengalese finch before and after an Area X lesion showing repeated stuttering on its syllable f (Kobayashi et al., 2001). (E) Song of an adult zebra finch after deafening (left) shows deterioration of acoustic structure and syntax, and this is prevented with lesions in lMAN (Brainard & Doupe, 2000).

1996) lead to poor imitation of new sounds with sparing of already learned, stereo-typed song or speech. Lesions in Broca's area and/or the adjacent DLPFC lead to poor syntax production in the transformation of phonemes into words and words into sentences, as well as uncontrolled echolalia imitation (Benson & Ardila, 1996). Lesions in the adjacent pre-SMA and anterior cingulate result in spontaneous-speech arrest, lack of spontaneous speech, and/or loss of emotional tone in speech, but with imitation preserved (Nielsen & Jacobs, 1951; Barris et al., 1953; Rubens, 1975; Valenstein, 1975; Jonas, 1981). Lesions in songbird mMAN adjacent to LMAN lead to a decreased ability to learn vocalizations and some disruption of syntax (Figure 4.4B) (Foster & Bottjer, 2001), as do lesions in Broca's area (Benson & Ardila, 1996). Interestingly, deafness in songbirds and humans causes learned vocal-izations to deteriorate, and lesions in LMAN in songbirds prevent this deteriora-tion, indicating that LMAN is always needed to change song, even in song disorders (Figure 4.4E) (Brainard & Doupe, 2000). In humans, these deficits are called verbal aphasias and verbal amusias (Benson & Ardila, 1996). In songbirds, no such distinc-tion is made, but it may be worth considering whether analogous deficits could be considered song aphasias.

Within the striatum of the anterior pathways, lesion studies have often led to ambiguous conclusions in both vocal-learning birds and humans, which in part appear to have to do with the exact location of the lesions and the nature of the behaviors measured. For example, lesions in songbird Area X and in the human anterior striatum do not lead to an inability to produce song or speech (Figure 4.4C) (Nottebohm et al., 1976; Nadeau & Crosson, 1997), and this is the original reason the nucleus was called Area X in songbirds (connected to the song pathway with an unknown function). But then it was later discovered that disruption of Area X dis-rupts song learning (Figure 4.4B) (Sohrabji, Nordeen, & Nordeen, 1990; Scharff & Nottebohm, 1991; Kobayashi, Uno, & Okanoya, 2001). In humans, disruption of the anterior head of the caudate and putamen leads to verbal aphasias (Mohr, 1976; Bechtereva, Bundzen, Gogolitsin, & Malyshev, 1979; Leicester, 1980; Damasio, Damasio, Rizzo, Varney, & Gersh, 1982; Alexander, Naeser, & Palumbo, 1987; Cum-mings, 1993; Speedie, Wertman, Ta'ir, & Heilman, 1993; Lieberman, 2000). Others call the deficits from striatal lesions a dyspraxia (Nadeau & Crosson, 1997). In their meta-analyses, Nadeau and Crosson (1997) argue that lesions in the head of the caudate-putamen in humans lead to verbal aphasia only if the lesions simultane-ously involve white-matter fibers that project from the cortex to the striatum; they thus believe that the deficits are really due to a lack of cortical input into the anterior striatum (or thalamus). A problem with these different views has to do with defini-tions. Dyspraxia is technically considered an aphasia, the latter term meaning a problem producing proper speech but not eliminating speech. The findings in song-birds are clearer. After Area X lesions during the song-learning critical period, the birds do not crystallize the correct syllable structure and syntax heard (Figure 4.4B). They instead show a persistent variability in the production of syllable acoustic structure and syntax, with the syllables being noisier than in normal birds (Sohrabji et al., 1990; Scharff & Nottebohm, 1991; Kobayashi et al., 2001). After the critical period is over, in adults, the Area X lesions have little effect. Lesions in adults can lead to stuttering, especially when the birds produce syllable repetitions before lesions (Figure 4.4D) (Kobayashi et al., 2001). A corollary of this finding is that damage to and developmental dysfunction in the anterior striatum are among the most common causes of stuttering in humans (Ludlow & Loucks, 2003; Giraud et al., 2008).

For the globus pallidus, lesions can lead to various types of vocal communication deficits in humans (Strub, 1989; Nadeau & Crosson, 1997). Well-defined lesions have been placed in the human globus pallidus in Parkinson's patients to alleviate their symptoms (Troster, Woods, & Fields, 2003). After the lesions, however, Parkinson's patients show lexical verbal deficits (spontaneous generation of words) but not verbal semantic deficits (generation of words with categorical meanings), indicating a problem in selecting spoken-language-specific tasks. The authors concluded that parts of the pallidum are necessary for selecting lexical verbal sounds during speech production. In songbirds, the role of pallidal neuron types has not yet been deciphered in inactivation studies, because one would have to lesion the specific cell types in Area X of the striatum. No one has so far lesioned the ventral pallidum and assessed the effects on song. I am not aware of studies testing the role of the striatum and pallidum in vocalizations of non-vocal learning species, except bilateral globus pallidus lesions in squirrel monkeys (MacLean, 1978). After such lesions, these animals still vocalize species typical calls.

In noncerebral areas, lesions in avian DM and the presumed homologous mammalian PAG in the midbrain result in muteness in both vocal learners and nonlearners; the same is true for avian 12th tracheosyringeal and mammalian Am vocal motor neurons (Brown, 1965; Nottebohm et al., 1976; Seller, 1981; Jurgens, 1994, 1998; Esposito, Demeurisse, Alberti, & Fabbro, 1999). Within the dorsal thalamus, damage to songbird aDLM leads to an immediate increase in song stereotypy (Halsema & Bottjer, 1991; Goldberg & Fee, 2011). Lesions in the anterior human thalamus (VA, VL, and A) can lead to temporary muteness followed by verbal aphasia deficits that are sometimes greater than after lesions to the anterior striatum or premotor cortical areas (Graff-Radford, Damasio, Yamada, Eslinger, & Damasio, 1985). A greater deficit may occur perhaps because there is convergence of inputs from the striatum into the thalamus (Beiser, Hua, & Houk, 1997). However, the interpretation of thalamic lesions in humans is controversial (Benson & Ardila, 1996), perhaps because of small but important differences in lesion locations among patients. The thalamus concentrates many functions into adjacent small nuclei, and thus a relatively small variance in the location of a lesion may lead to a large difference in the brain function affected.

The lesions in many of the brain areas mentioned in birds and humans can affect more than one modality. For example, lesions in LMAN or HVC in songbirds and in Broca's area or the anterior striatum in humans also lead to decreased abilities in song/speech perception and discrimination (Freedman, Alexander, & Naeser, 1984; Benson & Ardila, 1996; Scharff, Nottebohn, & Cynx, 1998; Burt, Lent, Beecher, & Brenowitz, 2000). The perceptual deficits, however, are usually not as great as the vocal motor deficits.

Taken together, the lesion findings are consistent with the presence in humans of a posterior-like vocal motor pathway and an anterior-like vocal premotor pathway that are similar to the production and learning pathways of song-learning birds (Figures 4.2 and 4.3). The clearest difference between birds and humans appears to be the greater complexity of the deficits that occur after lesions in humans, and a greater dependence on the anterior speech brain regions in human adults than in the most commonly studied songbird, the zebra finch. This dependence could be in part because humans are open-ended vocal learners, whereas zebra finches are closed-ended. A more appropriate comparative analysis for future investigations would be with open-ended vocal-learning avian species.

Brain Activation in Vocal Learners

Brain-activation studies support some of the parallels revealed by lesion and connectivity studies among vocal-learning birds and with humans. Brain activation includes changes in electrophysiological activity (recorded in both birds and humans during or after surgery on patients), electrical stimulation (birds and humans), motor- and sensory-driven gene expression (birds and nonhuman mammals), and PET and MRI imaging of activated brain regions (in anesthetized birds and awake humans).

In vocal-learning birds, such studies have revealed that all seven comparable cerebral song nuclei display singing-driven expression of immediate early genes (Figure 4.5A) (Jarvis & Nottebohm, 1997; Jarvis et al., 1998; Jarvis & Mello, 2000; Jarvis et al., 2000). These genes are responsive to changes in neural activity. The song nuclei still display singing-driven expression when songbirds are deaf (removal of the cochlea) or mute (removal of the 12th nerve to the vocal muscles), indicating that the singing-driven gene activation is independent of hearing and somatosensation from the syrinx (Jarvis & Nottebohm, 1997). That is, the singing-driven expression in the song nuclei is motor-driven. In support of this conclusion, singing-associated premotor neural firing is found in all five nuclei tested: HVC, RA, LMAN, Area X, NIf (Figure 4.5E) (McCasland, 1987; Yu & Margoliash, 1996; Hessler & Doupe, 1999;

Hahnloser, Kozhevnikov, & Fee, 2002). Similar singing-associated activity still occurs when deaf birds sing (tested for HVC and LMAN; McCasland & Konishi, 1981; Hessler & Doupe, 1999). The neural firing in HVC and RA correlates with sequencing of syllables and syllable structure, respectively, whereas firing in LAreaX and LMAN is more varied and correlates with song variability (Hessler & Doupe, 1999; Hahnloser et al., 2002; Kao et al., 2005; Olveczky, Andalman, & Fee, 2005). Moreover, the firing and immediate early gene activation in the anterior pathway (LAreaX and LMAN) and RA are dramatically different depending on the social context in which singing occurs (higher and more variable in undirected than in directed singing), and this difference is associated with more fine control of song behavior in different contexts (Jarvis et al., 1998; Hessler & Doupe, 1999). Electrical pulse stimulation to HVC and RA during singing temporarily disrupts song output (i.e., causing song arrest), and in HVC can further cause transient song degradation in syllable structure (Vu, Schmidt, & Mazurek, 1998; Hahnloser et al., 2002; Ashmore, Wild, & Schmidt, 2005). Stimulation in LMAN produces transient changes in amplitude and pitch (Kao et al., 2005).

Are there parallel types of patterns of activation in the human brain for vocal communication? Somewhat. In humans, the brain area with activation (as measured with PET and fMRI) most comparable to songbird HVC and RA is one that is always activated with all speech and singing tasks, which is in or near the face motor cortex, particularly the larynx representation (Figure 4.5C,D) (Petersen, Fox, Posner, Mintun, & Raichle, 1988; Rosen, Ojemann, Ollinger, & Petersen, 2000; Brown, Martinez, Hodges, Fox, & Parsons, 2004; Gracco, Tremblay, & Pike, 2005; Brown, Ngan, & Liotti, 2007). Other human vocal brain areas appear to be activated or not activated depending on the context in which speech or song is produced. Production of verbs and complex sentences can be accompanied by activation in all or a subregion of the strip of cortex anterior to the face motor cortex: the anterior insula, Broca's area, DLPFC, pre-SMA, and anterior cingulate (Figure 4.5D) (Petersen et al., 1988; Poeppel, 1996; Price et al., 1996; Crosson et al., 1999; Wise, Greene, Buchel, & Scott, 1999; Papathanassiou et al., 2000; Rosen et al., 2000; Palmer et al., 2001; Gracco et al., 2005). Activation in Broca's area, DLPFC, and pre-SMA is higher when speech tasks are more complex, including learning to vocalize new words or sentences, sequencing words into complex syntax, producing nonstereotyped sentences, and thinking about speaking (Figure 4.5D) (Hinke et al., 1993; Poeppel, 1996; Buckner, Kelley, & Petersen, 1999; Bookheimer, Zeffiro, Blaxton, Gaillard, & Theodore, 2000). The left-brain vocal areas show more activation than their right counterparts (Poeppel, 1996; Price et al., 1996; Papathanassiou et al., 2000; Rosen et al., 2000). Like song nuclei in birds, premotor speech-related neural activity has been found in Broca's area (Fried, Ojemann, & Fetz, 1981). Further, similar to HVC, low-threshold electrical stimulation to the face motor cortex, Broca's area,

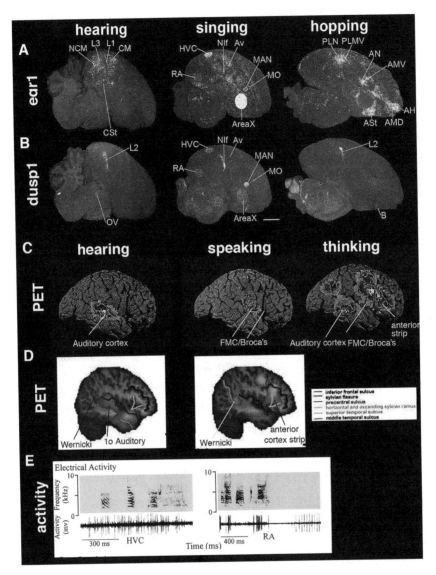

Figure 4.5
Hearing- and vocalizing-driven brain-activation patterns in songbirds and humans. (A) Brain expression patterns of the activity-dependent gene egr1 (white) in zebra finch males when they hear song (in the dark, standing still), sing, or hop (in a rotating wheel, when deaf, and in the dark). These are darkfield emulsion dipped sagittal sections reacted by in situ hybridizations. The hopping-induced activated regions are adjacent to song nuclei. Figures based on Mello et al. (1992); Jarvis and Nottebohm (1997); Feenders et al. (2008). (B) The same brain sections reacted with a probe to the dusp1 activity-dependent gene. Note the lack of hopping-induced dusp1 around song nuclei, but still the presence of singing-induced dusp1 in song

or the anterior supplementary areas causes speech arrest or generation of phonemes or words (Jonas, 1981; Fried et al., 1991; Ojemann, 1991, 2003).

In noncortical areas, speech production and singing in humans are accompanied by highest activation (fRMI and PET) of the anterior striatum and the thalamus (Wallesch, Henriksen, Kornhuber, & Paulson, 1985; Klein, Zatorre, Milner, Meyer, & Evans, 1994; Wildgruber, Ackermann, & Grodd, 2001; Brown et al., 2004; Gracco et al., 2005). Further, in songbirds and possibly in humans singing and speech are accompanied by dopamine release from the midbrain dopamine neurons (SNC-VTA) into the anterior striatum (Sasaki, Sotnikova, Gainetdinov, & Jarvis, 2006; Simonyan, Horwitz, & Jarvis, 2012). Low-threshold electrical stimulation to ventral lateral and anterior thalamic nuclei, particularly in the left hemisphere, leads to a variety of speech responses, including word repetition, speech arrest, speech acceleration, spontaneous speech, anomia, and verbal aphasia (but also auditory aphasia) (Johnson & Ojemann, 2000). The globus pallidus can also show activation during speaking (Wise et al., 1999). The PAG and Am in nonhuman mammals and the analogous DM and 12th motor neurons in birds display premotor vocalizing-associated neural firing (Yajima, Hayashi, & Yoshii, 1982; Larson, 1991; Larson, Yajima, & Ko, 1994; Zhang et al., 1995; Dusterhof, Hausler, & Jurgens, 2004) and/or vocalizing-driven gene expression (Jarvis et al., 1998; Jarvis & Mello, 2000; Jarvis et al., 2000). These findings demonstrate that it is not just cortical tissue that is involved in the production of learned vocalizations, but basal ganglia and thalamus tissue, as well as the brainstem premotor and motor neurons they control.

Forebrain vocal areas can also show action potential firing during hearing, depending on hearing task and species. In awake male zebra finches, firing is minimal in song nuclei (all the way down to the 12th) when a bird hears playbacks of song, but greater when he is anesthetized or asleep and presented with playbacks of his own song (Williams & Nottebohm, 1985; Dave & Margoliash, 2000; Nealen & Schmidt, 2002; Cardin & Schmidt, 2003). In song sparrows, the reverse occurs. Robust firing is observed in HVC when an awake bird hears playbacks of his own song, and this

---

nuclei (Horita et al., 2012). (C) Example brain-activation patterns in some brain regions on the surface of the cortex in humans, as seen with hearing-, speaking-, and thinking-driven PET signals minus rest. From Science Photo Library http://www.sciencephoto.com/media/307186/view. (D) PET signals superimposed on sagittal slices showing auditory and anterior strip of activation, including face motor cortex (FMC) during speaking. The shaded region is where higher activation occurs minus control conditions. Modified from Papathanassiou et al. (2000). (E) Neural activity in zebra finch HVC (interneuron) and RA (projection neuron) during singing (bottom plots), showing premotor neural firing milliseconds before song is produced (sonograph plots on top). Modified from Yu and Margoliash (1996). In panels A–D, not all activated brain areas are represented in these images; anterior is to the right, dorsal is up.

response is diminished when he is anesthetized (Nealen & Schmidt, 2002). In both species, the rate of firing or number of neurons that fire in song nuclei during hearing is lower than that during singing. Likewise, in humans, the face motor cortex, Broca's area, and/or the DLPFC often show increased activation when a person hears speech or is asked to perform a task that requires thinking in silent speech (Hinke et al., 1993; Poeppel, 1996; Price et al., 1996; Crosson et al., 1999; Wise et al., 1999; Papathanassiou et al., 2000; Rosen et al., 2000; Palmer et al., 2001). The magnitude of activation is usually lower during hearing than during speaking. The anterior insula, Broca's area, and DLPFC can also show activation due to other factors, such as by engaging working memory (MacLeod, Buckner, Miezin, Petersen, & Raichle, 1998; Zhang, Leung, & Johnson, 2003), which is short-term memory for future tasks. It is unclear, however, if working memory is a general property of the anterior cortex in which speech areas are located or if working memory and speech areas are separate.

It has long been assumed that non-vocal-learning species do not have forebrain regions that control the acoustic and syntactic structure of vocalizations. However, I believe that the evidence to support this conclusion has not been sought aggressively enough. I am not aware of studies that have attempted to record or stimulate regions of the forebrain to test for vocalization-elicited activity in non-vocal learning birds. A recent study of ours on a gene called dual sensitivity phosphatase 1 (dusp1) showed specialized singing-regulated expression in the forebrain song nuclei of vocal learners (songbirds, parrots, and hummingbirds) (Figure 4.5B) but no singing-activated forebrain regions in non–vocal learners (ring doves and suboscine songbirds) (Horita et al., 2012). This finding supports the conclusion of the connectivity and lesion studies, which is the absence of forebrain regions involved in direct vocal control in birds. But more investigation is necessary.

In mammals, the anterior cingulate cortex projects to the reticular formation surrounding the PAG, which in turn projects to Am (Jurgens, 2002). When the cingulate is lesioned, animals produce fewer calls, but the acoustic structure of the calls is normal (Kirzinger & Jurgens, 1982; Von Cramon & Jurgens, 1983). This has led to the conclusion that although the cingulate cortex is active in nonhuman primate vocalizations, it is not necessary for producing the acoustic structure of vocalizations but is necessary for the motivation to vocalize. Recently, two studies using immediate early genes in a small primate, marmosets, identified a region of the cingulate and adjacent cortex that is active in calling (Miller, Dimauro, Pistorio, Hendry, & Wang, 2010; Simoes et al., 2010). These studies challenged the notion that nonhuman primates do not have forebrain vocal-control regions outside of the cingulate cortex. However, both studies did not include silent controls or control for the effects of hearing oneself vocalize. Thus it is not clear if these are motor-control regions for vocalizations or regions associated with other aspects of the behavior, such as hearing oneself vocalize. In addition, this region was not the laryngeal premotor cortex identified in other primate species (Kirzinger & Jurgens, 1982; Jurgens, 2002). Humans, in contrast, are thought to have a premotor larynx region (possibly adjacent to or within Broca's area) as well as a primary motor cortex larynx region either within or adjacent to the face motor cortex (Figure 4.2E) (Jurgens, 2002; Brown et al., 2007), although Simonyan and Horwitz (2011) indicate that the presence of the premotor larynx cortex in humans needs to be validated. One possible explanation for the paradoxical findings in nonhuman primates is that the activated and indirectly connected forebrain regions to Am represent preadaptations for the evolution of vocal learning or that nonhuman primates may have lost part of the forebrain pathway necessary for vocal imitation (Simonyan & Horwitz, 2011).

Taken together, the brain-activation findings are consistent with the idea that songbird HVC and RA are more similar in their functional properties to the laryngeal-face motor cortex in humans than to any other human brain area, and that songbird MAN, Area X, and the anterior part of the dorsal thalamus are more similar in their properties to a strip of the anterior human premotor cortex, to part of the human anterior striatum, and to the human ventral lateral/anterior thalamic nucleus respectively (although Nadeau and Crosson (1997) argue for other human thalamic nuclei involved in speech). Non-vocal-learning birds as of now appear not to have any forebrain regions active with vocalizing, whereas non-vocal-learning primates may have some regions, but unlike in humans these are not required for producing species-specific vocalizations.

Molecular Specializations

Convergent behavioral and anatomical specializations for vocal learning in birds and humans might be expected to be associated with convergent molecular changes in those forebrain regions. Moreover, one would expect to find genes involved in neural connectivity that differs between vocal learners and non-learners. Findings in the last several years are starting to support this hypothesis. Through educated guesses and high-throughput gene expression investigations, our lab and that of Kazuo Okanoya have found convergent gene expression specializations of transcription factors, cell adhesion, and axon guidance molecules in song nuclei of all three vocal-learning orders. For example, the transcription factor *FoxP1* shows convergent up-regulation in the HVC analog of all vocal-learning bird lineages (Figure 4.6A) (Haesler et al., 2004; Feenders et al., 2008); its sister gene, *FoxP2*, is required for normal speech and song learning in humans and songbirds (via Area X) (Haesler et al., 2007; Fisher & Scharff, 2009). The axon guidance receptor neuropilin 1 shows enriched expression in the HVC and RA analogs of songbirds and parrots (not tested in hummingbirds) (Matsunaga, Kato, & Okanoya, 2008), whereas the axon

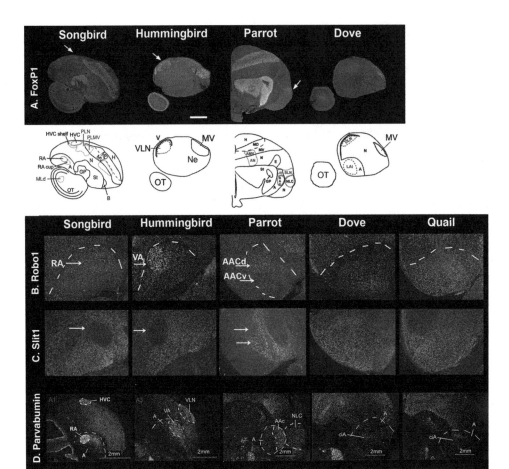

Figure 4.6
Molecular specializations in avian vocal-learning systems. (A). The FoxP1 mRNA (white) is overexpressed in the HVC analog (arrow) of all three vocal-learning bird groups (songbird, hummingbird, and parrot), but not in the analogous area of the dove brain. All sections are sagittal, except for parrot, which is frontal. The anatomical drawings below the gene expression images show brain regions. Images from Feenders et al. (2008). (B) Up-regulation of ROBO1 expression in the RA analog of vocal learners (highest in hummingbird), but not of nonlearners (dove and quail) relative to the surrounding arcopallium. (C) Down-regulation of Slit1 expression in the RA analog of vocal learners. Modified from Wang et al. (2012). (D) Up-regulation of parvalbumin expression in the RA and HVC analogs of vocal learners. Modified from Hara et al. (2012).

guidance receptor Robo1 shows enriched expression and its ligand Slit1 shows down-regulation in the RA analog of all three vocal-learning orders (Figure 4.6B,C) (Wang, Chen, Hara, & Jarvis, 2012). Interestingly, in parrots, the RA analog (AAC) has two divisions, a ventral division (AACv) that projects to other song nuclei and a dorsal division (AACd) that makes the specialized direct projection to brainstem vocal motor neurons. Only the dorsal division showed specialized Robo1 and Slit1 expression (Figure 4.6B,C). Different splice variants of the human *ROBO1* gene are also enriched in human fetal frontal and auditory brain areas (Johnson et al., 2009). Mutations of *ROBO1* in humans are associated with dyslexia, speech sound disorder, and other speech-language deficits (Hannula-Jouppi et al., 2005; Bates et al., 2010). Interestingly, the *FoxP2* gene appears to preferentially regulate axon guidance molecules, including specialized regulation of Slit1 by the human version of the gene but limited regulation by its chimpanzee ortholog (Konopka et al., 2009). Specialized expression of these genes was not present in the forebrains of non-vocal learning species (ring doves and quails) (Figure 4.6A,B,C). These findings indicate that there might be a link between the enriched gene regulation and the FoxP and Robo/Slit family of genes in vocal-learning species.

Axon guidance genes are not the only ones that show convergent gene expression changes in vocal-learning species. We have found that glutamate receptor subunits and several genes involved in neuroprotection also show convergent differential regulation in vocal-learning species. The NR2A, NR2B, and mGluR2 glutamate receptor subunits show either increased or decreased expression throughout the song nuclei of songbirds, parrots, and hummingbirds (Wada et al., 2004). The Ca2+ buffering protein, parvalbumin, shows differential up-regulation in the HVC analog and RA analog of all vocal-learning bird species (Figure 4.6D), and further surprisingly, also in the 12th vocal motor neurons relative to non-vocal-learning birds (Hara, Rivas, Ward, Okanoya, & Jarvis, 2012). Further, in a parallel fashion we found parvalbumin was differentially up-regulated in human but not monkey 12th motor tongue neurons (Hara et al., 2012). The dusp1 gene that shows specialized singing-induced up-regulation in song nuclei of vocal-learning species (Figure 4.5B), but does not show behaviorally driven expression outside the song system (Horita et al., 2010, 2012). This is strikingly different from other immediate early genes, which show behaviorally regulated expression (singing or movement) inside and outside the song nuclei (Figure 4.5A). Both dusp1 and parvalbumin are involved in neural protection. I hypothesize that their convergent up-regulation could be associated with the higher levels of activity found in song nuclei (of at least songbirds) relative to the surrounding brain subdivisions or possibly a higher rate of vocalizing relative to non–vocal learners.

In summary, these findings indicate that there is a convergent evolution mechanism that selects for similar genetic changes associated with the trait of vocal

neurons, these project to cochlea and lemniscal nuclei of the brainstem, which in turn project to midbrain (avian MLd, reptile torus, mammalian inferior colliculus) and thalamic (avian Ov, reptile reunions, mammalian medial geniculate) auditory nuclei. The thalamic nuclei in turn project to primary auditory cell populations in the cortex/pallium (avian Field L2, reptile caudal pallium, mammalian layer 4 cells of the primary auditory cortex; Figure 4.7). Avian Field L2 then projects to other pallial cells (L1, L3, NCM, CM) and striatal (CSt) populations that form a complex network. Mammalian layer 4 cells project to other layers of the primary auditory cortex and to secondary auditory regions, also forming a complex network. In birds, Field L1 and L3 neurons are similar in connectivity to mammalian layers 2 and 3 of the primary auditory cortex, the latter of which receive input, like Field L2, from layer 4 (Karten, 1991; Wild, Karten, & Frost, 1993). Avian NCM and CM are also similar to layers 2 and 3 in that they form reciprocal intrapallial connections with each other and receive some input from Field L2 (Figure 4.7) (Wang, Brzozowska-Prechtl, & Karten, 2010). Cerebral pathway connectivity is not well studied for nonavian reptiles.

Because an auditory forebrain pathway exist in all birds, reptiles, and mammals examined to date, I argue that the auditory forebrain pathway existed before vocal-learning pathways evolved and that the auditory pathway is not sufficient for vocal learning (Jarvis, 2004, 2006). Further, I hypothesize that the auditory pathway in vocal-learning birds and in humans was inherited from their common stem-amniote ancestor, thought to have lived ~320 million years ago (Jarvis, 2004; Jarvis et al., 2005). An alternative view proposed by Carr and colleagues is that the auditory pathway in each major tetrapod vertebrate group (amphibians, turtles, lizards, birds, and mammals) evolved independently of a common ancestor (Christensen-Dalsgaard & Carr, 2008). The rationale for this hypothesis is that in the different vertebrate groups, the cochlea nucleus in the midbrain develops from different neural rhombomeres, and therefore they cannot be homologous. The weaknesses of this hypothesis are that it is possible for homologous cell types to migrate and change rhombomere locations (Jacob & Guthrie, 2000), and there is no known vertebrate group that does not have an auditory forebrain pathway.

The source of auditory input into the vocal pathways of vocal-learning birds as well as humans is unclear. Proposed routes in songbirds include the HVC shelf into HVC; the RA cup into RA; Ov or CM into NIf; and from NIf dendrites in L2 (Wild, 1994b; Fortune & Margoliash, 1995; Vates et al., 1996; Mello, Vates, Okuhata, & Nottebohm, 1998; Bauer et al., 2008). One has to be cautious about what regions is considered an auditory pathway, because the CM input to NIf may have involved a mistaken identity of the song nucleus Av (surrounded by CM) input to NIf (Figure 4.3C) (Akutagawa & Konishi, 2010). The location of the song nuclei relative to the auditory regions also differs among vocal-learning groups. In songbirds, the posterior song nuclei are embedded in both auditory and motor regions; in hummingbirds, they are situated more laterally, but still adjacent to the auditory regions; in parrots, they are situated far laterally and physically separate from the auditory regions (Figure 4.2A–C). Humans are more similar to parrots, where the primary auditory cortex is topologically situated far caudal to the motor regions involved in the production of speech (Figure 4.2E). In humans, primary auditory cortex information is passed to secondary auditory areas, which include Wernicke's area (Figure 4.2E). Information from Wernicke's area has been proposed to be passed to Broca's area through arcuate fasciculus axons that traverse a caudal-rostral path (Geschwind, 1979), and this has been supported by diffusion tensor imaging (DTI) experiments in humans, which is a noninvasive method that allows paths of axons to be imaged (Glasser & Rilling, 2008). The presence of these arcuate fasciculus axons has been argued as an alternative hypothesis to the direct motor cortex–to–brainstem connection as the critical step in the evolution of spoken language (see the review by Fitch et al., 2010). The fiber pathway is present in some nonhuman primates, but is thinner in chimpanzees and is said to be separated by an indirect pathway in macaques (Rilling et al., 2008). One difficulty with this theory is that there is no known frontal cortex area that controls vocalizations in nonhuman primates, and the proponents of the theory tend to ignore the alternative explanation of the direct forebrain-to-brainstem connection. An alternative is that both hypotheses could be correct.

Bilateral damage to the primary auditory cortex and Wernicke's area leads to full auditory agnosia, the inability to consciously recognize sounds (speech, musical instruments, natural noises, etc.) (Benson & Ardila, 1996). Damage to Wernicke's area only leads to auditory aphasias, sometimes called fluent aphasia. A patient can speak well, but produces nonsense highly verbal speech. One reason for this symptom is that the vocal pathways may no longer receive feedback from the auditory system via the arcuate fasciculus, and thus run spontaneously on their own. In songbirds (at least in zebra finches), lesions in NCM and CM result in a significant decline in the ability to form auditory memories of songs heard, but do not affect the ability to sing or the acoustic structure of the songs (MacDougall-Shackleton, Hulse, & Ball, 1998; Gobes & Bolhuis, 2007). However, no one has yet tested whether lesions to these avian secondary auditory areas result in fluent song aphasias (which would be best tested in a species that produces more variable syntax) or in deficits of song learning. In macaques, the auditory cortex appears to be able to help form short-term auditory memories, but unlike humans, the animals have weak long-term auditory memories, and this has been argued as a potential difference between vocal-learning and non-vocal-learning species (Fritz, Mishkin, & Saunders, 2005). No one has yet tested if such differences in auditory memory occur in song-learning and non-song-learning birds.

In summary, the presence of cerebral auditory areas is not unique to vocal-learning species, which would explain why nonhuman animals, like dogs, exhibit auditory learning, including learning to understand the meaning of rudimentary human speech. It is possible that the primary and secondary auditory systems involved in speech perception in humans and song perception in birds represent an ancestral homologous system found at least in tetrapod vertebrates. Potential differences between vocal-learning and non-vocal-learning species may be in the weaker formation of long-term auditory memories, and in the weakness or absence of a direct projection from caudal auditory areas to frontal motor cortical areas. To support or refute these hypotheses, more comparative experiments are needed on vocal-learning and non-vocal-learning mammalian and avian species.

A Motor Theory for Vocal-Learning Origins

Whether the vocal-learning trait is dichotomous or continuous among species, the finding of remarkably convergent song/speech systems in distantly related birds and humans suggests that although brain pathways for vocal learning in different groups may have evolved independently from a common ancestor, they have done so under strong preexisting constraints. For many years, no experimental evidence suggested a convincing constraint. Recently, a possible constraint was revealed by identifying motor forebrain areas in birds (Feenders et al., 2008). Using behavioral molecular mapping, we discovered that in songbirds, parrots, and hummingbirds, all cerebral song nuclei are embedded in discrete adjacent brain regions that are selectively activated by limb and body movements (Figure 4.5A, hopping) (Feenders et al., 2008). Similar to the relationships between vocal nuclei activation and singing, activation in the adjacent regions correlates with the amount of movement performed and is independent of auditory and visual input. One exception is the motor areas adjacent to Nif and Av, called PLN and PLMV respectively, which showed both movement- and auditory-induced gene activation, independent of each other, and did so in both vocal learners and nonlearners (Figure 4.5A) (Feenders et al., 2008). These same movement-associated brain areas are also present in female songbirds that do not learn vocalizations and have atrophied cerebral vocal nuclei, and in non-vocal-learning birds such as ringdoves. It remains to be determined whether these areas function for motor output and/or somatosensory feedback from muscles during movement. However, the known somatosensory regions (AH and AMD) were active with movement behavior and they do not surround the song-learning nuclei (Figure 4.5A). Likewise, in humans, cortical areas involved in the production of spoken language are adjacent to or embedded in regions that control learned movement behavior, including dancing (Brown et al., 2006). Based on these findings, we proposed a motor theory for the origin of

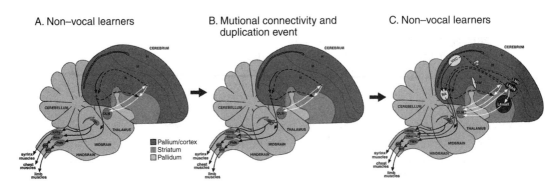

Figure 4.8
Proposed mechanism for the motor theory vocal-learning origin. (A) Non-vocal-learner brain with cerebrum nonvocal motor pathways, and midbrain and brainstem vocal innate pathways. (B) Proposed mutational event that led to descending cerebrum axons from the arcopallium to synapse onto vocal-motor (nXIIts) and respiratory (RAm) neurons (gray open arrows). (C) Vocal-learner brain now with a song-learning system with parallel connectivity to its cerebral motor pathway.

vocal learning: *Cerebral systems that control vocal learning in distantly related animals evolved as specializations of a preexisting motor system inherited from their common ancestor that controls movement and perhaps motor learning.* This preexisting forebrain motor pathway may represent a deep homology shared by vocal-learning systems.

A comparative analysis of the vocal-learning pathways and the surrounding motor systems may give clues to the mechanisms of how each of them work and how they evolved. The anatomical extent of the movement-associated areas in birds is larger than the song nuclei, which is consistent with a greater amount of musculature involved in the control of limb and body movements relative to that for the syrinx. The connectivity of the surrounding movement-associated areas in songbirds is similar to anterior and posterior song pathways (Figure 4.8A) (Iyengar, Viswanathan, & Bottjer, 1999; Bottjer, Brady, & Cribbs, 2000; Feenders et al., 2008). Like the songbird posterior vocal pathway, the posterior movement-associated regions are connected into a descending motor system that projects to premotor neurons (PMN) of the brainstem reticular formation (Figure 4.8A; black arrows); the PMN then projects to motor neurons that control body movements. The projection to PMN instead of directly to motor neurons represents a departure from vocal-learning systems, which may allow these systems to have more fine motor control of the musculature to produce vocalizations, since direct projections are usually associated with more fine motor control (Kuypers, 1958a; Lemon, Kirkwood, Maier, Nakajima, & Nathan, 2004).

Like the anterior song pathway, the anterior movement-activated regions are connected in a pallial-basal-ganglia-thalamic-pallial loop (Figure 4.8A; white arrows). Connectivity between posterior and anterior movement-associated regions has similarities to and differences from the song-learning systems. The differences are that unlike HVC's projection to Area X in songbirds, the adjacent nidopallium in zebra finches only sends a weak projection to the striatum (not shown), whereas the arcopallium adjacent to RA sends a strong projection to the striatum (Figure 4.8A) and many other areas besides the reticular PMN (Bottjer et al., 2000). These differences may reflect fewer constraints on interactions between posterior and anterior motor pathways. Mammalian nonvocal motor (posterior) and premotor (anterior) pathways follow a similar connectivity design (Figure 4.3G). So perhaps the evolution of vocal-learning brain areas for birds and humans exploited a more universal motor system that predates the split from the common ancestor of birds and mammals (i.e., stem amniotes).

An alternative view that has been suggested is that song nuclei evolved from adjacent auditory regions (Margoliash et al., 1994; Mello et al., 1998). Part of the regions adjacent to the posterior pathway songbird song nuclei (HVC shelf and RA cup) are also auditory. It has been further suggested that all regions adjacent to all song nuclei may be a parallel pathway involved in song learning and production, by bringing auditory input into the song system (Iyengar et al., 1999; Bottjer et al., 2000). This hypothesis was recently supported by lesion studies adjacent to RA (Bottjer & Altenau, 2011). However, the lesion studies have not acknowledged the adjacent motor areas to the song nuclei; further, parrots have such auditory regions in the same locations as songbirds, but the parrot posterior song nuclei (its NLC, AAC, LAN, and LAM) are situated far away from the auditory regions (Figure 4.2B), but still adjacent to the motor-activated regions (Feenders et al., 2008). Our results suggest that the latter areas are nonvocal motor, not involved in singing or hearing.

Our findings suggest that the three vocal-learning groups may have independently evolved similar cerebral vocal systems but that this was dependent (i.e., constrained) by a previous genetically determined motor system inherited from their common ancestor. If true, then such a posterior/anterior motor system could be used as a template for the evolution of a vocal motor/learning system that controls muscles of the syrinx or larynx by taking over control of midbrain and medulla vocal motor neurons. In this manner, a mutational event that caused descending projections of avian arcopallium neurons or human face motor cortex layer 5 neurons to synapse onto 12th tracheosyringeal and Am vocal motor neurons, respectively, may be the only major change that is needed to initiate a vocal-learning pathway (Figure 4.8B; gray arrows). Thereafter, other vocal brain regions could develop out of adjacent motor brain regions with preexisting connectivity.

Such a mutational event would be expected to occur in genes that regulate synaptic connectivity of upper pallial motor neurons to lower α-motor neurons. Not apparent in this view is the question of whether there is a genetic constraint for auditory information to enter vocal-learning pathways. I propose that the regions adjacent to NIf and Av (PLN and PLMV) may already provide auditory input into the motor system. This would represent another anatomical constraint from which to evolve a vocal-learning system. These hypotheses may be testable with fate mapping of different cell types across species and genetic manipulation studies of developing brain circuits.

At this point, I cannot say in our theory whether the forebrain vocal system developed using a preexisting part of a motor pathway as a scaffold or usurped a preexisting part of the pathway. However, I do not believe that a preexisting part of a motor pathway was lost. Rather, our idea is similar to gene evolution by duplication, where a gene is duplicated and one copy is used for a new function and the old copy maintains its old function (Ito, Ishikawa, Yoshimoto, & Yamamoto, 2007). In this regard, I propose one possible mechanism of vocal-learning pathway evolution, pathway duplication during embryonic development. I hypothesize that the vertebrate brain may have parallel posterior (motor) and anterior (premotor) pathways with sensory (auditory, visual, or somatosensory) input, which are connected to different brainstem and spinal cord premotor neuron groups to coordinate different behaviors. I argue that in vocal learners, this pathway could have been duplicated one more time, and then connected to brainstem neurons that control vocalizations and respiration. This is how similar brain pathways could have emerged independently from a common sensorimotor pathway.

We are not the first to implicate a motor origin for a learned vocal behavior. Based on a literature summary of studies conducted on humans, Robin Allott (1992) proposed a "motor theory for language origin" in a linguistic conference proceedings, where he argued that language brain areas evolved from a preexisting motor neural system. However, he did not provide experimental evidence or flesh out the anatomical or mechanistic details of this theory. Lieberman (2002) proposed that language areas evolved out of a preexisting cortical-basal-ganglia-thalamic loop, for which he identified the basal ganglia part as the reptilian brain. However, we now know that reptilian and avian cerebrums are not made up only of basal ganglia, that vocal-learning birds only have one nucleus of the vocal system in the basal ganglia, and that language areas may involve more than just this loop (Jarvis, 2004; Jarvis et al., 2005). Farries (2001) and Perkel (2004) proposed in birds and Jarvis (2004) in birds and humans that vocal-learning pathways in birds and humans may be similar to systems outside of the vocal pathways that logically could be motor pathways, but they did not have experimental evidence of these regions being motor to corroborate these suggestions. The findings of Feenders et al. (2008) provided evidence for a motor-origin theory. These findings have suggested that a cerebral motor system controls the vocal apparatus and creates the ability to translate auditory signals that everyone can process into vocal signals that only vocal learners can generate.

This theory is also concordant with the hypothesis of a gestural origin for spoken language, where the motor-learning ability to form gestures in human and nonhuman primates has been argued to be a precursor behavior for motor learning of speech/language (Pika, Liebal, & Tomasello, 2005; Gentilucci & Corballis, 2006; Pollick & De Waal, 2007). During child development, gesture production appears before speech production and is thought to enhance learning of speech, and adults use limb gestures automatically and often unconsciously during speech production (Galantucci, Fowler, & Turvey, 2006; Gentilucci & Corballis, 2006). This gestural hypothesis was one basis for the motor theory of language origins (Allott, 1992). We suggest that gesturing is controlled by a the motor system adjacent to the vocal system. Gesturing, although not a requirement in our theory, may be present in avian species too, as many avian species perform non-vocal gesturing-like movements such as a courtship dance or wing displays during vocalizing (Miller & Inouye, 1983; Zann, 1996; Prum, 1998; Cooper & Goller, 2004; Altshuler, 2006), and raven songbirds were recently discovered to communicate with gestures (Pika & Bugnyar 2011). Investigations into the behaviors and neural circuits for movement displays in birds may help shed light on these ideas. If verified in both birds and mammals, the evolution of vocal-learning brain systems as a specialization of a preexisting motor system could be a general mechanism of how brain pathways for complex traits evolve.

References

Akutagawa, E., & Konishi, M. (2010). New brain pathways found in the vocal control system of a songbird. *Journal of Comparative Neurology*, *518*, 3086–3100.

Alexander, G. E., & Crutcher, M. D. (1990). Functional architecture of basal ganglia circuits: Neural substrates of parallel processing. *Trends in Neurosciences*, *13*, 266–271.

Alexander, G. E., DeLong, M. R., & Strick, P. L. (1986). Parallel organization of functionally segregated circuits linking basal ganglia and cortex. *Annual Review of Neuroscience*, *9*, 357–381.

Alexander, M. P., Naeser, M. A., & Palumbo, C. L. (1987). Correlations of subcortical CT lesion sites and aphasia profiles. *Brain*, *110*, 961–991.

Allott, R. (1992). The motor theory of language: Origin and function. In J. Wind, B. H. Bichakjian, A. Nocentini, & B. Chiarelli (Eds.), *Language Origin: A Multidisciplinary Approach* (pp. 105–119). Dordrecht, The Netherlands: Kluwer.

Altshuler, D. L. (2006). Flight performance and competitive displacement of hummingbirds across elevational gradients. *American Naturalist*, *167*, 216–229.

Aroniadou, V. A., & Keller, A. (1993). The patterns and synaptic properties of horizontal intracortical connections in the rat motor cortex. *Journal of Neurophysiology*, *70*, 1553–1569.

Aronov, D., Andalman, A. S., & Fee, M. S. (2008). A specialized forebrain circuit for vocal babbling in the juvenile songbird. *Science*, *320*, 630–634.

Ashmore, R. C., Wild, J. M., & Schmidt, M. F. (2005). Brainstem and forebrain contributions to the generation of learned motor behaviors for song. *Journal of Neuroscience*, *25*, 8543–8554.

Avendano, C., Isla, A. J., & Rausell, E. (1992). Area 3a in the cat. II. Projections to the motor cortex and their relations to other corticocortical connections. *Journal of Comparative Neurology*, *321*, 373–386.

Baptista, L. F., & Schuchmann, K. L. (1990). Song learning in the anna hummingbird (*Calypte anna*). *Ethology*, *84*, 15–26.

Barris, R. W., Schuman, M. D., & Schuman, H. R. (1953). Bilateral anterior cingulated gyrus lesions: Syndrome of the anterior cingulate gyri. *Neurology*, *3*, 44–52.

Bates, T. C., Luciano, M., Medland, S. E., Montgomery, G. W., Wright, M. J., & Martin, N. G. (2010). Genetic variance in a component of the language acquisition device: ROBO1 polymorphisms associated with phonological buffer deficits. *Behavior Genetics*, *41*, 50–57.

Bauer, E. E., Coleman, M. J., Roberts, T. F., Roy, A., Prather, J. F., & Mooney, R. (2008). A synaptic basis for auditory-vocal integration in the songbird. *Journal of Neuroscience*, *28*, 1509–1522.

Bechtereva, N. P., Bundzen, P. V., Gogolitsin, Y. L., Malyshev, V. N., & Perepelkin, P. D. (1979). Neurophysiological codes of words in subcortical structures of the human brain. *Brain and Language*, *7*, 143–163.

Beiser, D. G., Hua, S. E., & Houk, J. C. (1997). Network models of the basal ganglia. *Current Opinion in Neurobiology*, *7*, 185–190.

Benson, D. F., & Ardila, A. (1996). *Aphasia: A clinical perspective*. New York: Oxford University Press.

Bolhuis, J. J., Okanoya, K., & Scharff, C. (2010). Twitter evolution: Converging mechanisms in birdsong and human speech. *Nature Reviews: Neuroscience*, *11*, 747–759.

Bookheimer, S. Y., Zeffiro, T. A., Blaxton, T. A., Gaillard, P. W., & Theodore, W. H. (2000). Activation of language cortex with automatic speech tasks. *Neurology*, *55*, 1151–1157.

Bottjer, S. W., & Altenau, B. (2011). Parallel pathways for vocal learning in basal ganglia of songbirds. *Nature Neuroscience*, *13*, 153–155.

Bottjer, S. W., Brady, J. D., & Cribbs, B. (2000). Connections of a motor cortical region in zebra finches: Relation to pathways for vocal learning. *Journal of Comparative Neurology*, *420*, 244–260.

Bottjer, S. W., & Johnson, F. (1997). Circuits, hormones, and learning: Vocal behavior in songbirds. *Journal of Neurobiology*, *33*, 602–618.

Bottjer, S. W., Miesner, E. A., & Arnold, A. P. (1984). Forebrain lesions disrupt development but not maintenance of song in passerine birds. *Science*, *224*, 901–903.

Brainard, M., & Doupe, A. (2000). Interruption of a basal ganglia-forebrain circuit prevents plasticity of learned vocalizations. *Nature*, *404*, 762–766.

Brown, J. (1965). Loss of vocalizations caused by lesions in the nucleus mesencephalicus lateralis of the Redwinged Blackbird. *American Zoologist, 5,* 693.

Brown, S., Martinez, M. J., Hodges, D. A., Fox, P. T., & Parsons, L. M. (2004). The song system of the human brain. *Brain Research: Cognitive Brain Research, 20,* 363–375.

Brown, S., Martinez, M. J., & Parsons, L. M. (2006). The neural basis of human dance. *Cerebral Cortex, 16,* 1157–1167.

Brown, S., Ngan, E., & Liotti, M. (2007). A larynx area in the human motor cortex. *Cerebral Cortex, 18,* 837–845.

Buckner, R. L., Kelley, W. M., & Petersen, S. E. (1999). Frontal cortex contributes to human memory formation. *Nature Neuroscience, 2,* 311–314.

Burt, J., Lent, K., Beecher, M., & Brenowitz, E. (2000). Lesions of the anterior forebrain song control pathway in female canaries affect song perception in an operant task. *Journal of Neurobiology, 42,* 1–13.

Caldwell, M. C., & Caldwell, D. K. (1972). Vocal mimicry in the whistle mode by an Atlantic bottlenosed dolphin. *Cetology, 9,* 1–8.

Capaday, C., Devanne, H., Bertrand, L., & Lavoie, B. A. (1998). Intracortical connections between motor cortical zones controlling antagonistic muscles in the cat: A combined anatomical and physiological study. *Experimental Brain Research, 120,* 223–232.

Cardin, J. A., & Schmidt, M. F. (2003). Song system auditory responses are stable and highly tuned during sedation, rapidly modulated and unselective during wakefulness, and suppressed by arousal. *Journal of Neurophysiology, 90,* 2884–2899.

Carr, C. E., & Code, R. A. (2000). The central auditory system of reptiles and birds. In R. J. Dooling, R. R. Fay, & A. N. Popper (Eds.), *Comparative hearing: Birds and reptiles* (pp. 197–248). New York: Springer.

Carrillo, G. D., & Doupe, A. J. (2004). Is the songbird Area X striatal, pallidal, or both? An anatomical study. *Journal of Comparative Neurology, 473,* 415–437.

Catchpole, C. K., & Slater, P. J. B. (1995). *Bird song: Biological themes and variations.* Cambridge: Cambridge University Press.

Christensen-Dalsgaard, J., & Carr, C. E. (2008). Evolution of a sensory novelty: Tympanic ears and the associated neural processing. *Brain Research Bulletin, 75,* 365–370.

Cooper, B. G., & Goller, F. (2004). Multimodal signals: Enhancement and constraint of song motor patterns by visual display. *Science, 303,* 544–546.

Crockford, C., Herbinger, I., Vigilant, L., & Boesch, C. (2004). Wild chimpanzees produce group-specific calls: A case for vocal learning? *Ethology, 110,* 221–243.

Crosson, B., Sadek, J. R., Bobholz, J. A., Gokcay, D., Mohr, C. M., Leonard, C. M., et al. (1999). Activity in the paracingulate and cingulate sulci during word generation: An fMRI study of functional anatomy. *Cerebral Cortex, 9,* 307–316.

Cummings, J. L. (1993). Frontal-subcortical circuits and human behavior. *Archives of Neurology, 50,* 873–880.

Damasio, A. R., Damasio, H., Rizzo, M., Varney, N., & Gersh, F. (1982). Aphasia with nonhemorrhagic lesions in the basal ganglia and internal capsule. *Archives of Neurology, 39,* 15–24.

Dave, A., & Margoliash, D. (2000). Song replay during sleep and computational rules for sensorimotor vocal learning. *Science, 290*, 812–816.

Derrickson, K. C. (1987). Yearly and situational changes in the estimate of repertoire size in northern mockingbirds (mimus polyglottos). *Auk, 104*, 198–207.

Dronkers, N. F. (1996). A new brain region for coordinating speech articulation. *Nature, 384*, 159–161.

Durand, S. E., Heaton, J. T., Amateau, S. K., & Brauth, S. E. (1997). Vocal control pathways through the anterior forebrain of a parrot (*Melopsittacus undulatus*). *Journal of Comparative Neurology, 377*, 179–206.

Dusterhoft, F., Hausler, U., & Jurgens, U. (2004). Neuronal activity in the periaqueductal gray and bordering structures during vocal communication in the squirrel monkey. *Neuroscience, 123*, 53–60.

Egnor, S. E., & Hauser, M. D. (2004). A paradox in the evolution of primate vocal learning. *Trends in Neurosciences, 27*, 649–654.

Esposito, A., Demeurisse, G., Alberti, B., & Fabbro, F. (1999). Complete mutism after midbrain periaqueductal gray lesion. *NeuroReport, 10*, 681–685.

Farabaugh, S. M., & Dooling, R. J. (1996). Acoustic communication in parrots: Laboratory and field studies of budgerigars, *Melopsittacus undulatus*. In D. E. Kroodsma & E. H. Miller (Eds.), *Ecology and evolution of acoustic communication in birds* (pp. 97–117). Ithaca, NY: Cornell University Press.

Farley, G. R. (1997). Neural firing in ventrolateral thalamic nucleus during conditioned vocal behavior in cats. *Experimental Brain Research, 115*, 493–506.

Farries, M. A. (2001). The oscine song system considered in the context of the avian brain: Lessons learned from comparative neurobiology. *Brain, Behavior and Evolution, 58*, 80–100.

Farries, M. A., Meitzen, J., & Perkel, D. J. (2005). Electrophysiological properties of neurons in the basal ganglia of the domestic chick: Conservation and divergence in the evolution of the avian basal ganglia. *Journal of Neurophysiology, 94*, 454–467.

Feenders, G., Liedvogel, M., Rivas, M., Zapka, M., Horita, H., Hara, E., et al. (2008). Molecular mapping of movement-associated areas in the avian brain: A motor theory for vocal learning origin. *PLoS ONE, 3*, e1768.

Ferreira, A., Smulders, T. V., Sameshima, K., Mello, C., & Jarvis, E. (2006). Vocalizations and associated behaviors of the sombre hummingbird (Trochilinae) and the rufous-breasted hermit (Phaethornithinae). *Auk, 123*, 1129–1148.

Fischer, J., & Hammerschmidt, K. (2011). Ultrasonic vocalizations in mouse models for speech and socio-cognitive disorders: Insights into the evolution of vocal communication. *Genes Brain & Behavior, 10*, 17–27.

Fisher, S. E., & Scharff, C. (2009). FOXP2 as a molecular window into speech and language. *Trends in Genetics, 25*, 166–177.

Fitch, W. T., Huber, L., & Bugnyar, T. (2010). Social cognition and the evolution of language: Constructing cognitive phylogenies. *Neuron, 65*, 795–814.

Foote, A. D., Griffin, R. M., Howitt, D., Larsson, L., Miller, P. J., & Hoelzel, A. R. (2006). Killer whales are capable of vocal learning. *Biology Letters, 2*, 509–512.

Fortune, E. S., & Margoliash, D. (1995). Parallel pathways converge onto HVc and adjacent neostriatum of adult male zebra finches (*Taeniopygia guttata*). *Journal of Comparative Neurology, 360,* 413–441.

Foster, E. F., & Bottjer, S. W. (2001). Lesions of a telencephalic nucleus in male zebra finches: Influences on vocal behavior in juveniles and adults. *Journal of Neurobiology, 46,* 142–165.

Foster, E. F., Mehta, R. P., & Bottjer, S. W. (1997). Axonal connections of the medial magnocellular nucleus of the anterior neostriatum in zebra finches. *Journal of Comparative Neurology, 382,* 364–381.

Freedman, M., Alexander, M. P., & Naeser, M. A. (1984). Anatomic basis of transcortical motor aphasia. *Neurology, 34,* 409–417.

Fried, I., Katz, A., McCarthy, G., Sass, K. J., Williamson, P., Spencer, S. S., et al. (1991). Functional organization of human supplementary motor cortex studied by electrical stimulation. *Journal of Neuroscience, 11,* 3656–3666.

Fried, I., Ojemann, G. A., & Fetz, E. E. (1981). Language-related potentials specific to human language cortex. *Science, 212,* 353–356.

Fritz, J., Mishkin, M., & Saunders, R. C. (2005). In search of an auditory engram. *Proceedings of the National Academy of Sciences of the United States of America, 102,* 9359–9364.

Gahr, M. (2000). Neural song control system of hummingbirds: Comparison to swifts, vocal learning (songbirds) and nonlearning (suboscines) passerines, and vocal learning (budgerigars) and nonlearning (dove, owl, gull, quail, chicken) nonpasserines. *Journal of Comparative Neurology, 426,* 182–196.

Galantucci, B., Fowler, C. A., & Turvey, M. T. (2006). The motor theory of speech perception reviewed. *Psychonomic Bulletin & Review, 13,* 361–377.

Gale, S. D., & Perkel, D. J. (2010). Anatomy of a songbird basal ganglia circuit essential for vocal learning and plasticity. *Journal of Chemical Neuroanatomy, 39,* 124–131.

Gardner, B. T., & Gardner, R. A. (1971). Two-way communication with an infant chimpanzee. In A. M. Schrier & F. Stollnitz (Eds.), *Behavior of Non-Human Primates* (pp. 117–184). New York: Academic Press.

Gentilucci, M., & Corballis, M. C. (2006). From manual gesture to speech: A gradual transition. *Neuroscience and Biobehavioral Reviews, 30,* 949–960.

Geschwind, N. (1979). Specializations of the human brain. *Scientific American, 241,* 180–199.

Giraud, A. L., Neumann, K., Bachoud-Levi, A. C., von Gudenberg, A. W., Euler, H. A., Lanfermann, H., et al. (2008). Severity of dysfluency correlates with basal ganglia activity in persistent developmental stuttering. *Brain and Language, 104,* 190–199.

Glasser, M. F., & Rilling, J. K. (2008). DTI tractography of the human brain's language pathways. *Cerebral Cortex, 18,* 2471–2482.

Glickstein, M., May, J. G., III, & Mercier, B. E. (1985). Corticopontine projection in the macaque: The distribution of labelled cortical cells after large injections of horseradish peroxidase in the pontine nuclei. *Journal of Comparative Neurology, 235,* 343–359.

Gobes, S. M., & Bolhuis, J. J. (2007). Birdsong memory: A neural dissociation between song recognition and production. *Current Biology, 17,* 789–793.

Goldberg, J. H., Adler, A., Bergman, H., & Fee, M. S. (2010). Singing-related neural activity distinguishes two putative pallidal cell types in the songbird basal ganglia: Comparison to the primate internal and external pallidal segments. *Journal of Neuroscience*, *30*, 7088–7098.

Goldberg, J. H., & Fee, M. S. (2011). Vocal babbling in songbirds requires the basal ganglia-recipient motor thalamus but not the basal ganglia. *Journal of Neurophysiology*, *105*, 2729–2739.

Gracco, V. L., Tremblay, P., & Pike, B. (2005). Imaging speech production using fMRI. *Neuro-Image*, *26*, 294–301.

Graff-Radford, N. R., Damasio, H., Yamada, T., Eslinger, P. J., & Damasio, A. R. (1985). Non-haemorrhagic thalamic infarction: Clinical, neuropsychological and electrophysiological findings in four anatomical groups defined by computerized tomography. *Brain*, *108*, 485–516.

Hackett, S. J., Kimball, R. T., Reddy, S., Bowie, R. C., Braun, E. L., Braun, M. J., et al. (2008). A phylogenomic study of birds reveals their evolutionary history. *Science*, *320*, 1763–1768.

Haesler, S., Rochefort, C., Georgi, B., Licznerski, P., Osten, P., & Scharff, C. (2007). Incomplete and inaccurate vocal imitation after knockdown of FoxP2 in songbird basal ganglia nucleus Area X. *PLoS Biology*, *5*, e321.

Haesler, S., Wada, K., Nshdejan, A., Morrisey, E. E., Lints, T., Jarvis, E. D., et al. (2004). FoxP2 expression in avian vocal learners and non-learners. *Journal of Neuroscience*, *24*, 3164–3175.

Hahnloser, R. H. R., Kozhevnikov, A. A., & Fee, M. S. (2002). An ultra-sparse code underlies the generation of neural sequences in a songbird. *Nature*, *419*, 65–70.

Halsema, K. A., & Bottjer, S. W. (1991). Lesioning afferent input to a forebrain nucleus disrupts vocal learning in zebra finches. *Neurosciences*, *17*, 1052.

Hannula-Jouppi, K., Kaminen-Ahola, N., Taipale, M., Eklund, R., Nopola-Hemmi, J., Kaari-ainen, H., et al. (2005). The axon guidance receptor gene ROBO1 is a candidate gene for developmental dyslexia. *PLoS Genetics*, *1*, e50.

Hara, E., Rivas, M. V., Ward, J. M., Okanoya, K., & Jarvis, E. D. (2012). Convergent differential regulation of parvalbumin in the brains of vocal learners. *PLoS ONE*, *7*, e29457.

Hauser, M. D., Chomsky, N., & Fitch, W. T. (2002). The faculty of language: What is it, who has it, and how did it evolve? *Science*, *298*, 1569–1579.

Hertz-Pannier, L., Chiron, C., Jambaque, I., Renaux-Kieffer, V., van de Moortele, P. F., Delalande, O., et al. (2002). Late plasticity for language in a child's non-dominant hemisphere: A pre- and post-surgery fMRI study. *Brain*, *125*, 361–372.

Hessler, N. A., & Doupe, A. J. (1999). Singing-related neural activity in a dorsal forebrain-basal ganglia circuit of adult zebra finches. *Journal of Neuroscience*, *19*, 10461–10481.

Hinke, R., Hu, X., Stillman, A., Kim, S., Merkle, H., Salmi, R., et al. (1993). Functional magnetic resonance imaging of Broca's area during internal speech. *NeuroReport*, *4*, 675–678.

Horita, H., Kobayashi, M., Liu, W.-C., Oka, K., Jarvis, E. D., & Wada, K. (2012). Repeated evolution of differential regulation of an activity-dependent gene dusp1 for a complex behavioral trait. *PLoS ONE*, *7*(8), e42173.

Horita, H., Wada, K., Rivas, M. V., Hara, E., & Jarvis, E. D. (2010). The dusp1 immediate early gene is regulated by natural stimuli predominantly in sensory input neurons. *Journal of Comparative Neurology*, *518*, 2873–2901.

Hosino, T., & Okanoya, K. (2000). Lesion of a higher-order song nucleus disrupts phrase level complexity in Bengalese finches. *NeuroReport, 11,* 2091–2095.

Ito, H., Ishikawa, Y., Yoshimoto, M., & Yamamoto, N. (2007). Diversity of brain morphology in teleosts: brain and ecological niche. *Brain, Behavior and Evolution, 69,* 76–86.

Iwatsubo, T., Kuzuhara, S., Kanemitsu, A., Shimada, H., & Toyokura, Y. (1990). Corticofugal projections to the motor nuclei of the brainstem and spinal cord in humans. *Neurology, 40,* 309–312.

Iyengar, S., Viswanathan, S. S., & Bottjer, S. W. (1999). Development of topography within song control circuitry of zebra finches during the sensitive period for song learning. *Journal of Neuroscience, 19,* 6037–6057.

Jacob, J., & Guthrie, S. (2000). Facial visceral motor neurons display specific rhombomere origin and axon pathfinding behavior in the chick. *Journal of Neuroscience, 20,* 7664–7671.

Jacobson, S., & Trojanowski, J. Q. (1975). Corticothalamic neurons and thalamocortical terminal fields: An investigation in rat using horseradish peroxidase and autoradiography. *Brain Research, 85,* 385–401.

Janik, V. M., & Slater, P. J. B. (1997). Vocal learning in mammals. *Advances in the Study of Behavior, 26,* 59–99.

Jarvis, E. D. (2004). Learned birdsong and the neurobiology of human language. *Annals of the New York Academy of Sciences, 1016,* 749–777.

Jarvis, E. D. (2006). Evolution of vocal learning systems in birds and humans. In J. Kaas (Ed.), *Evolution of nervous systems* (Vol. 2, pp. 213–228). New York: Academic Press.

Jarvis, E. D., Gunturkun, O., Bruce, L., Csillag, A., Karten, H., Kuenzel, W., et al. (2005). Avian brains and a new understanding of vertebrate brain evolution. *Nature Reviews: Neuroscience, 6,* 151–159.

Jarvis, E. D., & Mello, C. V. (2000). Molecular mapping of brain areas involved in parrot vocal communication. *Journal of Comparative Neurology, 419,* 1–31.

Jarvis, E. D., & Nottebohm, F. (1997). Motor-driven gene expression. *Proceedings of the National Academy of Sciences of the United States of America, 94,* 4097–4102.

Jarvis, E. D., Ribeiro, S., da Silva, M. L., Ventura, D., Vielliard, J., & Mello, C. V. (2000). Behaviourally driven gene expression reveals song nuclei in hummingbird brain. *Nature, 406,* 628–632.

Jarvis, E. D., Scharff, C., Grossman, M. R., Ramos, J. A., & Nottebohm, F. (1998). For whom the bird sings: Context-dependent gene expression. *Neuron, 21,* 775–788.

Johnson, M. B., Kawasawa, Y. I., Mason, C. E., Krsnik, Z., Coppola, G., Bogdanovic, D., et al. (2009). Functional and evolutionary insights into human brain development through global transcriptome analysis. *Neuron, 62,* 494–509.

Johnson, M. D., & Ojemann, G. A. (2000). The role of the human thalamus in language and memory: Evidence from electrophysiological studies. *Brain and Cognition, 42,* 218–230.

Jonas, S. (1981). The supplementary motor region and speech emission. *Journal of Communication Disorders, 14,* 349–373.

Jurgens, U. (1994). The role of the periaqueductal grey in vocal behaviour. *Behavioural Brain Research, 62,* 107–117.

Jurgens, U. (1995). Neuronal control of vocal production in non-human and human primates. In E. Zimmermann, J. D. Newman, & U. Jurgens (Eds.), *Current topics in primate vocal communication* (pp. 199–206). New York: Plenum Press.

Jurgens, U. (1998). Neuronal control of mammalian vocalization, with special reference to the squirrel monkey. *Naturwissenschaften, 85,* 376–388.

Jurgens, U. (2002). Neural pathways underlying vocal control. *Neuroscience and Biobehavioral Reviews, 26,* 235–258.

Jurgens, U., Kirzinger, A., & von Cramon, D. (1982). The effects of deep-reaching lesions in the cortical face area on phonation: A combined case report and experimental monkey study. *Cortex, 18,* 125–139.

Kao, M. H., Doupe, A. J., & Brainard, M. S. (2005). Contributions of an avian basal ganglia-forebrain circuit to real-time modulation of song. *Nature, 433,* 638–643.

Karten, H. J. (1991). Homology and evolutionary origins of the 'neocortex'. *Brain, Behavior and Evolution, 38,* 264–272.

Karten, H. J., & Shimizu, T. (1989). The origins of neocortex: Connections and lamination as distinct events in evolution. *Journal of Cognitive Neuroscience, 1,* 291–301.

Keizer, K., & Kuypers, H. G. (1989). Distribution of corticospinal neurons with collaterals to the lower brain stem reticular formation in monkey (Macaca fascicularis). *Experimental Brain Research, 74,* 311–318.

Kirzinger, A., & Jurgens, U. (1982). Cortical lesion effects and vocalization in the squirrel monkey. *Brain Research, 233,* 299–315.

Klein, D., Zatorre, R., Milner, B., Meyer, E., & Evans, A. (1994). Left putaminal activation when speaking a second language: Evidence from PET. *NeuroReport, 5,* 2295–2297.

Kobayashi, K., Uno, H., & Okanoya, K. (2001). Partial lesions in the anterior forebrain pathway affect song production in adult Bengalese finches. *NeuroReport, 12,* 353–358.

Konopka, G., Bomar, J. M., Winden, K., Coppola, G., Jonsson, Z. O., Gao, F., et al. (2009). Human-specific transcriptional regulation of CNS development genes by FOXP2. *Nature, 462,* 213–217.

Kroodsma, D. E. (1982). Song repertoires: Problems in their definition and use. In D. E. Kroodsma, E. H. Miller, & H. Ouellet (Eds.), *Acoustic communication in birds* (Vol. 2, pp. 125–146). New York: Academic Press.

Kuhl, P. K. (1986). Theoretical contributions of tests on animals to the special-mechanisms debate in speech. *Experimental Biology, 45,* 233–265.

Kuypers, H. G. J. M. (1958a). Corticobulbar connexions to the pons and lower brain-stem in man. *Brain, 81,* 364–388.

Kuypers, H. G. J. M. (1958b). Some projections from the peri-central cortex to the pons and lower brain stem in monkey and chimpanzee. *Journal of Comparative Neurology, 100,* 221–255.

Larson, C. R. (1991). On the relation of PAG neurons to laryngeal and respiratory muscles during vocalization in the monkey. *Brain Research, 552,* 77–86.

Larson, C. R., Yajima, Y., & Ko, P. (1994). Modification in activity of medullary respiratory-related neurons for vocalization and swallowing. *Journal of Neurophysiology, 71,* 2294–2304.

Lavenex, P. B. (2000). Lesions in the budgerigar vocal control nucleus NLc affect production, but not memory, of English words and natural vocalizations. *Journal of Comparative Neurology*, *421*, 437–460.

Leicester, J. (1980). Central deafness and subcortical motor aphasia. *Brain and Language*, *10*, 224–242.

Lemon, R. N., Kirkwood, P. A., Maier, M. A., Nakajima, K., & Nathan, P. (2004). Direct and indirect pathways for corticospinal control of upper limb motoneurons in the primate. *Progress in Brain Research*, *143*, 263–279.

Lieberman, P. (2000). *Human language and our reptilian brain: The subcortical bases of speech, syntax, and thought*. Cambridge, MA: Harvard University Press.

Lieberman, P. (2002). On the nature and evolution of the neural bases of human language. *American Journal of Physical Anthropology*, *119*(Suppl. 35), 36–62.

Lowndes, M., & Davies, D. C. (1995). The effect of archistriatal lesions on "open field" and fear/avoidance behaviour in the domestic chick. *Behavioural Brain Research*, *72*, 25–32.

Ludlow, C. L., & Loucks, T. (2003). Stuttering: A dynamic motor control disorder. *Journal of Fluency Disorders*, *28*, 273–295.

Luo, M., Ding, L., & Perkel, D. J. (2001). An avian basal ganglia pathway essential for vocal learning forms a closed topographic loop. *Journal of Neuroscience*, *21*, 6836–6845.

MacDougall-Shackleton, S. A., Hulse, S. H., & Ball, G. F. (1998). Neural bases of song preferences in female zebra finches (Taeniopygia guttata). *NeuroReport*, *9*, 3047–3052.

MacKay, B. K. (2001). *Bird sounds: How and why birds sing, call, chatter, and screech*. Mechanicsburg, PA: Stackpole Books.

MacLean, P. D. (1978). Effects of lesions of globus pallidus on species-typical display behavior of squirrel monkeys. *Brain Research*, *149*, 175–196.

MacLeod, A. K., Buckner, R. L., Miezin, F. M., Petersen, S. E., & Raichle, M. E. (1998). Right anterior prefrontal cortex activation during semantic monitoring and working memory. *NeuroImage*, *7*, 41–48.

Margoliash, D., Fortune, E. S., Sutter, M. L., Yu, A. C., Wren-Hardin, B. D., & Dave, A. (1994). Distributed representation in the song system of oscines: Evolutionary implications and functional consequences. *Brain, Behavior and Evolution*, *44*, 247–264.

Marler, P. (1970). Birdsong and speech development: Could there be parallels? *American Scientist*, *58*, 669–673.

Marshall, A. J., Wrangham, R. W., & Arcadi, A. C. (1999). Does learning affect the structure of vocalizations in chimpanzees? *Animal Behaviour*, *58*, 825–830.

Masataka, N., & Fujita, K. (1989). Vocal learning of Japanese and rhesus monkeys. *Behaviour*, *109*, 191–199.

Matsumura, M., & Kubota, K. (1979). Cortical projection to hand-arm motor area from postarcuate area in macaque monkeys: A histological study of retrograde transport of horseradish peroxidase. *Neuroscience Letters*, *11*, 241–246.

Matsunaga, E., Kato, M., & Okanoya, K. (2008). Comparative analysis of gene expressions among avian brains: A molecular approach to the evolution of vocal learning. *Brain Research Bulletin*, *75*, 474–479.

McCasland, J. S. (1987). Neuronal control of bird song production. *Journal of Neuroscience*, *7*, 23–39.

McCasland, J. S., & Konishi, M. (1981). Interaction between auditory and motor activities in an avian song control nucleus. *Proceedings of the National Academy of Sciences of the United States of America*, *78*, 7815–7819.

Mello, C. V., Vates, G. E., Okuhata, S., & Nottebohm, F. (1998). Descending auditory pathways in the adult male zebra finch (*Taeniopygia guttata*). *Journal of Comparative Neurology*, *395*, 137–160.

Mello, C. V., Vicario, D. S., & Clayton, D. F. (1992). Song presentation induces gene expression in the songbird forebrain. *Proceedings of the National Academy of Sciences of the United States of America*, *89*, 6818–6822.

Miller, C. T., Dimauro, A., Pistorio, A., Hendry, S., & Wang, X. (2010). Vocalization induced CFos expression in marmoset cortex. *Frontiers in Integrative Neuroscience*, *4*, 128.

Miller, S. J., & Inouye, D. W. (1983). Roles of the wing whistle in the territorial behavior of male broad-tailed hummingbirds (Selasphorus-platycercus). *Animal Behaviour*, *31*, 689–700.

Mohr, J. P. (1976). Broca's area and Broca's aphasia. In H. Whitaker & H. A. Whitaker (Eds.), *Studies in Neurolinguistics* (Vol. 1, pp. 201–235). New York: Academic Press.

Murphy, W. J., Eizirik, E., Johnson, W. E., Zhang, Y. P., Ryder, O. A., & O'Brien, S. J. (2001). Molecular phylogenetics and the origins of placental mammals. *Nature*, *409*, 614–618.

Murphy, W. J., Pringle, T. H., Crider, T. A., Springer, M. S., & Miller, W. (2007). Using genomic data to unravel the root of the placental mammal phylogeny. *Genome Research*, *17*, 413–421.

Nabholz, B., Kunstner, A., Wang, R., Jarvis, E. D., & Ellegren, H. (2011). Dynamic evolution of base composition: Causes and consequences in avian phylogenomics. *Molecular Biology and Evolution*, *28*, 2197–2210.

Nadeau, S. E., & Crosson, B. (1997). Subcortical aphasia. *Brain and Language*, *58*, 355–402, discussion 418–423.

Nealen, P. M., & Schmidt, M. F. (2002). Comparative approaches to avian song system function: Insights into auditory and motor processing. *Journal of Comparative Physiology, A: Neuroethology, Sensory, Neural, and Behavioral Physiology*, *188*, 929–941.

Nielsen, J. M., & Jacobs, L. L. (1951). Bilateral lesions of the anterior cingulated gyri. *Bulletin of the Los Angeles Neurological Society*, *16*, 231–234.

Nottebohm, F. (1972). The origins of vocal learning. *American Naturalist*, *106*(947), 116–140.

Nottebohm, F. (1977). Asymmetries in neural control of vocalizations in the canary. In S. Harnad, R. W. Doty, L. Goldstein, J. Jaynes, & G. Krauthamer (Eds.), *Lateralization in the nervous system* (pp. 23–44). New York: Academic Press.

Nottebohm, F., Alvarez-Buylla, A., Cynx, J., Kirn, J., Ling, C. Y., Nottebohm, M., et al. (1990). Song learning in birds: The relation between perception and production. *Proceedings of the Royal Society B: Biological Sciences*, *329*, 115–124.

Nottebohm, F., Nottebohm, M. E., & Crane, L. (1986). Developmental and seasonal changes in canary song and their relation to changes in the anatomy of song-control nuclei. *Behavioral and Neural Biology*, *46*, 445–471.

Nottebohm, F., Stokes, T. M., & Leonard, C. M. (1976). Central control of song in the canary, *Serinus canarius. Journal of Comparative Neurology, 165*, 457–486.

Ojemann, G. A. (1991). Cortical organization of language. *Journal of Neuroscience, 11*(8), 2281–2287.

Ojemann, G. A. (2003). The neurobiology of language and verbal memory: Observations from awake neurosurgery. *International Journal of Psychophysiology, 48*, 141–146.

Olveczky, B. P., Andalman, A. S., & Fee, M. S. (2005). Vocal experimentation in the juvenile songbird requires a basal ganglia circuit. *PLoS Biology, 3*, e153.

Owren, M. J., Dieter, J. A., Seyfarth, R. M., & Cheney, D. L. (1993). Vocalizations of rhesus (*Macaca mulatta*) and Japanese (*M. fuscata*) macaques cross-fostered between species show evidence of only limited modification. *Developmental Psychobiology, 26*, 389–406.

Palmer, E. D., Rosen, H. J., Ojemann, J. G., Buckner, R. L., Kelley, W. M., & Petersen, S. E. (2001). An event-related fMRI study of overt and covert word stem completion. *NeuroImage, 14*, 182–193.

Papathanassiou, D., Etard, O., Mellet, E., Zago, L., Mazoyer, B., & Tzourio-Mazoyer, N. (2000). A common language network for comprehension and production: A contribution to the definition of language epicenters with PET. *NeuroImage, 11*, 347–357.

Pepperberg, I. M. (1994). Vocal learning in grey parrots (Psittacus erithacus): Effects of social interaction, reference, and context. *Auk, 111*, 300–313.

Pepperberg, I. M. (1999). *The Alex studies: Cognitive and communicative abilities of grey parrots.* Cambridge, MA: Harvard University Press.

Perkel, D. J. (2004). Origin of the anterior forebrain pathway. *Annals of the New York Academy of Sciences, 1016*, 736–748.

Perkel, D., & Farries, M. (2000). Complementary "bottom-up" and "top-down" approaches to basal ganglia function. *Current Opinion in Neurobiology, 10*, 725–731.

Petersen, S. E., Fox, P. T., Posner, M. I., Mintun, M., & Raichle, M. E. (1988). Positron emission tomographic studies of the cortical anatomy of single-word processing. *Nature, 331*, 585–589.

Pika, S., & Bugnyar, T. (2011). The use of referential gestures in ravens (*Corvus corax*) in the wild. *Nature Communications, 2*, 560.

Pika, S., Liebal, K., & Tomasello, M. (2005). Gestural communication in subadult bonobos (*Pan paniscus*): Repertoire and use. *American Journal of Primatology, 65*, 39–61.

Plummer, T. K., & Striedter, G. F. (2002). Brain lesions that impair vocal imitation in adult budgerigars. *Journal of Neurobiology, 53*, 413–428.

Poeppel, D. (1996). A critical review of PET studies of phonological processing. *Brain and Language, 55*, 317–385.

Pollick, A. S., & de Waal, F. B. (2007). Ape gestures and language evolution. *Proceedings of the National Academy of Sciences of the United States of America, 104*, 8184–8189.

Poole, J. H., Tyack, P. L., Stoeger-Horwath, A. S., & Watwood, S. (2005). Animal behaviour: Elephants are capable of vocal learning. *Nature, 434*, 455–456.

Pratt, R. C., Gibb, G. C., Morgan-Richards, M., Phillips, M. J., Hendy, M. D., & Penny, D. (2009). Toward resolving deep neoaves phylogeny: Data, signal enhancement, and priors. *Molecular Biology and Evolution, 26*, 313–326.

Price, C. J., Wise, R. J., Warburton, E. A., Moore, C. J., Howard, D., Patterson, K., Frackowiak, R. S., & Friston, K. J. (1996). Hearing and saying: The functional neuro-anatomy of auditory word processing. *Brain, 119*, 919–931.

Prum, R. O. (1998). Sexual selection and the evolution of mechanical sound production in manakins (Aves: Pipridae). *Animal Behaviour, 55*, 977–994.

Ralls, K., Fiorelli, P., & Gish, S. (1985). Vocalizations and vocal mimicry in captive harbor seals, *Phoca vitulina. Canadian Journal of Zoology, 63*, 1050–1056.

Reiner, A., Jiao, Y., Del Mar, N., Laverghetta, A. V., & Lei, W. L. (2003). Differential morphology of pyramidal tract-type and intratelencephalically projecting-type corticostriatal neurons and their intrastriatal terminals in rats. *Journal of Comparative Neurology, 457*, 420–440.

Reiner, A., Laverghetta, A. V., Meade, C. A., Cuthbertson, S. L., & Bottjer, S. W. (2004). An immunohistochemical and pathway tracing study of the striatopallidal organization of area X in the male zebra finch. *Journal of Comparative Neurology, 469*, 239–261.

Rey, M., Dellatolas, G., Bancaud, J., & Talairach, J. (1988). Hemispheric lateralization of motor and speech functions after early brain lesion: Study of 73 epileptic patients with intracarotid amytal test. *Neuropsychologia, 26*, 167–172.

Rilling, J. K., Glasser, M. F., Preuss, T. M., Ma, X., Zhao, T., Hu, X., et al. (2008). The evolution of the arcuate fasciculus revealed with comparative DTI. *Nature Neuroscience, 11*, 426–428.

Rivas, E. (2005). Recent use of signs by chimpanzees (Pan Troglodytes) in interactions with humans. *Journal of Comparative Psychology, 119*, 404–417.

Rosen, H. J., Ojemann, J. G., Ollinger, J. M., & Petersen, S. E. (2000). Comparison of brain activation during word retrieval done silently and aloud using fMRI. *Brain and Cognition, 42*, 201–217.

Rubens, A. B. (1975). Aphasia with infarction in the territory of the anterior cerebral artery. *Cortex, 11*, 239–250.

Sanvito, S., Galimberti, F., & Miller, E. H. (2007). Observational evidences of vocal learning in southern elephant seals: A longitudinal study. *Ethology, 113*, 137–146.

Sasaki, A., Sotnikova, T. D., Gainetdinov, R. R., & Jarvis, E. D. (2006). Social context-dependent singing-regulated dopamine. *Journal of Neuroscience, 26*, 9010–9014.

Scharff, C., & Nottebohm, F. (1991). A comparative study of the behavioral deficits following lesions of various parts of the zebra finch song system: Implications for vocal learning. *Journal of Neuroscience, 11*, 2896–2913.

Scharff, C., Nottebohm, F., & Cynx, J. (1998). Conspecific and heterospecific song discrimination in male zebra finches with lesions in the anterior forebrain pathway. *Journal of Neurobiology, 36*, 81–90.

Seller, T. (1981). Midbrain vocalization centers in birds. *Trends in Neurosciences, 12*, 301–303.

Seyfarth, R. M., Cheney, D. L., & Marler, P. (1980). Vervet monkey alarm calls: Semantic communication in a free-ranging primate. *Animal Behaviour, 28*, 1070–1094.

Simoes, C. S., Vianney, P. V., de Moura, M. M., Freire, M. A., Mello, L. E., Sameshima, K., et al. (2010). Activation of frontal neocortical areas by vocal production in marmosets. *Frontiers in Integrative Neuroscience, 4*, 123.

Simonyan, K., & Horwitz, B. (2011). Laryngeal motor cortex and control of speech in humans. *Neuroscientist*, *17*(2), 197–208.

Simonyan, K., Horwitz, B., & Jarvis, E. D. (2012). Dopamine regulation of human speech and bird song: A critical review. *Brain & Language*. [Epub ahead of print]. PMID: 22284300.

Simonyan, K., & Jurgens, U. (2003). Efferent subcortical projections of the laryngeal motor-cortex in the rhesus monkey. *Brain Research*, *974*, 43–59.

Simonyan, K., & Jurgens, U. (2005). Afferent subcortical connections into the motor cortical larynx area in the rhesus monkey. *Neuroscience*, *130*, 119–131.

Simpson, H. B., & Vicario, D. S. (1990). Brain pathways for learned and unlearned vocalizations differ in zebra finches. *Journal of Neuroscience*, *10*, 1541–1556.

Sohrabji, F., Nordeen, E. J., & Nordeen, K. W. (1990). Selective impairment of song learning following lesions of a forebrain nucleus in the juvenile zebra finch. *Behavioral and Neural Biology*, *53*, 51–63.

Spaulding, M., O'Leary, M. A., & Gatesy, J. (2009). Relationships of Cetacea (Artiodactyla) among mammals: Increased taxon sampling alters interpretations of key fossils and character evolution. *PLoS ONE*, *4*, e7062.

Speedie, L. J., Wertman, E., Ta'ir, J., & Heilman, K. M. (1993). Disruption of automatic speech following a right basal ganglia lesion. *Neurology*, *43*, 1768–1774.

Springer, M. S., Murphy, W. J., Eizirik, E., & O'Brien, S. J. (2003). Placental mammal diversification and the Cretaceous-Tertiary boundary. *Proceedings of the National Academy of Sciences of the United States of America*, *100*, 1056–1061.

Striedter, G. F. (1994). The vocal control pathways in budgerigars differ from those in songbirds. *Journal of Comparative Neurology*, *343*, 35–56.

Strub, R. L. (1989). Frontal lobe syndrome in a patient with bilateral globus pallidus lesions. *Archives of Neurology*, *46*, 1024–1027.

Suh, A., Paus, M., Kiefmann, M., Churakov, G., Franke, F. A., Brosius, J., et al. (2011). Mesozoic retroposons reveal parrots as the closest living relatives of passerine birds. *Nature Communications*, *2*, 443.

Troster, A. I., Woods, S. P., & Fields, J. A. (2003). Verbal fluency declines after pallidotomy: An interaction between task and lesion laterality. *Applied Neuropsychology*, *10*, 69–75.

Valenstein, E. (1975). Nonlanguage disorders of speech reflect complex neurologic apparatus. *Geriatrics*, *30*, 117–121.

Vates, G. E., Broome, B. M., Mello, C. V., & Nottebohm, F. (1996). Auditory pathways of caudal telencephalon and their relation to the song system of adult male zebra finches. *Journal of Comparative Neurology*, *366*, 613–642.

Vates, G. E., & Nottebohm, F. (1995). Feedback circuitry within a song-learning pathway. *Proceedings of the National Academy of Sciences of the United States of America*, *92*, 5139–5143.

Vates, G. E., Vicario, D. S., & Nottebohm, F. (1997). Reafferent thalamo-"cortical" loops in the song system of oscine songbirds. *Journal of Comparative Neurology*, *380*, 275–290.

Von Cramon, D., & Jurgens, U. (1983). The anterior cingulate cortex and the phonatory control in monkey and man. *Neuroscience and Biobehavioral Reviews*, *7*, 423–425.

Vu, E. T., Schmidt, M. F., & Mazurek, M. E. (1998). Interhemispheric coordination of premotor neural activity during singing in adult zebra finches. *Journal of Neuroscience*, *18*, 9088–9098.

Wada, K., Sakaguchi, H., Jarvis, E. D., & Hagiwara, M. (2004). Differential expression of glutamate receptors in avian neural pathways for learned vocalization. *Journal of Comparative Neurology*, *476*, 44–64.

Wallesch, C. W., Henriksen, L., Kornhuber, H. H., & Paulson, O. B. (1985). Observations on regional cerebral blood flow in cortical and subcortical structures during language production in normal man. *Brain and Language*, *25*, 224–233.

Wang, R., Chen, C.-C., Hara, E., & Jarvis, E. D. (2012). Convergent signature of selection on the ROBO1 axon guidance pathway for vocal learning in both mammals and birds. Manuscript submitted for publication.

Wang, Y., Brzozowska-Prechtl, A., & Karten, H. J. (2010). Laminar and columnar auditory cortex in avian brain. *Proceedings of the National Academy of Sciences of the United States of America*, *107*, 12676–12681.

Webster, D. B., Popper, A. N., & Fay, R. R. (Eds.). (1992). *The mammalian auditory pathway: Neuroanatomy*. New York: Springer-Verlag.

Wild, J. M. (1994a). The auditory-vocal-respiratory axis in birds. *Brain, Behavior and Evolution*, *44*(4–5), 192–209.

Wild, J. M. (1994b). Visual and somatosensory inputs to the avian song system via nucleus uvaeformis (Uva) and a comparison with the projections of a similar thalamic nucleus in a nonsongbird, *Columba livia*. *Journal of Comparative Neurology*, *349*, 512–535.

Wild, J. M., Karten, H. J., & Frost, B. J. (1993). Connections of the auditory forebrain in the pigeon (*Columba livia*). *Journal of Comparative Neurology*, *337*, 32–62.

Wild, J. M., Li, D., & Eagleton, C. (1997). Projections of the dorsomedial nucleus of the intercollicular complex (DM) in relation to respiratory-vocal nuclei in the brainstem of pigeon (Columba livia) and zebra finch (Taeniopygia guttata). *Journal of Comparative Neurology*, *377*, 392–413.

Wildgruber, D., Ackermann, H., & Grodd, W. (2001). Differential contributions of motor cortex, basal ganglia, and cerebellum to speech motor control: Effects of syllable repetition rate evaluated by fMRI. *NeuroImage*, *13*, 101–109.

Williams, H., Crane, L. A., Hale, T. K., Esposito, M. A., & Nottebohm, F. (1992). Right-side dominance for song control in the zebra finch. *Journal of Neurobiology*, *23*, 1006–1020.

Williams, H., & Nottebohm, F. (1985). Auditory responses in avian vocal motor neurons: A motor theory for song perception in birds. *Science*, *229*, 279–282.

Wise, R. J., Greene, J., Buchel, C., & Scott, S. K. (1999). Brain regions involved in articulation. *Lancet*, *353*, 1057–1061.

Yajima, Y., Hayashi, Y., & Yoshii, N. (1982). Ambiguus motoneurons discharging closely associated with ultrasonic vocalization in rats. *Brain Research*, *238*, 445–450.

Yu, A. C., & Margoliash, D. (1996). Temporal hierarchical control of singing in birds. *Science*, *273*, 1871–1875.

Zann, R. A. (1996). *The zebra finch: A synthesis of field and laboratory studies* (pp. 196–247). New York: Oxford University Press.

Zhang, J. X., Leung, H. C., & Johnson, M. K. (2003). Frontal activations associated with accessing and evaluating information in working memory: An fMRI study. *NeuroImage, 20,* 1531–1539.

Zhang, S. P., Bandler, R., & Davis, P. J. (1995). Brain stem integration of vocalization: Role of the nucleus retroambigualis. *Journal of Neurophysiology, 74,* 2500–2512.

II ACQUISITION OF BIRDSONG AND SPEECH

5 Behavioral Similarities between Birdsong and Spoken Language

Sanne Moorman and Johan J. Bolhuis

Charles Darwin (1871, p. 55) noticed almost a century and a half ago that "the sounds uttered by birds offer in several respects the nearest analogy to language." Indeed, there are many parallels between the acquisition of spoken language in human infants and birdsong learning, at the behavioral, neural, genetic, and cognitive levels (Doupe & Kuhl, 1999; Bolhuis, Okanoya, & Scharff, 2010), as discussed in detail in the present book. In this chapter we introduce the behavioral parallels between birdsong and human spoken language. Most importantly, both human infants and juvenile songbirds imitate sounds that adult conspecifics (often their parents) make, a relatively rare ability in the animal kingdom. Until now, only a few mammalian taxa have been identified as vocal learners, namely humans, certain marine mammals, and bats, while vocal learning appears absent in our closest relatives, nonhuman primates (Hauser, Chomsky, & Fitch, 2002). In contrast, nearly half of the approximately 10,000 bird species are songbirds or oscine Passeriformes, which can vocally imitate species-specific communication sounds. In addition, parrots (Psittaciformes) and hummingbirds (Apodiformes) are also vocal imitators.

Avian Vocalizations

Birdsong Functions and Characteristics

The main function of song seems to be defense of territory and mate attraction (Collins, 2004). The importance of song in territory defense was shown in wild-living great tits (*Parus major*). Researchers caught and removed great tits from their territories, and placed speakers that broadcast songs in half of the empty territories. They found that empty territories were occupied faster than the territories where male songs were played (Krebs, 1977). However, gregarious species such as the zebra finch (*Taeniopygia guttata*) do not have a territory and consequently, their songs do not appear to have the function of territory defense. Singing males do not behave aggressively and their songs do not evoke aggressive behavior in other males (Zann, 1996).

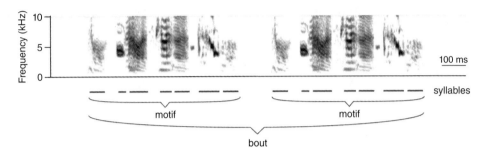

Figure 5.1
Sonogram of the song of a zebra finch, showing sound frequencies over time. One bout of a zebra finch song is depicted, consisting of two motifs, which are made up of seven syllables each (illustrated by the bars underneath the sonogram). Adapted, with permission, from Gobes, Zandbergen, and Bolhuis (2010).

Birdsongs consist of ordered strings of sounds (visualized in a sonogram in Figure 5.1). The separate sound elements are called syllables, and one complete string is a motif. Birds often sing multiple motifs that are together called a song bout. Some bird species, like the zebra finch, have a highly stereotyped song, where only one motif is repeated with just minor variability. Other species, such as Bengalese finches (*Lonchura striata* var. *domestica*) or European starlings (*Sturnus vulgaris*), sing several different motifs and thus have more variable songs (Figure 5.2).

Calls
Apart from producing songs, zebra finches also produce calls. Both males and females do this, and there are several call subtypes. The three most common calls are the distance call or long call, the tet, and the stack. The last two are used during hopping and take-off, respectively. The distance call is used often and in a variety of contexts. Male calls have a large learned component and there is considerable variation between individuals (Zann, 1996; Forstmeier, Burger, Temnow, & Derégnaucourt, 2009). In contrast, females have a much more stereotyped, highly heritable (Forstmeier et al., 2009) distance call, which they do not have to learn from their parents (Zann, 1996). Thus, male distance calls are more easily distinguishable from each other than female calls. It was found that juvenile zebra finches mistake unfamiliar female calls for their mother's call more often than they do for novel male calls and their father's call (Jacot, Reers, & Forstmeier, 2010). Both male and female zebra finches prefer female distance calls over male calls (Vicario, Naqvi, & Raksin, 2001; Gobes et al., 2009). Female budgerigars (*Melopsittacus undulatus*, a parrot species) recognize and prefer their mate's calls (Eda-Fujiwara et al., 2011). In black-capped chickadees (*Poecile atricapillus*), both males and females learn their songs

A

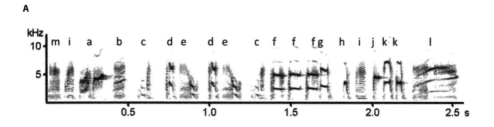

B

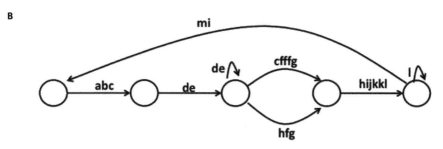

Figure 5.2
Sonogram of a song of a Bengalese finch. Individual syllables are indicated with a letter. (A) The sonogram of one song motif is shown. (B) State transition diagram of the song in A. See Okanoya (chapter 11, this volume) for further explanation. Figure courtesy of Kazuo Okanoya.

and calls. Interestingly, in this species, the call is more complex than the song, and calls are used in a variety of contexts (Avey, Kanyo, Irwin, & Sturdy, 2008).

Song Preferences

Female zebra finches recognize songs of male conspecifics. They have a preference for certain songs over others (reviewed in Riebel, 2009), and they may perform copulation solicitation displays in response to playbacks of male songs (Searcy, 1992). It was found that females acquire their song preferences as juveniles. For example, zebra finch females that were fostered by another songbird species preferred mates that sang the fostered songs rather than their own species' songs (Clayton, 1990). If song quality is related to fitness, it is expected that females would prefer high-quality songs. Indeed, female zebra finches prefer tutored over isolate song (Williams, Kilander, & Sotanski, 1993), and female song sparrows (*Melospiza melodia*) prefer songs of "good learners" over those of "poor learners" (Nowicki, Searcy, & Peters, 2002). However, it was shown that song preference is dependent on the physical quality of the female, which was manipulated by controlling brood size (large broods generate low-quality birds, while small broods generate high-quality birds): low-quality female zebra finches preferred low-quality males (Holveck & Riebel, 2010).

What are the specific song features that female songbirds attend to? It was demonstrated that female songbirds prefer complex songs (Okanoya, 2004; Leitão, ten Cate, & Riebel, 2006). Interestingly, in female budgerigars, neuronal activation in a brain region important for song perception and song memory was correlated with complexity of the stimulus song (Eda-Fujiwara, Satoh, Bolhuis, & Kimura, 2003). In the canary (*Serinus canaria*), females prefer syllables with a high trill rate and frequency bandwidth, the so-called sexy syllables (Vallet & Kreutzer, 1995; Draganoiu, Nagle, & Kreutzer, 2002). In swamp sparrows (*Melospiza georgiana*), this is true for trilled notes (Ballentine, Hyman, & Nowicki, 2004). Furthermore, females prefer songs with a large repertoire of song motifs, even in species whose songs would naturally have only one motif—such as the common grackle (*Quiscalus quiscula*) (Searcy, 1992) or the zebra finch (Collins, 1999)—which is not caused by a longer motif or song duration per se (Riebel, 2009). In addition, females prefer a high variety in syllables (Holveck & Riebel, 2007; Woodgate et al., 2011) and a high song rate (reviewed in Collins, 2004; Riebel, 2009).

Female Songs

In some species, such as the European starling, females also sing. The functions of female song are probably similar to those of males: territory defense and mate attraction. Furthermore, vocal duets may enhance pair bonds in those species (Collins, 2004). Zebra finch females do not sing, but Elie and colleagues (2010) demonstrated that zebra finch partners do communicate with each other through soft calls, including tets. These authors hypothesized that these private vocal duets also enhance pair-bond maintenance.

Some Avian Vocal Learners

The zebra finch is commonly used in laboratory studies. It is originally an Australian songbird that lives in large flocks. Only male zebra finches sing, and they sing only one song. Zebra finches form lifelong monogamous pair bonds (Zann, 1996).

Bengalese finches are a domesticated strain of the white-rumped munia (*Lonchura striata*), a bird native to Southeast Asia (Okanoya, chapter 25 this volume). They are closely related to zebra finches, and much like them, Bengalese finches are nonterritorial birds that learn one song type during life. However, they can repeat song elements so that a more flexible and complex song structure arises (Takahashi, Yamada, & Okanoya, 2010) (Figure 5.2).

The African grey parrot (*Psittacus erithacus*) is popular as a pet, and it can mimic human speech remarkably well. Its natural habitat is the West and Central African rainforests. Relatively little is known of the natural vocalizations of this species, but its human speech imitation abilities have been studied extensively (Pepperberg, 2010; chapter 13, this volume).

The European starling is native to most of temperate Europe and Western Asia. These birds can learn multiple songs and calls, and are able to mimic all kinds of sounds. Both male and female starlings sing, and they produce multiple vocalizations. Starlings sing long continuous bouts and possess unique repertoires (Gentner, 2004).

Behavioral Similarities between Speech Acquisition and Song Learning

Learning by Imitation

Inspection of sonograms of tutors and their tutees suggests that young songbirds copy many elements of the songs of their tutors (Figure 5.3). Research since the middle of the 20th century has demonstrated that songbirds indeed have to learn their songs. When young birds are reared in isolation, they will produce a highly abnormal song, the so-called isolate song (Marler, 1970; Fehér, Wang, Saar, Mitra, & Tchernichovski, 2009; Fehér & Tchernichovski, chapter 7, this volume). Isolate song does have some recognizable species-specific features, but it has a relatively simple structure and does not sound like the songs of socially raised birds. There is a predisposition to preferentially learn songs of conspecifics. When songbirds are exposed both to songs from their own and from another species, they will mainly imitate the conspecific song (Marler & Peters, 1977). However, if their preferred input is lacking, they will copy from songs available to them. If young songbirds are raised by parents of another species, they will imitate sounds of their foster parents (Clayton, 1988). Of course there are physical constraints on what sounds can be heard, remembered, and imitated. Physical constraints are formed on the one hand by neural mechanisms involved in the analysis, processing, and storage of auditory information, and on the other hand by the functioning of the ear and the sound-production organs (the larynx in humans, the syrinx in birds, and the vocal tract) (Bolhuis et al., 2010).

In parallel to songbirds, which learn by imitation, human infants are able to learn the language(s) to which they are exposed early in life, which could be any of more than 6,000 languages. They can do this with relative ease and without formal instruction (Kuhl & Rivera-Gaxiola, 2008; Bolhuis et al., 2010). Of course it would not be ethical to raise children without linguistic experiences to see how they will communicate. However, there are some stories of children that grew up separated from human contact, by themselves in the wild or in the presence of animals. Although most of these stories are probably fiction, there are some feral children that have been studied scientifically. One famous example is

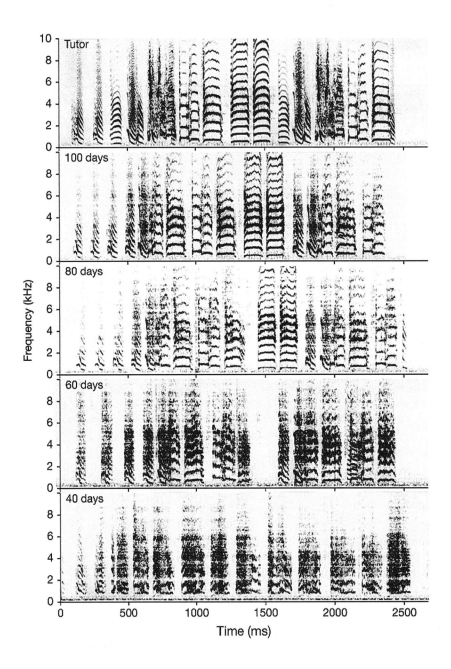

"Genie," a girl who grew up alone in her bedroom. Her father brought her food, but he never spoke to her, so she was never exposed to language. At the age of twelve, she was freed and people tried to socialize her and teach her English. Although she made some progress, she never learned to speak properly (Krashen, 1973).

Sensitive Periods

Feral children's failure to learn to speak suggests the existence of a sensitive period early in life, in which language learning is much easier than at a later time. More important evidence is that, in contrast to young children that can easily learn more than one language fluently, human adults have more trouble learning new languages and rarely learn to speak them fluently (Lenneberg, 1967; Doupe & Kuhl, 1999; Bolhuis et al., 2010).

By manipulating the auditory experience of songbirds early in life it was shown that there is a sensitive period in which songs are learned best (reviewed in Doupe & Kuhl, 1999). Some species, such as the zebra finch, sing only one song during their life, while others are able to learn multiple songs or adjust their songs every breeding season. Birds that learn one song that remains stable over the years are called "age-limited learners"; an example is the zebra finch. Birds that can learn songs throughout their lives, such as starlings, are called "open-ended learners" (Doupe & Kuhl, 1999; Bolhuis & Gahr, 2006).

Learning Phases

In the songbird species that have been studied, song learning has two phases: a memorization phase, during which a memory of the song of the tutor is formed, and a sensorimotor phase, in which the young bird learns to sing itself. The memorization phase can precede the sensorimotor phase by months. This is the case for many "seasonal breeders" such as the white-crowned sparrow (*Zonotrichia leucophrys*). In other species, such as the zebra finch, the two phases overlap (Figure 5.4). Konishi (1965) first introduced the concept of the "template," which is essentially a central

◀ Figure 5.3

Song imitation and development of a zebra finch male. Sonograms of two zebra finch males are shown. The top panel is the sonogram of the song of an adult tutor. Below it are sonograms of the song of its son, at different stages of development. When the son is 40 days old, he produces subsong, which sounds unstructured and does not yet resemble the tutor song. At 60–80 days the juvenile produces plastic song, which increasingly resembles the tutor song. Zebra finches develop their song until they are approximately 90 days old; from that age on their song is stable. When the son is 100 days old, he produces what is termed crystallized song. Modified and reproduced, with permission, from Bolhuis and Gahr (2006), copyright 2006 Nature Publishing Group. All rights reserved.

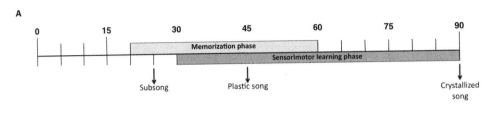

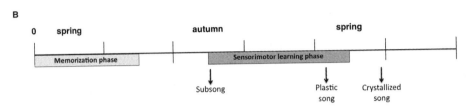

Figure 5.4
Timelines with phases in song learning in two songbird species. (A) Zebra finches have over-lapping memorization and production phases during which they learn their song. (B) In song sparrows there is no overlap between the memorization and sensorimotor phases. Vertical arrows indicate the approximate start of a new phase in song learning (see Figure 5.3 for an example in the zebra finch). See text for further explanation.

representation of the species-specific tutor song used for vocal learning. In this framework, songbirds are thought to be born with an elementary representation of their species' song, called a crude template. During the memorization phase, birds modify this crude template toward a more precise representation of the song of their tutor, resulting in a more exact template. During the sensorimotor phase, the young bird starts to produce sounds, much like children's babbling (Doupe & Kuhl, 1999). This is called subsong (see Figure 5.3 for a zebra finch example). Through auditory feedback, the bird matches its song to the template to improve its imitation of the parent's song, until the bird sings a more or less accurate copy of the tutor song when it reaches adulthood. This song is known as the crystallized song (Konishi, 1965; Nottebohm, 1968; Marler, 1970).

There is a remarkable developmental parallel between birdsong learning and speech acquisition in children. As early as in the uterus (DeCasper & Spence, 1986), but also in early stages after birth (Kuhl & Rivera-Gaxiola, 2008; Dehaene-Lambertz et al., 2010), human babies may learn characteristics of their parents' speech. From 6 or 7 months of age, babies start babbling, when they produce sounds that are still very different from adult speech. Then, after practice, the sounds develop into proper human speech from about the age of 3 (Doupe & Kuhl, 1999; Kuhl & Rivera-Gaxiola, 2008).

Auditory Feedback

Both humans and songbirds need to hear their own vocalizations for development and maintenance of their songs or speech (Doupe & Kuhl, 1999; Brainard & Doupe, 2000). Children born deaf cannot develop normal vocalizations; neither can birds surgically deafened early in life (Doupe & Kuhl, 1999). To imitate the parents' song, birds must hear their own song and match it to the template. Similarly, to learn to speak properly, children have to hear themselves speak to be able to compare their vocal output to the memorized representations of speech (Doupe & Kuhl, 1999). Auditory feedback is also important for maintenance of vocalizations. When humans or songbirds become deaf later in life—when normal speech or song has already developed—vocalizations will deteriorate slowly. However, their vocalizations will sound much more normal than those of people born deaf or of early-deafened birds (Nordeen & Nordeen, 1992; Doupe & Kuhl, 1999).

Syntactic Parallels?

Behaviorally, human language and birdsong both involve complex, patterned vocalizations (Doupe & Kuhl, 1999; Yip, chapter 9, this volume; Okanoya, chapter 11 this volume; ten Cate, Lachlan, & Zuidema, chapter 12, this volume). An essential aspect of human language is syntax, defined by Berwick et al. (2011, p. 113) as "the rules for arranging items (sounds, words, word parts or phrases) into their possible permissible combinations in a language." It is thought that human language can be distinguished from animal vocalizations by its syntactic complexity, where hierarchies can be assembled by combining words and words parts, a word-construction process called "morphology." Furthermore, in human syntax words can be organized into higher-order phrases and entire sentences (Berwick, Okanoya, Beckers, & Bolhuis, 2011). The songs of songbirds also consist of discrete acoustic elements that occur in a certain temporal order. Individual notes can be combined as particular sequences into syllables, syllables into "motifs," and motifs into complete song "bouts" (Figure 5.1). In principle, variable song-element sequences may be governed by sequential syntactic rules (Figure 5.2), as shown for example in the work of Okanoya (2004; chapter 11 this volume) and his collaborators. This is what Marler (1977) has termed "phonological syntax" (see also Yip, chapter 9, this volume). However, some authors have argued that there is a stronger linguistic parallel between songbirds and humans, and that the former may have the ability to acquire context-free syntactic rules (Gentner, Fenn, Margoliash, & Nusbaum, 2006; Abe & Watanabe, 2011). These claims have been challenged (Van Heijningen, Visser, Zuidema, & ten Cate, 2009; Beckers, Bolhuis, Okanoya, & Berwick, 2012), however, and Beckers et al. (2012) conclude that, although there is clearly evolutionary convergence between humans and songbirds regarding processes of auditory-vocal learning, birdsong is not a credible model for the study of the mechanisms of human language syntax. The principal difference between human language and birdsong is that the latter has neither a lexicon nor semantic complexity (Berwick et al., 2011). Thus, the production of sequential vocal elements is another behavioral parallel between birdsong and speech, but as yet there is no evidence for the kind of combinatorial complexity that is characteristic of human language syntax.

References

Abe, K., & Watanabe, D. (2011). Songbirds possess the spontaneous ability to discriminate syntactic rules. *Nature Neuroscience, 14*, 1067–1074.

Avey, M. T., Kanyo, R. A., Irwin, E. L., & Sturdy, C. B. (2008). Differential effects of vocalization type, singer and listener on ZENK immediate early gene response in black-capped chickadees (Poecile atricapillus). *Behavioural Brain Research, 188*, 201–208.

Ballentine, B., Hyman, J., & Nowicki, S. (2004). Vocal performance influences female response to male bird song: An experiment test. *Behavioral Ecology, 15*, 163–168.

Beckers, G. J. L., Bolhuis, J. J., Okanoya, K., & Berwick, R. C. (2012). Birdsong neurolinguistics: Songbird context-free grammar claim is premature. *NeuroReport, 23*, 139–145.

Berwick, R. C., Okanoya, K., Beckers, G. J. L., & Bolhuis, J. J. (2011). Songs to syntax: The linguistics of birdsong. *Trends in Cognitive Sciences, 15*, 113–121.

Bolhuis, J. J., & Gahr, M. (2006). Neural mechanisms of birdsong memory. *Nature Reviews: Neuroscience, 7*, 347–357.

Bolhuis, J. J., Okanoya, K., & Scharff, C. (2010). Twitter evolution: Converging mechanisms in birdsong and human speech. *Nature Reviews: Neuroscience, 11*, 747–759.

Brainard, M. S., & Doupe, A. J. (2000). Auditory feedback in learning and maintenance of vocal behaviour. *Nature Reviews: Neuroscience, 1*, 31–40.

Clayton, N. S. (1988). Song learning and mate choice in estrildid finches raised by two species. *Animal Behaviour, 36*, 1589–1600.

Clayton, N. S. (1990). Assortative mating in zebra finch subspecies Taeniopygia guttata guttata and T. g. castanotis. *Philosophical Transactions of the Royal Society B: Biological Sciences, 330*, 351–370.

Collins, S. A. (1999). Is female preference for male repertoires due to sensory bias? *Proceedings of the Royal Society B: Biological Sciences, 266*, 2309–2314.

Collins, S. A. (2004). Vocal fighting and flirting: The functions of birdsong. In P. Marler & H. Slabbekoorn (Eds.), *Nature's music: The science of birdsong* (pp. 39–79). San Diego: Elsevier Academic Press.

Darwin, C. (1871). *The descent of man, and selection in relation to sex*. London: Murray.

DeCasper, A. J., & Spence, M. J. (1986). Prenatal maternal speech influences newborns' perception of speech sound. *Infant Behavior and Development, 9*, 133–150.

Dehaene-Lambertz, G., Montavont, A., Jobert, A., Allirol, L., Dubois, J., Hertz-Pannier, L., et al. (2010). Language or music, mother or Mozart? Structural and environmental influences on infants' language networks. *Brain and Language, 114*, 53–65.

Doupe, A. J., & Kuhl, P. K. (1999). Birdsong and human speech: Common themes and mechanisms. *Annual Review of Neuroscience, 22*, 567–631.

Draganoiu, T. I., Nagle, L., & Kreutzer, M. (2002). Directional female preference for an exaggerated male trait in canary (*Serinusanaria*) song. *Proceedings of the Royal Society B: Biological Sciences, 269*, 2525–2531.

Eda-Fujiwara, H., Kanesada, A., Okamoto, Y., Satoh, R., Watanabe, A., & Miyamoto, T. (2011). Long-term maintenance and eventual extinction of preference for a mate's call in the female budgerigar. *Animal Behaviour, 82*, 971–979.

Eda-Fujiwara, H., Satoh, R., Bolhuis, J. J., & Kimura, T. (2003). Neuronal activation in female budgerigars is localized and related to male song complexity. *European Journal of Neuroscience, 17*, 149–154.

Elie, J. E., Mariette, M. M., Soula, H. A., Griffith, S. C., Mathevon, N., & Vignal, C. (2010). Vocal communication at the nest between mates in wild zebra finches: A private vocal duet? *Animal Behaviour, 80*, 597–605.

Fehér, O., Wang, H., Saar, S., Mitra, P. P., & Tchernichovski, O. (2009). De novo establishment of wild-type song culture in the zebra finch. *Nature, 459*, 564–569.

Forstmeier, W., Burger, C., Temnow, K., & Derégnaucourt, S. (2009). The genetic basis of zebra finch vocalizations. *Evolution: International Journal of Organic Evolution, 63*, 2114–2130.

Gentner, T. Q. (2004). Neural systems for individual song recognition in adult birds. *Annals of the New York Academy of Sciences, 1016*, 282–302.

Gentner, T. Q., Fenn, K. M., Margoliash, D., & Nusbaum, H. C. (2006). Recursive syntactic pattern learning by songbirds. *Nature, 440*, 1204–1207.

Gobes, S. M. H., ter Haar, S. M., Vignal, C., Vergne, A. L., Mathevon, N., & Bolhuis, J. J. (2009). Differential responsiveness in brain and behavior to sexually dimorphic long calls in male and female zebra finches. *Journal of Comparative Neurology, 516*, 312–320.

Gobes, S. M. H., Zandbergen, M. A., & Bolhuis, J. J. (2010). Memory in the making: Localized brain activation related to song learning in young songbirds. *Proceedings of the Royal Society B: Biological Sciences, 277*, 3343–3351.

Hauser, M. D., Chomsky, N., & Fitch, W. T. (2002). The faculty of language: What is it, who has it, and how did it evolve? *Science, 289*, 1567–1579.

Holveck, M.-J., & Riebel, K. (2007). Preferred songs predict preferred males: Consistency and repeatability of zebra finch females across three test contexts. *Animal Behaviour, 74*, 297–309.

Holveck, M.-J., & Riebel, K. (2010). Low-quality females prefer low-quality males when choosing a mate. *Proceedings of the Royal Society B: Biological Sciences, 277*, 153–160.

Jacot, A., Reers, H., & Forstmeier, W. (2010). Individual recognition and potential recognition errors in parent-offspring communication. *Behavioral Ecology and Sociobiology, 64*, 1515–1525.

Konishi, M. (1965). The role of auditory feedback in the control of vocalization in the white-crowned sparrow. *Zeitschrift für Tierpsychologie*, *22*, 770–783.

Krashen, S. D. (1973). Lateralization, language learning, and the critical period: Some new evidence. *Language Learning*, *23*, 63–74.

Krebs, J. R. (1977). Song and territory in the great tit, *Parus major*. In B. Stonehouse, & C. M. Perrins (Eds.), *Evolutionary Ecology* (pp. 47–62). London: Macmillan.

Kuhl, P., & Rivera-Gaxiola, M. (2008). Neural substrates of language acquisition. *Annual Review of Neuroscience*, *31*, 511–534.

Leitão, A., ten Cate, C., & Riebel, K. (2006). Within-song complexity in a songbird is meaningful to both male and female receivers. *Animal Behaviour*, *71*, 1289–1296.

Lenneberg, E. H. (1967). *Biological foundations of language*. Oxford: Wiley.

Marler, P. (1970). A comparative approach to vocal learning: Song development in white-crowned sparrows. *Journal of Comparative and Physiological Psychology*, *71*, 1–25.

Marler, P. (1977). The structure of animal communication sounds. In T. H. Bullock (Ed.), *Recognition of complex acoustic signals: Report of the Dahlem workshop on recognition of complex acoustic signals* (pp. 17–35). Berlin: Abakon-Verlagsgesellschaft.

Marler, P., & Peters, S. (1977). Selective vocal learning in a sparrow. *Science*, *198*, 519–521.

Nordeen, K. W., & Nordeen, E. J. (1992). Auditory feedback is necessary for the maintenance of stereotyped song in adult zebra finches. *Behavioral and Neural Biology*, *57*, 58–66.

Nottebohm, F. (1968). Auditory experience and song development in the chaffinch (*Fringilla coelebs*). *Ibis*, *110*, 549–568.

Nowicki, S., Searcy, W. A., & Peters, S. (2002). Quality of song learning affects female response to male bird song. *Proceedings of the Royal Society B: Biological Sciences*, *269*, 1949–1954.

Okanoya, K. (2004). The Bengalese finch: A window on the behavioral neurobiology of birdsong syntax. *Annals of the New York Academy of Sciences*, *1016*, 724–735.

Pepperberg, I. M. (2010). Vocal learning in grey parrots: A brief review of perception, production, and cross-species comparisons. *Brain and Language*, *115*, 81–91.

Riebel, K. (2009). Song and female mate choice in zebra finches: A review. *Advances in the Study of Behavior*, *40*, 197–238.

Searcy, W. A. (1992). Song repertoire and mate choice in birds. *Integrative and Comparative Biology*, *32*, 71–80.

Takahashi, M., Yamada, H., & Okanoya, K. (2010). Statistical and prosodic cues for song segmentation learning by Bengalese finches (*Lonchura striata* var. *domestica*). *Ethology*, *115*, 1–9.

Vallet, E., & Kreutzer, M. (1995). Female canaries are sexually responsive to special song phrases. *Animal Behaviour*, *49*, 1603–1610.

Van Heijningen, C. A. A., de Visser, J., Zuidema, W., & ten Cate, C. (2009). Simple rules can explain discrimination of putative recursive syntactic structures by a songbird species. *Proceedings of the National Academy of Sciences of the United States of America*, *106*, 20538–20543.

Vicario, D. S., Naqvi, N. H., & Raksin, J. N. (2001). Sex differences in discrimination of vocal communication signals in a songbird. *Animal Behaviour*, *61*, 805–817.

Williams, H., Kilander, K., & Sotanski, M. L. (1993). Untutored song, reproductive success and song learning. *Animal Behaviour*, *45*, 695–705.

Woodgate, J. L., Leitner, S., Catchpole, C. K., Berg, M. L., Bennett, A. T. D., & Buchanan, K. L. (2011). Developmental stressors that impair song learning in males do not appear to affect female preferences for song complexity in the zebra finch. *Behavioral Ecology*, *22*, 566–573.

Zann, R. (1996). *The zebra finch: A synthesis of field and laboratory studies*. Oxford: Oxford University Press.

6 Parametric Variation: Language and Birdsong

Neil Smith and Ann Law

What makes human language unique? The most influential recent suggestion is recursion, which has featured prominently since Hauser et al. (2002) proposed as an empirical hypothesis that what is unique to language and unique to humans is "(probably) recursion."[1] This answer is (probably) wrong: recursion is not unique to language, but is characteristic of the language of thought (in Fodor's 1975 sense; cf. also Smith, 2004; Fitch, Hauser, & Chomsky, 2005) and it may not be unique to humans given the hierarchical structure of canary song (Gardner, Naef, & Nottebohm, 2005), the vocal improvisation found in whales (Payne, 2000, p. 135), and perhaps the properties of animal navigation. Another potential answer that we investigate is parametric variation (PV) where this is a way of formulating and uniting two linguistic problems: the puzzle of first-language acquisition ("Plato's problem," Chomsky, 1986) and the limits of typological variation. Despite Hauser's (2006) spirited promotion of the claim that PV is characteristic of moral judgment, and Smith's (2007) parallel suggestion for music, we think it is plausible to suggest that PV is unique to human language and that the variation found in other cognitive domains and in animal vocalization is not "parametric." We do not exclude other possibilities: on the one hand, the putative uniqueness of PV to human language may be derivative from other characteristics, such as an immensely rich lexicon; on the other, human language may owe its uniqueness to a number of different properties.

The structure of the rest of the chapter is as follows: in the second section, we outline the properties of PV in language, concentrating on the differences between parametric and nonparametric variation. In the third section, we see if they generalize to birdsong. In the final section, we attempt to adjudicate among the conclusions in (1):

(1) a. PV is unique to human language.
   b. PV is unique to humans but not just to language.
   c. PV is common to human language and birdsong, but not the rest of cognition.

  d. PV is common to everything—language, cognition, birdsong . . .

  e. There is no coherent (or uniform) notion of PV.

For historical reasons we exclude logical possibilities such as (f):

  f. PV characterizes, for example, birdsong but *not* human language.

  We tentatively endorse (1a) and suspect that, if PV is unique to human language, it is because PV is a solution to Plato's problem and only human language confronts the learner with this problem in its full complexity.

Parametric Variation in the Language Domain

"Principles and Parameters" Theory

PV is part of "Principles and Parameters" theory (Chomsky, 1981, 2006; for overviews and history see Roeper & Williams, 1987; Roberts, 1997; Baker, 2001). The human language faculty is standardly described in terms of a contrast between the faculty of language in the "broad" sense (FLB) and a proper subpart of that faculty, referred to as the faculty of language in the "narrow" sense (FLN) (Hauser, Chomsky, & Fitch, 2002). The former includes a variety of performance mechanisms for parsing and producing utterances as well as our strictly grammatical ability. Many parts of FLB are shared with other organisms from bumblebees to bonobos, but FLN—which may be the empty set—is by hypothesis unique to humans and unique to language. FLN is characterized in terms of Universal Grammar (UG), the innate endowment that enables children to learn their first language and that defines the basic format of human language. It specifies that human languages consist of a lexicon and a "computational system" whose most important property is the possibility of recursion.[2] The lexicon consists of a set of lexical entries, each of which is a triple of phonological, morphosyntactic, and semantic features, and with a link to associated encyclopedic information. Every natural language has a lexicon containing tens of thousands of such entries whose essential function is to link representations of sound to representations of meaning.

  UG also provides a set of exceptionless principles such as structure dependence (Chomsky, 1971) and the extended projection principle (Chomsky, 1995), which constrain the operation of the computations. Structure dependence is a principle that states that all grammatical operations—phonological, morphological and syntactic—have to be defined in all languages over structures rather than simple linear sequences of elements. That is, the possibility of counting the number of words or constituents is excluded a priori. For instance, the formation of a question from a congeneric statement as in (2) can refer to syntactic categories, such as "auxiliary verb," and their movement to a particular position (the Infl node), but not to the "third word." In (2) the effect might appear to be the same: "Move the auxiliary verb *is* to the Infl position" or "Move the third word to initial position," but (3), with the putative questions in (4), shows that only one process is licit.

(2) The man is in the room — Is the man in the room?

(3) John is in the room.

(4) a. Is John in the room?
   b. *In John is the room?

This principle acts as a constraint on language acquisition: children learning their first language have their "hypothesis space" constrained with the result that they never make mistakes like that in (4b).

The extended projection principle (EPP) stipulates that all clauses must have a subject, so (5a) is acceptable but (5b) is impossible:

(5) a. John came home early.
   b. *Came home early.

The status of this principle is somewhat different from that of structure dependence in that "pro-drop" languages seem to be systematic exceptions. As we will see, this exceptionality is only apparent.

In addition to offering a set of universal principles, UG also provides a set of parameters that jointly define the limits of variation. This is typically conceptualized as the setting of a number of "switches" — on or off for particular linguistic properties. Typical examples of such parameters in syntax are the head-direction parameter (whether heads, such as Verb, Noun, and Preposition, precede or follow their complement), the null-subject (or "pro-drop") parameter (whether finite clauses can have empty pronominal subjects), and the null-determiner parameter (whether noun phrases can have empty determiners). English and Hindi have opposite values for each of these parametric choices, as illustrated in (6)–(8):

(6) Head-first — "on the table" — Head-last "mez par"
                                        table on

(7) Non-pro-drop — *"Is working" — Pro-drop — "Ø kaam kartaa hai"
                                                 work doing is

(8) Non-null D — *"boy has come" — Null D — "Ø laṛkaa aaya hai"
                                              boy come is

Typical examples of parametric choices in phonology are provided by the stress differences characteristic of English and French, and the possibility of complex consonant clusters found in English but not in Japanese. English stress is "quantity-sensitive" whereas French stress is "quantity-insensitive," with the result that words with the same number of syllables may have different stress in English but uniform stress in French, as shown in (9):

(9) English: *América / Manitóba*
       French: *endurcissement / sentimental*

There is a general phonological principle that syllables may have onsets—in no language do all syllables begin with a vowel. There is an associated parameter allowing some onsets to be complex. In English words may begin with clusters of consonants in a way that is impossible in Japanese, with the result that English loans into Japanese appear with the clusters separated by epenthetic vowels, as shown in (10):

(10) English: *screwdriver*
       Japanese: *sukuryūdoraibā*

The main discussion of PV has been in the domain of syntax. Originally, parameters were associated with the principles of UG, but following Borer's (1984) work, they were located in the lexicon (see Smith, 2004, for discussion). Moreover, for Chomsky and many others (e.g., Chomsky, 1995), the relevant domain is restricted to the functional lexicon, where this refers to that subset of the whole dealing with functional categories such as Tense, Complementizers, and Determiners, in contradistinction to the conceptual lexicon, which deals with substantive categories such as Noun, Verb, and Adjective. Thus *bumblebee* belongs to the conceptual lexicon, whereas an item such as *the* belongs to the functional lexicon. There is general consensus on the need for such a (traditional) distinction between functional and substantive categories, but little agreement on how to draw the boundary lines between the two (for discussion, see Muysken, 2008). For many theorists parameters are necessarily binary, but such a restriction is independent of the general conceptual claims of the theory.[3] There are many further distinctions within Principles and Parameters theory (see Smith & Law, 2007), but for our present purposes, what is important is providing identity criteria for distinguishing between variation that is parametric (PV) and variation that is nonparametric.

The domain of PV in syntax is the set of functional categories, but there is no comparable restriction in phonology. Rather, as indicated in the examples of stress and possible onset clusters above, parameters are associated with words, syllables, vowels, and so on (cf. Dresher & Kaye, 1990; Dresher, 1999). It follows that PV is not definable in terms of properties of the functional lexicon and suggests that we need to identify more abstract properties of the concept. One such crucial property of PV that follows from its being part of our innate endowment is that it gives rise to a situation in which language acquisition is cued or triggered rather than "learned." Learning in the traditional psychological sense (i.e., a process involving hypothesis formation and testing, association, generalization, and so on) plays little role in first-language acquisition.

Typology and Acquisition

"Principles and Parameters" theory unites—as an empirical claim—two domains: typology and first-language acquisition. By hypothesis, the principles do not vary from child to child or from language to language, and in first-language acquisition the child's task is reduced to setting the values of such parameters on the basis of the stimuli it is exposed to—utterances in the ambient language. Given the strikingly uniform success of such acquisition, the set of possibilities must be "narrow in range and easily attained by the first language learner" (Smith, 2004, p. 83). Variation among the world's languages is similarly defined in terms of parametric choices such as whether verbs precede their objects as in English, or vice versa as in Hindi. Thus PV is variation within a narrow range defined by universal principles; it facilitates the task of language acquisition and makes available a typology of the world's languages.

It is important to note that acquisition has conceptual priority, with the typological implication of PV being derivative. If it is correct to claim that first-language acquisition consists in making parametric choices, these have priority over any taxonomy based on them. Moreover, if acquisition is a matter of parameter setting and if there is no negative evidence available to the child, then all the possible alternatives are antecedently known or innate, and the child's task in learning its first language is a matter of selecting a grammar on the basis of the particular properties of the input, rather than needing instruction (cf. Piattelli-Palmarini, 1989).

This claim of "antecedent knowledge" or "knowledge without experience" has a number of implications. The first is that parametric choices may give rise to cascade effects: coming to know one fact (e.g., that verbs precede their objects) licenses knowledge by the learner of other facts (e.g., that prepositions precede their objects) without further exposure. "Cascades" have become unfashionable because of the dissociation of the properties associated with (notably) the pro-drop parameter (for discussion, see Ackema, Brandt, Schoorlemmer, & Weerman, 2006). This reaction may have been hasty: cascades could be operative in the domain of acquisition even if there is such dissociation. That is, the child leaps to the "cascade conclusion" (i.e., selects one parameter on the basis of the setting of a distinct but related parameter) unless there is evidence to the contrary, thereby solving Plato's problem (in part) (for discussion, see Smith & Cormack, 2002).

Principles and Parameters theory is at once "internalist" (i.e., it is a theory of states of the mind/brain) pertaining to knowledge that is largely unconscious, and "universalist," hence likely to take place in a critical period or periods.

Identity Criteria for Parametric Variation

The theory of PV hypothesizes that the range of choices is "antecedently known," and this basic property, our *first* criterion, correlates with a number of others that distinguish PV from nonparametric variation and allow us to provide identity criteria for it.

Our *second* criterion is that variants licensed by parametric choice must be cognitively represented. Consider by contrast acclimatization, specifically sweating. We have a critical period for setting our sweating switch: experiencing hot and humid weather in the first three years of life leads to a different setting from exposure to different conditions, and these settings cannot be significantly altered thereafter. Despite a certain superficial similarity, this is not PV because the different states are not (mentally) represented and have no cognitive effects.

Our *third* criterion is systematicity. A simple example is provided by irregular morphology of the type exemplified by the impossibility of *amn't* in (most varieties of) English, or the kind of defective paradigm seen in Latin *vis-vim-vi*. We do not consider this to be PV because it is by definition not systematic and hence we could not plausibly acquire knowledge of it by any process of triggering in the way that is plausible for systematic contrasts such as the possibility of null determiners.

Our *fourth* criterion is dependence on the input—that is, choices correspond to possible states of the adult language. The head-direction parameter clearly reflects properties of the ambient language in a way that is not characteristic of all variation. An example of systematic but nonparametric variation is provided by the individual differences in consonant harmony in phonological development (see Smith, 1973, p. 163), or the variation in the choice of initial or final negation in syntactic development (see Smith, 2005, p. 29). For instance, two children in essentially the same environment may produce the adult *duck* as [gʌk] and [dʌt] respectively. These are both manifestations of consonant harmony, but they do not count as PV because the particular variants chosen appear to be independent of the input. A comparable syntactic example is provided by the development of negation. All children typically go through a stage in which the negator is peripheral, either initial or final. Individual children then differ such that one child learning English may say *No like cabbage* and another *Like cabbage no*. We take such variation to be nonparametric because no language allows such peripheral negation. This universal exclusion enables us to differentiate this nonparametric variation from UG-licensed errors of the sort described by Crain and his colleagues (see Crain & Pietroski, 2002). A child may produce a form that never occurs in the input (e.g., *What do you think what pigs eat?*) because the structure is licensed by UG and so occurs as a parametric choice in other languages.[4] Despite this potential difficulty, the case of consonant harmony in phonology and negation in syntax makes the conceptual contrast between parametric and nonparametric variation clear.

Our *fifth* criterion is that PV must be deterministic—that is, the input to the child must be rich enough and explicit enough to guarantee that a parameter such as pro-drop or the presence of complex onsets in phonology can be set. If the input does not meet this requirement we are dealing with nonparametric variation. A syntactic example is provided by sequence of tense phenomena where individual variation verges on the random (see Smith & Cormack, 2002). A phonological example is provided by Yip (2003, p. 804), who argues that some people treat a postconsonantal glide as a secondary articulation of the consonant, others as a segment in its own right: "The rightful home of /y/ [is] underdetermined by the usual data, leaving room for variation." Her conclusion is that "speakers opt for different structures in the absence of conclusive evidence for either." Again that indicates for us that the variation is nonparametric.

Our *sixth* criterion is suggested by an observation of Dupoux and Jacob (2007) to the effect that PV in language is "discrete" (usually binary), whereas in moral judgment one typically finds continuous scales. An example of the contrast is provided by vowel height. Whether a language displays 2, 3, or 4 degrees of vowel height in its phonological system is a matter of parametric choice(s). The degree to which the particular articulation of some vowel is high—either randomly or as a matter of individual difference (maybe my articulations of [i] are systematically higher than yours)—is continuous and could not be parametric.

Our *seventh* and final criterion is "exclusivity." PV gives rise to mutually exclusive possibilities: languages are either [+Pro-drop] or [−Pro-drop]. The choice exhausts the conceptual space and leaves no room for compromise—no language is both. By contrast, the choice in a [+Pro-drop] language of using or not using a subject pronoun is nonparametric. The contrast is again with morality where moral diversity involves "different preference orderings among competing members of a finite set of universal moral values" (Dupoux & Jacob, 2007, p. 377).

In Table 6.1 we summarize and exemplify these criteria, all of which are common to syntax and phonology.

The criteria are intended to be individually necessary and jointly sufficient to identify PV. Some phenomena (e.g., cascade effects) may be sufficient to license the conclusion that there is PV even though it is not possible to make this a necessary condition. Other phenomena, such as occurring in a critical period, are compatible with PV, but are not evidence for it.

Assuming that the nature of PV is clear, we now see if the variation found in birdsong meets the identity criteria listed here and so counts as "parametric." A preliminary negative conclusion to the effect that PV is **not** characteristic of music, morality (or birdsong) can be found in Smith and Law (2007).

Parametric Variation in the Domain of Birdsong?

Is it plausible to generalize PV to the domain of animal cognition, in particular birdsong[5] in oscine birds? We begin by specifying some of the background

Table 6.1
Identity criteria for parametric variation

1. The range of choices must be antecedently known; hence acquisition is a matter of selection rather than instruction.	
Parametric	Nonparametric
±Pro-drop	Irregular morphology
2. Parametric choices must be mentally represented.	
Parametric	Nonparametric
Stress	Sweating
Word order	Consonant harmony
3. Choices must be systematic—variations are not accidents	
Parametric	Nonparametric
±Null determiner	Defective paradigms
4. Choices must be dependent on the input and hence correspond to a possible state of the adult language.	
Parametric	Nonparametric
Quantity-sensitivity	Consonant harmony
Word order—head direction	Word order—early negation
5. Choices must be deterministic.	
Parametric	Nonparametric
Pro-drop	Sequence of tense
Complex onsets in phonology	Postconsonantal glides
6. Choices must be discrete.	
Parametric	Nonparametric
Number of vowel heights	Realization of vowel height
7. Choices must be mutually exclusive.	
Parametric	Nonparametric
±Pro-drop	Choice of a pronoun (or not) in a pro-drop language

commonalities and differences between the two domains and listing the types of variation before going through the identity criteria one by one.

Parallels between Language and Birdsong

As pointed out, for example, by Kuhl (1999), Fitch (2005), and Doupe and Kuhl (1999), birdsong is parallel to language in being internalist, unconscious, reliant on auditory input, acquired in a sensitive period, and universal (for particular species).[6] Most strikingly, language and birdsong both manifest vocal learning (see, e.g., Nottebohm, 1999; Wilbrecht & Nottebohm, 2003) and hierarchical structure (see, e.g., Brenowitz, Margoliash, & Nordeen, 1997; Gardner et al., 2005), though this structure is defined over different units. The building blocks of birdsong are notes, syllables, phrases, motifs, types, and bouts, where typically "short stereotyped syllables are repeated to form phrases, which in turn are arranged to form songs" (Gardner et al., 2005, p. 1046). Assuming that PV operates over elements of the appropriate domain— as with syntax and phonology—this difference in "building blocks" is expected.

Like human infants, birds are sensitive to their own "language": "Young birds must hear the songs of their own species in order to learn them, but when faced with a potentially confusing array of songs, they are able to select the ones of their own species to serve as learning templates" (Whaling, 2000, p. 69). This is comparable to the sensitivity to their own language from intrauterine experience that newborn infants manifest but such sensitivity is presumably prior to, even if necessary for, any setting of parameters.

In later development songbirds appear to go through comparable stages to language-learning infants. Subsong, plastic song, and full (crystallized) song (Bolhuis, 2005) correspond to babbling and early and late phonological mastery, with "template memorization" and "vocalization matching" corresponding to the child's acquisition of the phonological representations of lexical items and the subsequent mastery of their production.

There are further parallels. Kuhl (1999, p. 424) observes that in humans "language input alters the brain's processing of the signal, resulting in the creation of complex mental maps." (This is an example of the perceptual magnet effect and the native-language magnet whereby infants lose some of their innate discriminatory abilities as a function of exposure to a specific ambient language. A parallel to this magnet effect is found in birdsong. Gardner et al. (2005, p. 1046) report on canaries' ability to imitate ill-formed song when young, and on how this ill-formed song is then "reprogrammed to form typical canary phrasing."

Finally, song development in oscine birds shows interesting parallels with the "selection" account of language acquisition (Piattelli-Palmarini, 1989). Thus, "Songbirds actually inherit much of the information required to generate a normal species-specific vocal repertoire . . . as though memorization is based not on instruction . . . but on selective processing, imposed on a fund of *innate* knowledge that is to some degree unique to each species" (Marler, 1999, p. 315).

Contrasts between Language and Birdsong

Despite these similarities, there are crucial differences. First, there are about 9,000 species of birds (including 4,000 species of songbirds) but only one human species and no one would describe our differences from other primates in terms of parametric choices. Moreover, there are "enormous between-species variation in song structure, as well as in the characteristics of song acquisition and production" (Bolhuis & Macphail, 2001, p. 429), so "each songbird species seems to go about the process of learning to sing in its own way" (Marler, 1999, p. 295; cf. Suthers, 1999).

This is important because birdsong typology is frequently cross-species, but comparability with language demands intraspecies treatment. The basic contrast between "open-ended learners" (such as canaries) and "age-limited learners" (such as zebra finches) could not usefully be viewed as parameter setting: there is no choice for the individual birds. We will anyway restrict our attention in what follows to acquisition rather than typology.

Second, in all species birdsong seems to lack any syntax or compositional semantics. Calls may have some minimal content such as indicating predators, and song has territorial and mate-selection functions, but it has no syntax. The richness of birdsong resides in what Marler (2000, p. 31) calls "phonocoding"—"the ability to create new sound patterns by recombination simply to generate signal diversity," and not in "lexicoding"—"when meaningful elements are syntactically joined." There are then no "sentences" in animal communication and when birdsong specialists talk of "syntax" they are referring to what linguists would call "phonotactics"—the syntax of phonology. It follows that starlings' apparent ability to learn recursive syntactic patterns (Gentner et al 2006; Marcus 2006) is unlikely to be comparable to the ability of humans to do so and hence not as remarkable as the authors suggest. For sobering discussion see Jackendoff et al (2006). The most relevant point for us is that the starlings' prowess appears to have no effect on their own song.

Is Variation in Birdsong Parametric?

The crucial consideration for us is whether any of the properties of birdsong satisfy the criteria for PV. Many, such as occurring in a critical period, having hierarchical structure, and illustrating "magnet" effects, are compatible with PV, but are not evidence for it. In an attempt to answer the question we list the types of variation observed and then go through the identity criteria.

Within any species, birdsong variation may be individual or dialectal, it may reflect differences in choice of tutor or model(s), in habitat, in the discrimination of self from other, and it may be the result of improvisation or reprogramming.

It is known that zebra finches can discriminate their own song from that of their tutors and can tell individuals apart (Nottebohm, 1999, p. 83). Moreover, Liu et al. (2004) demonstrated individual differences in the song learning of zebra finches, including cases of variation among siblings who were "members of the same clutch" all of whom "imitated the same model" (Liu, Gardner, & Nottebohm, 2004, p. 18178). Crucially, the juveniles gradually converge on the "same" adult song (Liu et al., 2004, p. 18180), making this kind of individual variation look comparable to that found in consonant harmony in the acquisition of phonology, with the selective choice characteristic of PV playing no role (cf. Doupe & Solis, 1999).

It is likewise well known that there are birdsong dialects (see, e.g., Catchpole, 1991), and one of the most suggestive parallels between birdsong and language is such regional variation. Searcy and Nowicki (1999) report that, as measured by courtship display to recordings, sparrows are sensitive to differences in the song of conspecifics from New York and Pennsylvania. Further, Nowicki, Searcy, Hughes, & Podos (2001) demonstrate that closely related species of sparrow imitate experimentally introduced variation—"broken syntax," characterized by artificially increased trill rate—so that it spreads through the population despite the conservatism of females. In language as in birdsong the development of dialects is presumably crucially dependent on the selection of particular patterns in the input, though in birdsong the choices seem continuous rather than discrete.

Further examples of variation are attributable to improvisation (the idiosyncratic variation characteristic of individual birds), and even to habitat. Improvisation is widespread but may be blocked by a live model (Nottebohm, 1999, p. 97), and more generally, "Motor song learning may restrict . . . the nature of further vocal learning" (Nottebohm, 1999, p. 95). Kopuchian, Lijtmaer, Tubaro, & Handford (2004) observe that songs of individual rufous-collared sparrows from closed habitat sites had trills with longer trill intervals and lower frequencies than those of individuals from open habitats.

With these scattered remarks on variation behind us, we now go through the identity criteria for PV that we proposed for language.

• Are the choices "antecedently known"? Where the song is independent of the input, it is clearly genetically determined, but if there is no choice, there is no variation and, a fortiori, no PV. Where there is a role for learning, we know of no examples where members of a single species manifest a choice among a significant range of possibilities. Rather, the variation seems exhausted by a combination of imitation and improvisation.

• Is the variation mentally represented? This criterion rests on the difference between phenomena like stress assignment and consonant harmony. The difficulty of determining the status of such a contrast in another species forces us to leave the issue open, but the ability of birds to recognize conspecific individuals and modify their behavior accordingly suggests a positive answer.

• Is the variation systematic? Again, a decision is not straightforward, but at least some variation is not. Thus, reporting on the perceptual magnet effect in canaries, Gardner et al. (2005) observe that "reprogramming" "occurred in the absence of any exposure to normal canary song," leading to the conclusion that "inferred innate rules forced a complete reprogramming of the imitated song" (p. 1047). None of this is comparable to PV in human learning of phonology.

• Is the variation dependent on the input? The huge literature on the role of tutor song shows that many details of the acquisition of birdsong are input-dependent, and the restriction on the nature of vocal learning reported by Nottebohm could result from the effect of the setting of a parameter. However, this is not a necessary conclusion and it is equally clear that some individual variation in song production is not parametric. The identifiably juvenile nature of the imitations cited by Liu et al. (2004) makes it reasonably clear that juvenile birdsong does not always correspond to the adult state but is more comparable to that found in consonant harmony. The conclusion bears elaboration. Salwiczek and Wickler (2004, p. 170) report that young chaffinches are unable to produce the chaffinch song unless they hear it, but naive individuals will unfailingly identify and copy chaffinch song presented among a sample of foreign birdsongs. We have already determined that PV should be restricted to a single species, so such species identification falls outside PV. Similarly, Houx, ten Cate, and Feuth (2000) write that in zebra finches song learning is a more active process than is generally assumed but that there was "no strong effect of tutor song characteristics" and that none of the variables (social context, tutor song, individual ability) had a profound effect (p. 1387). That is, there were no discontinuities of the kind expected under a PV account—even good learners and poor learners seemed roughly comparable.

• Is the variation deterministic? In general the bird's final state seems deterministic, but the example of "broken syntax" shows the possibility of deviations.

• Is the variation discrete? In general the choices birds make are continuous ones. Typical evidence comes from the discussion of the role of habitat mentioned earlier. Kopuchian et al. (2004, p. 551) report finding consistent differences between the songs of birds occupying different habitats. Revealingly, they interpret this pattern of variation as "a song cline that correlates with the environmental gradient."

• Is the variation mutually exclusive? The answer seems to be negative. Some young songbirds are able to incorporate material from different tutors: "Zebra finches exposed to many models may imitate syllables from more than one model . . . [thereby] achieving a more unique song" (Nottebohm, 1999, p. 97). The closest linguistic parallel to such a phenomenon is dialect mixture where a single individual uses pronunciations from different sources in the same utterance, giving for example [pla:nt ðə plænts] for "plant the plants." Such mixture reflects multilingualism rather than PV.

It is time to take stock. The unification of typology and acquisition in birdsong is moot and we know of no cascade effects in its development that would make the case for PV cogent. The existence of critical periods and universals may be necessary but is clearly not sufficient to motivate postulating PV in birdsong. The best case for it resides in the existence of regional birdsong dialects. These dialects appear to be learned rather than inherited (Catchpole, 1991, p. 288; cf. Searcy & Nowicki, 1999) and are clearly functional. However, even here there is no evidence (that we know of) that the choices are antecedently given, and the complexity of what is learned may not be sufficient to motivate the need for PV: there may be no avian equivalent of Plato's problem.

We began by expressing our skepticism toward Hauser et al.'s claim that what is unique to human language is recursion. Can we provide an alternative conclusion?

Conclusion

The claim that PV is unique to human language could be trivially true—PV presupposes UG and, by hypothesis, birds do not have UG. We attempted to make the claim nontrivial by providing putative identity criteria for PV in language and seeing if they generalize to birdsong. That such an extension is plausible is suggested by the commonalities between language and birdsong. In both areas there are putative universals, suggesting an innate basis; there are simultaneously clear effects of the environment, suggesting interplay between genes and learning; and there are parallels in the various stages that organisms pass through in mastering the complexity of the system they are acquiring.

Accordingly, we are now in a position to make a tentative selection among the possibilities in (1). As far as typology is concerned, PV can be adduced harmlessly to describe the limits of variation in both the relevant domains. We view this, however, as somewhat banal and somewhat misleading in view of the species complexity among oscine birds. The core interest of PV in linguistics lies in its solution to Plato's problem and in its unification of typology and acquisition. Neither of these seems to generalize to birdsong and it follows that (1a) is correct.

But our conclusion is muted. We reject the pessimistic (1e), though on bad days it looks fairly persuasive, but our reasoning is less decisive than we had hoped, leaving unanswered the question of why PV should be unique to language. We suspect that the answer is going to be messy and complex. Human language is complicated and it is that complexity that makes Plato's problem so hard and so interesting. Other domains, including birdsong, despite their interest and richness, are not complex in the same way. So we envisage multiple answers, partly, as Jackendoff, Liberman, Pullum, and Scholz (2006) put it: What is "unique to language is a very large learned vocabulary consisting of long-term memory associations between meanings and structured pronunciations plus varied phrasal syntax." And partly PV, where crucially typology and acquisition are united and the central factor distinguishing language from all the rest is the antecedently available knowledge of the possible choices.

Acknowledgments

For comments, conversation, correspondence and criticism we are grateful to David Adger, Misi Brody, Noam Chomsky, Annabel Cormack, Tecumseh Fitch, Paul Harris, Rita Manzini, Gary Marcus, Ian Roberts, Amahl Smith, and Deirdre Wilson. None of these should be taken as agreeing with anything we say. Some indication of relevant developments in our thinking since we submitted this chapter can be found in Smith and Law (2009; in press).

Notes

1. It is not easy to interpret such claims precisely. Tomalin (2007) points out that *recursion* is used in at least five different ways, so that evaluating the claim by Hauser, Chomsky, and Fitch (2002) is problematic.

2. Recursion is generally restricted to syntax and not extended to the phonological domain.

3. Marten, Kula, and Thwala (2007, p. 257) give binarity as one of their criteria for parameters but offer no reason.

4. We take it that such overgeneralization is a sign that the child has, temporarily, mis-set the relevant parameter.

5. We assume the standard distinction among birdsong, bird calls, and bird mimicry, because PV is of potential relevance only to the first of these.

6. Though often only the male sings.

References

Ackema, P., Brandt, P., Schoorlemmer, M., & Weerman, F. (Eds.). (2006). *Arguments and agreement*. Oxford: Oxford University Press.

Baker, M. (2001). *The atoms of language: The mind's hidden rules of grammar*. Oxford: Oxford University Press.

Bolhuis, J. (2005). Development of behavior. In J. Bolhuis & L.-A. Giraldeau (Eds.), *The behavior of animals: Mechanisms, function, and evolution* (pp. 119–145). Oxford: Blackwell.

Bolhuis, J., & Macphail, E. (2001). A critique of the neuroecology of learning and memory. *Trends in Cognitive Sciences, 5*, 426–433.

Borer, H. (1984). *Parametric syntax: Case studies in Semitic and Romance languages*. Dordrecht: Foris.

Brenowitz, E., Margoliash, D., & Nordeen, K. (1997). An introduction to birdsong and the avian song system. *Journal of Neurobiology, 33*, 495–500.

Catchpole, C. (1991). Songs. In M. Brooke & T. Birkhead (Eds.), *The Cambridge encyclopaedia of ornithology* (pp. 283–291). Cambridge: Cambridge University Press.

Chomsky, N. (1971). On interpreting the world. *Cambridge Review, 92*, 77–93.

Chomsky, N. (1981). *Lectures on government and binding*. Dordrecht: Foris.

Chomsky, N. (1986). *Knowledge of language: Its nature, origin and use*. New York: Praeger.

Chomsky, N. (1995). *The Minimalist Program*. Cambridge, MA: MIT Press.

Chomsky, N. (2006). *Language and mind* (3rd ed). Cambridge: Cambridge University Press.

Crain, S., & Pietroski, P. (2002). Why language acquisition is a snap. *Linguistic Review, 19*, 163–183.

Doupe, A., & Kuhl, P. (1999). Birdsong and speech: Common themes and mechanisms. *Annual Review of Neuroscience, 22*, 567–631.

Doupe, A., & Solis, M. (1999). Song- and order-selective auditory responses emerge in neurons of the songbird anterior forebrain during vocal learning. In M. Hauser & M. Konishi (Eds.), *The design of animal communication* (pp. 343–368). Cambridge, MA: MIT Press.

Dresher, E. (1999). Charting the learning path: Cues to parameter setting. *Linguistic Inquiry, 30*, 27–67.

Dresher, E., & Kaye, J. (1990). A computational learning model for metrical phonology. *Cognition, 34*, 137–195.

Dupoux, E., & Jacob, P. (2007). Universal Moral Grammar: A critical appraisal. *Trends in Cognitive Sciences, 11*, 373–378.

Fitch, T. (2005). The evolution of language: A comparative review. *Biology and Philosophy, 20*, 193–230.

Fitch, T., Hauser, M., & Chomsky, N. (2005). The evolution of the language faculty: Clarifications and implications. *Cognition, 97*, 179–210.

Fodor, J. (1975). *The language of thought*. New York: Crowell.

Gardner, T., Naef, F., & Nottebohm, F. (2005). Freedom and rules: The acquisition and reprogramming of a bird's learned song. *Science, 308*, 1046–1049.

Gentner, T., Fenn, K., Margoliash, D., & Nusbaum, H. (2006). Recursive syntactic pattern learning by songbirds. *Nature, 440*, 1204–1207.

Hauser, M. (2006). *Moral minds*. New York: HarperCollins.

Hauser, M., Chomsky, N., & Fitch, W. T. (2002, November 22). The faculty of language: What is it, who has it, and how did it evolve? *Science, 298*, 1569–1579.

Houx, B., ten Cate, C., & Feuth, E. (2000). Variations in zebra finch song copying: An examination of the relationship with tutor song quality and pupil behaviour. *Behaviour, 137*, 1377–1389.

Jackendoff, R., Liberman, M., Pullum, G., & Scholz, B. (2006). Starling study: Recursion. http://linguistlist.org/issues/17/17-1528.html.

Kopuchian, C., Lijtmaer, D. A., Tubaro, P. L., & Handford, P. (2004). Temporal stability and change in a microgeographical pattern of song variation in the rufous-collared sparrow. *Animal Behaviour, 68*, 551–559.

Kuhl, P. (1999). Speech, language and the brain: Innate preparations for learning. In M. Hauser & M. Konishi (Eds.), *The design of animal communication* (pp. 419–450). Cambridge, MA: MIT Press.

Liu, W., Gardner, T., & Nottebohm, F. (2004). Juvenile zebra finches can use multiple strategies to learn the same song. *Proceedings of the National Academy of Sciences of the United States of America, 101*, 18177–18182.

Marcus, G. (2006). Startling starlings. *Nature, 440,* 1117–1118.

Marler, P. (1999). On innateness: Are sparrow songs "learned" or "innate"? In M. Hauser & M. Konishi (Eds.), *The design of animal communication* (pp. 293–318). Cambridge, MA: MIT Press.

Marler, P. (2000). Origins of music and speech: Insights from animals. In N. Wallin, B. Merker, & S. Brown (Eds.), *The origins of music* (pp. 31–48). Cambridge, MA: MIT Press.

Marten, L., Kula, N., & Thwala, N. (2007). Parameters of morphosyntactic variation in Bantu. *Transactions of the Philological Society, 105,* 253–338.

Muysken, P. (2008). *Functional categories.* Cambridge: Cambridge University Press.

Nottebohm, F. (1999). The anatomy and timing of vocal learning in birds. In M. Hauser & M. Konishi (Eds.), *The design of animal communication* (pp. 63–110). Cambridge, MA: MIT Press.

Nowicki, S., Searcy, W., Hughes, M., & Podos, J. (2001). The evolution of birdsong: Male and female response to song innovation in swamp sparrows. *Animal Behaviour, 62,* 1189–1195.

Payne, K. (2000). The progressively changing songs of humpback whales: A window in the creative process in a wild animal. In N. L. Wallin, B. Merker, & S. Brown (Eds.), *The origins of music.* Cambridge, MA: MIT Press.

Piattelli-Palmarini, M. (1989). Evolution, selection and cognition: From learning to parameter setting in biology and in the study of language. *Cognition, 31,* 1–44.

Roberts, I. (1997). *Comparative syntax.* London: Arnold.

Roeper, T., & Williams, E. (1987). *Parameter setting.* Dordrecht: Reidel.

Salwiczek, L., & Wickler, W. (2004). Birdsong: An evolutionary parallel to human language. *Semiotica, 151,* 163–182.

Searcy, W., & Nowicki, S. (1999). Functions of song variation in song sparrows. In M. Hauser & M. Konishi (Eds.), *The design of animal communication* (pp. 577–595). Cambridge, MA: MIT Press.

Smith, N. V. (1973). *The acquisition of phonology: A case study.* Cambridge: Cambridge University Press.

Smith, N. V. (2004). *Chomsky: Ideas and ideals* (2nd ed.). Cambridge: Cambridge University Press.

Smith, N. V. (2005). *Language, frogs and savants: More linguistic problems, puzzles and polemics.* Oxford: Blackwell.

Smith, N. V. (2007). Unification and the unity of Chomsky's thought. *Chomskyan Studies, 2,* 1–21.

Smith, N. V. & Cormack, A. (2002). Parametric poverty. *Glot International, 6,* 285–287. (Reprinted in N. V. Smith, 2005, *Language, frogs and savants: More linguistic problems, puzzles and polemics,* Oxford: Blackwell)

Smith, N. V., & Law, A. (2007). Twangling instruments: Is parametric variation definitional of human language? *UCL Working Papers in Linguistics, 19,* 1–28.

Smith, N., & Law, A. (2009). On parametric (and non-parametric) variation. *Biolinguistics, 3,* 332–343.

Smith, N., & Law, A. (in press). On parametric variation and how it might have evolved. In A. Kalokerinos & I. Tsimpli (Eds.), *Language and evolution.* [in Greek] Heraklion, Greece: Crete University Publishers.

Suthers, R. (1999). The motor basis of vocal performance in songbirds. In M. Hauser & M. Konishi (Eds.), *The design of animal communication* (pp. 37–62). Cambridge, MA: MIT Press.

Tomalin, M. (2007). Reconsidering recursion in syntactic theory. *Lingua, 117,* 1784–1800.

Whaling, C. (2000). What's behind a song? The neural basis of song learning in birds. In N. Wallin, B. Merker, & S. Brown (Eds.), *The origins of music* (pp. 65–76). Cambridge, MA: MIT Press.

Wilbrecht, L., & Nottebohm, F. (2003). Vocal learning in birds and humans. *Mental Retardation and Developmental Disabilities Research Reviews, 9,* 135–148.

Yip, M. (2003). Casting doubt on the onset-rime distinction. *Lingua, 113,* 779–816.

7 Vocal Culture in Songbirds: An Experimental Approach to Cultural Evolution

Olga Fehér and Ofer Tchernichovski

Spoken language might define us as humans, but it is an outcome of vocal learning and vocal culture—two phenomena not unique to humans. Vocal culture is a most striking case of cultural evolution, because as opposed to other instances of culture (e.g., tool use), where certain functional behaviors spread nongenetically, vocal culture is purely communicative—there is no function to vocalization except for the communication itself. Hence the transition from hardwired (innate) vocal communication into a plastic, learned vocal repertoire establishes a biological foundation for a potentially limitless evolution of an arbitrarily complex communication system, where human speech constitutes an extreme case within a continuum (Bolhuis, Okanoya, & Scharff, 2010).

At the mechanistic level, some equivalence between genetic evolution and the evolution of vocal culture is apparent: genetic transmission can be equated to vocal imitation (by social learning), and genetic mutations parallel errors in the learning process. However, as opposed to genetic evolution, the units of cultural evolution are not as clearly defined as genes are. Since vocal behavior does not fossilize, direct evidence for the evolution of vocal cultural traits is difficult to find in nature. Although culture often evolves more rapidly than genes, the timescales are still usually too long for observational studies. Because of these issues, changes occurring over time and across geographic distances are difficult to follow. Field studies have focused on song dialects, which sometimes change fairly quickly—within a few years—due to changing ecological conditions (Baptista, 1977; Chilton & Lein, 1996; Baker, Baker, & Baker, 2001). Recent studies have overcome some of the obstacles of controlled laboratory studies by developing experimental systems that can control and monitor the evolution of vocal culture with sufficient precision to allow the investigation of vocal culture at the mechanistic level (Fehér, Wang, Saar, Mitra, & Tchernichovski, 2009; Belzner, Voigt, Catchpole, & Leitner, 2009). This chapter focuses on such experimental approaches.

Imitation of Isolate Song as a Model for Rapid Cultural Evolution

About 10 years ago, Partha P. Mitra raised an interesting question: Is it experimentally feasible to have a songbird colony established by an isolate founder and then test if and how the improvised song produced by such isolate (ISO) birds would evolve toward wild-type (WT) song over generations without any external influence? To test this question, our group established an "island" bird colony with an isolate founder, and, in addition, performed a series of experiments where exposure to songs was controlled across "generations" of song tutoring: ISO songs were imitated by unrelated juvenile birds who, when adults, trained another generation of birds. This recursive training allowed us to observe changes in the ISO song as it was passed down across a few generations of song learners, in a situation where the social environment (e.g., the presence of females) can be controlled. If any of these experimental conditions would result in some tendency to change or normalize the ISO songs toward WT songs, one can explore this evolutionary process mechanistically and ask what the minimal and necessary conditions are that allow the evolution of vocal culture. What is the role of sexual selection, if any? Is it a Darwinian process or perhaps a different evolutionary mechanism? What are the proximate mechanisms of the transition? What is the role of random errors in song imitation? Are there any perceptual mechanisms involved?

The zebra finch, *Taeniopygia gutatta*, is an excellent model organism to study the evolution of vocal culture. This songbird has a short ontogeny of about 90 days, by the end of which the adult male sings one crystallized song. He then continues to sing this song until the end of his life. When raised in complete acoustic and social isolation, zebra finches improvise a song that has different structural characteristics than normal, WT song (Price, 1979; Williams, Kilander, & Sotanski, 1993). It has been known for a long time that pupils of isolates imitate their tutor's song, and despite their relative lower reproductive success, their prominence as song tutors can be higher than that of WT males (Williams et al., 1993). The studies that had described differences between ISO and WT songs relied on subjective tools to do this, and we felt that an objective, quantitative description of ISO song was needed before we could attempt to recursively train with these ISO songs. We did this by using song features that describe zebra finch songs on different timescales: moment-to-moment (millisecond-level) features such as pitch, amplitude, and frequency modulation; on the level of the song syllable by measuring song note duration; and on the level of the whole song bout using song rhythm as a measure.

After we had a multilevel descriptive tool, we trained birds with ISO song tutors, and we established the island colony with an ISO founder and observed song evolution over generations of song learners. As previously presented (Fehér et al., 2009), we found that ISO song evolved toward WT song features on all timescales within three or four generations in both the socially deprived (one-to-one) setting and in the colony. A description of the behavioral details will follow, but first we briefly discuss some possible cultural evolutionary mechanisms that could have accounted for the rapid evolution to WT songs.

Mechanisms of Cultural Evolution

If we follow Boyd and Richerson's (1985) definition of culture as acquired information inherited through social learning and expressed in behavior and artifacts, the vocal behavior and its transmission in songbirds can be categorized as cultural behavior. It has high variation in learned behavioral patterns, a structured geographic distribution (Marler & Tamura, 1962), and a well-documented mechanism for transmission by social learning (Marler & Tamura, 1964; Payne, 1981; Baker & Cunningham, 1985; Williams, 1990). Such treatment of birdsong gives us the option of considering ideas from cultural evolutionary theory when asking questions about the evolutionary mechanisms underlying the rapid multigenerational song changes in our experiments.

One of the debates that dominate the field of cultural evolutionary theory is how much of a parallel can be drawn between Darwinian biological evolution (Darwin, 1859) and cultural evolution and to what extent Darwinian ideas can explain cultural evolution. On the one hand, the memeticists argue for the existence of cultural replicators, "memes" (Dawkins, 1976), which, equivalent to biological replicators or genes, allow Darwinian processes to result in cultural evolution (Blackmore, 1999; Aunger, 2002). According to this theory, genetic and cultural evolution can be directly compared by making use of identical replication and inheritance mechanisms. On the other hand, inferences about nongenetic replicators remain elusive, which questions the appropriateness of applying Darwinian ideas to the study of cultural evolution (Atran, 2001; Boyer, 1994). More recently, some cultural evolutionary theorists have argued that Darwinian ideas can be applied to study cultural evolution if interpreted more loosely (Henrich, Boyd, & Richerson, 2008; Mesoudi, Whiten, & Laland, 2004). Here, we consider three mutually nonexclusive scenarios based on Darwinian ideas and theories of cultural evolution.

One possibility is that sexual selection is the driving force behind the rapid cultural changes that we observed in our isolate songs (Figure 7.1a). In this scenario, random errors ("mutations") occur in the imitation of songs, generating the diversity of song phenotypes (Figure 7.1b). Females prefer a subset of those altered songs, and therefore males that sing them are most likely to become fathers. Since chicks tend to imitate more from their fathers (Bohner, 1983; Williams, 1990; Mann & Slater, 1994; Zann, 1996), the combination of random copying errors with female preferences constitutes a nongenetic Darwinian process, gradually altering the

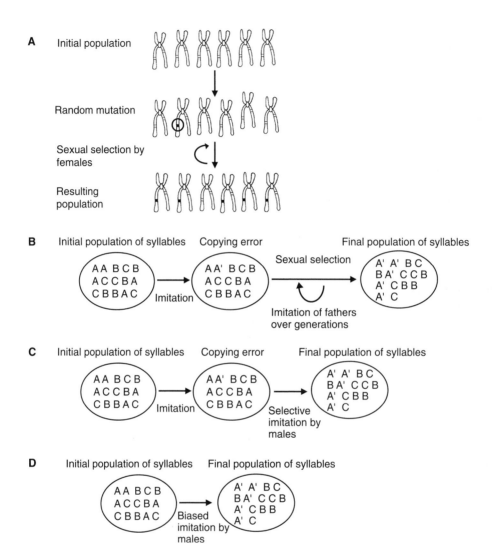

Figure 7.1
Different cultural evolutionary mechanisms can drive the rapid cultural change of ISO song.
(a) Genetic evolution by sexual selection. (b) Cultural evolution by sexual selection. Syllables
(letters) are imitated by successive generations, and when copying errors (A') occur that are
preferred by females, they put a selective pressure on the imitation process that will result in
the spreading of the mutated syllables. (c) Cultural evolution by selective imitation. The
frequency of syllables undergoing copying errors is increased by the preferential imitation of
young males. (d) Cultural evolution by biased imitation. Young males change the song features of syllable A by applying imitation biases.

distribution of song phenotypes in the population. The WT song might be approximated by the subspace of the most sexually attractive songs that males are capable of producing and learning while imitating ISO tutor songs.

The other alternative mechanisms are selective imitation and biased imitation. These concepts are not differentiated in cultural evolutionary theory, because, in general, it is the imitation biases that give rise to selective imitation (Boyd & Richerson, 2005). We use these terms slightly differently. By selective imitation, we mean that young birds preferentially imitate certain parts of the tutor song (Figure 7.1c). This can be the result of variability in perceptual saliency across song elements, or of some social reinforcement and action-based learning where the juveniles initially produce all of the song syllables but prune them during development (Marler & Nelson, 1992). We will use the term "biased imitation" to describe modification in song structure that stems from the process of imitation alone. Here, song elements are *not being selected but are transformed* (Figure 7.1d). Each cycle of imitation is a recursion of this process, which might eventually lead to a new equilibrium as imitation biases accumulate. Changes in song structure due to the process of imitation alone can be thought of as a null model for testing for a Darwinian cultural evolution (Prum, 2010), because the outcome of the imitation of ISO songs will necessarily reflect the preferences and innate biases of the pupils.

Tracking the Evolution of Song Culture

We raised zebra finch males in complete social and acoustic isolation and recorded their crystallized songs when they reached adulthood. Then we introduced juvenile males into their cages, and let them train the young males one to one. We recorded the songs of the pupils and when they reached adulthood at around 120 days, we separated them from their tutors. After making sure that their songs were stable, we introduced new juvenile pupils into the cages of the former pupils, who now acted as the tutors for the next generation of song learners. We continued this recursive training for about six generations, training multiple pupils with the same tutors (Figure 7.2a).

The island colony was housed in a large acoustically isolated sound box with three chambers connected by windows that allowed free movement between them. Females were not attracted to the isolate males, and it took us several trials before one isolate male was able to pair up with one of them and they laid a clutch, and, as much as we can tell by visual observations, this was the only clutch that male ever fathered. His offspring then paired up with the remaining females and the island colony was born, with no external song-culture input, except perhaps in the perception of the female founders.

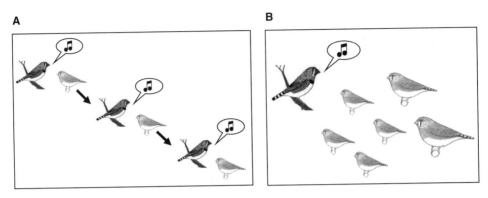

Figure 7.2
Schematic diagram of tutoring paradigm. (a) Recursive one-to-one training paradigm. The pupils (small gray birds), on reaching adulthood, become another juvenile's tutor (black breast patch and light cheek patch). The tutoring continues recursively over several learning generations. (b) In the colony, juvenile males learn to sing from adult males in the presence of their siblings (other small birds) and adult females (larger gray birds).

The two settings differed both in the strength of cultural isolation and in the extent of social deprivation: in the one-to-one setting, cultural isolation was complete with no females or siblings present, and the only social context was the male tutor and his one (genetically unrelated) pupil. In the colony, animals were genetically related, and females, who do not sing but had been exposed to WT songs, could interact with the pupils during song learning, and were making mating choices that, as mentioned before, can indirectly affect song imitation because zebra finches tend to imitate more syllables from their fathers (see, however, Williams, 1990). In a rich acoustic environment, the juvenile birds tend to learn from multiple tutors, so even by verifying paternity we could not exclude other males as potential tutors.

Our qualitative assessment and analysis of song structure over generations indicated that normalization of song occurred in both the one-to-one tutoring situation (Figure 7.3a,b) and in the colony (Figure 7.3c), indicating that isolate song can evolve toward WT song in a few generations *even in an impoverished social environment*. Moreover, the song features evolved toward a WT distribution quickly, in only three or four generations.

While the approximation of WT song was observed in both social settings, we noticed some interesting differences. One of the pronounced differences was that in the one-to-one training experiment, the song motifs and introductory notes (and the notes derived from them) were surrounded by long silence intervals (Figure 7.4a). In the colony, however, the song bouts were tight, the silence intervals short,

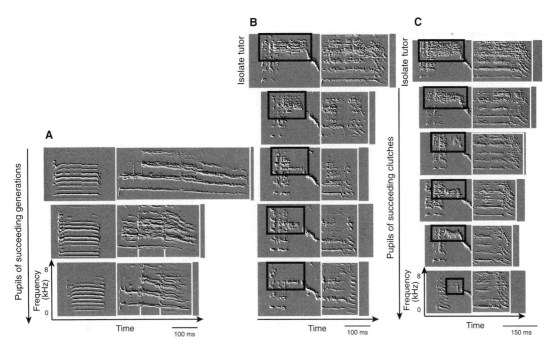

Figure 7.3

Isolate song features evolve toward WT in just three to five learning generations. (a) Abnormally long ISO syllables (white squares) are shortened by the pupil and even further shortened by the pupil's pupil in a recursive one-to-one training experiment. (b) and (c) The same ISO tutor was used in a one-to-one training experiment (b) and as a founder of an ISO colony (c), but the changes made to his song by the succeeding learning generations are very similar. The broadband white-noise-like note (black squares) was reduced in frequency band and duration, and the long harmonic syllable (white squares) was shortened and differentiated into two acoustically distinct notes by pupils in both social environments.

and there were more numerous and diverse song notes with many novel harmonic syllables (improvised or adopted from female calls, Figure 7.4b).

Song syntax can be very unusual in ISO songs. Some isolates sing a large number of syllable repetitions, but WT birds hardly ever do. In the case of ISO songs that contained such stuttered syllables, we observed an immediate and extreme reduction of syllable repetitions in the imitations of the pupils (Figure 7.5). The ISO song in Figure 7.5 was very unusual both because of the spectral features of the dominant syllable (B) and because it was repeated consecutively (15 times on average) by the ISO tutor (top panel). Two other syllable types of similar duration (C and D), not repeated, were nested among these repetitions. Interestingly, the first-generation pupil (second panel from top) copied all the syllables of his isolate tutor's song, but

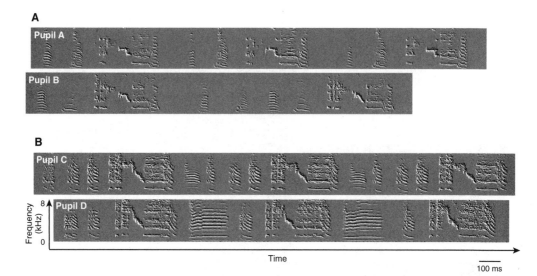

Figure 7.4
Song bouts are less diverse and more spread out in a deprived social setting (a) than in the island colony (b). (a) Song bouts of birds in the one-to-one training experiment. Song bouts of a bird representing learning generation 3 (top panel) and his pupil, representing learning generation 4 (bottom panel), are shown. (b) Song bouts of two birds in the island colony, siblings from the same clutch representing learning generation 3.

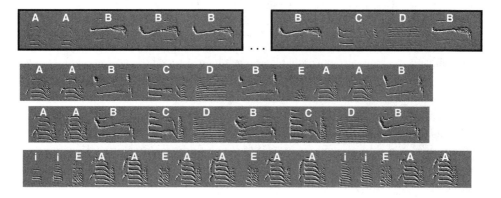

Figure 7.5
Song syntax evolution. The ISO song contains many repetitions of syllable B (on average 15, top panel), but its imitation in the first-generation learner does not have any (second panel). The second-generation learner (third panel) further modifies the syntax by adding a repetition of the new syntactically complex song. The next-generation learner selects the most WT-like syllable and does not imitate the rest (bottom panel).

strongly altered the syntax to avoid the repetitions of syllable B. He constructed his motif out of the three long syllable types, singing them serially once and ending with the first one. Another syntactic reorganization happened in the second generation (third panel), when the song went from one rendition introduced by two renditions of syllable A (AABCDB) to AABCDBCDB. This motif repetition and bout lengthening is reminiscent of the syntax changes that take place during development in an individual juvenile bird (personal observations). The generation 3 pupil completely omitted the long syllables and sang a simple song whose spectral features were WT-like. This was one of the very few instances of selective imitation, where the pupil only imitated the more WT-like syllables and did not imitate the more abnormal syllables at all.

As mentioned before, we examined song imitation at three timescales: moment-to-moment song features, durations of notes, and song rhythm. To be able to compare across many birds and levels, we computed cumulative histograms of features at these three levels: song-phonology features (e.g., pitch), durations of notes, and rhythm spectrum (an estimate of periodicity in the overall structure of the song bout). We used principal component analysis (PCA) to reduce the dimensionality of the feature distribution data and retained the first two principal components, PC1 and PC2. The song structure of each bird is then reduced to a single dot, as shown in Figure 7.6, and the gray cloud represents the distribution boundaries of WT song features across many birds. To contrast how the evolutionary dynamics played out differently in different social contexts, we plotted vectors showing the multigenerational progression to the WT distribution: each vector represents the transition from the tutor's feature (the tail of the vector) to that of his pupil (the head of the vector). The head-to-tail strings of vectors show trajectories over a few generations, starting from an isolate tutor and across generations of pupils (Figure 7.6a,b).

We found that spectral features changed most and fastest in the impoverished social setting (Figure 7.6a). Rhythm evolution was very inconsistent and the song rhythms of successive learners, though somewhat closer to the WT rhythm, did not improve very much (Figure 7.6b). In the colony, the opposite picture materialized. Spectral features, though closer to WT in later generations than in the ISO founder's song, hover at the edge of the WT distribution cloud (Figure 7.6c). However, song rhythm gets closer and closer in every clutch and by the end it is in the middle of the WT distribution (Figure 7.6d). This was confirmed by ear as well, because listening to the songs of the birds in the colony, we could easily tell that by the third clutch, the males were singing songs that sounded exactly like WT songs. A probable reason for this difference in the two social settings is the previously described tightening of song bouts (Figure 7.4b), which are much more characteristic of WT zebra finch songs than the long silence intervals that characterized the songs of the pupils in the one-to-one experiment (Figure 7.4a).

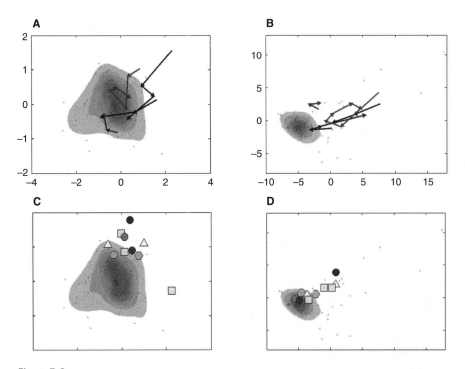

Figure 7.6
The evolution of song features (a, c) and rhythm (b, d) in the one-to-one experiment (a, b) and the island colony (c, d) across a few generations. The gray blobs represent the center of the WT distribution of song features. Arrows point from tutors toward pupils in the top panels. In the bottom panels, increasingly circular and darkening symbols represent successive clutches in the island colony. The dark circles far from the WT cloud represent the ISO founder. In both conditions, successive generations get closer and deeper into the WT distribution, but in the one-to-one training, the results for spectral features (a) are much more conclusive, whereas in the colony song rhythm (d) shows a much stronger effect.

Song Development in Pupils of Isolates

Pupils of ISO tutors clearly imitate most and often all of their ISO tutors' song syllables, but change them in a nonrandom fashion. These changes accumulate over learning generations and produce WT-like songs within as little as three generations. Biases in song imitation that produce such rapid progression toward WT songs can emerge at different stages of song development: they might be present as early as we can identify the onset of song imitation, and perhaps even prior to that, during subsong. The latter would mean that at the beginning of development, the song production is already biased toward the WT-like song. Gradually emerging biases, on the other hand, would suggest a stronger role of developmental changes (e.g., in hormonal levels and perception) in the normalization process. In this case, young birds would initially show fewer biases in their imitation, but as they grow, the biases would kick in and guide the emerging song toward WT-like song features. A similar scenario has been uncovered in the canary by Gardner, Naef, & Nottebohm (2005), who trained young birds with abnormally organized synthetic song and found that the young birds achieved a very good imitation in early development but deviated from it toward a species-typical program in late development by getting rid of some notes and reorganizing the rest.

Having examined some of the developmental trajectories of our birds, we suspect that at least some imitation biases appear very early—as soon as imitation becomes apparent. At the onset of subsong, individual vocal sounds are highly unstructured, but the distribution of their features might still have some structure. If we consider duration, for example, at the beginning of development the birds often vary syllable duration, singing short and long versions of the same syllable with little stability. After a few days, they lock on to a specific length, which they later may gradually modify. The exploration may include typical ISO durations (which tend to be longer), but it quickly shifts in the direction of more regular, WT-like durations. Figure 7.7 shows the developmental trajectory of one of our birds trained by an ISO tutor. Each dot represents the duration and frequency modulation of one song syllable. The scatterplots include all of the tutor as well as the pupil syllables (since they were housed together during the development of the pupil when we continuously recorded their vocalizations), but the cluster for the main motif (tutor's surrounded by dashed line, pupil's by dotted line in top-right panel) can be fairly easily distinguished. The tutor's cluster does not change over time, because he is singing a crystallized song, but we can track the pupil's syllable cluster over the whole development. The motif's FM does not seem to change too much, but its duration does. In early development, there is a diffuse cluster that splits into two at some point (second panel from bottom) and then stabilizes before the motif duration undergoes a gradual lengthening until quite late in development. The vocal exploration in early development, though containing long versions of the motif, still does not attempt a faithful copy of the tutor's motif duration.

The developmental trajectory shown in Figure 7.7 differs from Gardner et al.'s findings, because here the young bird appears to have been starting out imitating a much shorter version of the tutor song, and only later increases the duration to approximate the tutor song (up to a point). In addition, we noticed that pupils, when imitating ISO song that contains stuttered syllables, never at any point sing multiple repetitions like their tutors. Rather, from the beginning, they sing WT-like syntax.

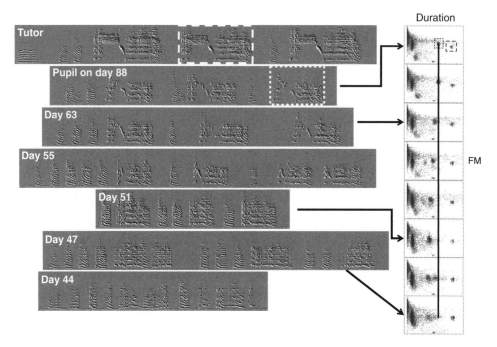

Figure 7.7
Developmental trajectory of a pupil of an ISO tutor. The tutor song is shown in the top
sonogram. The sonograms under the tutor's are the pupil's, with the latest copy on top and
the earliest at the bottom. On the right, syllable FMs are plotted against their durations, again
with the earliest plots at the bottom and the latest on top. A dashed line surrounds the tutor's
motif and a dotted line the pupil's. A line extends vertically from the center of the pupil's
motif cluster to show the gradual lengthening of his motif over song development.

Conclusions

Because rapid cultural evolution from ISO song features toward WT features
appeared in both social settings, it seems that sexual selection is not necessary for
the song culture to evolve toward WT-like song, although both selective imitation
and sexual selection are likely to play a significant role in the natural establishment of song culture. Our findings could provide some clues to how they might
do this. Although song rhythm, which no doubt could be a salient ecological
feature, did not progress toward WT in the absence of females and siblings, other
acoustic features, such as note duration and frequency modulation, evolved just as
much or more in the one-to-one training situation. Therefore, at least to some
degree, juveniles are born with implicit imitation biases that direct them toward
the WT zebra finch song. We call these imitation biases and not production biases
because for the WT phenotype to emerge, the birds must be exposed to songs

that they can imitate. Moreover, it takes a few recursions before the WT pheno-type fully emerges, because the imitation is always some compromise between the model the bird is copying and its internal tendency to shape the song in a certain way. The recursions are amplifying the expression of this internal tendency over generations.

Selective imitation supposes that some syllables are preferentially imitated. However, we observed very little of such selective copying. In fact, almost all ISO syllables were imitated regardless of their degree of "WT-ness." Pupils imitated nearly everything but dynamically transformed the syllable features into more WT-like features. This process presumes the employment of biased imitation in driving the rapid multigenerational cultural evolution. The nature of these biases can be perceptual (reminiscent of the Weber-Fechner law), such that in some feature ranges (for example, when durations are too long), perception is inaccurate and consequently, the auditory templates the juvenile males might form in early devel-opment are already biased toward WT features. Alternatively, ISO songs may be perceived accurately but biases may be employed during sensorimotor conversion. It is also possible that there are limitations on the matching performance to ISO song when it is imitated. Some sounds might be easy to improvise but difficult to copy (for example, singing a long note with lots of pitch fluctuations may not be difficult for a person to produce but it may be extremely hard for another to copy). In addition, subsong—the starting point of developmental learning—may impose biases on the imitation process, and top-down processes could be responsible as well. We hope future studies focusing on the development of song perception in male and female birds could answer some of these questions.

References

Atran, S. (2001). The trouble with memes. *Human Nature (Hawthorne, NY)*, *12*, 351–381.

Aunger, R. (2002). *The electric meme*. New York: Free Press.

Baker, M. C., Baker, E. M., & Baker, M. S. A. (2001). Island and island-like effects on vocal repertoire of singing honeyeaters. *Animal Behaviour*, *62*, 767–774.

Baker, M. C., & Cunningham, M. A. (1985). The biology of bird-song dialects. *Behavioral and Brain Sciences*, *8*, 85–133.

Baptista, L. F. (1977). Geographical variation in song and dialects of the Puget Sound white-crowned sparrow. *Condor*, *79*, 356–370.

Belzner, S., Voigt, C., Catchpole, C. K., & Leitner, S. (2009). Song learning in domesticated canaries in a restricted acoustic environment. *Proceedings. Biological Sciences*, *276*, 2881–2886.

Blackmore, S. (1999). *The meme machine*. Oxford: Oxford University Press.

Bohner, J. (1983). Song learning in the zebra finch (*Taeniopygia guttata*): Selectivity in the choice of a tutor and accuracy of song copies. *Animal Behaviour*, *31*, 231–237.

Bolhuis, J. J., Okanoya, K., & Scharff, C. (2010). Twitter evolution: Converging mechanisms in birdsong and human speech. *Nature Reviews. Neuroscience, 11*, 747–759.

Boyd, R., & Richerson, P. J. (1985). *Culture and the evolutionary process*. Chicago: University of Chicago Press.

Boyd, R., & Richerson, P. J. (2005). *The origin and evolution of cultures*. Oxford: Oxford University Press.

Boyer, P. (1994). *The naturalness of religious ideas*. Berkeley: University of California Press.

Chilton, G., & Lein, M. R. (1996). Song repertoires of Puget Sound white-crowned sparrows *Zonotrichia leucophrys pugetensis*. *Journal of Avian Biology, 27*, 31–40.

Darwin, C. (1859). *On the origin of species*. London: John Murray.

Dawkins, R. (1976). *The selfish gene*. Oxford: Oxford University Press.

Fehér, O., Wang, H., Saar, S., Mitra, P. P., & Tchernichovski, O. (2009). De novo establishment of wild-type song culture in the zebra finch. *Nature, 459*, 564–568.

Gardner, T. J., Naef, F., & Nottebohm, F. (2005). Freedom and rules: The acquisition and reprogramming of a bird's learned song. *Science, 308*, 1046–1049.

Henrich, J., Boyd, R., & Richerson, P. J. (2008). Five misunderstandings about cultural evolution. *Human Nature (Hawthorne, NY), 19*, 119–137.

Mann, N. I., & Slater, P. J. B. (1994). What causes young male zebra finches, Taeniopygia guttata, to choose their father as song tutor? *Animal Behaviour, 47*, 671–677.

Marler, P., & Nelson, D. A. (1992). Action-based learning: A new form of developmental plasticity in bird song. *Netherlands Journal of Zoology, 43*, 91–103.

Marler, P., & Tamura, M. (1962). Song "dialects" in three populations of white-crowned sparrows. *Condor, 64*, 368–377.

Marler, P., & Tamura, M. (1964). Culturally transmitted patterns of vocal behavior in sparrows. *Science, 146*, 1483–1486.

Mesoudi, A., Whiten, A., & Laland, K. N. (2004). Is human cultural evolution Darwinian? Evidence reviewed from the perspective of *The origin of species*. *Evolution; International Journal of Organic Evolution, 58*, 1–11.

Payne, R. B. (1981). Song learning and social interaction in indigo buntings. *Animal Behaviour, 29*, 688–697.

Price, P. H. (1979). Developmental determinants of structure in zebra finch song. *Journal of Comparative and Physiological Psychology, 93*, 260–277.

Prum, R. O. (2010). The Lande-Kirpatrick mechanism is the null model of evolution by intersexual selection: Implications for meaning, honesty, and design in intersexual signals. *Evolution; International Journal of Organic Evolution, 64*, 3085–3100.

Williams, H. (1990). Models for song learning in the zebra finch: Fathers or others? *Animal Behaviour, 39*, 745–757.

Williams, H., Kilander, K., & Sotanski, M. L. (1993). Untutored song, reproductive success and song learning. *Animal Behaviour, 45*, 695–705.

Zann, R. (1996). *The zebra finch: A synthesis of field and laboratory studies*. Oxford: Oxford University Press.

8 Acquisition of Linguistic Categories: Cross-Domain Convergences

Frank Wijnen

Categorization is an essential process in language acquisition. To be able to construct and analyze words and sentences, the child has to build a system of phonemes (i.e., abstract representations of speech sounds), as well as a system of grammatical categories. Categorization in language acquisition hinges on induction. Even if we assume that notions such as "phoneme," "syntactic category," or even "noun" and "verb," are somehow hardwired, the language-learning child needs to assign units segmented from the input speech stream to the correct classes. A rapidly growing number of studies underscore the crucial contribution to this process of distributional analysis of spoken language, both for phonology and syntax. However, studies in the two domains—phonology and grammar—place different accents. Phoneme acquisition is associated with a process of "counting" occurrences of phonetic units, whereas grammatical categorization would appear to depend primarily on the detection of sequential co-occurrence relations.

In this chapter, I argue that the properties of distributional learning in syntax and phonology are basically the same (i.e., they are specific implementations of a generic computational scheme). As a corollary, I put forward the conjecture that basic acquisitional processes in the two domains are implemented by a single neurocognitive mechanism. In addition, I provide some arguments for the speculative hypothesis that category induction in language is a particular instantiation of a general system subserving implicit statistical learning in various domains. By way of evaluating these proposals, I briefly explore two predictions, namely that category acquisition at the level of phonology is not dissociable from category formation in syntax, and that delays or deficits in language acquisition are associated with implicit learning deficiencies, specifically in the motor domain.

Categories in Language

The ability to produce and understand verbal utterances is based on stored knowledge consisting of two components: basic linguistic "building blocks," and computational procedures that control the assembly of structured strings from the basic elements (Pinker, 2000). The building blocks at different levels of linguistic structure are assigned to categories, and linguistic rules refer to these categories, rather than to their variable instantiations. At the sound level (phonology), the building blocks are *phonemes:* some 40 distinct units from which syllables and words are constructed. Phonemes represent classes of speech sounds — "phonetic units" (Kuhl, 2004) — that are recognized as the same despite considerable physical variation due to speaker characteristics, speaking style, and assimilation to neighboring units. At the level of sentence structure (syntax), the building blocks are words, which are assigned to discrete grammatical categories such as "noun," "verb," or "determiner."

There are fundamental differences between phonemes and grammatical categories. For one, phonemes can in principle be described in terms of (idealized) physical features, whereas grammatical categories cannot. There are some important similarities as well. Phonemes as well as grammatical categories represent infinite sets of linguistic units that are functionally equivalent. Just as the phonemic identity of a phonetic unit, rather than its physical realization, determines where it can and cannot occur in a syllable (as specified by phonotactic rules), the grammatical identity of a word determines where it can and cannot occur in a phrase or sentence (as specified by syntactic rules). In this sense, both category types are strictly cognitive entities.

The Acquisition of Linguistic Categories

Phonemes

Evidence collected over the past 20 years indicates that children begin acquiring the phonology of their native language long before the onset of word production (Gerken, 2002; Kuhl, 2004). The crucial evidence concerns infants' discrimination of native and nonnative speech-sound contrasts. Before the age of roughly six months, children respond to differences between phonetic units that instantiate different phonemes in the native language just as well as to physically different units that instantiate one phoneme in the native language, but that may span a phoneme boundary in some other language. After the 10th month, children begin to ignore differences that are nonphonemic in the native language. Thus, a 6-month-old infant growing up in a Japanese-speaking environment discriminates the phonetic units [r] and [l] just as well as a child exposed to English does, even though the two units are functionally equivalent in Japanese. After 10 months, however, the Japanese child will ignore the [r]–[l] difference. By contrast, the child growing up in an English-speaking environment retains the ability to discriminate [r] and [l]. A study by Kuhl et al. (2006)

shows that American children even respond more strongly to this contrast at 11 months than they do at 6 months. Such developmental patterns are believed to reflect the emergence of abstract phoneme categories (Best, McRoberts, & Sithole, 1988; Gerken, 2002; Kuhl, 2004; Werker & Pegg, 1992; Werker & Tees, 1984).

A rapidly growing body of evidence strongly suggests that language-specific phonological categorization results from analyzing the distributional properties of perceived sequences of speech sounds (phonetic units). As pointed out above, the realizations of a phoneme vary greatly, but these realizations nonetheless cluster around a "point of gravity" defined by the relevant acoustic/articulatory dimensions (see Kuhl, 2004, and references cited therein). Children can apparently make use of a system that detects the distribution of physically varying phonetic units in a relevant (multidimensional) acoustic space. This allows them to recognize the "points of gravity" and, hence, the (boundaries between) phonemes of their native language (Gerken, 2002; Kuhl, Ramus, & Squire, 1992). Direct evidence for the existence of such a statistical system is provided by experiments in which speech-sound discrimination is shown to be affected by frequency distributions of (minimally different) phonetic units taken from a continuum spanning a phonemic contrast, such as (voiced) [b] to (voiceless) [p]. If, in a repetitive sample played during a few minutes, the frequency distribution is bimodal, with the two modi close to the two endpoints of the continuum, infants will after exposure (continue to) discriminate the contrast. If the distribution is unimodal, however, with the modus corresponding to a point in the middle of the continuum, the result of the exposure is that children (begin to) ignore the acoustic differences between units corresponding to the endpoints of the continuum (Maye, Werker, & Gerken, 2002; Maye, Weiss, & Aslin, 2008; Capel, De Bree, Kerkhoff, & Wijnen, 2008, 2011).

In addition to registering the relative frequencies of phonetic units, children also detect statistical patterns in sequences of such units, which helps them learn what types of phoneme sequences are allowed, as well as detect the boundaries between words (Bonte, 2005; Chambers, Onishi & Fisher, 2003; Jusczyk, Friederici, Wessels, Svenkerud, & Jusczyk, 1993; Jusczyk, Luce, & Charles-Luce, 1994; Saffran et al., 1996). There is growing consensus that registering and encoding transitional probabilities is at the basis of these learning processes.

Grammatical Categories
Acquiring grammatical categories entails assigning class labels to individual words that determine their combinatorial properties in phrases and sentences. A classic hypothesis, the *semantic bootstrapping hypothesis* (Pinker, 1984), holds that during the initial phase of this process, children make use of universal (and therefore possibly innate), unidirectional meaning-to-grammar associations. For instance, a linking

rule stating that words referring to objects belong to the category "noun" assists the child in setting up a noun category. At a later stage, when the child learns nonobject labels such as *love* or *peace*, other principles must come into play to guide their classification. It is assumed that the child is capable of detecting that a word like *peace* shares sequential co-occurrence patterns with prototypical nouns such as *box*, and will consequently classify *peace* as a noun as well.

The *distributional bootstrapping* hypothesis (Braine, 1987; Maratsos & Chalkley, 1980) holds that distributional regularities alone are sufficient for the initial categorization of some (types of) words. Words are assigned to categories on the basis of overlapping lexical co-occurrence patterns. Computational and simulation studies have demonstrated that major grammatical categories (e.g., noun, verb) can be induced from the distributional regularities in word sequences in child-addressed language (Mintz, 2002), and that the algorithms required are relatively simple (Redington, Chater, & Finch, 1998). Experiments in artificial and natural language learning show that learners are sensitive to sequential co-occurrence patterns, and use them to induce grammatical categories as well as combinatorial rules (Gerken, Wilson, & Lewis, 2005; Gómez & Gerken, 1999; Marcus, Vijayan, Bandi Rao, & Vishton, 1999; Saffran, 2002).

Commonalities in the Acquisition of Phonemes and Grammatical Categories

The overview above suggests that while statistical learning plays a role in the acquisition of phonemes as well as grammatical categories, different types of computations appear to be at stake. Phoneme acquisition would seem to rest on establishing the frequencies of occurrence of phonetic units (tokens), while the induction of grammatical categories depends on establishing sequential co-occurrence patterns. However, a moment's thought reveals that "counting units" as well as registering sequential patterns must come into play in both types of categorization.

Grammatical categories are relational entities, and detecting which words belong together within one class hinges on registering co-occurrence relations with other words.[1] In an extensive computational study, Redington et al. (1998) demonstrate that even though better-than-chance grammatical categorization is attainable by an algorithm that registers mere lexical co-occurrences, the performance of the system improves markedly when the frequencies of such co-occurrences are taken into account. In other words, counting the occurrences of individual words in particular distributional contexts would seem to be necessary to *efficiently* categorize them. In addition, behavioral evidence shows that increasing the number of different lexical items X facilitates detecting a dependency between lexical items A and B in A-X-B sequences (Gómez, 2002). Conversely, the frequency of occurrence of a lexical frame A_B is a relevant factor in assigning various X tokens to a single class (Mintz, 2003). In order to see that A_B is a frame, it is necessary to register the high probability of A and B occurring together (or the conditional probability linking them), which presupposes keeping track of the frequencies of occurrence of both individual items A and B, as well as bigram statistics. In summary, counting frequencies of occurrence of word tokens is indispensible in the calculation of sequential regularities that can be used to induce grammatical categories.

Conversely, and perhaps less obviously, detecting sequential structure must be an integral part of phonological formation. One case in point concerns the interplay between acquiring phonemes and detecting allophonic variation (see Gerken, 2002). In numerous languages, the default realization of a phoneme may be altered in certain, distributionally defined contexts. For example, English voiceless stop consonants such as /p/, /t/, and /k/ are aspirated when they occur in initial position in a stressed syllable. The default (nonaspirated) variant occurs in other positions. Given this complementary distribution, it is to be expected that infants will assign the aspirated and nonaspirated variants to separate categories. However, mature speakers of English recognize aspirated and nonaspirated variants as belonging to the same category. This implies that the (young) learner at some point collapses the two, and recognizes the aspiration alternation as a regular phonological process. Peperkamp, Le Calvez, Nadal, & Dupoux (2006) demonstrate that this "collapsing" of allophones can be based on a statistical algorithm that detects complementarity in distributional patterns. To ensure that spurious allophone pairs are avoided (i.e., pairs that have complementary distribution but are insufficiently similar in terms of their phonetic makeup), the algorithm needs to be supplemented by a constraint defining articulatory or acoustical similarity. This type of similarity can in principle be gleaned from the distance between the relevant segments (putative allophones) in a multidimensional acoustic (or articulatory) space as described above.

More fundamentally, it should be noted that the frequency-analysis hypothesis for phoneme acquisition as outlined above presupposes that speech is perceived as a string of phonetic units, while in reality discovering what constitutes a unit is one of the problems the learner must solve. Phonetic units are the acoustic result of the interleaved movements of relatively independent articulatory systems. One of the properties of speech is that in any sequence A–X–B, the realization of X is affected by A and/or B. Therefore, to grasp the systematicity of the co-occurrence of articulatory events correlated with X or their audible consequences, one has to take the effects of A and B into account. In other words, recognizing the properties that define X as a phonetic unit depends on encoding the properties of X's neighbors, and this rests on sequential pattern recognition.

What is proposed here, then, is that grammatical and phonological category formation rest on the same statistical learning mechanisms, and can in fact be seen as instantiations of the same computational processes. Without pretending to be exhaustive in this respect (see Peperkamp et al., 2006), it can be said that the core properties of the learning system include the following:

1. A capacity to keep track of *quantities* of elements (tokens such as phonetic units or words) defined by a relevant set of dimensions—that is, keeping track of their (numerical) distributions.
2. A capacity to detect dependencies among such elements when they occur in series. This implies that the system is capable of representing *serial order* (i.e., transitions from element X to element X+i, as well as the probabilities of occurrence of such transitions).

Notice that there is a fuzzy boundary between characteristics 1 and 2, because dimensions that define a representational space used for tracking units' frequencies of occurrence may themselves be distributional in nature. This is the case for instance in frequent frames and other string-dependent co-occurrences that are critical for grammatical categorization (Redington et al., 1998; Mintz, 2002, 2003).

The Neurocognition of Sequential Learning

A pertinent question is how and where the computational system outlined above is implemented in the central nervous system. There are several clues to guide our search. First, because the capacity to detect and represent serial order is a distinctive property of this neural substrate, it would need to comprise *recurrent* (reciprocal) circuitry, allowing for a state i to be associated with a preceding state $i-1$ (or states $i, i-1, \ldots, i-x$). That recurrent connections are necessary for the detection and representation of serial structure has been demonstrated by artificial neural network studies (Cleeremans, Destrebecqz, & Boyer, 1998; Clegg, DiGirolamo, & Keele, 1998; Dominey, 2005; Elman, 1990). A second clue is that natural primary language acquisition is quite clearly implicit learning. Moreover, artificial language-learning experiments confirm that people past the critical period for language acquisition pick up on sequential regularities without conscious awareness (Saffran, Newport, Aslin, Tunick, & Barrueco, 1997). Consequently, it makes sense to link language acquisition to neural structures that are implicated in implicit learning and memory.

There is converging evidence from various sources suggesting that a network comprising parts of the inferior frontal cerebral cortex, the basal ganglia, and the cerebellum fulfills these two criteria (DeLong, 2000; Dominey, 2005; Doyon, Penhune, & Ungerleider, 2003; Ghez & Thach, 2000; Ullman 2004). The structural connections between frontal lobe regions and basal ganglia, as well as between frontal lobe and the cerebellum, are recurrent. Both the basal ganglia and the cerebellum receive projections from cortical areas to which they in turn project back (via the thalamus). Different parts of the basal ganglia and the cerebellum interface with different cortical structures, forming segregated loops or circuits. The different corticobasal ganglia circuits have a parallel synaptic organization, and the same holds for the corticocerebellar circuits. It is a plausible conjecture, therefore, that these circuits perform analogous computations on different types of cortical information, corresponding to different perceptual or cognitive domains. The following overview will show that the inferior frontal lobe—basal ganglia—cerebellum (IFLBC) network is involved in learning and reproducing sequences of sensory or motor events. Interestingly, there are also various indications that the basal ganglia and the cerebellum play a role in language processing. Conversely, the areas of the inferior frontal lobe associated with language processing at various levels of structure (see Hagoort, 2005) appear to be implicated in nonlinguistic sequence learning.

Neuropsychological studies indicate that densely amnesic patients (i.e., individuals whose explicit memory system shows significant deterioration) are still capable of learning in (nonlinguistic) serial reaction time tasks (which test the implicit acquisition of serial order; see, e.g., Squire, Knowlton, & Musen, 1993). They also perform relatively well in artificial language learning, which involves the induction of structural patterns from strings of nonexistent words (Knowlton, Ramus, & Squire, 1992). This is suggestive of a connection between nonlinguistic sequence learning and language learning, as well as their shared independence from the explicit memory system. Functional decline of the basal ganglia, as witnessed in neurodegenerative diseases such as Parkinson's and Huntington's, is typically associated with disorders of motor programming and execution. In addition to this, patients perform poorly on serial reaction time tasks (Clegg, DiGirolamo, & Keele, 1998; Smith & McDowall, 2004), and (mild) grammatical disruptions have been observed in spontaneous and elicited language production (Ullman, 2004; Ullman, Corkin, Coppola, Hickok, & Korosnetz 1997). Notably, Teichmann and coauthors (Teichmann, Dupoux, Kouider, & Bachoud-Lévi, 2006; Teichmann, et al., 2008) showed that patients with early-stage Huntington's disease have difficulties computing regular and subregular verbal inflections and (noncanonical) syntactic structure. Isolated acquired damage to the basal ganglia has been linked to deficits in motor and cognitive sequencing, as well as to linguistic comprehension difficulties (Lieberman, 2000). These neuropsychological observations are corroborated and extended by neuroimaging studies. The encoding of motor sequences as well as the retrieval of learned sequences of movements are associated with increased activity in the basal ganglia, specifically the striatum. The striatum as well as the cerebellum also become activated by tasks that involve implicit serial-order learning (Doyon, Penhune, & Ungerleider, 2003; Goschke, Friederici, Kotz, & van Kampen, 2001), and some studies document activation of these areas in (complex) sentence processing (Stowe, Paans, Wijers, & Zwarts, 2004).

The cerebellum has long been known to be involved in the acquisition and execution of motor skills (De Smet, Baillieux, De Deyn, Mariën, & Paquier, 2007; Ghez & Thach, 2000). Sequence learning appears to be impaired in individuals with cerebellar dysfunctions. Interestingly, dyslexia, a condition widely assumed to be based in fundamental language processes, is associated with deficits in motor learning (de Kleine & Verweij, 2009; Nicolson, Fawcett, & Dean, 2001). Nicolson, Fawcett, Berry, Jenkins, & Dean (1999) observed reduced activation of the cerebellum in dyslexic adults (as compared to nondyslexic controls) in a motor sequence-learning task. However, MR images obtained by Menghini, Hagberg, Caltagirone, Petrosini, & Vicari (2006) in a serial reaction time task, suggested sustained high cerebellar activation in dyslexic participants, which the authors associate with these individuals' difficulty in learning the task, as indicated by the behavioral results. Cerebellar lesions have been observed to occasionally result in grammatical disorders. Imaging studies have demonstrated increased activation of the right cerebellar hemisphere in (complex) sentence processing in healthy participants (De Smet et al., 2007).

Patients with Broca's aphasia, which typically results from damage to the left hemisphere inferior frontal cortex and surrounding gray as well as white matter, and is characterized by grammatical processing problems, have been reported to show difficulties in learning nonlinguistic sequences (Dominey, Hoen, Lelekov, & Blanc, 2003; but see Goschke et al., 2001). Hoen, Pachot-Clouard, Segebarth, & Dominey (2006) report that both sentence processing and abstract (nonlinguistic) sequence processing (well-formedness judgment task) activate Brodmann's area 44 (BA) (*pars opercularis,* part of Broca's area), as well as adjacent BA's 6 (roughly: premotor cortex and supplementary motor area) and 46 (roughly: dorsolateral prefrontal cortex). Moreover, Dominey et al. (2003) report that training of nonlinguistic sequential structures improved comprehension of complex sentences with an analogous grammatical structure.

In summary, although the evidence is not unequivocal, the neuropsychological and functional neuroimaging studies with adults strongly suggest that the IFLBC network is involved in learning and representing sequential structure (serial order) in (motor) output as well as sensory input, and appears to be engaged in language processing and the implicit induction of structural patterns (grammar) from artificial languages. Given these observations, I propose that it is this network that implements the sequential-distributional learning processes necessary for the acquisition of phonological and grammatical categories in primary-language acquisition. To my knowledge, this hypothesis has not yet been addressed directly. One piece of indirect evidence comes from neural network simulations by Dominey (2005; Dominey & Ramus, 2000), which indicate that a recurrent network modeled on frontostriatal circuitry in the primate brain is capable of mimicking learning in serial reaction time tasks, and emulates the results of infant studies on serial and abstract structure in linguistic strings (Saffran, Aslin, & Newport, 1996; Marcus et al., 1999). In the next section, I discuss two predictions derived from the hypothesis put forth here: (1) deficits in language acquisition are associated with deficits in (motor) skill acquisition and nonlinguistic sequence learning; (2) there cannot be a dissociation between phonological category acquisition and grammatical category acquisition.

Language Acquisition and Nonlinguistic Implicit Learning

If, as hypothesized, categorization in language acquisition depends on the IFLBC network outlined above, we can expect that language acquisition in children is strongly correlated with ability in nonlinguistic skill acquisition, particularly the acquisition of motor (movement) skills, as well as learning serial-order patterns in sequences of (nonlinguistic) sensory events.

Reports on correlations between motor development and language acquisition in typically developing children are scarce. Siegel (1982) reports that scores on perceptuomotor subscales of the Bayley Scales of Infant Development at as early as 4 months are predictive of language test scores (Reynell Developmental Language Scales) at 2 years and older. A study by Alcock (2006) shows that at the age of 21 months, oral (nonverbal) motor skills, particularly pertaining to complex movements, are significantly correlated with vocabulary size, total function words used, and sentence complexity. Cheng, Chen, Tsai, Chen, and Cherng (2009) found significant correlations between motor development and language test scores in a large sample of 5- to 6-year-olds. Such associations are, however, open to multiple interpretations. In particular, general maturation (as indicated by age) can explain correlated development in different domains. In this connection it is interesting to note, however, that Cheng et al. still found motor–language correlations when IQ—which can be viewed as an index of general cognitive maturation—was partialed out.

Children diagnosed with Developmental Coordination Disorder (DCD), or "motor clumsiness," have been reported to score lower than typically developing peers on various measures of language and literacy skills (Kadesjo & Gillberg, 1999; Kaplan, Dewey, Crawford, & Wilson, 2001). Gillberg (2003) found that 50% of children with deficits in motor control as well as attention and perception had difficulties in speech and language. The motor discoordination disorder appears to be more predictive than the attention deficit of a language disorder (Rasmussen & Gillberg, 2000). In the same vein, Archibald and Alloway (2008) report that almost 50% of a sample of children diagnosed with DCD showed language profiles similar to those of an age-matched group of children with specific language impairment (SLI). On the whole, language difficulties and motor discoordination are more often comorbid than expected on the basis of prevalence figures of each of these dysfunctions separately.[2] Also, cerebral palsy (which goes under different names), a condition characterized by severe difficulty in motor praxis, has been associated with language deficits (Redmond & Johnston, 2001).

Specific language impairment (SLI) appears to be strongly associated with motor discoordination and slow acquisition as well as execution of motor movement patterns (e.g., Bishop, 2002; Owen & McKinlay, 2003; Powell & Bishop, 1992; Visscher, Houwen, Scherder, Moolenaar, & Hartman, 2007; Webster, Majnemer, Platt, & Shevell, 2005; Webster et al., 2006; for extensive reviews, see Hill, 2001; Ullman & Pierpont, 2005). Interestingly, balancing difficulties—typically seen as a cerebellar symptom—are mentioned as a prominent feature of SLI children's motor deficit. On the whole, the motor profile of children with language disorders overlaps with that of children diagnosed with DCD. There are indications that the severity of comprehension difficulties in language-delayed children is correlated with performance on fine motor tasks (Schwartz & Regan, 1996). Like children with SLI, many children with dyslexia display deficits in learning and executing movements and movement sequences (i.e., dyspraxia; Nicolson & Fawcett, 2007). Correlations between poor orofacial, nonverbal motor praxis and language (i.e., morphosyntactic deficiences) have also been found in the KE family (e.g., Watkins, Dronkers, & Vargha-Khadem, 2002). There are indications as well that children with language impairments (SLI, dyslexia) perform poorly on implicit (nonlinguistic) sequential learning tasks (Howard, Howard, Japikse, & Eden, 2006; Tomblin, Mainela-Arnold, & Zhang, 2007; Vicari, Marotta, Menghini, Molinari, & Petrosini, 2003; Vicari et al., 2005). Notably, Lum, Gelgec, & Conti-Ramsden (2009) report a dissociation between serial reaction time task performance (poor) and nonverbal paired associate learning (normal) in 7- to 8-year-old-children with SLI.

Anatomical studies indicate that developmental language disorders and dyslexia are associated with abnormalities in, among others, the inferior frontal cortex, basal ganglia (particularly caudate nucleus), and cerebellar structures (Brown et al., 2001; Webster & Shevell, 2004). A structural neuroimaging study with members of the KE family (Watkins et al., 2002) found gray matter reductions in caudate nucleus (part of the striatum), sensorimotor cortex, and cerebellum in the affected family members, as compared to unaffected family members and matched controls. In a structural MRI study, Eckert et al. (2003) found volume reductions of the right cerebellar anterior lobe and the pars triangularis in the inferior (cerebral) frontal lobe of dyslexic children, as compared to age-matched controls. These reductions were significantly correlated with scores on language and literacy tests. Conceivably, anatomical and functional abnormalities such as these compromise the functioning of the corticostriatal and corticocerebellar loops involved in the sequential-statistical analysis that is assumed to underlie phonological and grammatical categorization.

The (Non)dissociability of Grammatical and Phonological Categorization

If phonological and grammatical category formation share one neurocognitive mechanism, a prediction is that efficacy of category formation at one level (e.g., phonology) is predictive of the efficacy at another (e.g., grammar). Interindividual differences in speed of language acquisition (within the normal range) are well documented (Bates, Dale, & Thal, 1995). Such differences generally occur across the board; onset and rate of phonological and grammatical development are correlated. According to some, this is due to conditional relations between learning processes at different (hierarchically related) levels of linguistic structure. Thus, a delay in, for example, phonological acquisition would lead to a delay in lexical development (Newman, Bernstein Ratner, Jusczyk, Jusczyk, & Dow, 2006), which would in its turn delay grammar. To adjudicate between such "cascadic" accounts and the present proposal, one of the things that need to be done is to experimentally assess the categorization performance at different levels of structure within the same individuals. Such assessments should address the responsivity to identical distributional manipulations at the levels of phonology and grammar. Nonadjacent dependencies are an example in point, because they have been studied at the level of grammar (e.g., Gómez, 2002) as well as phonology (Newport & Aslin, 2004). The hypothesis proposed here predicts, for instance, that the learnability of A–B dependencies in A–X–B sequences should be similarly affected by variations in the set size of X, and that sensitivity to such variations should be parallel within individuals, for phonology and grammar. In the same vein, children's detecting of phonotactic allophony (such as the aspirated–nonaspirated contrast discussed above) should be predictive of their detecting complementary distributions in syntax (e.g., the complementary distribution of finite and nonfinite verb forms in languages such as Dutch and German).

Because interindividual differences in phonological and grammatical learning in typically developing children are likely to be subtle and therefore difficult to pinpoint statistically, it is useful to look at nontypical language acquisition. Specific language impairment is prototypically associated with grammatical (morphosyntactic) deficits, but in a majority of cases, phonological delays are observed as well, while lexical learning and interpretation (semantics) appear to be relatively spared (de Bree, 2007; Leonard, 1998; Rescorla & Bernstein Ratner, 1996; Roberts, Rescorla, Giroux, & Stevens, 1998). Also, children diagnosed with dyslexia at school age have typically shown delays in both phonological and grammatical development at a younger age (Bishop & Adams, 1990; Catts & Kamhi, 1999; Scarborough 1990, 1991, 2005; Snowling & Hayiou-Thomas, 2006). Children with a familial risk of dyslexia—of whom approximately 40–60% will develop manifest dyslexia—have a mild but broad language delay, affecting the perception and production of phonological as well as grammatical structure (de Bree, Wijnen, & Zonneveld, 2006; de Bree, Van

Alphen, Fikkert, & Wijnen, 2008; De Jong, Wijnen, & de Bree, 2012; Van Alphen et al., 2004; Wilsenach, 2006). The specific reading/spelling problems in children who are (or will be) manifestly dyslexic stem from difficulties in associating alphabetical symbols (graphemes) with phonemes, which is assumed to result from an incomplete or deficient phonological development, resulting in underspecified phoneme representations (Vellutino, Fletcher, Snowling, & Scanlon, 2004).

Research on speech-sound perception in children with (a familial risk of) dyslexia as well as in children with specific language impairment is quite abundant. Numerous studies show that identification and discrimination of phonemes in these children are deviant (e.g., Gerrits & de Bree, 2008; Joanisse, Manis, Keating, & Seidenberg, 2000; Serniclaes, Van Heghe, Mousty, Carré, & Sprenger-Charolles, 2004; Tallal, 1980, to name just a few). However, so far there are no data on nontypical children's sensitivity to distributional properties of phonetic input. The hypothesis proposed here predicts that these children are less sensitive to distributional information. Consequently, they will need more extensive exposure to speech in order for categorization (as well as rule extraction) to occur. Similarly, sensitivity to statistical sequential patterns in phoneme strings is expected to be reduced in language-delayed children. Evans, Saffran, & Robe-Torres's (2009) results confirm this. Children with SLI turned out to be less sensitive to transitional probabilities (TPs) as cues to word boundaries in strings of consonant-vowel (CV) syllables than typically developing children. This difference decreased with longer exposure to the CV strings. A similar difference between children with SLI and controls was found in a structurally identical task that employed musical tones instead of CV syllables. This similarity suggests that TP-based segmentation is subserved by a domain-general mechanism.

Also at the level of grammar (i.e., the detection of sequential patterns across and dependencies among words), progress is expected to be slower in language-impaired children than in controls. Pertinent data are fairly scarce. Wilsenach and Wijnen (2004) showed that 19-month-old children with a familial risk of dyslexia, in contrast to age-matched controls, have not yet detected the dependency between the Dutch auxiliary *heeft* and the past participle prefix *ge-*. This is suggestive of a lowered sensitivity to grammatical co-occurrence relations, which, as argued above, is a precondition for categorization (and grammatical rule learning). Congruently, results on artificial language learning from our lab indicate that 18-month-old at-risk infants are less sensitive to nonadjacent lexical dependencies than controls (Kerkhoff, de Bree, de Klerk, & Wijnen, in press). In line with these infant results, Pavlidou and colleagues (Pavlidou, Williams, & Kelly, 2009; Pavlidou, Kelly, & Williams, 2010) showed that 9- to 12-year-old dyslexic children perform worse than age-matched controls in an implicit artificial grammar learning task. Importantly, the strings from which the grammar was to be induced did not consist of words, but of geometric shapes. Thus, Pavlidou et al.'s results cannot be accommodated by a cascadic account, according to which grammatical difficulty or delay is due to difficulty processing or representing phonological information. Clearly, this work needs to be extended in order to fully test the predictions formulated here.

Conclusion

What is proposed here is that the emergence of linguistic categories, both at the level of phonology and grammar, is subserved by a learning mechanism sensitive to statistical patterns in linguistic strings. The neurological substrate for this mechanism is argued to be associated with the circuitry underlying procedural (implicit) learning, which comprises parts of the inferior frontal cerebral cortex, the basal ganglia (specifically the striatum), and the cerebellum, and their manifold reciprocal connections. Since this substrate is critically involved in learning and representing (nonlinguistic) sequential structure, particularly in the context of motor skills, it is predicted that language acquisition processes are strongly associated with performance in nonlinguistic sequential learning tasks and motor skill acquisition. Such associations have been reported in the (developmental) psychological, neuropsychological, and neuroimaging literature. However, the available literature does not allow incontrovertible conclusions. Developmental language delays, for example, are not 100% comorbid with motor deficiencies. It is not inconceivable, though, that more fine-grained analyses, using critical tasks and encompassing subclinical deficiencies, might change the picture. Particularly, to further corroborate the hypothesis put forth here, we need to demonstrate that language acquisition—notably the processes of phonological and grammatical category formation, nonlinguistic sequential learning, and motor skill acquisition—are associated within individuals. Also, within an individual, a deficit or delay in one of these three domains is expected to predict deficits in the other two, and such correlated deficits are expected to be associated with functional and/or anatomical abnormalities in the hypothesized neural substrate. Crucially, other types of learning, in which detecting statistical-sequential structure does not play a role—"declarative learning," for short—should be unaffected.

It should be noted that the proposal made in this chapter is not the first to associate language processes and language acquisition with the neural circuitry that is thought to comprise the procedural learning system (Lieberman, 2000; Ullman & Pierpont, 2005; Nicolson & Fawcett, 2007). Also, in each of the proposals cited, (developmental) language disorders (SLI, dyslexia) are seen as the result of some subtle malfunctioning of this procedural system (possibly due to genetically conditioned abnormalities in brain tissue architecture and/or connectivity). However, these models fall short of connecting the association between a dysfunctional procedural system and language deficiencies to a viable account of (normal) language acquisition. I believe that the proposal outlined in this chapter fills this gap, by arguing that the procedural learning system is crucially involved in the statistical-sequential analysis that underlies the construction of a system of linguistic categories (as well as rule acquisition, which is not discussed here). Thus, the hypothesis sketched opens up an avenue of research through which language acquisition during infancy, statistical learning, and the neurocognition of learning and language can fruitfully interact. This will contribute to a better understanding of what is, and what is not, unique about human language.

Notes

1. For reasons of brevity, I am ignoring here that grammatical category distinctions may be correlated with systematic phonological differences, so that a statistical learner could map out individual words in a multidimensional space defined by a set of phonological attributes (see Monaghan, Chater, & Christiansen, 2005).

2. However, the same can be said of comorbidities of other neurodevelopmental disorders, which is one of the reasons some researchers argue that the diagnostic distinctions, though clinically useful, may be etiologically obsolete (e.g., Pennington, 2006; Nicolson & Fawcett, 2007).

References

Alcock, K. (2006). The development of oral motor control and language. *Down Syndrome Research and Practice, 11*, 1–8.

Archibald, L. M. D., & Alloway, T. P. (2008). Comparing language profiles: Children with specific language impairment and developmental coordination disorder. *International Journal of Language & Communication Disorders, 43*, 165–180.

Bates, E., Dale, P., & Thal, D. (1995). Individual differences and their implications for theories of language development. In P. Fletcher & B. MacWhinney (Eds.), *Handbook of child language* (pp. 96–151). Oxford: Blackwell.

Best, C. T., McRoberts, G. W., & Sithole, N. M. (1988). Examination of perceptual reorganization for non-native speech contrasts: Zulu click discrimination by English-speaking adults and infants. *Journal of Experimental Psychology, 14*, 345–360.

Bishop, D. V. M. (2002). Motor immaturity and specific speech and language impairment: Evidence for a common genetic basis. *American Journal of Medical Genetics, 114*, 56–63.

Bishop, D. V. M., & Adams, C. (1990). A prospective study of the relationship between specific language impairment, phonological disorders, and reading retardation. *Journal of Child Psychology and Psychiatry, and Allied Disciplines, 31*, 1027–1050.

Bonte, M. (2005). *Between sounds and words: Neurophysiological studies on speech processing in adults, normally reading children, and children with developmental dyslexia.* Unpublished doctoral dissertation, Universiteit Maastricht.

Braine, M. D. S. (1987). What is learned in acquiring word classes—a step toward an acquisition theory. In B. MacWhinney (Ed.), *Mechanisms of language acquisition* (pp. 65–88). Hillsdale, NJ: Erlbaum.

Brown, W. E., Eliez, S., Menon, V., Rumsey, J. M., White, C. D., & Reiss, A. L. (2001). Preliminary evidence of widespread morphological variations of the brain in dyslexia. *Neurology, 56,* 781–783.

Capel, D., de Bree, E., Kerkhoff, A., & Wijnen, F. (2008). Nederlandse baby's gebruiken statistische informatie om spraakklanken te leren onderscheiden. *Toegepaste Taalwetenschap in Artikelen, 79,* 21–30.

Capel, D., de Bree, E., Kerkhoff, A., & Wijnen, F. (2011). Distributional cues affect phonetic discrimination in Dutch infants. In W. Zonneveld, H., Quené, & W. Heeren (Eds.). *Sound and sounds. Festschrift for Bert Schouten* (pp. 33–44). Utrecht, Netherlands: LOT Publications.

Catts, H. W., & Kamhi, A. G. (Eds.). (1999). *Language and reading disabilities.* Boston: Allyn & Bacon.

Chambers, K. E., Onishi, K. H., & Fisher, C. (2003). Infants learn phonotactic regularities from brief auditory experience. *Cognition, 87,* B69–B77.

Cheng, H.-C., Chen, H.-Y., Tsai, C.-L., Chen, Y.-J., & Cherng, R.-J. (2009). Comorbidity of motor and language impairments in preschool children of Taiwan. *Research in Developmental Disabilities, 30,* 1054–1061.

Cleeremans, A., Destrebecqz, A., & Boyer, M. (1998). Implicit learning: News from the front. *Trends in Cognitive Sciences, 2,* 406–416.

Clegg, B. A., DiGirolamo, G. J., & Keele, S. W. (1998). Sequence learning. *Trends in Cognitive Sciences, 2,* 275–281.

de Bree, E. (2007). *Dyslexia and phonology: A study of the phonological abilities of Dutch children at-risk of dyslexia.* Doctoral dissertation, Utrecht University (LOT Dissertation Series 155).

de Bree, E., van Alphen, P., Fikkert, P., & Wijnen, F. (2008). Metrical stress in comprehension and production of Dutch children at risk of dyslexia. In H. Chan, H. Jacob, & E. Kapia (Eds.), *Proceedings of the 32nd Annual Boston University Conference on Language Development* (pp. 60–71). Somerville, MA: Cascadilla Press.

de Bree, E., Wijnen, F., & Zonneveld, W. (2006). Word stress production in three-year old children at risk of dyslexia. *Journal of Research in Reading, 29,* 304–317.

de Jong, J., Wijnen, F., & de Bree, E. (2012) *Morphosyntax in children at risk of dyslexia: Noun and verb inflection.* Unpublished manuscript.

de Kleine, E., & Verweij, W. B. (2009). Motor learning and chunking in dyslexia. *Journal of Motor Behavior, 41,* 331–337.

DeLong, M. R. (2000). The basal ganglia. In E. R. Kandel, J. H. Schwartz, & T. M. Jessell (Eds.), *Principles of neural science* (pp. 853–872). New York: McGraw-Hill.

De Smet, H. J., Baillieux, H., De Deyn, P. P., Mariën, P., & Paquier, P. (2007). The cerebellum and language: The story so far. *Folia Phoniatrica et Logopaedica, 59,* 165–170.

Dominey, P. F. (2005). From sensorimotor sequence to grammatical construction: Evidence from simulation and neurophysiology. *Adaptive Behavior, 13,* 347–361.

Dominey, P. F., Hoen, M., Lelekov, T., & Blanc, J. M. (2003). Neurological basis of language in sequential cognition: Evidence from simulation, aphasia and ERP studies. *Brain and Language*, *86*, 207–225.

Dominey, P. F., & Ramus, F. (2000). Neural network processing of natural language: I. Sensitivity to serial, temporal and abstract structure of language in the infant. *Language and Cognitive Processes*, *15*, 87–127.

Doyon, J., Penhune, V., & Ungerleider, L. (2003). Distinct contribution of the cortico-striatal and cortico-cerebellar systems to motor skill learning. *Neuropsychologia*, *41*, 252–262.

Eckert, M. A., Leonard, C. M., Richards, T. L., Aylward, E. H., Thomson, J., & Berninger, V. W. (2003). Anatomical correlates of dyslexia: Frontal and cerebellar findings. *Brain*, *126*, 482–494.

Elman, J. L. (1990). Finding structure in time. *Cognitive Science*, *14*, 179–211.

Evans, J. L., Saffran, J. R., & Robe-Torres, K. (2009). Statistical learning in children with specific language impairment. *Journal of Speech, Language, and Hearing Research*, *52*, 321–335.

Gerken, L. (2002). Early sensitivity to linguistic form. *Annual Review of Language Acquisition*, *2*, 1–36.

Gerken, L., Wilson, R., & Lewis, W. (2005). Infants can use distributional cues to form syntactic categories. *Journal of Child Language*, *32*, 249–268.

Gerrits, E., & de Bree, E. (2008). Early language development of children at familial risk of dyslexia: Speech perception and production. *Journal of Communication Disorders*, *42*, 180–194.

Ghez, C., & Thach, W. T. (2000). The cerebellum. In E. R. Kandel, J. H. Schwartz, & T. M. Jessell (Eds.), *Principles of neural science* (pp. 832–852). New York: McGraw-Hill.

Gillberg, C. (2003). Deficits in attention, motor control, and perception: A brief review. *Archives of Disease in Childhood*, *88*, 904–910.

Gómez, R. L. (2002). Variability and detection of invariant structure. *Psychological Science*, *13*, 431–436.

Gómez, R. L., & Gerken, L. (1999). Artificial grammar learning by 1-year-olds leads to specific and abstract knowledge. *Cognition*, *70*, 109–135.

Goschke, T., Friederici, A. D., Kotz, S. A., & van Kampen, A. (2001). Procedural learning in Broca's aphasia: Dissociation between the implicit acquisition of spatio-motor and phoneme sequences. *Journal of Cognitive Neuroscience*, *13*, 370–388.

Hagoort, P. (2005). On Broca, brain, and binding: A new framework. *Trends in Cognitive Sciences*, *9*, 416–423.

Hill, E. L. (2001). Non-specific nature of specific language impairment: A review of the literature with regard to concomitant motor impairments. *International Journal of Communication Disorders*, *36*, 149–171.

Hoen, M., Pachot-Clouard, M., Segebarth, C., & Dominey, P. F. (2006). When Broca experiences the Janus syndrome: An ER-fMRI study comparing sentence comprehension and cognitive sequence processing. *Cortex*, *42*, 605–623.

Howard, J. H., Howard, D. V., Japikse, K. C., & Eden, G. F. (2006). Dyslexics are impaired on implicit higher-order sequence learning, but not on implicit spatial context learning. *Neuropsychologia*, *44*, 1131–1144.

Joanisse, M. F., Manis, F. R., Keating, P., & Seidenberg, M. S. (2000). Language deficits in dyslexic children: Speech perception, phonology, and morphology. *Journal of Experimental Child Psychology*, *77*, 30–60.

Jusczyk, P., Friederici, A., Wessels, J., Svenkerud, V., & Jusczyk, A. (1993). Infants' sensitivity to the sound patterns of native language words. *Journal of Memory and Language*, *32*, 402–420.

Jusczyk, P., Luce, P., & Charles-Luce, J. (1994). Infants' sensitivity to phonotactic patterns in the native language. *Journal of Memory and Language*, *33*, 630–645.

Kadesjo, B., & Gillberg, C. (1999). Developmental Coordination Disorder in Swedish 7-year-old children. *Journal of the American Academy of Child and Adolescent Psychiatry*, *38*, 820–828.

Kaplan, B. J., Dewey, D. M., Crawford, S. G., & Wilson, B. N. (2001). The term comorbidity is of questionable value in reference to developmental disorders: Data and theory. *Journal of Learning Disabilities*, *34*, 555–565.

Kerkhoff, A., de Bree, E., de Klerk, M., & Wijnen, F. (in press). Non-adjacent dependency learning in infants at familial risk of dyslexia. *Journal of Child Language*.

Knowlton, B. J., Ramus, S. J., & Squire, L. R. (1992). Intact artificial grammar learning in amnesia: Dissociation of classification learning and explicit memory for specific instances. *Psychological Science*, *3*, 172–179.

Kuhl, P. K. (2004). Early language acquisition: Cracking the speech code. *Nature Reviews. Neuroscience*, *5*, 831–843.

Kuhl, P. K., Stevens, E., Hayashi, A., Deguchi, T., Kiritani, S., & Iverson, P. (2006). Infants show a facilitation effect for native language phonetic perception between 6 and 12 months. *Developmental Science*, *9*, F13–F21.

Kuhl, P. K., Williams, K. A., Lacerda, F., Stevens, K. N., & Lindblom, B. (1992). Linguistic experience alters phonetic perception in infants by 6 months of age. *Science*, *255*, 606–608.

Leonard, L. B. (1998). *Children with Specific Language Impairment*. Cambridge, MA: MIT Press.

Lieberman, P. (2000). *Human language and our reptilian brain: The subcortical basis of speech, syntax, and thought*. Cambridge, MA: Harvard University Press.

Lum, J. A. G., Gelgec, C., & Conti-Ramsden, G. (2009). Procedural and declarative memory in children with and without specific language impairment. *International Journal of Language and Communication Disorders*, *45*, 96–107.

Maratsos, M., & Chalkley, M. (1980). The internal language of children's syntax. In K. Nelson (Ed.), *Children's language* (Vol. 2). New York: Gardner Press.

Marcus, G. F., Vijayan, S., Bandi Rao, S., & Vishton, P. M. (1999). Rule learning by seven-month-old infants. *Science*, *283*, 77–80.

Maye, J., Weiss, D., & Aslin, R. (2008). Statistical phonetic learning in infants: Facilitation and feature generalization. *Developmental Science, 11*, 122–134.

Maye, J., Werker, J., & Gerken, L. (2002). Infant sensitivity to distributional information can affect phonetic discrimination. *Cognition, 82*, B101–B111.

Menghini, D., Hagberg, G. E., Caltagirone, C., Petrosini, L., & Vicari, S. (2006). Implicit learning deficits in dyslexic adults: An fMRI study. *NeuroImage, 33*, 1218–1226.

Mintz, T. (2002). Category induction from distributional cues in an artificial language. *Memory & Cognition, 30*, 678–686.

Mintz, T. (2003). Frequent frames as a cue for grammatical categories in child directed speech. *Cognition, 90*, 91–117.

Monaghan, P., Chater, N., & Christiansen, M. H. (2005). The differential role of phonological and distributional cues in grammatical categorisation. *Cognition, 96*, 143–182.

Newman, R., Bernstein Ratner, N., Jusczyk, A. M., Jusczyk, P. W., & Dow, K. A. (2006). Infants' early ability to segment the conversational speech signal predicts later language development: A retrospective analysis. *Developmental Psychology, 42*, 643–655.

Newport, E. L., & Aslin, R. N. (2004). Learning at a distance I. Statistical learning of non-adjacent dependencies. *Cognitive Psychology, 48*, 127–162.

Nicolson, R. I., & Fawcett, A. (2007). Procedural learning difficulties: Reuniting the developmental disorders? *Trends in Neurosciences, 30*, 135–141.

Nicolson, R. I., Fawcett, A. J., Berry, E. L., Jenkins, I. H., & Dean, P. (1999). Association of abnormal cerebellar activation with motor learning difficulties in dyslexic adults. *Lancet, 353*, 1662–1667.

Nicolson, R. I., Fawcett, A. J., & Dean, P. (2001). Developmental dyslexia: The cerebellar deficit hypothesis. *Trends in Neurosciences, 24*, 508–511.

Owen, S. E., & McKinlay, I. A. (2003). Motor difficulties in children with developmental disorders of speech and language. *Child: Care, Health and Development, 23*, 315–325.

Pavlidou, E. U., Kelly, M. L., & Williams, J. M. (2010). Do children with developmental dyslexia have impairments in implicit learning? *Dyslexia, 16*, 143–161.

Pavlidou, E. U., Williams, J. M., & Kelly, M. L. (2009). Artificial grammar learning in primary school children with and without developmental dyslexia. *Annals of Dyslexia, 59*, 55–77.

Pennington, B. (2006). From single to multiple deficit models of developmental disorders. *Cognition, 101*, 385–413.

Peperkamp, S., Le Calvez, R., Nadal, J.-P., & Dupoux, E. (2006). The acquisition of allophonic rules: Statistical learning with linguistic constraints. *Cognition, 101*, B31–B41.

Pinker, S. (1984). *Language learnability and language development*. Cambridge: Harvard University Press.

Pinker, S. (2000). *Words and rules*. New York: Perennial.

Powell, R. P., & Bishop, D. V. M. (1992). Clumsiness and perceptual problems in children with specific language impairment. *Developmental Medicine and Child Neurology, 34*, 755–765.

Rasmussen, P., & Gillberg, C. (2000). Natural outcome of ADHD with developmental coordination disorder at age 22: A controlled, longitudinal, community-based study. *Journal of Child Psychology and Psychiatry, and Allied Disciplines, 44*, 712–722.

Redington, M., Chater, N., & Finch, N. (1998). Distributional information: A powerful cue for acquiring syntactic categories. *Cognitive Science, 22,* 425–469.

Redmond, S. M., & Johnston, S. S. (2001). Evaluating the morphological competence of children with severe speech and physical impairments. *Journal of Speech, Language, and Hearing Research: JSLHR, 44,* 1362–1375.

Rescorla, L., Bernstein Ratner, N. (1996). Phonetic profiles of toddlers with specific expressive language impairment (SLI-E). *Journal of Speech, Language, and Hearing Research: JSLHR, 39,* 153–165.

Roberts, J., Rescorla, L., Giroux, J., & Stevens, L. (1998). Phonological skills of children with specific expressive language impairment (SLI-E): Outcome at age 3. *Journal of Speech, Language, and Hearing Research: JSLHR, 41,* 374–384.

Saffran, J. (2002). Constraints on statistical language learning. *Journal of Memory and Language, 47,* 172–196.

Saffran, J. R., Aslin, R. N., & Newport, E. L. (1996). Statistical learning by 8-month old infants. *Science, 274,* 1926–1928.

Saffran, J. R., Newport, E. L., Aslin, R. N., Tunick, R. A., & Barrueco, S. (1997). Incidental language learning: Listening and learning out of the corner of your ear. *Psychological Science, 8,* 101–105.

Scarborough, H. S. (1990). Very early language deficits in dyslexic children. *Child Development, 61,* 1728–1734.

Scarborough, H. S. (1991). Antecedents to reading disability: Preschool language development and literacy experiences of children from dyslexic families. *Reading and Writing, 3,* 219–233.

Scarborough, H. S. (2005). Developmental relationships between language and reading: Reconciling a beautiful hypothesis with some ugly facts. In H. W. Catts & A. G. Kamhi (Eds.), *The connections between language and reading disabilities* (pp. 3–22). Mahwah, NJ: Erlbaum.

Schwartz, M., & Regan, V. (1996). Sequencing, timing, and rate relationships between language and motor skill in children with receptive language delay. *Developmental Neuropsychology, 12,* 255–270.

Serniclaes, W., Van Heghe, S., Mousty, P., Carré, R., & Sprenger-Charolles, L. (2004). Allophonic mode of speech perception in dyslexia. *Journal of Experimental Child Psychology, 87,* 336–361.

Siegel, L. S. (1982). Early cognitive and environmental correlates of language development at 4 years. *International Journal of Behavioral Development, 5,* 433–444.

Smith, J. G., & McDowall, J. (2004). Impaired higher order implicit sequence learning on the verbal version of the serial reaction time task in patients with Parkinson's disease. *Neuropsychology, 18,* 679–691.

Snowling, M., & Hayiou-Thomas, M. E. (2006). The dyslexia spectrum: Continuities between reading, speech and language impairments. *Topics in Language Disorders, 26,* 110–126.

Squire, L. R., Knowlton, B., & Musen, G. (1993). The structure and organization of memory. *Annual Review of Psychology, 44,* 453–495.

Stowe, L., Paans, A. M. J., Wijers, A. A., & Zwarts, F. (2004). Activations of "motor" and other non-language structures during sentence comprehension. *Brain and Language, 89,* 290–299.

Tallal, P. (1980). Auditory temporal perception, phonics, and reading disabilities in children. *Brain and Language*, *9*, 182–198.

Teichmann, M., Dupoux, E., Kouider, S., & Bachoud-Lévi, A.-C. (2006). The role of the striatum in processing language rules: Evidence for word perception in Huntington's disease. *Journal of Cognitive Neuroscience*, *18*, 1555–1569.

Teichmann, M., Gaura, V., Démonet, J.-F., Supiot, F., Delliaux, M., Verny, C., . . . & Bachoud-Lévi, A.-C. (2008). Language processing within the striatum: Evidence from a PET correlations study in Huntington's disease. *Brain*, *131*, 1046–1056.

Tomblin, J. B., Mainela-Arnold, E., & Zhang, X. (2007). Procedural learning in adolescents with and without specific language impairment. *Language Learning and Development*, *3*, 269–293.

Ullman, M. (2004). Contributions of memory circuits to language: The declarative/procedural model. *Cognition*, *92*, 231–270.

Ullman, M. T., Corkin, S., Coppola, M., Hickok, J. H., & Korosnetz, W. J. (1997). A neural dissociation within language: Evidence that the mental dictionary is part of declarative memory, and that grammatical rules are processed by the procedural system. *Journal of Cognitive Neuroscience*, *9*, 266–276.

Ullman, M. T., & Pierpont, E. I. (2005). Specific language impairment is not specific to language: The procedural deficit hypothesis. *Cortex*, *41*, 399–433.

Van Alphen, P., De Bree, E., Gerrits, E., De Jong, J., Wilsenach, C., & Wijnen, F. (2004). Early language development in children with a genetic risk of dyslexia. *Dyslexia*, *10*, 265–288.

Vellutino, F. R., Fletcher, J. M., Snowling, M. J., & Scanlon, D. M. (2004). Specific reading disability (dyslexia): What have we learned in the past four decades? *Journal of Child Psychology and Psychiatry, and Allied Disciplines*, *45*, 2–40.

Vicari, S., Finzi, A., Menghini, D., Marotta, L., Baldi, S., & Petrosini, L. (2005). Do children with developmental dyslexia have an implicit learning defect? *Journal of Neurology, Neurosurgery, and Psychiatry*, *76*, 1392–1397.

Vicari, S., Marotta, L., Menghini, D., Molinari, M., & Petrosini, L. (2003). Implicit learning deficit in children with developmental dyslexia. *Neuropsychologia*, *41*, 108–114.

Visscher, C., Houwen, S., Scherder, E. J. A., Moolenaar, B., & Hartman, E. (2007). Motor profile of children with developmental speech and language disorders. *Pediatrics*, *120*, e158–e163.

Watkins, K. E., Dronkers, N. F., & Vargha-Khadem, F. (2002). Behavioural analysis of an inherited speech and language disorder: Comparison with acquired aphasia. *Brain*, *125*, 452–464.

Watkins, K. E., Vargha-Khadem, F., Ashburner, J., Passingham, R. E., Connelly, A., Friston, K. J., Fracowiak, R. S., Mishkin, M., Gadian, D. G. (2002). MRI analysis of an inherited speech and language disorder: Structural brain abnormalities. *Brain*, *125*, 465–478.

Webster, R. I., Erdos, C., Evans, K., Majnemer, A., Kehayia, E., Thordardottir, E., Evans, A., & Shevell, M. I. (2006). The clinical spectrum of developmental language impairment in school-aged children: Language, cognitive, and motor findings. *Pediatrics*, *118*, e1541–e1549.

Webster, R. I., Majnemer, A., Platt, R., & Shevell, M. (2005). Motor function at school age in children with a preschool diagnosis of developmental language impairment. *Journal of Pediatrics*, *146*, 80–85.

Webster, R. I., & Shevell, M. I. (2004). Neurobiology of specific language impairment. *Journal of Child Neurology*, *19*, 471–481.

Werker, J. F., & Pegg, J. E. (1992). Infant speech perception and phonological acquisition. In C. Ferguson, L. Menn, & C. Stoel-Gammon (Eds.), *Phonological development: Models, research, implications* (pp. 285–311). Timonium, MD: York Press.

Werker, J. F., & Tees, R. C. (1984). Cross-language speech perception: Evidence for perceptual reorganization during the first year of life. *Infant Behavior and Development*, *7*, 49–63.

Wilsenach, A. C. (2006). *Syntactic processing in children at risk of dyslexia and children with Specific Language Impairment.* Doctoral dissertation, Utrecht University (LOT Dissertation Series 128).

Wilsenach, A. C., & Wijnen, F. (2004). Perceptual sensitivity to morphosyntactic agreement in language learners: Evidence from Dutch children at risk for developing dyslexia. In A. Brugos, L. Micciula, & C. E. Smith (Eds.), *Proceedings of the 28th Boston University Conference on Language Development.* Somerville, MA: Cascadilla Press.

III PHONOLOGY AND SYNTAX

9 Structure in Human Phonology and in Birdsong: A Phonologist's Perspective

Moira Yip

Why Birdsong, Not Primate Calls?

For a phonologist, birdsong is clearly the most intricate and tantalizing place to look for analogs of human phonology in nonhuman species. There are many reasons for this, three of which seem especially compelling.

First, like human language, birdsong involves vocal learning, guided by innate templates of varying degrees of strictness (Fehér, Wang, Saar, Mitra, & Tchernichovski, 2009; Gardner, Naef, & Nottebohm, 2005; Marler, 2000; Tchernichovski et al., 2001; Liu, Gardner, & Nottebohm, 2004; and many others). Early stages of birdsong have analogs of vocal babbling known as subsong (see Aronov, Andalman, & Dee, 2008). In both humans and some birds, learning is subject to a critical period but also continues in adulthood, with the brain retaining sufficient plasticity to learn new accents (Evans & Iverson, 2005) or new songs (Margoliash, Staicer, & Inouet, 1991; Mountjoy & Lemon, 1995; Brenowitz & Beecher, 2005) and monitoring itself via lifelong error correction (Sober & Brainard, 2009). In many species, innate templates do not completely specify the song form, and the final adult song is determined by "a synergy between innate . . . and experience-based forces" (Rose et al., 2004). Like humans, some birds have quite large repertoires of song elements, and vary their order and arrangement so as to provide a rich variety of songs. For example, the brown thrasher (*Toxostoma rufum)* has about 1,800 different song types (Kroodsma & Parker, 1979). They are built up by "ubiquitous copying, re-arrangement, and innovation through processes that presuppose an emancipation of vocal learning from constraints imposed by innate templates" (Merker & Okanoya, 2006, p. 410). About a third of the song units are "reduplications" (one syllable repeated twice). Birdsong requires auditory feedback, even after initial learning (Okanoya & Yamaguchi, 1997; Leonardo & Konishi, 1999). Finally, it is interesting that the language areas in the human and avian brains may show some similarities, and FoxP2 seems to play a role in both language and birdsong (Bolhuis, Okanoya, & Scharff, 2010).

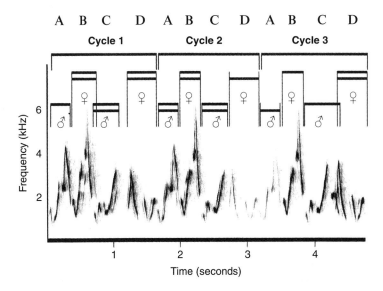

Figure 9.1
Male and female duet of the plain-tailed wren, reprinted with permission from Mann et al. (2005).

Second, unlike many other species (such as humpback whales), some birds not only sing male solos, but also interact vocally in reciprocal, competitive, or cooperative ways (Brumm & Slater, 2007; Todt & Naguib, 2000). The phenomena known as duetting, countersinging, and chorusing are all instances of this. A remarkable example of duetting is found in the plain-tailed wren (*Thryothorus euophrys*) (Mann, Dingess, & Slater, 2005). Males and females take turns, and their coordination is so precise that it sounds like a single song. In Figure 9.1, males sing sections A and C (lower bars) and females sing B and D (upper bars). The sequence ABCD then repeats.

Groups of birds have repertoires of phrases—for example, one group used 27 different phrases in position A. However, they are not combined at random, so certain AC, BD, and ABCD combinations occur and others do not. Note how each bird picks up exactly where the other left off, or the note that the other bird would have reached.. If bird A ends having just started a rise, bird B completes the rise, either seamlessly, or, if there is a brief pause, starting at the pitch bird A would have reached if it was still singing. Visual inspection of 142 "handovers" in sonograms kindly provided by Nigel Mann and Peter Slater shows this tight coordination in 92% of cases. Duetting also takes practice: coordination improves over time in bonded pairs (Hall & Magrath, 2007). The parallels in human language can be found in the literature on turn-taking, particularly on collaborative completions and on simultaneous "choral" performances (Ford & Thompson, 1996; Lerner, 1996, 2002;

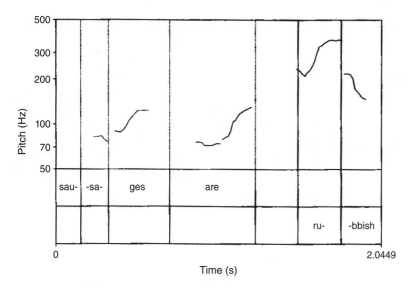

Figure 9.2
A pitch contour continued across two speakers, reprinted with permission from Szczepek-Reed (2006, p. 161).

Local, 2005; Szczepek Reed, 2006). For example, in human conversations, prosodic "projection" foreshadows how the next item will continue prosodically. Consider Figure 9.2. The first speaker sets up a prosodic pattern on *Sausages are*, and this "projects" the same pitch and timing for a new item, *rubbish*, produced by an incoming speaker.

An analogy can also be drawn between the ability of some birds to sing in chorus (Wingfield & Lewis, 1993; Seddon, 2002) and with humans' ability to speak synchronously (Cummins, 2003).

Third (and this is the particular aspect of birdsong that will mainly concern us here) birdsong has internal structure. Although this is sometimes called "song syntax," it is in fact much closer to phonology and is sometimes called "phonocoding" (Marler, 2000). In human syntax, the pieces of structure have meaning. In phonology, they don't. For example, *academic* is structured into four syllables, *a.ca. de.mic*, and two stress feet, *(áca)(démic)*, but none of these pieces has independent meaning. As far as we know, the structures in bird songs don't have meanings either. Even if changing one aspect of the song (a note or a motif) changes its meaning (e.g., identity of singer), it does not mean the changed element itself has a meaning: changing *bill* to *fill* does not mean [b] and [f] have a meaning.

The structures of birdsong play a role in learning, with songs made up of separately learned "chunks" (Williams & Staples, 1992; Hultsch & Todt, 1989); they control where song is interrupted (Cynx, 1990); and they have been argued in many species to be hierarchical (Hultsch & Todt, 2004b; Okanoya, 2004).

This chapter takes a look at the extensive literature on birdsong through the eyes of a phonologist. It is a hunt for analogs of human phonological skills in the avian kingdom. Not surprisingly, there is a deep gulf between the complexity of human phonology and that of birdsong. But more surprisingly, for a phonologist, is how extensive the parallels are. While they are not known to be under cognitive control, nor manipulable with consequences for the message to be conveyed, they look very much like the basic building blocks that would be needed to create the kind of code that could subsequently be mapped onto meanings. For the linguist, this chapter should serve to introduce, in outline, some of the more striking properties of birdsong, and hopefully make it clear that we should pay attention to these commonalities.

The chapter unfolds as follows. In the next section I list some human phonological skills, and summarize which are or are not found in songbirds. In the third section I look more closely at structures, first in humans and then in birds. In the fourth section I propose five mechanisms by which structure could arise and be passed on. The fifth section sums up.

Human vs. Avian Phonological Skills

Humans bring to their phonology a collection of skills, some auditory/perceptual, some articulatory/productive, and some purely cognitive and organizational. For a brief summary, see Yip (2006). Some of these skills have been shown to have analogs in one or more bird species. Table 9.1 illustrates some of these, with selected references. The term *phonological skill* is used here for any tool that plays into our human phonological competence, whether clearly cognitive or not. Other skills are less well documented, but may exist (Table 9.2). Other skills seem at present to be limited to humans. References in Table 9.3 are to chapters in De Lacy (2007).

In this chapter, I focus on one aspect only, the evidence for structure in birdsong, and the extent to which it shows any similarities with the structures found in human language. The one structural issue I do not address at all is the controversy surrounding recursion, since that is treated in some detail in chapter 10. See also Abe and Watanabe (2011) and Berwick, Okanoya, Beckers, & Bolhuis (2011) for a useful recent overview.

Table 9.1
Phonological skills definitely found in at least one bird species

Phonological skill	Sample species	Sample references
Imitation of conspecifics	Many species, including mockingbird, European starling, parrots, and all vocal learners	Allard (1939); Pepperberg (2005)
Conspecific categorical perception, with learned component	Swamp sparrows	Nelson & Marler (1989); Prather et al. (2009)
Perceptual magnet effect	European starling	Kluender et al. (1998)
Preference for learning "natural" patterns	Canary, white-crowned sparrow	Gardner et al. (2005); Rose et al. (2004); Fehér et al. (2009)
Natural classes of sounds	European starling: clicks vs. whistles	Mountjoy & Lemon (1995)
Detection of adjacent transitional probabilities	European starling	Gentner & Hulse (1998, 2000)
Rhyme	Mockingbirds	Thompson et al. (2000); Wildenthal (1965)
Internal structural groupings within the song	Nightingale, Bengalese finch, and many others	Todt & Hultsch (1998); Honda & Okanoya (1999); Okanoya (2004)

Focus on Structure

Structure and Cues to Structure in Humans

Human utterances are well known to be highly structured, and the hierarchy includes at least the categories of intonation phrase, phonological phrase, prosodic word, stress foot, and syllable. The only one of these that may not be familiar to nonlinguists is the stress foot, which consists of a stressed syllable, and adjacent unstressed syllables if present. So the English word *innovation* contains two feet (ìnno)(vátion). The situation is diagrammed in Figure 9.3.

Table 9.2
Phonological skills possibly found in at least one bird species

Phonological skill	Sample species	References
Detection of recursive embedding	European starling (in lab)	Gentner et al. (2006), but cf. Corballis (2007); on recursion in human phonology see Ladd (1986)
Prosodic effects	Zebra finch: harmonic emphasis and lengthening; but cooling of HVC slows entire song	Williams et al. (1989); Scharff and Jarvis (personal communication); Long & Fee (2008);
	Chaffinch: trill/flourish length Final length	Leitao et al. (2004); Tierney et al. (2008);
	Collared dove: silent beats[a]	Ballintijn & ten Cate (1999);
	Great tits: "drift" (slowing down during songs)	Lambrechts (1988, 1996);
	Cockatoo: rhythmic synchronization to a beat	Patel et al. (2009)
Stuttering	Collared dove	Ballintijn & ten Cate, (1999);
	Bengalese finch	Okanoya (2004);
	Zebra finch	Helekar et al. (2003)
Copying	White-crowned sparrow	Baptista (1977);
	Brown thrasher	Kroodsma & Parker (1979)

[a]Note that the collared dove is not a vocal learner and has calls rather than songs.

Table 9.3
Phonological skills possibly found only in humans

Phonological skill	Example	References
Systematic, lawful alternations	Final devoicing, place assimilation	Bakovic, chap. 14; Rice, chap. 4
Prosodic distinctions	Heavy vs. light syllables Initial vs. final stress	Kager, chap. 9
Nonadjacent total/ partial identity (or nonidentity) computation	Vowel harmony Restrictions on homorganic consonants in roots	Archangeli & Pulleyblank, chap. 15; Alderete & Frisch, chap. 16
Abstract structures	Syllables composed of any consonant followed by any vowel; bimoraic foot composed of one heavy or two light syllables	Zec, chap. 8; Kager, chap. 9

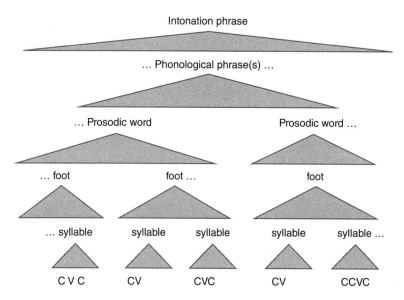

Figure 9.3
Human prosodic hierarchy.

For example, one possible structure for the sentence *After lunch, children love gin-gerbread* is as follows:

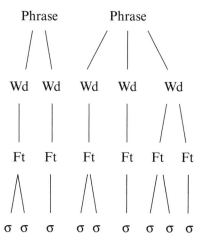

After lunch children love gingerbread

It is important to note that structures are not defined by their contents: the contents of each foot/word/phrase are different and are not a closed set. [ta] and [strɪŋ] are both syllables, even though they differ considerably in size and contents. The number of actually occurring syllables in a language ranges from about 400 in a language like Mandarin (not counting tonally distinct versions) to nearly 10,000 in English (Ke, 2006). The number of bisyllabic feet in Mandarin is therefore 400^2, or 160,000, and the number in English is 10^8, or 100 million.[1]

The structures repeat, but the contents usually don't. It is of course true that repetition (reduplication) and rhyme are found in human phonology and morphology, but typically with semantic content, so that the repeated form carries a different meaning from its unrepeated components.

These structures typically obey certain principles, such as the Strict Layer Hypothesis: "Each level is composed entirely of constituents of the next level down" (Selkirk, 1984). As a result, all words contain feet, and all syllables are grouped into feet. Many structures are preferentially binary, most clearly the foot and the prosodic word (Downing, 2006).

The existence of these structures is motivated by their role in phonology. They may be domains for processes such as tone association and vowel harmony (Pearce, 2006) or lenition (Harris, 2004). The edges of these structures are locations that allow pauses, anchor pitch accents, trigger lengthening, or authorize repetition.

A final property of these structures is that they are "headed": each has one element that is more prominent in some way than the others. So phrases have an accented syllable; words have word stress; feet are left- or right-prominent; and syllables have a sonority peak (typically a vowel). Taken together with the preference for binarity, the result is an alternating strong-weak pattern that shows up as . . . CVCVCV . . . strings, and as stress on every other syllable in many languages.

There is also acoustic evidence for the existence of these heads, since they show clusters of acoustic properties including increased duration, amplitude, and higher pitch or a sharp pitch change.

What about acoustic cues to the structures themselves? It is known that speech pauses play an important role in speech production and perception. However, even the highest-level units, such as phrases, are not necessarily separated by silent intervals. Conversely, some, but not all, silent intervals are perceived as pauses: even nondetectable 25 ms silent intervals are interpreted as word boundaries in artificial language learning (Peña, Bonatti, Nespor, & Mehler, 2002; Onnis, 2003), but others, such as the silence during a stop consonant, are not. Finally, pauses may be perceived when no silent interval is present (Nooteboom, Brokx, & De Rooij, 1978), typically after final lengthening, or in the presence of a melodic boundary marker, or at major phrasal constituent boundaries. In sum, there is a tendency for silent intervals and structural boundaries to co-occur, but they cannot be used as diagnostic.

Other cues may be more useful, but vary from language to language. They include the position of prominence (e.g., word-initial/final stress in Finnish/French (Vroomen, Tuomainen, & De Gelder, 1998)); final lengthening (e.g., phrase-finally in English); initial strengthening (Quené, 1993; Fougeron & Keating, 1997); more coarticulation within than across domains; the domain of phonological spreading (e.g., vowel harmony in Finnish words (Suomi, McQueen, & Cutler, 1997; Vroomen et al., 1998)); and the domain of tone sandhi (e.g., tone sandhi in S.Min phrases (Chen, 1987)).

This then is the background against which I will now examine some of the bird-song literature.

Structure and Cues to Structure in Birds

Comparison to Human Equivalents

Although we should really be comparing our species to a single bird species, not to the entire bird kingdom, I will continue the common practice of spreading my net wide, and looking for the most humanlike examples wherever I can find them. Work on birdsong usually decomposes the song into four levels: the song bout, the motif, the syllable, and the note. In Figure 9.4 I show an example of two songs. (a) is the

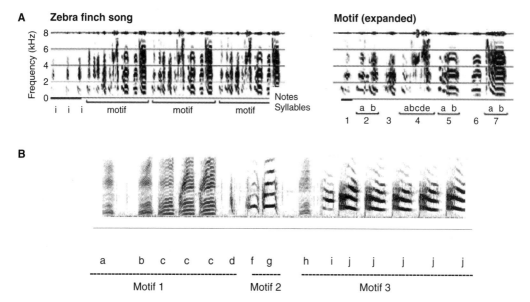

Figure 9.4
(a) Song of the zebra finch, reprinted with permission from Williams (2004). (b) Song of the Bengalese finch, reprinted with permission from Suge and Okanoya (2010). Terminology has been changed.

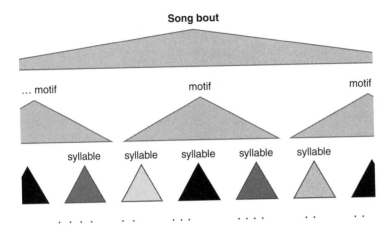

Figure 9.5
Structure of zebra finch song bout (after Doupe and Kuhl, 1999).

song of the zebra finch (*Taeniopygia guttata*). The left-hand sonogram shows an entire song bout containing several motifs, and the right-hand sonogram looks inside a motif to show the individual syllables and notes (terminology changed from Williams 2004 to allow cross-species comparisons). (b) is the song of the domesticated Bengalese finch (*Lonchura striata*), which has more than one motif (also called chunk; see Suge & Okanoya (2010) for a discussion of the terminology).

In Figure 9.5 I diagram the structure of the zebra finch song more abstractly, after Doupe and Kuhl (1999); the single fixed motif repeats several times. For a linguist, birdsong terminology needs some explanation. The key terms are defined as follows:

- *Notes*   Segments of song separated by rapid transitions in the spectrogram
- *Syllable*   Defined as "acoustic productions separated by gaps of silence"[2]
- *Motif*   Stereotyped sequence of syllables; also called "song type" and "phrase"

These terms, especially the term *syllable*, imply direct analogies with human language, but this is dangerous. Consider the motif. What might its human equivalent be? In birdsong, it is the name for a collection of syllables, but it is not like a foot, for many reasons. The foot in human language is defined rather strictly by its form, not its contents, but for some bird species the motif's form and length can vary with apparent freedom, as shown by the different European starling (*Sturnus vulgaris*) motifs in Figure 9.6.

Unlike feet, there are no obvious other restrictions on motif form: there is no evidence for headedness or for binarity, and they may contain at least 11 syllables, even in birds with large repertoires (Devoogd, Krebs, Healy, & Purvis, 1993). Dif-

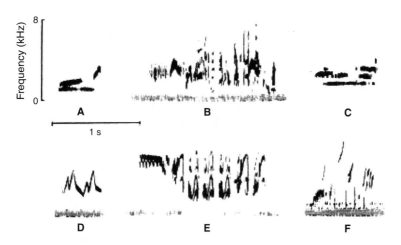

Figure 9.6
European starling motifs, reprinted with permission from Mountjoy and Lemon (1995).

ferent motifs do not necessarily start or end with syllables of the same type, or on the same pitch. Identical motifs are frequently repeated, unlike syllables, feet, or words in humans, and this repetition is a major diagnostic for motif-hood, and indeed it is encoded in the word *stereotypical* in the definition above. By comparison with the number of feet (10^8 for English), or words (much larger), the number of different motifs mostly ranges from 1 (zebra finch) to 300 (canary), with the champion being the brown thrasher, reported to have about 1,800 (Kroodsma & Parker, 1979). Finally, by definition (and again unlike feet in human language), each motif is separated by a silent interval from the next motif.

So if the motif does not resemble the foot, what is it? It is not like a word, because it lacks meaning. It is more like a phrase, in its tendency to be surrounded by "pauses" (i.e., silent intervals), but in that case, since it directly dominates the "syllable" level and has no other internal structure of its own, we see that the entire structure is distinctly flattened by comparison with human language.

There are a few interesting exceptions to my statement above that there are no restrictions on motif form. The most striking is probably the nightingale, *Luscinia megarynchos*, which is famous for its intricate song. A male may have up to 200 "song types," and a "song bout" consists of a series of these, separated by silences of around 3 seconds. In the work on nightingales, the terminology is slightly different. Every "song type" conforms to a template that has four sections, denoted by Greek letters in Figure 9.7, and the most elaborate of these sections is called the "motif." It is preceded by two or three quieter notes, and followed by a single syllable repeated multiple times as well as a distinctive final note (Todt & Hultsch, 1998).

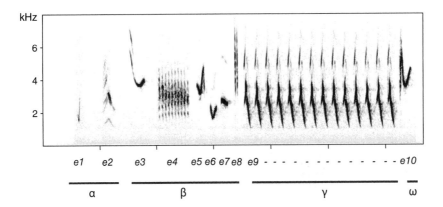

Figure 9.7
Nightingale "song type," showing four sections (Hultsch, personal communication).

This template is the closest thing to an abstract structure that can be filled by any suitable content that I have encountered in the birdsong literature, although the chaffinch's trill-flourish pattern (Riebel & Slater, 1998; Leitao, Van Dooren, & Riebel, 2004) and the blackbird's whistle part followed by twitter part may show something similar, on a simpler scale.

Now let us look at how the motifs combine into songs. The European starling (*Sturnus vulgaris*) has a repertoire of up to 70 motifs, of four kinds (Adret-Hausberger & Jenkins, 1988; Eens, Pinxten, & Verheyen, 1989; Eens, 1997; and others). If W = whistle, Wb = warble, R = rattle, H = high frequency, the song has the shape $W^i Wb^j R^k H^l$. A section of such a song is shown schematically in Figure 9.8; note that it has one more layer of structure than the zebra finch song.

Starlings memorize large inventories of motifs, and specialized neurons respond only to these familiar motifs (Gentner, 2004). New motifs may be added over time (even for adults), but this structure apparently does not change (Mountjoy & Lemon, 1995). Interestingly, then, they appear to sort their motifs into four "natural classes" reminiscent of human phonology.[3] Within this overall structure, individuals vary their song considerably. Gentner and Hulse (1998, 2000) show that the ability to recognize other starlings depends partly on their repertoire of motifs, and on the order in which they occur. For example, within the warble section, the relative order of different warble motifs matters. Note that this sensitivity to ordering is like phonology (as well as syntax): *tack* and *cat* have the same phonemes in different orders, and we detect the change. Gentner (2008) further shows that starlings learn and recognize motifs as whole auditory objects, but that they also have access to sub-motif elements and when these are extracted from learned motifs and rearranged, recognition is still somewhat above chance.

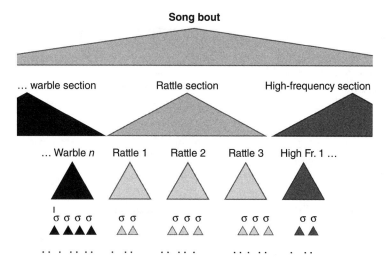

Figure 9.8
Structure of European starling song. σ stands for "syllable."

Although the details vary, it should be clear that there is considerable agreement that these various species have songs with internal structure, based largely on distributional evidence.

Acoustic and Behavioral Cues to Structure in Birds

I now move on to ask whether there is evidence of other kinds for the existence of these structures.

Starting with the note, the lowest level of analysis, Yu and Margoliash (1996) show that RA neurons have motor-activity histograms (MAHs) that match individual notes, grounding them in the neuroanatomy.

There is much more good acoustic, neurological, and behavioral evidence for the existence of the syllable. First, a syllable is separated from the next syllable by a short silent interval. Second, if a strobe light induces stopping in the single-motif zebra finch song (Cynx, 1990), breaks are far more likely between than within syllables (in 331/334 cases). However, no such effect was found for nightingale song by Riebel and Todt (1997). Third, Yu and Margoliash (1996) also show that zebra finch "syllables" of the same type have identical MAHs for HVc neurons.

Moving up to groupings of syllables, within the motif (since no other motif-internal structure is usually assumed) all syllable breaks should be equally likely stopping points, but this is false: while breaks are possible between any two syllables, there is a significant effect of length of silence (Williams & Staples, 1992). The human analogy is that (1) silent intervals are heard as word boundaries (Peña et al.,

2002; Onnis, 2003) and (2) clicks are heard at the boundaries between perceptual units (Fodor & Bever, 1965). We will see later that the length of the silent interval also has effects in learning.

It is interesting that there is a slight hint of internal "chunk" structure to the motif whereby a motif may be broken into two or more chunks in acquisition (see below for details), and one might ask whether any other acoustic cues support this view. Evidence from recent work suggests that chunk boundaries are more likely to occur at junctions of lower statistical probability (Takahashi, Yamada, & Okanoya, 2010), reminiscent of the way infants apparently recognize "words" in the speech stream (Saffran, Newport, & Aslin, 1996). It also appears that the detection of clicks during a bird's own song is delayed until a song chunk is processed (Suge & Okanoya, 2010). However, Williams and Staples (1992) state that melodic cues (i.e., note type, downsweep versus stack, etc.) do not seem to play a role in chunk boundaries, and Hultsch and Todt (2004a) have shown the same for nightingales.

Rather surprisingly, I have not found any experiments testing for whether induced breaks are more likely between than within motifs, or for whether the length of the silent interval is a reliable cue to motif boundaries.

Other evidence for structures could come from positional variants of syllables at structure edges. Williams and Staples's (1992) data includes a slightly different "allo-syllable" when a song restarts after interruption, thereby placing the usually medial syllable in initial position. In contrast, Yu and Margoliash (1996) show that zebra finch "syllables" of the same type in different contexts have identical MAHs for HVc neurons, and that there is little variation in the duration of the syllables, though more variation in the intervals between the syllables.

To pursue this line of inquiry, a linguist would ask whether the two best-motivated structures, syllables and motifs, are subject to restrictions on their form. Do we find equivalents of initial strengthening, final weakening, final lengthening, and so on? There is evidence that suggests some songs tend to "drift" over time (Lambrechts, 1988, 1996; Chi & Margoliash, 2001), getting gradually slower, and Glaze and Troyer (2006) argue that the details of this process support a hierarchical model of song structure, with silent intervals stretching more than syllables, and syllable onsets aligning with global song structure. Other manipulations reminiscent of prosody in human language include "harmonic emphasis" in zebra finch song (Williams, Cynx, & Nottebohm, 1989), in which some harmonics (formants) are strengthened and others suppressed. This may be done only on certain syllables, and appears to be under vocal control. Zebra finches are also reported to display final lengthening (Scharff and Jarvis, personal communication). Finally, Tierney, Russo, and Patel (2008) look at 56 songbird families, and find some evidence of a statistically signifi-cant tendency for final notes to be longer than non-final ones. Their data do not allow one to distinguish between a preference for selecting longer note types in final position versus different durations for a single note type depending on its position in the song.

In the chaffinch (*Fringilla coelebs*), portions of a motif may be differentially lengthened. The song consists of a trill section followed by a flourish. Playback experiments suggest that females prefer longer flourishes (Riebel & Slater, 1998). Leitao et al. (2004), noting that an individual may show considerable variation in the length of the trilled section, suggest that the length of the trill section (i.e., the number of repetitions of the trilled syllable) may be used as a graded aggression signal, but this has not yet been tested in playback experiments.

A very interesting phenomenon is reported by Ballintijn and ten Cate (1999) in the collared dove. The collared dove (*Streptopelia decaocto*) is not a songbird, but it is nonetheless of interest here. It has a simple call that usually contains three coos. Some birds have only two coos in some calls, and the authors show that the duration of the call is essentially the same as the three-coo call, but with the final coo silent. In human language, silent beats of this kind are common in such things as poetic recitation (or indeed rap). More generally, timing is often preserved in speech even when segments are deleted, and this is typically achieved by compensatory lengthening of a neighboring segment. A second mechanism with human analogs is also of interest here: a few birds have calls of four coos, and Ballintijn and ten Cate argue that this results when the first coo is "stuttered."

In sum, there is acoustic, neurological, and behavioral evidence for the syllable, and for some higher-order constituents. There are some tantalizing glimpses of prosodic-like phenomena that manipulate these structures, but there is disappointingly little work on this question, which is ripe for future research.

How Could Structures Arise, Evolve, and Be Passed on to a New Generation?

The birdsong research provides evidence of at least five mechanisms that could, singly or in conjunction, give rise to structure when none existed before, or be used to create more complex structures out of simpler ones:

1. Genetic mutation plus sexual selection
2. Rate increases that can result in novel silent breaks
3. Interpretation of silent intervals as indicating constituent boundaries: learning of chunks/packages
4. Concatenation of motifs usually sung in isolation: A and B on their own have no structure, but AB has the beginnings of structure, since A precedes B
5. Addition of extra copies: AB has little "structure," but AABB is A^nB^n

I will deal with each in turn. The first case is a little different from the others because it posits genetic mutation molded by natural selection. The remaining four cases

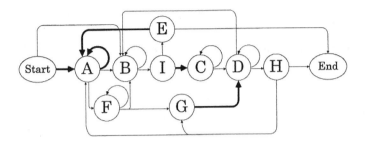

Figure 9.9
Transition diagrams of a Bengalese finch song. Reprinted with permission from Figure 4B of Honda and Okanoya (1999), *Zoological Science*, *16*, 319–326 .

involve an error in copying that is then culturally transmitted. These latter four cases are somewhat analogous to many cases of historical change in human phonology resulting from a misperception perpetuated during acquisition, which is then passed on to the next generation, as discussed in Blevins (2004).

Genetic Mutation Results in Novel Structured Song
Given that many songs are demonstrably innate, we might expect that genetic change could result in song change. Okanoya and colleagues have studied the difference in the songs of the wild and domesticated varieties of the Bengalese finch, and they suggest that a genetic mutation resulted in the appearance of structure in song. The wild Bengalese finch or white-rumped munia has a simple song (a b c d e f g h j a b c d e f g h j a b c d e . . .), but the domesticated variety has a complex song syntax, describable by a second-order Markov model (Honda and Okanoya 1999). (See Figure 9.9.)

This complex song is preferred by females, so it is sexually selected for. If one of these birds is lesioned in NIf (the higher-order song control nucleus), it causes simplification of this syntax. Honda and Okanoya (1999), Okanoya (2004), and Sasahara and Ikegami (2003) thus conclude that the change started with a genetic mutation, resulting in changes in the relevant brain structures, and then stabilized as a result of sexual selection.

The other four scenarios do not assume any genetic change; they would arise in the course of acquisition and then be culturally transferred to the next generation.

Rate Increases Can Result in Novel Silent Intervals
As songs speed up, they may begin to fragment, breaking a unitary song into two or more pieces. These can form the beginnings of internal structure.

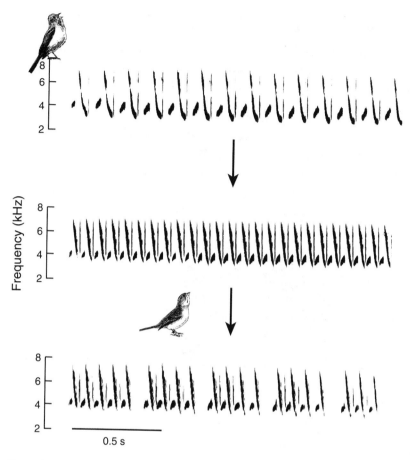

Figure 9.10
Normal song of swamp sparrow (top), speeded-up version (middle), and broken response (bottom). Reprinted with permission from Podos et al. (1999).

If the swamp sparrow (*Melospiza georgiana*) is presented with an artificially speeded up training song, they find this hard to learn, at least partly for motoric reasons caused by the excessively fast trill rate. One response is to produce the faster trill rate, but only in brief spurts separated by pauses, resulting in "broken syntax." If this novel song is then used to train the next generation, some will revert to an unbroken copy, but some learn the new "structured" song (Podos, Nowicki, & Peters, 1999).

The same mechanism could arise naturally and be responsible for the first internal structure in a species' song.

Interpretation of Silent Intervals as Constituent Boundaries: Learning of Chunks and Packages

Perhaps the most interesting source of new structures is found in the course of acquisition. Many species of birds do not seem to learn entire songs from their tutors, but smaller sections. These are then put together to form their own adult song, and may persist as structural entities in that song.

Young zebra finch males copy "chunks" from the single motif of their tutors' songs (Williams & Staples, 1992). Different birds copy different chunks of the same tutor song, but chunk boundaries are strongly correlated with the length of the silent interval. Figure 9.11 shows an adult spectrogram. The bars show the chunks copied

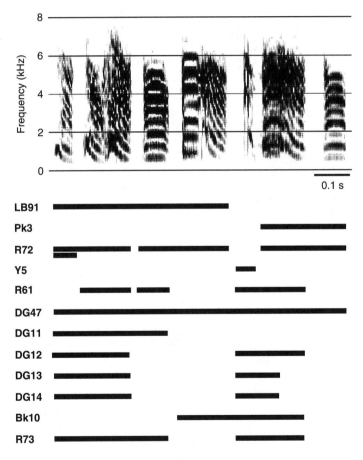

Figure 9.11
Chunks copied by 12 different young males from the same tutor song. Reprinted with permission from Williams and Staples (1992).

by 12 different young males. Most birds break between syllables 6 and 7, the longest silent interval, whereas almost no birds break between syllables 2 and 3, 5 and 6, or 8 and 9, where the silent interval is extremely short.

In their own song, subsequent production breaks are highly correlated with chunk boundaries, suggesting that these motif-internal chunks acquire some structural status in the new song. In human infants, we know that acquisition pays close attention to prosody, including pauses, in the process known as prosodic bootstrapping (Jusczyk, Hohne, & Mandel, 1995).

What about more complex songs? Nightingales *(Luscinia megarhynchos;* Hultsch & Todt, 1989) also learn their tutors' songs in continuous sections, but these are two to seven "song types" long, and are known as "packages." The breaks coincide with motif boundaries, but different birds pick different sequences, suggesting that any cues in the input song are ignored. Unfortunately for our purposes, the artificially prepared stimuli had a constant 3-second interval between motifs, meaning that it is not possible to know if package size is related to timing, or to information content, nor whether package breaks might correlate with the length of the silent interval in natural song. The packages are subsequently kept together in production (Todt & Hultsch, 1998, p. 495), suggesting that they form a new level of structure intermediate between song type and song bout.

Concatenation of Two Lone Song Types into a New Complex One

In the previous section, structure was described as arising because long songs were broken down into smaller pieces (and then reconstituted). The inverse of this can also happen: short songs are concatenated into new longer songs, and these could in principle retain their historical origins as structures within the new song.

Banded wrens (*Thryothorus pleurostictus)* have a repertoire of 20–25 motifs, 80% of which are shared between neighbor birds (Molles & Vehrencamp, 1999). These song types vary in form; there is no obvious "template." Occasionally, a bird will create a new song by putting together two of the original motifs into a new longer motif. If this new motif is composed of songs shared by other birds, it is a "compound" motif roughly analogous to the creation of a compound term like *ice cream* in human language, although presumably without meanings being attached to each subpart. As such it now has a minimal AB structure. More speculatively, if the next generation were to combine this with a third motif, the result would be a new motif with the structure (AB)C.

Addition of Extra Copies

A song like the compound song of the banded wren has minimal structure, AB. If these notes are copied, creating AABB, the structure emerges more obviously, and we end up eventually with something more like the song of the starling, with repeated motifs. An example from two dialects of one species is instructive.

Baptista (1977) looks at two white-crowned sparrow (*Zonotrichia leucophrys*) dialects. They differ in several respects, but two of the most prominent differences are the presence of an early buzz in the northern dialect that is missing in the southern dialect, and the presence of an extra copy of the medial complex syllable. So the southern dialect has Whistle-**Complex syllable**-Buzz-Trill, and the northern has Whistle-**Buzz**-**Comp.syll**-**Comp.syll**-Buzz-Trill. Some northern birds expand this sequence of complex syllables to three. One way to view this is that the only real structural change here is the addition of the new early buzz. Both songs contain a complex syllable section, which in the northern dialect is expanded by copying, but it is still one section. Such copying is common. For the brown thrasher, about a third of the motifs are "reduplications," with one syllable repeated twice (Kroodsma & Parker 1979).

Most of the mechanisms discussed here give rise only to structures that are much simpler than those of human language, but in full combination they start to build up a more complex picture with up to six levels of structure:

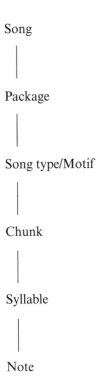

Song

Package

Song type/Motif

Chunk

Syllable

Note

What Should We Look for Next?

Although the preceding discussion makes it clear that the structures in birdsong are generally shallower than in human phonology, and also that they are neither headed nor binary, there are enough similarities to make it a plausible model for the structure of the earliest language (see also Sereno, 2005). We can also find suggestive evidence of mechanisms by which structure could arise, persist, and evolve.

However, there are two significant differences. First, the structures of human phonology are subject to strict conditions that control their size, shape, and realization. Take two examples. First, the syllable. The syllable always contains a vowel, and usually begins with a consonant. In some languages, the resulting CV sequence is the only well-formed syllable. VC, V, or CC won't do. In others, CVC is also permitted, and in some such languages the final consonant may be obligatorily unreleased. So far, the most complex analog of these sorts of well-formedness conditions in birdsong structures is the nightingale song-type template. What is more, the nightingale's distinctive final notes are found only in final position of the song type, rather reminiscent of human languages that restrict unreleased stops to syllable-final position. As a second example, consider the foot. In English a stress foot contains one to two syllables, and the first of these carries the stress. The vowel in the unstressed syllable may be reduced to a schwa. In many dialects phonemes have different allophones depending on their location in the foot: [t] is pronounced as aspirated at the start of this foot, but as flapped or glottalized in the middle (consider the initial versus medial [t] in U.S. English *titter*). I have not found any clear analogs of this situation in birdsong.

What might we look for? The equivalents would be well-defined limits on syllable or motif size; evidence for headedness within motifs, perhaps in terms of the pitch, amplitude, or duration of the component syllables; positional requirements such as an expectation that all syllables begin with a sharp onset and end with a gradual finish; and allophonic variation, such that the same syllable is subtly different when it begins a motif from when it ends a motif. As far as I know, there has been no systematic attempt to search for such things, and so we cannot yet know if they are out there in the song of some species.

The second difference is an apparent lack of real innovation within the confines of the template. In human language, a syllable template cannot only be filled by any suitable material, but speakers can create new sequences of syllables that they have never heard before. It is true that some species of birds are lifelong learners, but it is not clear that they create their own new songs. Nightingales, for example, sing new songs in their second year of life, but they appear to be songs that they heard as juveniles but did not sing in the first year (Kiefer et al., 2009).

The exchange of detailed ideas and information between phonologists and bird-song experts is a recent development and a healthy one. Each discipline has its own traditions, and we have each asked very different questions. We are now beginning to learn from each other. For example, we know much more about the neuroanatomy of birdsong than the neuroanatomy of human language, because the techniques used are not acceptable on humans. But we know much more about the cognitive organization of human language, because scientists can expect subjects to answer questions rather than simply having to watch reactions in playback experiments. Although we frequently cannot share experimental methods, we can share results, and as we do so more and more, the fields will only benefit.

Acknowledgments

I would like to acknowledge all the participants in the "Workshop on Birdsong, Speech and Language: Converging Mechanisms" in Utrecht in April 2007. Thanks especially to the organizers, Martin Everaert, Johan Bolhuis, and Peter Hagoort, and the community of birdsong scholars, who were unfailingly patient and courteous to a phonologist trying to understand their field, and who have been generous with their time and materials ever since. Special thanks to Tecumseh Fitch, Erich Jarvis, Gary Marcus, Daniel Margoliash, Kazuo Okanoya, Constance Scharff, and Carel ten Cate. I would also like to thank Fred Cummins, Henrike Hultsch, Nigel Mann, Peter Slater, and Robert Truswell for corresponding with me and supplying me with papers and even data. Finally, I would like to thank two anonymous reviewers, whose questions and suggestions greatly improved this chapter. None of these is responsible for any errors or omissions.

Notes

1. I ignore vowel reduction in unstressed syllables, and weight-related restrictions on syllable position. These would reduce the number of actual feet.

2. In some species, the syllable has been argued to correspond to the pattern of respiration (Franz & Goller, 2002).

3. Examples of natural classes are the labials [p,b,m,f,v], the stops [p,t,k,b,d,g], and the high vowels [i,u].

References

Abe, K., & Watanabe, D. (2011). Songbirds possess the spontaneous ability to discriminate syntactic rules. *Nature Neuroscience, 14*, 1067–1074.

Adret-Hausberger, M., & Jenkins, P. (1988). Complex organization of the warbling song in starlings. *Behaviour, 107*, 138–156.

Allard, H. A. (1939). Vocal mimicry of the starling and the mockingbird. *Science, 90*, 370–371.

Aronov, D., Andalman, A., & Dee, M. (2008). A specialized forebrain circuit for vocal babbling in the juvenile songbird. *Science, 320*, 630–634.

Ballintijn, M., & ten Cate, C. (1999). Variation in number of elements in the perch-coo vocalization of the collared dove (Streptopelia Decaocto) and what it may tell about the sender. *Behaviour, 136*, 847–864.

Baptista, L. (1977). Geographic variation in song and dialects of the Puget Sound white-crowned sparrow. *Condor, 79*, 356–370.

Berwick, R. C., Okanoya, K., Beckers, G. J. L., & Bolhuis, J. J. (2011). Songs to syntax: The linguistics of birdsong. *Trends in Cognitive Sciences, 15*, 113–121.

Blevins, J. (2004). *Evolutionary phonology*. New York: Cambridge University Press.

Bolhuis, J. J., Okanoya, K., & Scharff, C. (2010). Twitter evolution: Converging mechanisms in birdsong and human speech. *Nature Reviews: Neuroscience, 11*, 745–759.

Brenowitz, E. A., & Beecher, M. D. (2005). Song learning in birds: Diversity and plasticity, opportunities and challenges. *Trends in Neurosciences, 28*, 127–132.

Brumm, H., & Slater, P. (2007). Animal communication: Timing counts. *Current Biology, 17* (13), R522.

Chen, M. (1987). The syntax of Xiamen tone sandhi. *Phonology Yearbook, 4*, 109–150.

Chi, Z., & Margoliash, D. (2001). Temporal precision and temporal drift in brain and behavior of zebra finch song. *Neuron, 32*, 899–910.

Corballis, M. C. (2007). Recursion, language, and starlings. *Cognitive Science, 31*, 697–704.

Cummins, F. (2003). Practice and performance in speech produced synchronously. *Journal of Phonetics, 31*, 139–148.

Cynx, J. (1990). Experimental determination of a unit of song production in the zebra finch *(Taeniopygia gunata). Journal of Comparative Psychology, 104*, 3–10.

Devoogd, T. J., Krebs, J. R., Healy, S. D., & Purvis, A. (1993). Relations between song repertoire size and the volume of brain nuclei related to song: Comparative evolutionary analyses amongst oscine birds. *Proceedings of the Royal Society B: Biological Sciences, 254*, 75–82.

Doupe, A., & Kuhl, P. (1999). Birdsong and human speech: Common themes and mechanisms. *Annual Review of Neuroscience, 22*, 567–631.

Downing, L. (2006). *Canonical forms in prosodic morphology*. Oxford: Oxford University Press.

Eens, M. (1997). Understanding the complex song of the European starling: An integrated ethological approach. *Advances in the Study of Animal Behavior, 26*, 355–434.

Eens, M., Pinxten, M., & Verheyen, R. F. (1989). Temporal and sequential organization of song bouts in the European starling. *Ardea, 77*, 75–86.

Evans, B. G., & Iverson, P. (2005). Plasticity in speech production and perception: A study of accent change in young adults. *Journal of the Acoustical Society of America, 117*, 2426.

Fehér, O., Wang, H., Saar, S., Mitra, P., & Tchernichovski, O. (2009). *De novo* establishment of wild-type song culture in the zebra finch. *Nature, 459*, 564–568.

Fodor, J. A., & Bever, T. G. (1965). The psychological reality of linguistic segments. *Journal of Verbal Learning and Verbal Behavior, 4*, 414–420.

Ford, C. E., & Thompson, S. A. (1996). Interactional units in conversation: Syntactic, intonational, and pragmatic resources for the management of turns. In E. Ochs, E. A. Schegloff, & S. A. Thompson (Eds.), *Interaction and grammar* (pp. 134–184). Cambridge: Cambridge University Press.

Fougeron, C., & Keating, P. (1997). Articulatory strengthening at edges of prosodic domains. *Journal of the Acoustical Society of America, 101*, 3728–3740.

Franz, M., & Goller, F. (2002). Respiratory units of motor production and song imitation in the zebra finch. *Journal of Neurobiology, 51*, 129–141.

Gardner, T., Naef, F., & Nottebohm, F. (2005). Freedom and rules: The acquisition and reprogramming of a bird's learned song. *Science, 308*, 1046–1049.

Gentner, T. Q. (2004). Neural systems for individual song recognition in adult birds. *Annals of the New York Academy of Sciences, 1016*, 282–302.

Gentner, T. Q. (2008). Temporal scales of auditory objects underlying birdsong vocal recognition. *Journal of the Acoustical Society of America, 124*, 1350–1359.

Gentner, T., Fenn, K., Margoliash, D., & Nussbaum, H. (2006). Recursive syntactic pattern learning by songbirds. *Nature, 440*, 1204–1207.

Gentner, T. Q., & Hulse, S. H. (1998). Perceptual mechanisms for individual vocal recognition in European starlings, Sturnus vulgaris. *Animal Behaviour, 56*, 579–594.

Gentner, T. Q., & Hulse, S. H. (2000). Perceptual classification based on the component structure of song in European starlings. *Journal of the Acoustical Society of America, 107*, 3369–3381.

Glaze, C., & Troyer, T. (2006). Temporal structure in zebra finch song: Implications for motor coding. *Journal of Neuroscience, 26*, 991–1005.

Hall, M., & Magrath, R. (2007). Temporal coordination signals coalition quality in a duetting songbird. *Current Biology, 17*, R406–R407.

Harris, J. (2004). Release the captive coda: The foot as a domain of phonetic interpretation. In J. Local, R. Ogden, & R. Temple (Eds.), *Phonetic interpretation: Papers in Laboratory Phonology 6* (pp. 103–129). Cambridge: Cambridge University Press.

Helekar, S. A., Espino, G. G., Botas, A., & Rosenfield, D. B. (2003). Development and adult phase plasticity of syllable repetitions in the birdsong of captive zebra finches (*Taeniopygia guttata*). *Behavioral Neuroscience, 117*, 939–951.

Honda, E., & Okanoya, K. (1999). Acoustical and syntactical comparisons between songs of the white-backed munia Lonchura striata and its domesticated strain, the Bengalese finch Lonchura striata var. domestica. *Zoological Science, 16*, 319–326.

Hultsch, H., & Todt, D. (1989). Memorization and reproduction of songs in nightingales (*Luscinia megarhynchos*): Evidence for package formation. *Journal of Comparative Physiology A: Neuroethology, Sensory, Neural, and Behavioral Physiology, 165*, 197–203.

Hultsch, H., & Todt, D. (2004a). Approaches to the mechanisms of song memorization and singing provide evidence for a procedural memory. *Anais da Academia Brasileira de Ciencias, 76*, 219–230.

Hultsch, H., & Todt, D. (2004b). Learning to sing. In P. Marler & H. Slabbekoorn (Eds.), *Nature's music—the science of birdsong* (pp. 80–107). London: Elsevier Academic Press.

Jusczyk, P. W., Hohne, E., & Mandel, D. (1995). Picking up regularities in the sound structure of the native language. In W. Strange (Ed.), *Speech perception and linguistic experience: Theoretical and methodological issues in cross-language speech research* (pp. 91–119). Timonium, MD: York Press.

Ke, J. (2006). A cross-linguistic quantitative study of homophony. *Journal of Quantitative Linguistics, 13,* 129–159.

Kiefer, S., Sommer, C., Scharff, C., Kipper, S., & Mundry, R. (2009). Tuning towards tomorrow? Common nightingales Luscinia megarhynchos change and increase their song repertoires from the first to the second breeding season. *Journal of Avian Biology, 40,* 231–236.

Kluender, K., Lotto, A., Holt, L., & Bloedel, S. (1998). Role of experience for language-specific functional mappings of vowel sounds. *Journal of the Acoustical Society of America, 104,* 3568–3582.

Kroodsma, D., & Parker, L. (1979). Vocal virtuousity in the brown thrasher. *Auk, 94,* 783–785.

De Lacy, P. (Ed.). (2007). *The Cambridge handbook of phonology.* Cambridge: Cambridge University Press.

Ladd, D. R. (1986). Intonational phrasing: The case for recursive prosodic structure. *Yearbook of Phonology, 3,* 311–340.

Lambrechts, M. M. (1988). Great tit song output is determined both by motivation and by constraints in singing ability: A reply to Weary et al. *Animal Behaviour, 36,* 1244–1246.

Lambrechts, M. M. (1996). Organization of birdsong and constraints on performance. In D. E. Kroodsma & E. H. Miller (Eds.), *Ecology and evolution of acoustic communication in birds* (pp. 305–320). Ithaca, NY: Cornell University Press.

Leitao, A., van Dooren, T., & Riebel, K. (2004). Temporal variation in chaffinch *Fringilla coelebs* song: Interruptions between the trill and flourish. *Journal of Avian Biology, 35,* 199–203.

Leonardo, A., & Konishi, M. (1999). Decrystallization of adult birdsong by perturbation of auditory feedback. *Nature, 399,* 466–470.

Lerner, G. H. (1996). On the "semi-permeable" character of grammatical units in conversation: Conditional entry into the turn space of another speaker. In E. Ochs, E. A. Schegloff, & S. Thompson (Eds.), *Interaction and grammar* (pp. 238–276). Cambridge: Cambridge University Press.

Lerner, G. H. (2002). Turn-sharing: The choral co-production of talk in interaction. In C. Ford, B. Fox, & S. Thompson (Eds.), *The language of turn and sequence.* (pp. 225–256). Oxford: Oxford University Press.

Liu, W., Gardner, T., & Nottebohm, F. (2004). Juvenile zebra finches can use multiple strategies to learn the same song. *Proceedings of the National Academy of Sciences of the United States of America, 101,* 18177–18182.

Local, J. (2005). On the interactional and phonetic design of collaborative completions. In W. Hardcastle & J. Beck (Eds.), *A figure of speech: A Festschrift for John Laver* (pp. 263–282). Mahwah, NJ: Erlbaum.

Long, M., & Fee, M. (2008). Using temperature to analyse temporal dynamics in the songbird motor pathway. *Nature*, *456*, 189–194.

Mann, N. I., Dingess, K. A., & Slater, P. J. B. (2005). Antiphonal four-part synchronized chorusing in a neotropical wren. *Biology Letters*, *2*, 1–4.

Margoliash, D., Staicer, C. A., & Inouet, S. A. (1991). Stereotyped and plastic song in adult indigo buntings, *Passerina cyanea*. *Animal Behaviour*, *42*, 367–388.

Marler, P. (2000). Origins of music and speech: Insights from animals. In N. L. Wallin, B. Merker, & S. Brown (Eds.), *The origins of music* (pp. 31–48). Cambridge, MA: MIT Press.

Merker, B., & Okanoya, K. (2006). The natural history of human language: Bridging the gaps without magic. In C. Lyon, C. Nehaniv, & A. Cangelosi (Eds.), *Emergence of communication and language* (pp. 403–420). Berlin: Springer.

Molles, L., & Vehrencamp, S. (1999). Repertoire size, repertoire overlap, and singing modes in the banded wren (*Thryothorus Pleurostictus*). *Auk*, *116*, 677–689.

Mountjoy, D. J., & Lemon, R. E. (1995). Extended song learning in wild European starlings. *Animal Behaviour*, *49*, 357–366.

Nelson, D., & Marler, P. (1989). Categorical perception of a natural stimulus continuum: Birdsong. *Science*, *244*, 976–978.

Nooteboom, S. G., Brokx, J. P. L., & de Rooij, J. J. (1978). Contributions of prosody to speech perception. In W. J. M. Levelt & G. B. Flores d'Arcais (Eds.), *Studies in the perception of language* (pp. 75–107). New York: Wiley.

Okanoya, K. (2004). Song syntax in Bengalese finches: Proximate and ultimate analyses. *Advances in the Study of Behavior*, *34*, 297–345.

Okanoya, K., & Yamaguchi, A. (1997). Adult Bengalese finches require real-time auditory feedback to produce normal song syntax. *Journal of Neurobiology*, *33*, 343–356.

Onnis, L. (2003). *Statistical language learning*. Doctoral dissertation, University of Warwick.

Patel, A., Iversen, J., Bregman, M., & Schulz, I. (2009). Experimental evidence for synchronization to a musical beat in a nonhuman animal. *Current Biology*, *19*, 827–830.

Pearce, M. (2006). The interaction between metrical structure and tone in Kera. *Phonology*, *23*, 259–286.

Peña, M., Bonatti, L., Nespor, M., & Mehler, J. (2002). Signal-driven computations in speech processing. *Science*, *298*, 604–607.

Pepperberg, I. M. (2005). Grey parrots do not always "parrot": Roles of imitation and phonological awareness in the creation of new labels from existing vocalizations. *Language Sciences*, *29*, 1–13.

Podos, J., Nowicki, S., & Peters, S. (1999). Permissiveness in the learning and development of song syntax in swamp sparrows. *Animal Behaviour*, *58*, 93–103.

Prather, J., Nowicki, S., Anderson, R., Peters, S., & Mooney, R. (2009). Neural correlates of categorical perception in learned vocal communication. *Nature Neuroscience*, *12*, 221–228.

Quené, H. (1993). Segment durations and accent as cues to word segmentation in Dutch. *Journal of the Acoustical Society of America*, *94*, 2027–2035.

Riebel, K., & Slater, P. (1998). Testing female chaffinch song preferences by operant conditioning. *Animal Behaviour*, *56*, 1443–1453.

Riebel, K., & Todt, D. (1997). Light flash stimulation alters the nightingale's singing style: Implications for song control mechanisms. *Behaviour, 134,* 789–808.

Rose, G., Grolier, F., Gritton, H. J., Plamondon, S. L., Baugh, A. T., & Cooper, B. G. (2004). Species-typical songs in white-crowned sparrows tutored with only phrase pairs. *Nature, 432,* 753–757.

Saffran, J., Newport, E., & Aslin, R. (1996). Statistical learning by 8-month old infants. *Science, 274,* 1926–1928.

Sasahara, K., & Ikegami, T. (2003). Coevolution of birdsong grammar without imitation. In W. Banzhaf, J. Ziegler, T. Christaller, P. Dittrich, J. T. Kim (Eds.), *Advances in artificial life: Lecture notes in artificial intelligence* (pp. 482–490). Berlin: Springer.

Seddon, N. (2002). The structure, context and possible functions of solos, duets and choruses in the subdesert mesite *(Monias benschi). Behaviour, 139,* 645–676.

Selkirk, E. (1984). *Phonology and syntax: The relation between sound and structure.* Cambridge, MA: MIT Press.

Sereno, M. (2005). Language origins without the semantic urge. *Cognitive Science, 3,* 1–12.

Sober, S. J., & Brainard, M. S. (2009). Adult birdsong is actively maintained by error correction. *Nature Neuroscience, 12,* 927–931.

Suge, R., & Okanoya, K. (2010). Perceptual chunking in the self-produced songs of Bengalese finches (Lonchura striata var. domestica). *Animal Cognition, 13,* 515–523.

Suomi, K., McQueen, J. M., & Cutler, A. (1997). Vowel harmony and speech segmentation in Finnish. *Journal of Memory and Language, 36,* 422–444.

Szczepek-Reed, B. (2006). *Prosodic orientation in English conversation. Houndmills.* Hampshire, UK: Palgrave Macmillan.

Takahashi, M., Yamada, H., & Okanoya, K. (2010). Statistical and prosodic cues for song segmentation learning by Bengalese finches (*Lonchura striata var. domestica). Ethology, 116,* 481–489.

Tchernichovski, O., Mitra, P. P., Lints, T., & Nottebohm, F. (2001). Dynamics of the vocal imitation process: How a zebra finch learns its song. *Science, 291,* 2564–2569.

Thompson, N. S., Abbey, E., Wapner, J., Logan, C., Merritt, P. G., & Pooth, A. (2000). Variation in the bout structure of northern mockingbird (*Mimus polyglottos*) singing. *Bird Behaviour, 13,* 93–98.

Tierney, A. T., Russo, F. A., & Patel, A. D. (2008, July). *Empirical comparisons of pitch patterns in music, speech and birdsong.* Paper presented at Acoustics '08 (155th meeting of the Acoustical Society of America), Paris.

Todt, D., & Hultsch, H. (1998). How songbirds deal with large amounts of serial information: Retrieval rules suggest a hierarchical song memory. *Biological Cybernetics, 79,* 487–500.

Todt, D., & Naguib, M. (2000). Vocal interactions in birds: The use of song as a model of communication. *Advances in the Study of Behavior, 29,* 247–296.

Vroomen, J., Tuomainen, J., & de Gelder, B. (1998). The roles of word stress and vowel harmony in speech segmentation. *Journal of Memory and Language, 38,* 133–149.

Wildenthal, J. L. (1965). Structure in primary song of the mockingbird (*Mimus polyglottos). Auk, 82,* 161–189.

Williams, H. (2004). Birdsong and singing behavior. *Annals of the New York Academy of Sciences, 1016,* 1–30.

Williams, H., Cynx, J., & Nottebohm, F. (1989). Timbre control in zebra finch (*Taeniopygia guttata*) song syllables. *Journal of Comparative Psychology, 103,* 366–380.

Williams, H., & Staples, K. (1992). Syllable chunking in zebra finch (Taeniopygia guttata) song. *Journal of Comparative Psychology, 106,* 278–286.

Wingfield, J. C., & Lewis, D. M. (1993). Hormonal and behavioural responses to simulated territorial intrusion in the cooperatively breeding white-browed sparrow weaver, Plocepasser mahali. *Animal Behaviour, 45,* 1–11.

Yip, M. (2006). The search for phonology in other species. *Trends in Cognitive Sciences, 10,* 442–446.

Yu, A., & Margoliash, D. (1996). Temporal hierarchical control of singing in birds. *Science, 273,* 1871–1875.

10 Recursivity of Language: What Can Birds Tell Us about It?

Eric Reuland

Most researchers have their fascination. For me this fascination is the computational system of human language, briefly syntax, the way it relates to the interpretive system, and how it is embedded in our general cognitive system. Why choose a biological/evolutionary perspective? Like perhaps for many others of my colleagues, this perspective is not a main line in my research. But I got intrigued by the subject some nine years ago when I was asked to be a commentator in a workshop on the evolution of language, and this led me further to this symposium, where the net has been cast so wide that we discuss possibly converging mechanisms in birdsong, speech, and language. So, indeed, what does this perspective contribute?

Language looks like a very complex system. And in much of the discussion this keeps being stressed. But the more we focus on this complexity the harder it will be to understand it, and to relate what we know about language to what we know about very different biological systems. As the discussions in evolutionary biology show us, the evolution of complex organs can only be understood if we focus on the evolutionary origins of their component parts, and realize that the chances for the emergence of a complex organ by a number of evolutionary steps that require each other in order to have any selective advantage can become vanishingly small. Conversely, adopting an evolutionary perspective forces us to consider factors simple enough to have actually occurred in the events that gave rise to language.

In general, complexity can be intrinsic. However, complexity can also arise from the interaction between different components that are each simple. If we wish to relate the complexity of language to other biological systems, it seems to me that we have no choice but to decompose it, following the latter course.

What Makes Language Special?

What sets humans apart from cats, chimps, starlings, parrots, ants, lice, cod, squid, bees, seals, and other competitors for the resources of our world? Of course, we look different, we move different, we eat different, but above all we talk different. We have language, and they don't. But, one might say: Well, bees have language, don't they? And some dogs perfectly well understand when their boss says: *Sit!* Believe it or not, even my cat reacts, when I say: *Naar bed!* 'to bed!' And, of course, Hoover the harbor seal was able to say *Get outta here*! And look at parrots! So what's the difference with us? In their thought-provoking article on the evolution of language, Hauser, Chomsky, and Fitch (2002) put forward the challenging idea that what really sets us humans apart is our ability to create and handle discrete infinity. They propose that for a good perspective on the evolution of language we must distinguish between the faculty of language broadly conceived (FLB), and the faculty of language narrowly conceived (FLN). FLB contains aspects of language we share with our fellow creatures—for instance, the ability to produce sounds or gestures, and a system to organize an internal representation of the world around us (a conceptual system). The really novel and distinctive property of human language is that it reflects our ability to recursively combine minimal form-meaning combinations into interpretable expressions.

The notion of "recursion" used in Hauser et al. (2002) occasionally sparks discussion. Which of the uses of the term found in the literature is intended? If put into the perspective of Chomsky (2008), no more is intended than the notion of recursion underlying the definition of natural number:

(1) i.   1 is a natural number;
     ii. if n is a natural number, its successor n+1 is a natural number.
     iii. These are all the natural numbers.

So, we have an instruction that applies to its own output: if you want to create a natural number, take something that is a natural number and add 1 to it. More generally, we can say that *recursion* is the calling of an instruction while that instruction is being carried out.

In current theory natural language expressions are formed as in (2):

(2) i.   given a finite vocabulary V of lexical items, every e ∈ V is a natural language expression;
     ii. if e_1 and e_2 are natural language expressions $M(e_1, e_2)$ is a natural language expression;
     iii. These are the natural language expressions.

M is the Merge operation. Merge in turn is defined as follows:

(3) For any two natural language expressions e_1 and e_2: $M(e_1, e_2) = \{e_i, \{e_1, e_2\}\}$, with $i \in \{1,2\}$.

The Merge operation stipulates that the combination of two natural language expressions results in a set-theoretical object that represents, in addition to the combination itself, which of the objects combined is the "head" of the composite expression. In this way Merge reflects the pervasive asymmetry of natural language expressions; this asymmetry is nicely illustrated in the venerable textbook examples in (4):

(4) a. Merge (squadron, commander) →
{commander, {squadron, commander}}
Merge (top, (squadron, commander)) →
{{commander, {squadron, commander}}, {top, {commander, {squadron, commander}}}}
Simplified:
{commander, {top, {commander, {squadron, commander}}}}

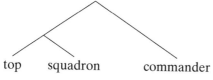

    top   squadron   commander

b. {{top, squadron}, commander}

    top   squadron   commander

In (4a) the result of merging *squadron* and *commander* is merged with *top*; in (4b) the result of merging *top* and *squadron* is merged with *commander*. In both cases the noun *commander* is the head of the construction. This order of Merge is reflected in the interpretation: in (4a) the commander is "top"; in (4b) the squadron is "top."

In earlier stages of the theory a natural language grammar was taken to consist of a set of recursively applying rules. For instance, as in (5),

(5) S → NP VP
NP → Det N′
N′ → A N′
N′ → N S′
N′ → N
A → bloody, sad, unbelievable, ….
N → fact, news, ….
S′ → that S
VP → V NP
V → suppressed, hid, …

Intuitively, if you want to construct a sentence, construct a nominal expression (a Noun Phrase — NP) and a predicate (a Verb Phrase — VP) and combine them. If you want to construct a nominal expression follow the construction rules for NP, and if you want to construct a predicate follow the rules for VP. It is easily seen that these rules allow you to construct an infinity of sentences, including *The fact suppressed the news*, *The bloody fact that the news hid the fact suppressed the sad, unbelievable news that the fact hid the news*, etc. The rules allow for any number of adjectives modifying a noun, and for the "content" of nouns such as *fact* or *news* to be specified by a sentence, which in turn may contain a noun whose content is specified by a sentence, and so on.

The selection of rules and lexical items given here is such that a "grammatical" outcome is guaranteed, but it is easily seen that any broadening of the set of nouns, verbs, and adjectives will lead to strange outcomes. So, any set of rules of this type must be supplemented by lexical properties of nouns, verbs, and adjectives, specifying their *selection restrictions* — for instance, facts may suppress news, but cannot hit, sadden, burn, worry, etc., news. We can have *the fact* that the news is sad, but not *the table* or *the rabbit* that the news is sad. We can have a bloody fact, but not a red or square fact.

A move in the development of linguistic theory culminating in Chomsky (2008) can be understood in the following way: we have selection restrictions anyway. Every item you select from the lexicon puts restrictions on what expressions it can be combined with. These semantic restrictions are part of our conceptual structure, and largely invariant across languages. We can ease the burden on the syntax if we avoid duplication, and let such dependencies be taken care of by conceptual principles and not incorporate them in the syntax. As Jim McCawley once noted in the 1960s, if someone maintains that his toothbrush is pregnant, you should not send him to a language course. If so, syntax is just the basic combinatory mechanism, not caring about pregnancy of toothbrushes or redness of facts. In Chomsky (2008) the notion of combinability is encoded in a minimal feature combinable lexical items have: their *edge feature*. The edge feature on a lexical element just expresses that it can be combined with a natural language expression, yielding an expression that in turn can be combined. Thus, the edge feature is what underlies (2).

This is almost all, but not quite. There is a type of dependency syntax should care for, namely *agreement*. Natural language is rife with requirements of the following type: if expression *a* is of a certain type, an expression *b* it is construed with must be of the same type.

In many languages predicates and NPs, used as their subjects, must *agree* in properties such as *person* and *number*, sometimes also gender. Even in English, impoverished as it is, we can see this in the contrast between ***the rabbit is*** running and ***the rabbits are*** running, but not *****the rabbit are** running and *****the rabbits is** running. Also, NPs show their relation to the verb by a *Case form*, as in *I/*me saw him/*he*.

In languages like German or Russian, agreement and Case requirements are pervasive. So, even if we dispense with semantic matching requirements as part of syntax, there are formal requirements that syntax must handle. We see another instance in question formation. If I know that John has been reading something, and I want to know what it was, I can say *What did John read?* *What* serves here in two capacities. It signals a question, but it also acts as the object of *read*. It even fills the object position, since we cannot say **What did John read it?* We can understand this as *For which object x is it true that John read x?* This dependency can wait for its satisfaction over some distance, as in *What do you think that Bill told me that Annie discovered that John read?*

From a processing perspective, this agreement phenomenon is just the possibility to put some formal property of an expression in store and look for a match. Or, from a production perspective, to put a property in store and make sure there will be a match. In neutral terms, agreement is just the property of matching formal features. The matching requirement can vary from "there must be an empty position that *what* can relate to" to "as for a third-person singular predicate, make sure that there is a third-person singular subject to be combined with." What we end up with, then, as constitutive of the syntactic combinatory system is just the *edge feature* as the embodiment of combinability, and *agreement/match* as the syntactic expression of dependency. Whereas combinability is free, matching is sensitive to the hierarchy that Merge produces. This can be seen on the basis of (6):

(6) i.

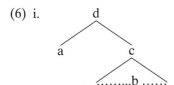

ii.

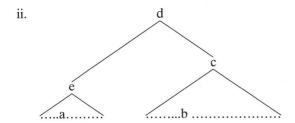

In (6i) *a* has been merged with *c*, forming *d*. That is, *a* is a *sister of c*. *a* can be in a dependency relation with *b* contained in *c*. The technical term for this is c-command:

(7) *c-command*
    a c-commands b iff b is contained in c, c a sister of a.

In (6ii) *a* cannot be in a dependency relation with *b*, since *a* does not c-command *b*. A simple illustration of the effects of c-command on dependencies is given in (8):

(8) a. [Mary's father]$_a$ [$_c$ admires himself$_b$]
    b. *[Mary$_a$'s father]$_e$ [$_c$ admires herself$_b$]

In (8a) *Mary's father* = *a* as a whole can *provide the value*/serve as the *antecedent* of *himself* = *b*, since *a* is a sister of the phrase *c* that contains *himself*. In (8b) *Mary* = *a* cannot serve as the antecedent of *herself* since *Mary* is too far embedded. It is an empirical claim that this relation of c-command is the *sole basis of syntactic dependency relations* in natural language grammars.[1]

One may wonder how this agreement relation is related to Merge: Is it an independent mechanism, or can it ultimately be reduced? For a subclass of dependencies it is uncontroversial that they can be reduced. As argued by Chomsky (1995) and subsequence work, the double duty seen with question words, and other cases of *dislocation*, can be explained if we assume that the Merge mechanism can access expressions already merged at a previous step. So, (9a) is derived by the following steps (using an embedded clause for ease of exposition, and omitting the set-theoretic notation):

(9) a. (I wonder) what John read
    b. Merge (read, what) → read what
    c. Merge (John, (read what)) → John read what
    d. Merge (what, (John read what)) → what John read ~~what~~
       (The strikethrough indicates that a phrase is not pronounced.)

The element *what* first inserted as an object of *read* in step b is reused and merged anew in the initial position of the embedded clause. This is what Chomsky refers to as *Internal Merge*. Reuland (2011) shows how this mechanism can be extended to dependencies of the type in (8) (discussion here would carry us too far afield). If so, we have one general mechanism both for building structure and for the syntactic encoding of dependencies.

This leads to the following general conception of the language system (Chomsky, 1995), where the syntax makes available objects that can be represented as forms at the Form interface, and can be interpreted as meanings at the Meaning interface:

(10)

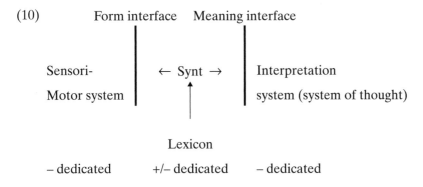

Mechanism available in syntax: Merge (external, internal)

Our sensorimotor system performs a variety of tasks (from eating, to drinking, to grasping and locomotion), hence is not dedicated to language (although it may well have certain properties that are special to language). Our system of thought may be used to set up and evaluate mental models of the world around us, is used in reasoning, and so on. Hence it is not dedicated to language, although it too may have properties that depend on language. Our lexicon contains elementary combinations of forms and meaning. To the extent that meanings reflect our conceptual system, it is not entirely dedicated. However, the very fact that its elements are elementary form-meaning combinations is special to language. This must be a dedicated feature, in fact even essential for language as we know it.

Merge was defined as an operation on lexical items and expressions formed of lexical items. If we define it as an operation on dedicated elements it is certainly dedicated to language. However, strictly speaking this is not necessary. We could take Merge to be an operation that can apply to any type of mental object resulting in complex, structured objects in that domain. In fact, Chomsky (2008) shows how the Merge operation in a simplified form applying in the domain of natural numbers can be used to implement the successor function; however, no role is played by the notion of headedness in this implementation. If so, the version of Merge with the expression of headedness would still be a candidate for a dedicated linguistic operation. Since it is hard to see how headedness can play a role in other domains, I will tentatively assume Merge to be indeed an operation dedicated to language. The next question is whether it is also constitutive of language.

Setting the Stage

As I argued in Reuland (2005) in discussing the "evolution of language" it is important to observe the following distinctions:

(11) i. The evolution of the human lineage up to the emergence of language
      ii. The event that gave rise to the faculty of language as we know it
      iii. The subsequent evolution of humans and the emergence of language

This sketch does not necessarily presuppose a clearcut distinction between human and nonhuman in the course of our evolution. Such a distinction will presumably be hard to draw in a nonarbitrary way. At what point was some ancestor nonhuman and did its offspring become human? In this form, the question is not sufficiently defined to warrant discussion. However, there is a clearcut distinction between us and our chimpanzee cousins, and there is nothing "in between." Hence the two lineages can be distinguished, and the distinction between (11i) and (11ii) must be real, provided (11ii) is real. The question is, then, whether it makes sense to talk about "the event" that gave rise to the faculty of language as we know it. I claim that, unlike in the case of (11i), it is possible to identify a trait that "clinches the matter," when added to whatever one might think of as a protolanguage and without which an ancestor could have had no "language as we know it." This trait can be present independently of whether an ancestor actually had developed the use of language.

My discussion is against the background of the current more general debate about what is special about language, as in Hauser et al. (2002), Pinker and Jackendoff (2005a), Fitch, Hauser, and Chomsky (2005), and Pinker and Jackendoff (2005b), as well as many of the contributors to this book. I will not recapitulate this discussion. Some of the issues will come up as the discussion goes on. The main question is what ultimately underlies the vast difference in linguistic abilities that separates us from even our most closely related nonhuman primate family members.

As schematized in (10), the computational system of human language enables a mapping between forms and interpretations. The mapping is based on an inventory of lexical elements representing elementary form-meaning pairs, and a combinatory system. The lexical elements represent the access channel to our conceptual system. Further requirements for language include memory systems (Baddeley, 2000, 2001, 2003; Ullman, 2004; Coolidge & Wynn, 2005, 2006). Their structure must be such that they allow planning (not just needed for language). Planning in turn involves imagination: the ability to construct models of the world that reflect how it should be in view of some goals, and to compare these with a model reflecting how it is (see Reuland, 2005). Having planning and imagination entails having at least a rudimentary theory of mind (ToM, imagining how it would look if I were you, etc.). We can summarize these requirements for language in (12):

(12) *Cognitive faculties underlying human language*
      Language requires:
      • Expressive system

-Inventory of elements representing elementary form/meaning pairs
-Conceptual system
-Combinatory system
- Interpretation system
Requirements for language use
- Memory systems (declarative/procedural)
- Planning
- Theory of mind (propositional attitude reports)
- ....

Let's compare this to what nonhuman primates have access to. They have a realization system that is not incompatible with language (either gestures or sound—see Fitch (2006) for arguments that the role of anatomy should not be overrated). They have a conceptual system, memory systems, and some forms of rudimentary planning, and ToM (see Bischof-Koehler, 2006, for discussion). But they do not have language in anything like our sense. So, there is an asymmetry. Whereas chimpanzees show some planning capacity under natural conditions, and some ToM under experimental conditions—both rudimentary as compared to humans—their language is not just rudimentary under natural conditions, but absent. Their language under experimental conditions stays far behind given what one would expect if planning and ToM were the decisive factors. The situation is summarized in (13):

(13) Non-humans may have functional homologues/analogues of:
  - Expression system
  - Inventory of elements representing elementary form/meaning combinations (under experimental conditions)
  - Conceptual system
  - Interpretation system
  - Memory systems
  - Planning
  - Theory of Mind (rudimentary, under experimental conditions)
  But no rudimentary form of a language in the relevant sense.

Why would this be so? Hauser et al. (2002) and Fitch et al. (2005) explore the idea that the core of the issue resides in the combinatorics of the syntactic system. To put it simply: humans have *recursion*, nonhumans do not. This brings us back to the previous discussion. It is the property of recursion that gives rise to the discrete infinity characteristic of human language.

Pinker and Jackendoff (2005a, 2005b), by contrast, argue that the differences between human and nonhuman functional homologs in cognitive functions are pervasive. They argue that there is no reason to single out recursion and the human *syntactic* combinatory system.

This leads us to the question of what types of evolutionary events are needed for language to have arisen. Let's assume some kind of protohuman in the "final stage" before language: What change could then lead to language as we know it? *Gradual changes* in what was already there? Adaptive, continuous, quantitative, etc.[2] Or rather a *discontinuity*?

It is important to note that there can be no evolutionary pressure on a trait without that trait already being *there* as a "target." Hence, at each turning point in the evolution of a species there must have been an evolutionary event that cannot be understood in adaptive terms. To understand the origins of language one must focus on those changes that are (i) constitutive of language and (ii) cannot be gradual, and then look for possible concomitant genetic changes.

Comparing the cognitive faculties underlying language and their functional homologs in nonhumans, there is indeed an important asymmetry. With all the caveats about adaptive value, one can imagine a gradual increase in working memory, accuracy of articulation, suppression of breathing, vocal range, speed in lexical access, etc., that could be selected for. But there can be no gradual increase in *recursivity*: recursivity is a yes-no property. This implies that the transition from a system without recursion to a system that has it is necessarily discontinuous. Therefore, there is a very good reason to single it out, pace Pinker and Jackendoff: it is a property that is nongradual by necessity.

There is another nongradual property underlying language: our ability to produce and deal with *arbitrary form-meaning combinations*. Many authors (from Deacon (1997) to Arbib (2005)) view protolanguage as a collection of *Saussurean signs* (De Saussure, 1916) with (14) being the traditional example:

(14)

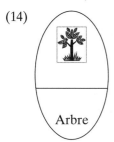

Such a Saussurean sign is a pair:

(15) a. <significant, signifié>, or
      b. <f,i>, where *f* is a form in a medium (sound, gesture), and *i* its
          interpretation as a concept.

As discussed in Reuland (2010), viewing the Saussurean sign as typical of the words of language is an oversimplification. Natural language has instructions for interpretation that cannot be immediately associated with a form in a medium, such as the empty subject in languages such as Italian (– *sono arrivati* '**they** have arrived'). It has forms that cannot be interpreted on their own, such as *of* in *the destruction of Rome* or *There* in ***There** arrived a man*. The relation between form and meaning is far more complex than is expressible in this notion of a sign.

Even so, we are able to handle arbitrary form-meaning combinations. What is rarely appreciated is that the emergence of our ability to handle arbitrary signs must reflect a discontinuity as well. Just as there cannot be only a bit of recursivity, there cannot be only a bit of arbitrariness. So, we have identified two yes-no properties underlying language. It is an open question whether they emerged from one and the same evolutionary event, or separately.[3]

Using this notion of a sign as a starting point, recursion/combinability can be implemented by a simple change in its format. Instead of (15) we need (16):

(16) <f, g, i>

with *f* and *i* as above, and *g* a formal instruction driving the computation.

The addition is *g*, as a formal instruction representing *combinability*. It is effectively Chomsky's "edge feature" that I discussed earlier. It is the addition of *g* that leads us beyond the Saussurean sign. It is a minimal change, but it is *qualitative* in nature. Adding this property to the sign opens the door to the development of the purely grammatical "machinery" such as "null subjects," purely formal prepositions, and so on, as I discussed above. In this sense, then, Merge is indeed constitutive of language.

Recursion and Birdsong

Hauser et al.'s (2002) proposal that recursivity is the core feature of FLN and represents what makes human language unique gave rise to three types of reactions: (i) there is much more in language that is unique; (ii) recursivity is not just the basis of syntax, but recursivity is also—or even primarily—a property of the other components of the language system, notably the conceptual system; and (iii) manifestations of recursivity are also found in other species. Pinker and Jackendoff's (2005a, 2005b) responses represent the lines in (i) and (ii). I will not discuss them here. Instead, I will focus on Marcus (2006), who represents the line in (iii).

Marcus discusses the consequences of a finding reported in Gentner, Fenn, Margoliash, and Nusbaum (2006). As Marcus puts it: "Gentner *et al.* showed that at least one non-human species, the European starling, can be trained to acquire complex recursive grammars. . . . This is strong evidence that humans are not alone in their capacity to recognize recursion."

To evaluate this claim we have to bring together two issues. What are the complex recursive grammars starlings are argued to have acquired, and what does the experiment tell us about the starlings' actual achievement?

Recursion and Formal Grammar Types

The discussion is conducted against the background of a theory of formal grammar types developed in Chomsky (1959); see Hopcroft and Ulmann (1969) for a by-now classic overview. Chomsky showed how the formal properties of rule systems correlate with the complexity of dependencies they can capture. For precision's sake I will recapitulate the main definitions characterizing this class of formal grammars.

In general, a grammar consists of a finite set of terminal symbols V_T (one may think of these as the "words" of the grammar), a finite set of nonterminal symbols V_N (symbols that if one wishes can be interpreted categories such as Noun, Noun Phrase, Verb, Verb Phrase, etc.), a start symbol (S) that belongs to the set of nonterminal symbols, and a finite set of rules relating expressions consisting of terminal and nonterminal symbols. The sets of terminal and nonterminal symbols are disjoint. Using standard notation, a grammar is represented as in (17):

(17) $G = (V_T, V_N, R, S)$, $S \in V_N$, $V_N \cap V_T = \emptyset$
   R is a set of rules of the general form $x \rightarrow y$, $x, y \in (V_T \cup V_N)^*$, and x is nonempty.

The notation $(V_T \cup V_N)^*$ stands for the set of all strings one can form concatenating the terminal and nonterminal symbols of the grammar. So, if one has a grammar in which V_T is {a,b}, and V_N is {S,A,B}, $(V_T \cup V_N)^*$ is an infinite set containing, among others the strings S,A,B,a,b, SA,SB,Sa,Sb, AS,AB, Aa,Ab, etc., including the empty string. Possible rules are all the rules given in (5), but also rules such as ABABA $\rightarrow$ aSb, ASa $\rightarrow$ SBbbbbbbS comply with the general format.

A derivation D is a sequence of strings $s_1 \ldots s_n$ over $(V_T \cup V_N)^*$ such that, for all i, $1 \leq i < n$, $s_i = uxv$, $s_{i+1} = uyv$, $x \rightarrow y \in R$, and $u,v \in (V_T \cup V_N)^*$.

A derivation D of s_n is terminated iff there is no s_{n+1} such that $s_n R s_{n+1}$.

The language generated by G is the set of all $z \in V_T^*$ such that there is a terminated derivation D with $s_1 = S$ and s_n is z.

These formal definitions just capture what is intuitively clear if we apply the rules in (5) to obtain English sentences.

While (17) is general, different classes of grammars arise by putting various restrictions on the format of the rules in R.

For instance, Chomsky (1959) discussed the following restriction on R:

(18) R is a set of rules of the general form $x \rightarrow y$, $x \in V_N$, and y has the form uA or u, with $u \in V_T^*$ and $A \in V_N$

Grammars of this type are also known as *finite-state grammars*. Consider the grammar in (19):

(19) V_T is {a,b}, V_N is {S, A,B}

R is {S→ aA, A→ aA, A→ a, S→ bB, B→ bB, B→ b}

As is easily seen, (19) derives the language consisting of all the strings of the form a^n, and all the strings of the form b^n, but no mixed forms. One of the questions occasionally raised is whether such grammars exhibit recursion. Clearly, what we see in a rule like B→bB is that it applies to its own output. It entails that we have subsequent lines in the derivation xB, xbB, xbbB, xbbb, each connected by an application of this rule. So, what we have here does fall under the standard notion of recursion. Systems of this type are highly restricted in their expressive power, though. This is because the recursive step only occurs in the periphery of the expression being derived: finite-state grammars only exhibit *peripheral recursion*. And in fact it does not matter whether recursion uniformly applies in the right periphery or the left periphery—that is, whether y in (18) has the form uA or Au.

As shown in Chomsky (1957, 1959), finite-state grammars (peripheral recursion only) lack the expressive power needed for dependencies found in natural language. Their essential limitation is that they are unable to represent unbounded dependencies. A simple formal example is the language of (20a), which is in fact the language Gentner used in his starling experiments.

(20) a. $a^n b^n$

b. R = {S → aSb; S→ ab}

(20a) stands for the set of strings consisting of a string of a's followed by an equal number of b's (all and only the strings meeting that requirement).

It has been formally proven that there is no finite-state grammar that can generate this language. Intuitively, it is easy to see why. Category symbols in a derivation encode "what the system has to do on the basis of what it has been doing." So, in a derivation by the grammar of (19), the category symbol A in a string aaaaaaA encodes that the system has been producing strings of a's, and the options are to go on or stop, and either is fine. Suppose, however, that a finite-state grammar is attempting to generate the language (20a), and suppose we have the string aaaaaaZ, where Z must encode how to go on. The options are to go on producing a's or to switch to producing b's. If the latter option is chosen, the system must also know how many a's have been produced, since the number of b's has to match it. In the current case Z must encode that six a's have been produced, hence six b's must follow. If instead another a is produced, another final category symbol Z+1 must be introduced to encode that property. By definition the total number of category symbols is finite. However, there is no upper limit on n. So, however large we choose the number of category symbols, there will always be a number n such that the system cannot encode for numbers equal to or higher than that n, that a^n must be followed by b^n.

What is impossible to represent in a finite-state grammar becomes easy under the seemingly trivial enrichment in (20b).

The crucial step is that (20b) shows *embedded recursion*. Under embedded recursion material is not necessarily added at the periphery, but in any position in the string to which the rule applies. So, given some space in which the computation is being carried out one can say: Write down ab and put in the middle another pair ab, etc., and stop at some point. This is what allows one to match a's and b's without an upper bound on the number of paired elements.

The rules in (20) exemplify a *context-free phrase structure grammar* (CFG). The one given here is very simple. The general format of CFGs is given in (21):

(21) R is a set of rules of the general form $x \rightarrow y$, $x \in V_N$, and $y \in (V_T \cup V_N)^+$
      where + stands for an arbitrary string formed over that vocabulary provided
      it is not empty.

All the rules in (5) obey the format of (21). Hence that grammar is a CFG. The example given has recursion over S and N'. By expanding it, more complicated patterns can be generated (I just note this, without going into detail). Consider again the first steps of (5), repeated in (22):

(22) $S \rightarrow NP\ VP, NP \rightarrow \ldots; VP \rightarrow \ldots$

In the expansion of VP one may encounter NPs (e.g., when the sentence has a direct object); in the expansion of NP one may encounter Ss and VPs (e.g., when the NP has a relative clause), etc. (23) illustrates this pattern:

(23)

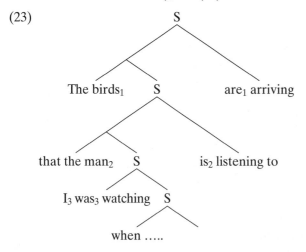

(23) shows a dependency between *the birds* and *are*, separated by a dependency between *the man* and *is*, in turn separated by a dependency between *I* and *was*. Note that at many points there can be further expansion. Adjectives can be added to *man* and *birds*, adverbial modifiers to *arriving*, *listening*, and *watching*, both as single adverbs and in the form of full clauses like the *when*-clause indicated.[4] This too requires embedded recursion, the power of a CFG.

The Starlings' Achievements

Let's now see how starlings fare. What did Gentner and his colleagues do? They rewarded European starlings for pressing a bar in response to strings of starling-produced sounds that were of the form $a^n\ b^n$—for instance, *rattle rattle warble warble*—and gave no reward for responses to strings of the form $(ab)^n$ (such as *rattle warble rattle warble rattle warble*) (and vice versa for another group of starlings). Learning required considerable training, but nine of eleven birds eventually (after 10,000–50,000 trials) learned to discriminate reliably between instances produced by the two grammars. By itself this is interesting, since these starlings succeeded, where cotton-top tamarin monkeys in an earlier experiment carried out by Fitch and Hauser (2004) had failed.

The question is what the success of the experiment tells us. First of all, the starlings were able to extend the $a^n\ b^n$ pattern only to new sequences of familiar sounds. But more importantly, their ability did not really go beyond $n = 3$. And, as we saw from the intuitive representation of the proof above, if n is maximized, there is a finite-state grammar that can generate the language. Can we really say that the starlings acquired embedded recursion? By raising this doubt, one might say, I open the door for doubting that human language requires recursion. There is certainly an upper limit to the length and complexity of sentence we humans can handle. However, in the case of human language it is clear that as soon as we wish to characterize "what is going on," our metalanguage requires recursive structures. We know that in the position between *the* and *birds* we can insert an arbitrary number of adjectives as modifiers of *birds*. We know that we can modify the predicate *are arriving* by as many modifiers as we would feel necessary; the same holds true for the NP and VP in the embedded clauses.

Even if, for the sake of the argument, one were to assume some practical upper limit on the number of natural language sentences, it is sure to be very large. It has been calculated that the number of grammatical English sentences of 20 words and less is 10^{20}. (Note that this very normal sentence is already over this limit, being exactly 21 words and costing 9 seconds to pronounce.) At an average of 6 seconds per sentence it will take 19,000,000,000,000 years to say (or hear) them all. A finite-state grammar, rather than being simple and practical, would in fact be of an unwieldy complexity, precisely since introducing recursive steps in intermediate positions would—by assumption—have to be avoided.[5]

The upshot is that recursivity is not a property of a system that can be assessed on the basis of a small finite sample of its output as such. If the output of a system is all we have, saying that the system producing this output uses recursive processes is an empirical hypothesis about this system, which has to be evaluated and put to the test like any empirical hypothesis. There is a difference between saying that the grammar in (20b) generates the language in (20a) and that no grammar without recursive procedures can do so, and saying that their being able to tell *rattle warble*, *rattle rattle warble warble*, and *rattle rattle rattle warble warble warble*, apart from *rattle warble rattle warble rattle warble*, or *rattle rattle rattle warble warble*, shows that European starlings are able to "acquire complex recursive grammars." The latter step is not warranted on the basis of such limited data.

This is not to deny that the experiment is very clever and interesting. The difference it shows between starlings and cotton-top tamarins is already important by itself. What the experiment as it is reported really demonstrates is that starlings are able to differentiate between patterns that tamarins aren't. I use *differentiate*, rather than *recognize*. The phrase *recognize a pattern* is ambiguous. If *recognize* is used in the meaning "see that *rattle rattle rattle warble warble warble* is different from *rattle rattle rattle warble warble*," its use is innocent and fine. If it is used with the meaning *recognize an underlying regularity* instantiated in *rattle warble*, *rattle rattle warble warble*, and *rattle rattle rattle warble warble warble*, but not in *rattle warble rattle warble rattle warble*, this would be misleading, since this is what has not been shown. But still, this ability to differentiate between patterns is a prerequisite for various types of learning. So it is truly interesting that the presence of this ability is not commensurate with the genetic distance from us.

The question is, then, what it would take to show that starlings or other songbirds have a system with embedded recursion. Consider the system they have been shown to have. Songbirds have an inventory of motifs. For instance, in the case of zebra finches there are stereotyped sequences of four to seven syllables. These constitute the basic repeated phrase of zebra finch song (Hessler & Doupe, 1999). Their and other songbirds' song is best described by a very simple finite-state grammar schematically given in (24):

(24) $S \rightarrow m_{1-n} S; S \rightarrow m_{1-n}$

   where m_{1-n} are members of M, M a finite inventory of motifs

What would be evidence of embedded recursion in this or any other bird species? Suppose we identified a number of motifs, just like themes in Western music. And suppose we found the pattern in (25a): a sequence of first halves of motifs followed by (in reverse order to keep the grammar simple) their respective second halves. So the structure could be as in (25b):

(25) a. $m1_1$ $m2_1$ $m3_1$ .... mn_{1+2} .... $m3_2$ $m2_2$ $m1_2$

b.

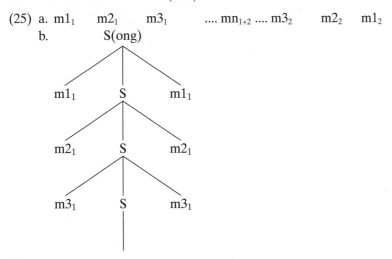

If something like this were a regularly occurring pattern in the songs of some type of bird, one would indeed be inclined to represent the regularity by means of a system with embedded recursion. Doing so would express the empirical hypothesis that this is how the internal system works. However, to my knowledge no such patterning has been observed to date. Consequently only the more modest interpretation of the starlings' achievement I presented above is warranted:

(26) Starlings can differentiate between patterns in sequences. They need not internally represent such patterns with systems exhibiting embedded recursion.

To differentiate between patterns, it is enough to have a good memory system. Consequently, no far-reaching claims about the classes of languages starlings are able to acquire can be supported.

By Way of Conclusion: Grammars and the Structure of Expressions

The discussion in the previous section only concerned what is called *weak generative capacity*: whether or not there is a grammar in a particular class of grammars that can generate all and only the strings fitting a particular description, such as $a^n b^n$, $a^n b^n c^n$, etc. However, as we saw in the first section, it is not only the string, but also the structure it is assigned that is interpreted. After the initial discussions about generative capacity in the 1950s and 1960s, the focus of the investigation shifted toward the structure the grammar assigns to expressions, and the question of what structure is needed for the interpretation system to operate on. As we saw, the Merge operation expresses more than just concatenation. Rather it creates a structure, and it is this structure that is essential for interpretation. From a current perspective the goal of grammatical theory is not so much to derive/generate the set of well-formed expressions in a language as to specify the form-meaning pairings of a language, reflecting the interpretation a speaker would assign to each form. For instance, if a is merged with b, it is crucial for the interpretation system whether a or b is to be the head. It is crucial for the interpretation system to "see" whether or not a depends on b or vice versa. If a is internally remerged to the structure containing it, this encodes a formal dependency between two positions, which the interpretation system will see as a real, interpretive dependency.

Having Merge (see (2)) as the basic operation defining linguistic structure embodies the hypothesis that the recursive definition of interpretable structures, not strings, is the "core business" of the human language faculty. This by itself is already vastly different from what songbirds have to offer, even in the most optimistic estimates of their capacities. However, Marcus (2006) does make a very important point. What he shows is that the notion of recursion in Hauser et al. (2002) was badly in need of refinement. This refinement is presented in Chomsky (2008).

In short, what birds can tell us about the processes of evolution is truly fascinating, but there is little reason to expect that they will tell us much about recursivity in the human language faculty.

Notes

1. One may wonder whether there is a further rationale for the c-command requirement. Reinhart (1983) already provided such a rationale, elaborated in Reuland (1998). An element such as *himself* or *herself* in (8) is defective. The interpretation of the expression *admired himself* cannot be completed unless *himself* has a value. We may assume that the syntactic computation carries along as little information as possible for reasons of economy; however, carrying over onto *c* that a value is needed is inevitable. Expressions like *Mary* or *Mary's father* do not need anything. They are both complete as they are. Consequently, at the level where we have *Mary's father* there is no intrinsic need to keep *Mary* accessible, hence, by economy it is not accessible, and the current phrase *Mary's father*, headed by *father*, is all the procedure has access to.

2. There is an important caveat about adaptive value. From the primordial soup evolved species as diverse as squid, E. Coli, jacaranda trees, lichens, sloths, us, ants, bonobos, cats, corn, . . .

This relativizes any story about adaptive values. The crucial notion is a niche, a particular type of environment. Any explanation of an evolutionary development in terms of adaptive value must be relativized to a particular niche, which may not be easy to reconstruct.

3. It is unclear whether arbitrariness carries some selective advantage with it (or was parasitic on another property with an advantage), or whether it just arose in part of the population as

a consequence of a minor mutation and stayed dormant until the emergence of recursion enabled its being put to use.

4. Although the type of dependency illustrated here falls within the power of CFG, natural languages show other dependencies that require a yet more powerful type of grammar (Chomsky 1957). For instance, the use of *respectively* in English, as in *John, Bill, Charles, . . . , and George are respectively blond, gray, red, . . . , and bald*, requires a crossing dependency that is beyond CFG. Note that in this case only the mapping onto the interpretation is beyond CFG. It requires more effort to construct natural language examples that show this at the level of the strings (Dutch bare infinitival complements of perception verbs are often cited). But for current purposes this example should suffice. Another instance of a formal language that is beyond CFG is $a^n b^n c^n$. For this pattern no natural language instantiation has been offered as far as I know.

5. To some colleague who might wish to argue, "Well, but finite is finite, and it is finite after all," one could respond saying that no principle requires a sentence to have only one author. If so, there is no upper limit on the length of sentences other than what is imposed by the age to be reached by the universe. No clear limit seems warranted here.

References

Arbib, M. A. (2005). From monkey-like action recognition to human language: An evolutionary framework for neurolinguistics. *Behavioral and Brain Sciences*, *28*, 105–124.

Baddeley, A. D. (2000). The episodic buffer: A new component of working memory? *Trends in Cognitive Sciences*, *4*, 417–423.

Baddeley, A. D. (2001). Is working memory still working? *American Psychologist*, *11*, 851–864.

Baddeley, A. D. (2003). Working memory and language: An overview. *Journal of Communication Disorders*, *36*, 189–208.

Bischof-Koehler, D. (2006). *Theory of mind and mental time travel: Specifically human abilities.* Paper presented at the Second Biennial Conference on Cognitive Science, St. Petersburg, Russia, June 9–13.

Chomsky, N. (1957). *Syntactic structures*. The Hague: Mouton.

Chomsky, N. (1959). On certain formal properties of grammars. *Information and Control*, *2*, 137–146.

Chomsky, N. (1995). *The Minimalist Program*. Cambridge, MA: MIT Press.

Chomsky, N. (2008). On phases. In C. Otero and M.-L. Zubizarreta (Eds.), *Foundational issues in linguistic theory: Essays in honor of Jean-Roger Vergnaud* (pp. 133–166). Cambridge, MA: MIT Press.

Coolidge, F. L., & Wynn, T. (2005). Working memory, its executive functions, and the emergence of modern thinking. *Cambridge Archaeological Journal*, *15*, 5–26.

Coolidge, F. L., & Wynn, T. (2006). *Recursion, pragmatics, and the evolution of modern speech.* Paper presented at the Cradle of Language Conference, November 9, University of Stellenbosch, South Africa, November.

Deacon, T. (1997). *The symbolic species*. New York: Norton.

De Saussure, F. (1916). *Cours de linguistique générale* (C. Bally & A. Sechehaye, Eds., with Albert Riedlinger). Paris: Payot.

Fitch, T. (2006). *Comparative data and fossil cues to speech*. Paper presented at the Cradle of Language Conference, University of Stellenbosch, South Africa, November 9.

Fitch, T., & Hauser, M. (2004). Computational constraints on syntactic processing in a nonhuman primate. *Science, 303*, 377–380.

Fitch, T., Hauser, M., & Chomsky, N. (2005). The evolution of the language faculty: Clarifications and implications. *Cognition, 97*, 179–210.

Genter, T. Q., Fenn, K. M., Margoliash, D., & Nusbaum, H. C. (2006). Recursive syntactic pattern learning by songbirds. *Nature, 440*, 1204–1207.

Hauser, M., Chomsky, N., & Fitch, W. T. (2002). The faculty of language: What is it, who has it, and how did it evolve? *Science, 298*, 1569–1579.

Hessler, N. A., & Doupe, A. J. (1999). Social context modulates singing-related neural activity in the songbird forebrain. *Nature Neuroscience, 2*, 209–211.

Hopcroft, J. E., & Ullman, J. D. (1969). *Formal languages and their relation to automata*. Reading, MA: Addison-Wesley.

Marcus, G. F. (2006). Startling starlings. *Nature, 440*, 1117–1118.

Pinker, S., & Jackendoff, R. (2005a). The faculty of language: What's special about it? *Cognition, 95*, 201–236.

Pinker, S., & Jackendoff, R. (2005b). The nature of the language faculty and its implications for evolution of language. *Cognition, 97*, 211–225.

Reinhart, T. (1983). *Anaphora and semantic interpretation*. London: Croom Helm.

Reuland, E. (1998). Structural conditions on chains and binding. In P. N. Tamanji & K. Kusumoto (Eds.), *Proceedings of NELS 28* (pp. 341–356). Amherst: GLSA, University of Massachusetts.

Reuland, E. (2005). On the evolution and genesis of language: The force of imagination. *Lingue e Linguaggio, 1*, 81–110.

Reuland, E. (2010). Imagination, planning, and working memory: The emergence of language. [Special Issue: *Working memory: Beyond language and symbolism*]. *Current Anthropology, 51* (S1), S99–S110.

Reuland, E. (2011). *Anaphora and language design*. Cambridge, MA: MIT Press.

Ullman, M. (2004). Contributions of neural memory circuits to language: The declarative/procedural model. *Cognition, 92*, 231–270.

11 Finite-State Song Syntax in Bengalese Finches: Sensorimotor Evidence, Developmental Processes, and Formal Procedures for Syntax Extraction

Kazuo Okanoya

Birdsong conveys the motivational, nutritional, cultural, and developmental condition of the singer (Nowicki, Peters, & Podos, 1998). Each birdsong note has specific acoustical properties; these song notes are ordered according to rules that are typically referred to as "syntax." However, they are *not* connected to any referential meanings (Hauser, 1997). Birdsong does not have syntax, but only ordering of the elements. Songs are sung by fully motivated individuals in response to potential mates or rivals (Catchpole & Slater, 2008). Singing itself establishes a context for mating or fighting, and the style of singing governs the intensity of the signal. For example, European blackbirds change the arrangement of song notes depending on how many rivals are singing. The order of song notes becomes increasingly jumbled as the aggressiveness of the singer increases (Dabelsteen & Pedersen, 1990). In this sense, the song is a graded or holistic system of communication rather than a compositional system like language, in which different combinations convey different meanings (Berwick, Okanoya, Beckers, & Bolhuis, 2011; Bolhuis, Okanoya, & Scharff, 2010).

Thus, it is impossible to directly compare the syntax of birdsong with that of human language. Instead, we need to identify a strategy for drawing valid comparisons between the two systems. One possible approach involves restricting the definition of syntax. Indeed, formal language theory addresses only the form and not the content of language (Hopcroft, Motwani, & Ullman, 2001). This theory uses the Chomsky hierarchy to characterize different degrees of linguistic complexity. At the lowest end of the hierarchy is finite-state grammar, in which sequences of language are determined only by local transitional relationships. This class of grammar has been described in terms of finite-state syntaxes in which state transitions produce certain strings. Context-free grammar, in which the production rule can contain itself and thus enables the recursive production of strings (Berwick et al., 2011), forms the next rung in the hierarchy. Higher classes of grammar have been described but are beyond the scope of this study.

After it is assigned to the appropriate level of the hierarchy, birdsong "syntax" can be compared with the different structural rules underlying human speech.

Analyses of song sequences have suggested that local transition rules adequately describe the order of the structures that contain the elements of birdsong. For example, the sequences of the elements in starling songs were analyzed with the first-order (e.g., A to B), second-order (e.g., AB to C), and third-order (e.g., ABC to D) transitional matrices, and the resulting entropy states were compared (Gentner, 2007). Results indicated that the first-order analyses provided full descriptions of the starling songs. The same analyses have been performed in Bengalese finches, and the second- or third-order matrix was necessary for a full description, but no self-embedding structures were found (Okanoya, 2004). Thus, finite-state syntax, a level of syntax that can be expressed by local transition rules alone, is an appropriate level of description for birdsong because no evidence of recursive structures has been identified with regard to this phenomenon.

To describe the finite-state syntax characterizing birdsong, both the terminal symbols and the structure by which these are ordered need to be identified. The first issue relates to the basic unit of song production and perception. Although birdsong consists of a set of elements that are delimited by silent intervals, these elements do not necessarily correspond to the unit of perception and production. Statistical analysis of co-occurring elements and behavioral analysis of song productions and perceptions are necessary in this regard. The second issue is related to the statistical and rule-based relationships between the units and the resulting topography. These relationships can be expressed in several ways, including traditional transition diagrams, state notations, sets of derivation rules, and more linguistically oriented branching diagrams.

In the following analyses, I used the Bengalese finch, a songbird, as a model to study the perceptual and motor aspects of song segmentation and chunking (Okanoya, 2004). Unlike other experimental birds, such as zebra finches, Bengalese finches sing complex songs that have a hierarchical structure and that are learned via segmentation and chunking. The smallest unit of song production is the "note." In most species, several notes are serially ordered to form a song, and the order becomes more or less fixed. However, Bengalese finches combine notes to constitute a "chunk" and then further combine several chunks in multiple ways to constitute song phrases (Figure 11.1, upper). Thus, Bengalese finch songs can be expressed by finite-state song syntax (Figure 11.1, lower). Because of their hierarchical and syntactic nature, Bengalese finch songs provide a suitable model for studying segmentation and chunking in perception and behavior (Okanoya, 2004). Furthermore, because songbird songs are learned, Bengalese finch songs represent an interesting parallel to the acquisition of human language (Okanoya, 2002).

As a note of caution, the term *syllable* is also used in birdsong literature. A syllable is a smallest unit of continuous repetitions in birdsong. If a song contains the

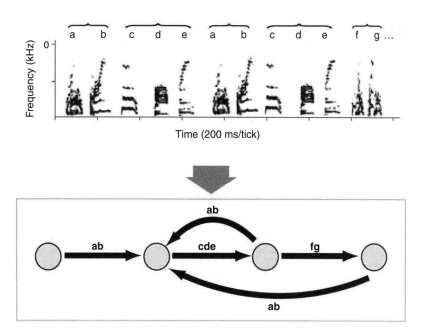

Figure 11.1
Upper: Song notes are organized into a "chunk," a set of co-occurring song notes. In this song, chunks ab, cde, and fg are identified. Lower: Transitional relationship among these chunks is expressed by the finite-state syntax.

sequence $(AB)^n$, then notes AB forms a syllable. I try to avoid this term, because a single note can also be a syllable if repeated. I prefer the term *chunk* because it does not imply whether it is continuously repeated or not.

Segmentation and Chunking

In human language, multiple syllables are chunked to form a word, and these words are arranged to form a sentence. Words have a perceptual and motor reality for humans in that we can use discrete memories to both perceive and produce a particular word, and a focal brain lesion can obliterate the mental correlates of a specific category of words (Hills & Caramazza, 1991). Do birdsongs have such units of production and perception? An examination of the developmental process of birds constitutes one approach to this question. Juvenile birds with more than one tutor often learn parts of songs from different tutors and merge the parts to form one song. Two species of songbirds, the zebra finch and the Bengalese finch, have been examined in detail with regard to their development of song segmentation (Williams & Staples, 1992; Takahasi, Yamada, & Okanoya, 2010).

A second approach to the question of whether birds have units of production and perception involves studying real-time song production and perception. This perspective underscores the motor aspect of birdsong insofar as it involves disrupting singing with a mild disturbance; the point at which a song terminates is considered to represent a boundary of motor chunking. If a motor program actually chunks together several elements, the break points in songs might correspond to the boundaries between these chunks. When zebra finches (Cynx, 1990) were subjected to experiments focused on this issue, the minimum unit of song production in this species was found to correspond to song notes (Seki, Suzuki, Takahasi, & Okanoya, 2008).

A study inspired by Chomsky's generative syntax (Chomsky, 1957) is very suggestive with regard to the perceptual aspects of this issue. When an extraneous stimulus such as a click was introduced while humans were processing a sentence, the part of the sentence that had perceptual unity could be processed prior to the click (Fodor & Bever, 1965). Indeed, humans can retain units of a higher order than words in their store of mental representations. Noun and verb phases are among these higher-order entities. The psychological existence of such abstract constructs has been demonstrated with a behavioral experiment. When subjects were asked to identify the positions of clicks while listening to a sentence, they often displaced the position of the click to the direction of the boundaries of phrases. Similar procedures can be applied to examine perceptual chunking in birdsong (Suge & Okanoya, 2010).

In the following sections, I present detailed accounts of a series of studies that addressed developmental, perceptual, and motor chunking in Bengalese finch songs. Ideally, if all of these experiments were done on a set of the same birds, integrative interpretations would be possible. However, in reality, each experiment was done on different sets of individuals.

Developmental Segmentation

I hypothesized that the complexity of songs sung by Bengalese finches might have developed from the perceptual and motor segmentation/chunking of available model songs by young birds. When chicks are raised by only one tutor, they learn the elements and order of songs primarily by copying the songs sung by that tutor. However, chicks with more than one tutor might learn pieces from each tutor and recombine them to form original songs. Our purpose was to clarify the developmental process by which Bengalese finches use chunking and segmentation to learn songs. The following is a short description of our recent work (Takahasi et al., 2010).

Eleven adult male and female Bengalese finches were kept in a large aviary with 11 nest pots where they raised a total of 40 male chicks within 6 months. When the finches reached adulthood, we analyzed their songs and assessed their similarity to the songs sung by tutors by examining which part of the song came from which tutor. Using second-order transition analysis, we also analyzed the songs sung by tutors and chicks in terms of transitional structures; we measured the note-to-note transition probabilities as well as the internote intervals and jumps in pitch.

Bengalese finches raised with multiple tutors learned songs from several tutors. Approximately 80% of the birds learned from between two and four tutors (Figure 11.2), implying the operation of three underlying processes. First, juvenile finches segmented the continuous singing of an adult bird into smaller units. Second, these units were learned as chunks when the juveniles practiced singing. Third, juvenile birds recombined the chunks to create an original song. As a result, the chunks copied by juveniles had higher transition probabilities and shorter intervals of silence than did those characterizing the boundaries of the original chunks. When one tutor song was copied by more than one individual pupil, the pattern of

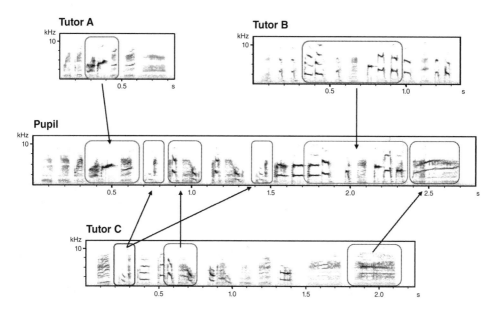

Figure 11.2
In a multitutor environment, chicks learn pieces of songs from multiple tutors and recombine these in certain orders to form original songs. This pupil learned segments of songs from three different tutors and combined these to make his own song.

chunking was similar among them. These processes suggest that Bengalese finches segmented songs by using both statistical and prosodic cues during learning.

Perceptual Segmentation

Following the experiment by Fodor and Bever (1965), we demonstrated the existence of a hierarchy characterizing song perception by Bengalese finches. The unit of song perception in this hierarchy is larger than the note, which is the smallest unit of song production.

Using operant conditioning, we trained male Bengalese finches to peck a key when they heard a short burst (10 ms) of white noise. The birds were trained to initiate the trial by pecking the observation key. After a random delay of 0–3 sec, a short burst of noise was presented, and the birds were trained to peck the report key within a limited time interval (4 sec initially, gradually decreasing to 1 sec). Correct detection of the noise burst within the time interval was reinforced with a reward of yellow millet seeds. We gradually shortened the time available for an effective response to train the birds to respond as soon as they heard the burst of noise. We then introduced the subject bird's own song as a background stimulus. After the bird responded to the observation key, we played a short segment (~2 sec) of the bird's song and superimposed a noise burst within the song, either inside or outside a "chunk" (two to five successive song notes that always appeared together). The song was played back either forward or in reverse.

We predicted that the reaction time for detecting the noise burst would be shorter when the noise burst occurred outside the chunk and longer when it was buried within the chunk. We also predicted that the reaction times would not differ when the song was reversed because a chunk would be processed before the noise burst within that chunk when the song was played forward. In addition, we hypothesized that reaction times to noise bursts that were outside a chunk would not differ from those to songs without a superimposed noise. Finally, we predicted that reaction times would not differ according to whether the noise burst was inside or outside the chunk when the song was reversed.

As predicted, the reaction times were significantly longer when the noise burst was within than when it was outside the chunk. We found no differences in reaction times when the song was reversed (Figure 11.3). Our results supported the notion that the perceptual unit of Bengalese finch songs is greater than an isolated note. The birds perceive songs by chunking notes and processing the chunk as a unit.

Motor Chunking

We next demonstrated the existence of the motor chunk, a higher-order unit of song production. We subjected singing Bengalese finches to a strobe light flash and deter-

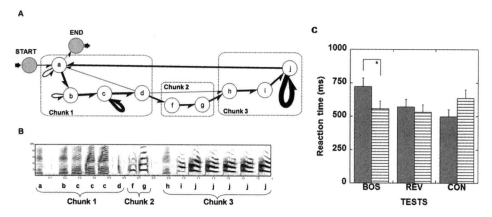

Figure 11.3
Perceptual correlates of chunk structure in Bengalese finch songs. (A) An example of a transition diagram based on a Markov analysis of a Bengalese finch song. (B) Sonogram indicating chunk structure of the song. (C) Operant reaction time detecting a short noise burst with a background of species-specific songs. BOS—bird's own song; REV—reversal of BOS; CON—conspecific song. Filled bar indicates the target was inside the chunk, and hatched bar indicates the target was outside the chunk.

mined the position in the song at which the bird stopped singing (Seki et al., 2008). Thirty song episodes were recorded for each of the eight birds. We then used the recordings to count the note-to-note transitions and calculated the transition probability of song notes for each bird. When a note transitioned to the same specific note with a >90% probability (i.e., they tended to appear together), the notes were regarded as a "within-chunk" sequence. After the songs were recorded, each bird was placed in a small cage in a sound-attenuated box.

A light was flashed when the bird began a song, and this usually resulted in the immediate termination of the song, implying that the fundamental unit of song production was the note. However, when the light interruption occurred within a chunk, the subsequent notes of that chunk tended to be produced, whereas interruptions presented between chunks tended to cause instantaneous song termination. This difference was statistically significant. These results suggested that the associations among sequences of song notes within chunks were more resistant to interruption than were those between chunks. This confirmed that Bengalese finch songs are characterized by a motor hierarchy (Figure 11.4).

Higher-order song units might exist not only as the unit of perception but also as a unit of motor production. In both cases, the unit should also have a cognitive ground. Thus, to produce songs, the brain of the bird might require a representation not only of each song note, but also of each group of song notes that serve as segmented parts of the song. Our results suggest that similar mechanisms of stimulus

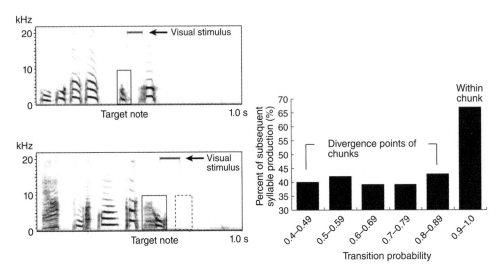

Figure 11.4
Motor correlates of chunk structure. (A) Song continued after the visual stimulation. (B) Song terminated after the visual stimulation. (C) Percentage of song continued after the visual stimulus.

segmentation operate in Bengalese finches and humans and that these result from complicated interactions among perception, motor, segmentation, and chunking processes.

Procedures to Extract Song Syntax

Two primary approaches have been proposed for extracting the syntax from a sequence of birdsong: the descriptive approach and the explanatory approach. The former attempts to produce a shorthand description of the birdsong sequence to enable comparisons among experimental manipulations or developmental processes, and the latter attempts to use certain mechanistic assumptions about a mental or a brain representation of the birdsong to provide an account of why a particular sequence emerged.

Descriptive models use actual data on song strings and try to condense these into a set of rules or probabilities. Earlier attempts primarily utilized the transition diagram based on element-to-element transitional probabilities (also known as a first-order Markov model or bigram). The weaknesses of such probabilistic representations involve their sensitivity to subtle fluctuations in data sets and their inability to address long-distance dependency. N-gram models (also called as $[N–1]$ th-order Markov models) try to overcome the second weakness by addressing probabilistic relationships longer than immediate adjacency and predicting the Nth elements based on the preceding $N-1$ elements. A variant of this model involves changing the length of N according to the data so that N is always optimal. However, these modifications are not free from the fluctuation problem (reviewed in Okanoya, 2004).

The hidden-Markov model (HMM) is among the most applicable models in terms of its explanatory power. This model assumes that actual elements of a song represent output from hidden states (Kogan & Margoliash, 1998). The task is to estimate the number of these hidden states, the transitional relationships between the states, and the probability with which each song element emerges from each of the states. This task requires prior knowledge or assumptions about the nature of the states, as well as inferences about the transitional relationships between the states (Katahira, Okada, & Okanoya, 2007; Katahira, Nishikawa, Okanoya, & Okada, 2010. The established neuroanatomical and neurophysiological characteristics of the song control system (Figure 11.5) suggest that HMM represents one candidate for modeling the valid representations involved in the generative aspects of birdsong. The HVC probably stores hidden states (Weber & Hahnloser, 2007), and the RA is probably responsible for producing each song element (Leonardo & Fee, 2005).

In the service of overcoming the drawbacks of the two approaches and of providing a model for both descriptive and explanatory purposes, we developed a set of procedures for automatically producing a deterministic finite-state automaton (Kakishita et al., 2009). We first constructed an N-gram representation of the sequence data. Based on this representation, song elements were chunked to yield a hierarchically higher-order unit. We then developed a diagram that mapped transitions among these units, and subjected this diagram to further processing for k-reversibility, where k-reversibility referred to the property of the resulting automaton that was able to determine the state that existed k steps back from the present state (Angluin, 1982). According to our experience with Bengalese finch songs, N was usually between 3 and 4, and k was often between 0 and 2. This set of procedures provided a robust estimation of automaton topography and has been useful in evaluating the effects of developmental or experimental manipulations on birdsong syntax. Furthermore, the resulting finite-state automaton can be used to guide further neurophysiological studies.

Discussion

Using analyses of Bengalese finch songs as examples, we demonstrated ways to advance studies in birdsong syntax and to clarify the relationship between the syntax of birdsong and human language. This approach offers both limitations and advantages. The limitations involve our original assumption that birdsong is a holistic

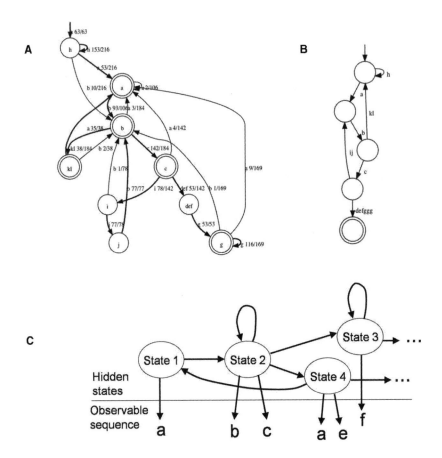

Figure 11.5
Song syntax representation. (A) Bigram model based on actual song episodes of a Bengalese finch. (B) The same data set as A, but analyzed with 4-gram model followed by a 0-reversible automaton. (C) A hypothetical song sequence and its hidden Markov representation. A and B are derived from Kakishita et al. (2009).

signal with constituents that do not correspond to particular meanings; thus, bird-song is not characterized by compositionality. As a result, birdsong can serve as only a formal model for human syntax rather than as a detailed model for this phenom-enon. However, the absence of referential content in birdsong does not imply its inadequacy as a model of human language if we assume that the mechanisms under-lying form and content are independent. It has been shown that human inferior frontal areas respond differently to strings derived from finite-state grammar than they do to those derived from context-free grammar. Area 45 was active for both types of grammar, whereas Area 44 was active only in response to context-free grammar (Bahlmann, Schubotz, & Friederici, 2008). Thus, satisfactory linguistic form, even without linguistic content, activated neural systems involved in language. In this regard, recent demonstrations that starlings and Bengalese finches could parse context-free grammar in an operant discrimination task (Gentner, Fenn, Mar-goliash, & Nusbaum, 2005; Abe & Watanabe, 2011) is somewhat unexpected. Although no evidence for the use of recursive mechanisms to produce birdsong has been reported, birds might be able to perceptually process such structures. However, reanalyses of stimulus properties suggest alternative explanations: phonetic cues might be sufficient to account for such results (Beckers, Bolhuis, Okanoya, & Berwick, 2012). Thus the issue remains unsolved and a breakthrough in conceptual and experimental design is expected to clarify this area of research (Berwick et al., 2011).

For present purposes, birdsong syntax should serve as an attractive biological model for human language acquisition because key features are common to both systems (Bolhuis et al., 2010). For example, the development of both birdsong and human language is characterized by a period of babbling, premature vocalizations similar to adult phonology but less stereotyped. A recent neuroanatomical study demonstrated that lesioning the higher-order vocal control nucleus (analogous to Broca's area) in zebra finches abolished normal song development; birds having such treatment remained at the babbling stage of song production. Thus, the higher vocal area is absolutely necessary for the crystallization of song syntax in birds (Aronov, Andalman, & Fee, 2008).

More direct comparisons can be drawn from recent results in our own laboratory. Using artificial sound sequences that were segmented based solely on statistical cues, we found that when human adults continued listening to auditory streams, an area adjunctive to the anterior cingulate cortex was activated during the initial learning phase, and this activity gradually disappeared as learning advanced (Abla, Katahira, & Okanoya, 2008). When the subject completely learned the segmentation task, inferior frontal areas, including Broca's area, were activated (Abla & Okanoya, 2008). It will be interesting to test whether juvenile birds with lesions performed on a structure analogous to the anterior cingulate cortex (Okanoya, 2007) and reared in a multitutor environment learn by segmentation and chunking. In Bengalese finches, lesioning one of the higher-order song control nuclei, the NIf, substantially reduced the syntactic complexity of songs by eliminating some of the note-to-note transitions (branches) from the original song (Hosino & Okanoya, 2000); thus, this structure seems to contribute to the development of complexity in songs. In addition, we used single-unit electrophysiology to show that note-to-note transitional information is population-coded in the HVC neurons of Bengalese finches (Nishikawa et al., 2008). The accumulation of neurophysiological evidence on chunking, segmentation, and the operations governing the stochastic transitions between song notes should enable understanding of the mechanisms responsible for sequential processing in language and birdsong.

References

Abe, K., & Watanabe, D. (2011). Songbirds possess the spontaneous ability to discriminate syntactic rules. *Nature Neuroscience*, *14*, 1067–1074.

Abla, D., Katahira, K., & Okanoya, K. (2008). Online assessment of statistical learning by event-related potentials. *Journal of Cognitive Neuroscience*, *20*, 952–964.

Abla, D., & Okanoya, K. (2008). Statistical segmentation of tone sequences activates the left inferior frontal cortex: A near-infrared spectroscopy study. *Neuropsychologia*, *46*, 2787–2795.

Angluin, D. (1982). Inference of reversible languages. *Journal of the Association for Computing Machinery*, *29*, 741–765.

Aronov, D., Andalman, A. S., & Fee, M. S. (2008). A specialized forebrain circuit for vocal babbling in the juvenile songbird. *Science*, *320*, 630–634.

Bahlmann, J., Schubotz, R. I., & Friederici, A. D. (2008). Hierarchical artificial grammar processing engages Broca's area. *NeuroImage*, *42*, 525–534.

Beckers, G. J. L., Bolhuis, J. J., Okanoya, K., & Berwick, R. C. (2012). Birdsong neurolinguistics: Songbird context-free grammar claim is premature. *NeuroReport*, 23, 139-145.

Berwick, R. C., Okanoya, K., Beckers, G. J. L., & Bolhuis, J. J. (2011). Songs to syntax: The linguistics of birdsong. *Trends in Cognitive Sciences*, *16*, 113–121.

Bolhuis, J. J., Okanoya, K., & Scharff, C. (2010). Twitter evolution: Converging mechanisms in birdsong and human speech. *Nature Reviews: Neuroscience*, *11*, 747–759.

Catchpole, C., & Slater, P. J. B. (2008). *Bird song*. Cambridge: Cambridge University Press.

Chomsky, N. (1957). *Syntactic structures*. The Hague: Mouton.

Cynx, J. (1990). Experimental determination of a unit of song production in the zebra finch (Taeniopygia guttata). *Journal of Comparative Psychology*, *104*, 3–10.

Dabelsteen, T., & Pedersen, S. B. (1990). Song and information about aggressive responses of blackbirds, *Turdus merula*: Evidence from interactive playback experiments with territory owners. *Animal Behaviour*, *40*, 1158–1168.

Fodor, J., & Bever, T. (1965). The psychological reality of linguistic segments. *Journal of Verbal Learning and Verbal Behavior*, *4*, 414–420.

Gentner, T. (2007). Mechanisms of temporal auditory pattern learning in songbirds. *Language Learning and Development*, *3*, 1–22.

Gentner, T., Fenn, K., Margoliash, D., & Nusbaum, H. (2005). Recursive syntactic pattern learning by songbirds. *Nature*, *440*, 1204–1207.

Hauser, M. D. (1997). *Evolution of communication*. Cambridge, MA: MIT Press.

Hills, A., & Caramazza, A. (1991). Category-specific naming and comprehension impairment: Double dissociation. *Brain*, *114*, 2081–2094.

Hopcroft, J. E., Motwani, R., & Ullman, J. D. (2001). *Introduction to automata theory, languages, and computation*. Boston: Addison-Wesley.

Hosino, T., & Okanoya, K. (2000). Lesion of a higher-order song nucleus disrupts phrase level complexity in Bengalese finches. *NeuroReport*, *11*, 2091–2095.

Kakishita, Y., Sasahara, K., Nishino, T., Takahasi, M., & Okanoya, K. (2009). Ethological data mining: An automata-based approach to extracting behavioral units and rules. *Data Mining and Knowledge Discovery*, *18*, 446–471.

Katahira, K., Nishikawa, J., Okanoya, K. & Okada, M. (2010). Extracting state transition dynamics from multiple spike trains using Hidden Markov Models with correlated Poisson distribution. *Neural Computation*, *22*, 2369–2389.

Katahira, K., Okada, M., & Okanoya, K. (2007). A neural network model for generative complex birdsong syntax. *Biological Cybernetics*, *97*, 441–448.

Kogan, J. A., & Margoliash, D. (1998). Automated recognition of bird song elements from continuous recordings using dynamic time warping and hidden Markov models: A comparative study. *Journal of the Acoustical Society of America*, *103*, 2185–2196.

Leonardo, A., & Fee, M. S. (2005). Ensemble coding of vocal control in birdsong. *Journal of Neuroscience*, *25*, 652–661.

Nishikawa, J., Okada, M., & Okanoya, K. (2008). Population coding of song element sequence in the Bengalese finch HVC. *European Journal of Neuroscience*, *27*, 3273–3283.

Nowicki, S., Peters, S., & Podos, J. (1998). Song learning, early nutrition and sexual selection in songbirds. *American Zoologist*, *38*, 179–180.

Okanoya, K. (2002). Sexual display as a syntactical vehicle: The evolution of syntax in birdsong and human language through sexual selection. In A. Wray (Ed.), *Transition to language* (pp. 46–63). Oxford: Oxford University Press.

Okanoya, K. (2004). Song syntax in Bengalese finches: Proximate and ultimate analyses. *Advances in the Study of Behavior*, *34*, 297–345.

Okanoya, K. (2007). Language evolution and an emergent property. *Current Opinion in Neurobiology*, *17*, 271–276.

Seki, Y., Suzuki, K., Takahasi, M., & Okanoya, K. (2008). Song motor control organizes acoustic patterns on two levels in Bengalese finches (*Lonchura striata* var. *domestica*). *Journal of Comparative Physiology A: Neuroethology, Sensory, Neural, and Behavioral Physiology*, *194*, 533–543.

Suge, R., & Okanoya, K. (2010). Perceptual chunking in the self-produced songs of the Bengalese finches, *Lonchura striata* var. *domestica. Animal Cognition, 13*, 515–523.

Takahasi, M., Yamada, H., & Okanoya, K. (2010). Statistical and prosodic cues for song segmentation learning by Bengalese finches. *Ethology, 119*, 481–489.

Weber, A., & Hahnloser, R. (2007). Spike correlations in a songbird agree with a simple Markov population model. *PLoS Computational Biology, 3*, e249.

Williams, H., & Staples, K. (1992). Syllable chunking in zebra finch (*Taeniopygia guttata*) song. *Journal of Comparative Psychology, 106*, 278–286.

12 Analyzing the Structure of Bird Vocalizations and Language: Finding Common Ground

Carel ten Cate, Robert Lachlan, and Willem Zuidema

Virtually all birds produce vocalizations, but bird species differ wildly in the complexity of their vocal repertoire. In some species we only find a handful of peeps or calls; in others, like the blackbird or nightingale, we find elaborate songs. These songs are built up from building blocks ("elements") that each require a sophisticated motor program to be produced as well as advanced auditory analysis to be recognized. Moreover, many species maintain a large repertoire of possible elements, and follow complex rules when sequencing elements into songs, or sequencing songs into "song bouts."

In the last 50 years or so, many researchers have on various occasions pointed at similarities between the production, development, and perception of bird vocalizations and that of speech sounds and language. Although there are equally interesting parallels in the complexity of their structure, only few studies have attempted to see whether this can be tackled in a similar way. Only occasionally, popular models that were first developed for describing language have been used to describe "song syntax"; these include the use of n-gram models (Lemon & Chatfield, 1971, 1973), finite-state machines (Okanoya, 2004), and hierarchical models (Hultsch, Mundry, & Todt, 1999). It seems fair to say, however, that none of these attempts has really found its way to the standard toolbox of the birdsong researcher. A more direct comparison of linguistic and song syntax would be very useful, because it can reveal how the complexities of each relate to those of the other and make clear if and where fundamental differences are present. In recent years there has been increasing interest in analyzing the organizational rules underlying birdsong from the perspective of whether song organization shares any features with human language syntax patterns. Such parallels have received new impetus from the view that one of the most important differences between the human language system and the vocal structure in other animals is in the level of syntactic complexity (Hauser, Chomsky, & Fitch, 2002). Examining such parallels requires comparable approaches to analyzing birdsong and language, and thus the definition of units and sequencing rules in similar ways.

In this chapter, we review several findings about birdsong syntax and discuss their relation to human language and the various techniques developed to formalize and quantify syntactic structure. Such efforts start from the observation that in many species of songbirds we can recognize both a limited set of basic elements and a precise set of rules that seem to regulate the order of those elements in a phrase, song, or song bout. Figure 12.1, for instance, shows spectrograms of four very similar songs recorded from a single blackbird (unpublished data, Monique Gulickx, Leiden University). Blackbird song follows a stereotyped scheme, where a low-pitched *motif* part consisting of several elements is followed by a high-pitched *twitter* part consisting of two or more elements from a different set. Examples A and B only differ in the motif, where B's motif contains one additional element. Examples A and C only differ in the twitter, where A contains one repetition and C a second repetition of what appear to be identical elements. Song D is yet another variation, with many repetitions of a second twitter element, following the two productions of the earlier twitter element. Blackbird song often shows such variations. This makes it unlikely that each song represents a single and independent representation.

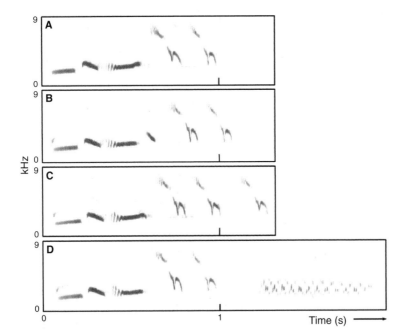

Figure 12.1
Four similar songs from a blackbird.

More likely, the similarities between these four different songs are a consequence of the singing bird using the same set of elements/motor programs, but slightly altering their sequencing between the four instances. The aim of this chapter is to analyze how researchers have tried to come to grips with this and other types of variation in song output. To this end, we will review the state of the art in birdsong research on the syntax of songs and compare it to observations from linguistics, with the aim of identifying the types of tools that would potentially be helpful in these fields.

Vocalizations, Birdsong, and Language

Although the songs of songbirds are only one of a range of vocalizations produced by birds, they are most conspicuous and complex and for that reason invite comparison with features of human speech and language. Songs are defined as the vocalizations that birds make in the context of mate attraction and competition; most species also have other types of vocalizations, commonly known as "calls." Songbirds form the taxonomic unit of *oscines*, and are by far the largest bird order, comprising about 4,000 of the 9,000 bird species. Songbirds are special in that they need exposure to song, usually at an early age, in order to sing this song later on. In this respect the development of song differs from that of calls, which often (though not always) develop without previous exposure to calls. As highlighted in other chapters in this book, the song-learning process shares many features with that of vocal development in humans. For this reason songbirds are the most popular bird model for comparative research.

Like birds, all humans produce communicative signals, using the vocal-auditory or other modalities, and they do so very frequently. Not all of that signaling, however, counts as language. Laughter, for instance, falls outside of what linguists like to call "natural language." In this chapter, we will compare findings in birdsong to findings in natural language phonology (concerning the units of sound and the rules governing sequences of sounds) as well as in morphosyntax (concerning the structure of words and the stringing together of words into sentences). Throughout the chapter, we sometimes talk about song or language as a property of populations and sometimes as a property of individuals. It is important to distinguish these levels. In humans, we can assume that the language of individuals is a relatively large subset of the language of populations in both phonology and syntax. In songbirds, this differs between species and between aspects of the song: it is unusual for an individual bird to produce in its own repertoire a large proportion of the phonological variation within a population, although exceptions exist. In contrast, syntactic structure is often quite rigid within a species, and a single bird's repertoire may express the entire range of syntactic structure found in a population.

Identifying the Units of Song and Language

The first major problem one encounters when trying to describe the structure of birdsong (or language) is the difficulty of deciding on elementary building blocks and the level of analysis. The basis for classifying units is usually how they appear on sonograms. An uninterrupted trace on a sonogram is typically taken as the smallest unit, commonly indicated as a *note* or *element* (we will use the term *element*). At the other end of the spectrum, the term *song* is usually applied to the longest unit.

One of the earliest methods used to distinguish between different levels within songs has been to examine the frequency distribution of the duration of the pauses between elements (e.g., Isaac & Marler, 1963). A song can then be defined as a series of elements separated by brief pauses. Songs are usually separated by audible pauses that are substantially longer, during which a bird might show other behaviors. Between the level of the element and that of the song one can often distinguish intermediate levels of structure. Thus, the intervals that separate subsequent elements may show a nonuniform frequency distribution, with two or more clear peaks. If so, this can be used to identify larger units. This is how Isaac and Marler (1963) distinguished between elements (units separated by pauses < 45 msec) and *syllables* (groups of elements separated by pauses > 45 msec) in the song of the mistle thrush. If there is a second peak of intervals within songs, this might be called *phrases* (e.g., see Figure 12.2), although that term is also often used for recurrent series of elements or syllables, as in zebra finch song.

Even apart from such confusing differences in terminology, assigning an acoustic unit to a specific category is rarely unambiguous between or even within species. For instance, if one compares the songs of zebra finch tutors with those of pupils in song-learning experiments, one may find that what seems like one unit in one of them is a combination of separate ones in the other (e.g., Franz & Goller, 2002). Thus, researchers often use additional criteria or are forced to make arbitrary or intuitive decisions when delineating the different units. For example, a continuous trace of a sonogram showing sudden transitions in character, like switching from a noiselike structure to a more tonal one, might be used to label each separate part as an element, or the full unit as a syllable. Researchers have also tried to arrive at classifications based on how acoustic structures are produced by birds. For the zebra finch, for instance, Cynx (1990) showed that flashing a light during singing resulted in song interruptions that were mostly (but not always) in the brief pauses between rather then within the elements, supporting the notion of visually identified elements as units of production. In another study, Franz and Goller (2002) examined the pattern of expiration and inspiration during singing. This showed that while single expiratory (and sometimes inspiratory) pressure pulses can coincide with single elements identified on the sonogram, others underlie what look like several

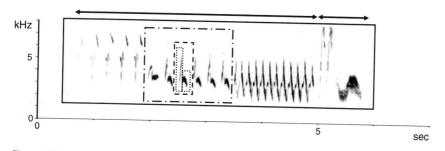

Figure 12.2

Sonagram of a chaffinch song, showing the different levels of segmentation. The smallest units (dotted lines) are single *elements*, separated by a brief gap. Two of these form one unit (the *syllable*—hatched line) that is repeated several times. A series of identical syllables form a *phrase*. In this song (enclosed by the unbroken line) there are three phrases consisting of brief syllables, repeated several times. This song ends, like many chaffinch songs, with a few elements that show a different spectrographic structure and are often not repeated. The first three phrases together form what is known as the "trill" part of the chaffinch song (indicated by the left arrow), which is followed by what is known as the "flourish" part (right arrow). The sonogram also shows some of the problems with identifying the units. Whereas most researchers might agree that the first phrase consists of syllables of two elements, the subdivision of the third phrase might pose more problems: Is there a syllable consisting of two elements or rather a single element with an upsweep and a downsweep? The decision to make it two elements gives rise to a phrase with a structure ABABAB . . ., whereas making it a single element results in an AAA . . . structure (sonogram courtesy of Katharina Riebel).

separate elements, suggesting a syllable structure. Such methods help to provide a better picture of the units of production, although they do not always correspond.

An alternative approach to segmenting songs into their constituent units (applicable only when there are few elements and many combinations) is to examine patterns of recombination within a bird's repertoire (Podos, Peters, Rudnicky, Marler, & Nowicki, 1992), based on identifying (arbitrarily long) parts of the song that always occur together. Although perhaps closer to how birds themselves categorize acoustic input, it is often not clear with this method how many songs need to be sampled (e.g., how to treat rare divisions and coincidental similarities).

So, in all existing approaches arbitrary decisions seem unavoidable and this is why different researchers can differ in their descriptions of songs, with both "splitters" and "lumpers" being present. In many types of analyses this does not matter, but it might when analyzing song structure based on sequences of labeled units. For instance, a "splitter" may label a song as consisting of a sequence of elements ABCABCABCABC, with ABC forming a syllable, thus arriving at a song structure in which elements are never repeated, while syllables are. On the other hand, a "lumper" might consider ABC to form one unit, thus arriving at the sequence AAAA, a simple repetition (Figure 12.2).

A related issue concerns the decision about categorization of similar units. In many species, the differences between the elements or syllables shown by a particular individual are so clearcut that they allow unambiguous classification in discrete categories. In others, however, the phonological structure of elements varies in a more gradual way, again giving rise to different interpretations. This is even more true at the species level. Whether at this level elements are clustered into categories or vary continuously is a contentious issue. In some species, like the swamp sparrow (*Melospiza georgiana*), researchers have claimed that all elements in the species can be assigned to one of a small number of relatively invariant categories (Marler & Pickert, 1984; Clark, Marler, & Beeman, 1987). In others, like the chaffinch (*Fringilla coelebs*; Figure 12.2), no such categorization by researchers has been used. In this species, although a clear distinction can be drawn between the structures of the terminal flourish syllables and the trills that precede it, there is little sign that the considerable variation between trill syllables falls into discernible categories. Finally, in a third group of species, like the zebra finch (*Taenopygia guttata*), there have been various efforts to categorize universal element types (e.g., Zann, 1993; Sturdy, Phillmore, & Weisman, 1999), but these schemes differ considerably and clearly encompass much intracategory variation.

Based on these findings, it is tempting to conclude that there are clear differences between bird species in the pattern of acoustic variation in their songs. However, the accounts are largely based on subjective human interpretation of spectrograms. Humans have a tendency to group stimuli ranging along a continuum into categories, even where none exist in the input (Harnad, 1987), so it is difficult to conclude from this evidence alone that there is any objective basis for these categories. On the other hand, spectrographs are an imperfect method of visualizing acoustic variation, and it is possible that clear categories are obscured for species with harmonically complex song, like zebra finches.

Clearly a computational approach to comparing song units would improve matters, but it is only recently that statistical tools developed to measure the degree and structure of clustering in data sets (see Tan, Steinback, & Kumar, 2006) have been deployed in the analysis of birdsong variation. Lachlan, Peters, Verhagen, and ten Cate (2010) used a dynamic time-warping algorithm to compare song elements and next used a hierarchical clustering algorithm and a statistical analysis to examine the clustering tendency for song elements in chaffinches, song sparrows, and zebra finches. The study revealed very broad categories of song units in all three species, at least at the population level if not the species level. One interpretation of this finding is that at least to some extent and for some species, it is possible to identify a population-wide repertoire of elements, somewhat like an inventory of speech sounds characteristic of a language, but less specific. Although the categories found only explained about half of the total variation between elements (swamp sparrows), or even much less (chaffinches, zebra finches), they do appear to correspond to biologically meaningful categories. For instance, in the case of chaffinches, the major distinction found is between trill and flourish syllables, corresponding to a suggested functional difference in whether these syllables are directed particularly at males or at females (Leitao & Riebel, 2003; Leitao, ten Cate, & Riebel, 2006). Recent work on different perceptual boundaries in two populations of swamp sparrows (Prather, Nowicki, Anderson, Peters, & Mooney, 2009) also matches the computational analysis of note-type structure (Lachlan et al., 2010).

In the study of human language, similar issues have arisen in the identification of the basic units, even though introspection and the fact that words and sentences carry semantic meaning have sometimes made the problem slightly easier. For instance, in phonology the basic unit is traditionally the *phoneme*, defined using the minimal difference in sounds that can signify a difference in semantic meaning. Thus, the English words /pin/ and /pen/ differ in one sound and can be used to define the phonemes /i/ and /e/. Using such a minimal-pair test, phonologists have described the phoneme repertoire of thousands of languages.

Phonemes are defined with reference to meaning, which has obvious methodological disadvantages and makes the comparison with birdsong less obvious. The minimal unit of sound that can be defined without such reference is usually called a *phone*. However, it has turned out to be very difficult to unambiguously label phones based on directly observable information alone. Acoustic information provides unreliable cues: in the acoustic signal there are no clear equivalents of phoneme, syllable, or word boundaries, and what we perceive as the same phoneme or syllable can be realized in rather different ways. The classic demonstration of this fact, from Liberman, Cooper, Shankweiler, and Studdert-Kennedy (1967), is the difference between adding the phoneme /d/ before an /i/ or /u/: in the case of an articulated /di/ we see a rising second formant, while /du/ shows a falling second formant preceding the original vowel (see Figure 12.3). The two realizations of /d/ thus differ radically in terms of objective, acoustic measures, even if humans perceive them as very similar and linguists like to label them with the same phoneme. Articulatory information (concerning the exact movements of *articulators* such as the larynx, tongue, and lips) provides perhaps a closer match to cognitively real motor programs associated with specific units of sound (Liberman et al., 1967), but is in turn much harder to obtain.

However, even when we accept the use of semantics in identifying the units of language, the traditional notion of a phoneme does not provide an unproblematic atomic unit of linguistic form as we might hope. When applying the definition, one encounters examples where no minimal pair can be given, even though distinguishing phonemes seems appropriate (e.g., in English, word-initial /h/ and word-final /ng/ have no minimal pair to stop us from considering them a single phoneme).

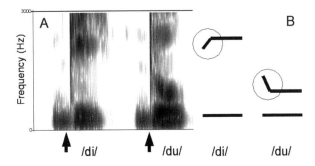

Figure 12.3
A single phoneme can correspond to acoustically very different sounds. Panel A shows a spectrogram of the syllables /di/ and /du/. The /d/ phoneme corresponds in both cases to a temporary blocking of the vocal tract (indicated with the arrow), followed by a rising second formant in the case of /di/ but a falling second formant in the case of /du/. (Time is on the horizontal axis; clip duration is 1 sec.) Panel B illustrates this difference schematically, as in Liberman et al. (1967). (Recording courtesy of Bart de Boer.)

Moreover, it is not even clear at all that the found phonemes indeed correspond to cognitively real categories. Some researchers—the "splitters"—have argued for a finer level of description (e.g., "distinctive features"; Chomsky & Halle, 1968), thus observing that /d/ and /t/ only differ in whether the vocal folds vibrate (the *voicing* feature) and are much more similar to each other than to, for example, a /k/. However, other researchers—the "lumpers"—have argued for larger basic units (e.g., the syllable; Levelt & Wheeldon, 1994), based on experimental evidence on speech production that best fits a model where humans have access to a redundant repertoire of motor programs associated with particular syllables. Current consensus in phonology seems to be that humans must have access to both sub- and supra-phonemic levels of representation.

Similar debates exist about the basic units of meanings, words, and sentences. The chosen definition of building block is very much dependent on what the ultimate research questions are. Investigations of production, perception, acquisition, language change, and so on might all need different atomic units of description. Also in these domains it seems likely that humans have access to cognitive representations at various degrees of granularity, for some tasks relying more on smaller units and for others on larger memorized chunks. Although there is some work on trying to identify the basic units (e.g., Borensztajn, Zuidema, & Bod, 2008), linguists have generally, like birdsong researchers, taken a pragmatic approach and made intuitive choices about the appropriate level of analysis of a given utterance. A crucial difference with birdsong is that such choices have been mostly guided by native speaker intuitions about meanings. Such information is necessarily somewhat subjective (even if disagreements are rare), and of course not available in the case of birdsong. Because we must conclude that no objective unambiguous classification schedule for acoustic units exists, be it at the level of elements, syllables, or songs for birdsong or phoneme, morpheme, or word compounds in language, any attempt to examine the structuring rules of smaller units into larger ones must to some degree be buffered against variation in classification.

Describing Sequence Structure and Complexity

Once a pragmatic choice of the units of description is made, we can focus our attention on the sequencing of these units. The pioneering theoretical work on describing sequences is that of Claude Shannon (1948). Among Shannon's many contributions was the introduction of *entropy* as a measure of the degree of uncertainty about every next element when observing a sequence. Shannon further defined what are now called "n-gram models," where the probability of the next element to be generated is based on the last $n-1$ elements. For instance, in a *bigram* model ($n = 2$), the probability of generating the next element is assumed to only depend on the last element. In a *trigram* model ($n = 3$), that probability depends on the last two elements. Shannon demonstrated that increasingly accurate approximations of written English could be obtained by increasing this "window" of $n-1$ previous letters/words. The value of n (or $n-1$) is also called the "Markov order."

Most early studies addressing birdsong structure focused on the sequencing of song units in one or a few particular individuals, and also addressed the issue for one particular level of analysis (e.g., that of the sequencing of syllables and not that of songs or elements). For instance, Isaac and Marler (1963) analyzed the sequence of syllables, as identified from the temporal distribution of interelement pauses, for one individual mistle thrush. The authors looked at syllable-to-syllable transitions; in current terminology, this is a *bigram* analysis. Their contingency table, counting which syllable followed which other, showed that there were far fewer pair combinations than would be expected if the transitions were random, and that the transitions were almost deterministic.

Other results from this analysis were that songs were begun by only a few of the possible syllable types; that most syllables were never repeated; and that the same syllable type never both preceded and followed another one. Also, the transition probability between two successive syllables was higher (more predictable) the shorter their interval was. Finally, they also showed that the syllables ending one song did not predict with which syllable the next one would start.

A next step in analyzing sequences was done by Chatfield and Lemon (1970; Lemon & Chatfield 1971, 1973), who pioneered in this context the use of entropy as a measure for how well a specific model described the observed data. Their approach laid the foundation for many subsequent studies. Among the first species they examined was the cardinal, concentrating on two individuals. Like Isaac and Marler, they distinguished the elements, syllables, and songs by the duration of pauses between them. They noted that each individual had a few simple songs, mostly consisting of repetitions of the same syllable, but sometimes of a few different syllables. The number of syllables was inversely related to their duration. But rather than examining the sequencing of elements or syllables within songs, as Isaac and Marler did, they examined the sequencing of songs.

Again using transition matrices, they first calculated a number of parameters, like the transition probability between songs; the conditional probability (i.e., the chance of repeating a particular song type given that the previous utterance was of the same type); and the relation between the number of "bouts" (here: series of songs of the same type) and the mean number of utterances per bout. They used these calculated parameters to construct a model describing the sequencing of songs of the same type. Next they examined the sequence of switching from bouts of one song type to another. They compared n-gram models with increasing n, and measured the entropy of each model to decide how large an n was needed to obtain an adequate model. For the sequences of songs of cardinals, Chatfield and Lemon (1970) found that a bigram model already obtained a very good fit. Since this pioneering work, several later studies have been using transition matrices and Markov chain models to examine the sequencing of songs in birds.

While Markovian short-term dependencies seem sufficient to predict how certain species (or better: certain individuals of a species) sequence their songs, for other species the sequencing seems more complex. Todt (1968, 1970, 1975, 1977) provides one of the most compelling examples, with the study of the song of the European blackbird. As mentioned in the introduction, blackbirds sing quite variable songs, usually described as consisting of an initial motif part, followed by a twitter part. Each part consists of a variety of different elements. When comparing different songs it becomes clear that there is a "branching" structure to the songs: a limited number of elements are used to begin a song, but each initial element may give rise to several song varieties because either this or subsequent elements may be followed by two or three different element types, resulting in several song variants per initial element (Todt, 1968, 1970).

In his analyses, Todt concentrated on the rules underlying the sequencing of songs starting with different initial elements (here indicated as "song types"), irrespective of which other elements were in those songs. He noted five separate types of dependencies in the singing behavior of the blackbirds (Todt & Wolffgramm, 1975): (1) some sequences of song types showed a close association between the types in that each could precede or follow the other (this could be sequences of the type A-A or A-B); (2) some sequences (e.g., A>B) occurred above chance in one direction, but below chance in the other direction (B>A); (3) when a particular sequence A-B was already present during a series of songs, this increased the likelihood that a later occurrence of A was also followed by B; (4) the occurrence of a given song type showed a periodicity in that it had a high chance of reappearing after the bird had sung a particular number of other songs; and (5) singing a particular song type suppressed its reappearance for some time.

Based on these five components, Todt and Wolffgramm (1975) constructed a cybernetic computer program including the different processes and tested whether it was possible to simulate a song pattern similar to that of a real blackbird. This turned out to be the case, albeit dependent on the parameter settings, but only when all five processes were included in the model. It is obvious that in blackbirds the sequencing of songs is more complex than in cardinals, and that n-grams do not suffice. Nevertheless, the pattern remains quite deterministic. So, the question is whether models exist that can cope with such complexities, but are still simpler and more generic than Todt and Wolffgramm's program. Various researchers have explored the use of such richer formalisms (e.g., Okanoya, 2004). We also suggest these models exist, and to illustrate this we return to our initial example of blackbird song (Figure 12.1).

The sequences represented in the four songs in Figure 12.1 can be analyzed by making a transition matrix in the conventional way. This results in a description of the song sequence in terms of the probability that a particular element "A" would be followed by an element "B"; for the four recordings of Figure 12.1 this is illustrated in Figure 12.4a, with motif elements labeled M_1–M_4 and twitter elements T_1–T_2. However, it is also possible to analyze the songs in a different way. While the bird was producing the songs, it presumably went through a sequence of functionally different "brain states." Importantly, it seems that these brain states do not correspond one to one to the produced elements. For instance, in our example twitter element T_1 is repeated exactly two or three times, not one or four times. The bigram analysis has no way to represent how many times an element may be produced. Paradoxically, it considers an unobserved song with a single T_1 more likely (probability $p_1 \times p_5$) than the *observed* song with two repetitions (e.g., the song in Figure 12.1a, with probability $p_1 \times p_3 \times p_5$). Thus, although it is possible to analyze these songs in the classical way by tracking transition probabilities, the bigram description implicitly assumes that the "brain states" are sufficiently characterized by the last element produced, which is clearly not the case.

Although many details cannot be decided on with so little information, it does seem clear that an adequate description of these songs will at least consist of the following components: a set of song elements and a *different* set of potential brain states, probabilities of transitioning from one state to another, and probabilities of producing each element in every state. If we assume the set of states and elements

	\multicolumn{7}{c}{Next element}						
	M_1	M_2	M_3	M_4	T_1	T_2	#
0	1	0	0	0	0	0	0
M_1	0	1	0	0	0	0	0
M_2	0	0	1	0	0	0	0
M_3	0	0	0	p_2	p_1	0	0
M_4	0	0	0	0	1	0	0
T_1	0	0	0	0	p_3	p_4	p_5
T_2	0	0	0	0	0	0	1

(a) Transition probabilities

State $\rightsquigarrow$ Sound		Probability
0 $\rightsquigarrow$ M_1		1
M_1 $\rightsquigarrow$ M_2		1
M_2 $\rightsquigarrow$ M_3		1
M_3 $\rightsquigarrow$ T_1		p_1
M_3 $\rightsquigarrow$ M_4		p_2
M_4 $\rightsquigarrow$ T_1		1
T_1 $\rightsquigarrow$ T_1		p_3
T_1 $\rightsquigarrow$ T_2		p_4
T_1 $\rightsquigarrow$ #		p_5
T_2 $\rightsquigarrow$ #		1

(b) Bigram analysis

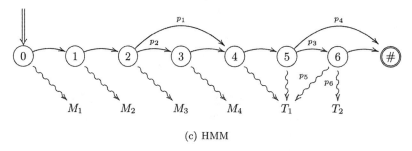

(c) HMM

Figure 12.4
Three models for the production of a blackbird song.

is finite, this means such a description corresponds to a class of mathematical models
known as *hidden Markov models* (HMM; the term *hidden* means we cannot directly
observe the brain states, and *Markov* means we assume the relevant probabilities
are only influenced by the current brain state; hidden Markov models are essentially
"finite-state machines" with probabilities on the transitions). A graphic representation of an HMM is given in Figure 12.4b (see also Kogan & Margoliash, 1998, for
another application of HMMs to birdsong). We see six states (labeled 1–6), and six
different elements, corresponding to four different observed motif elements (labeled
M1–M4) and two observed twitter elements. Solid arrows indicate transitions
between states; curly arrows indicate the production of elements and whether there
are multiple choices at a particular state. Finally, little p's mark different probabilities of following that arrow, and double circles mark possible stop states. It is easy
to check that this HMM can produce six different sequences: <M1 M2 M3 T1 T1>,

<M1 M2 M3 M4 T1 T1>, <M1 M2 M3 T1 T1 T1>, <M1 M2 M3 T1 T1 T2>, <M1 M2 M3 M4 T1 T1 T1>, and <M1 M2 M3 M4 T1 T1 T2>. The first four of these correspond to the observed spectrograms in Figure 12.1.

There is a simple relation between *n*-grams and the HMM: *n*-grams are restricted versions of HMMs where the relevant "brain states" are assumed to correspond to the last *n–1* elements (based on the "brain state" we predict the next element; *n*-grams thus model a relation between *n–1 + 1 = n* elements, hence the *n* in the name). *N*-grams are therefore quite restricted models that cannot, for instance, explicitly model that element M4 seems optional. That is, in *n*-grams states are tied to actually produced elements; hence, we cannot return to the same state after producing M4 that we would have been in had we not produced it. In Figures 12.5 and 12.6, we illustrate the difference between *n*-grams and HMMs by giving a simple bigram and HMM analysis of the sentence *A man sees the woman with the telescope*. In the bigram model, every word forms a "state," which then produces the following word. Such a model cannot make the generalization that the words *man* and *woman* are in fact substitutable: in every sentence where *man* is used, we could replace it with *woman* without making the sentence ungrammatical. HMMs do allow us to define a state for many different words that are equivalent in this sense; in Figure 12.5b we see that state 1 permits any of the three nouns to be produced.

Whereas to date there is no convincing example of a song, or even any nonhuman vocalization, that might not be described with an HMM, in human morphosyntax, Chomsky (1957) famously demonstrated that finite-state machines—and thus *n*-grams and HMMs—are inadequate models of natural language syntax. Chomsky proposed a richer class of models, now known as *context-free grammars* (CFGs). The relation between CFGs and HMMs is now well understood. Recall that HMMs assume a *finite* set of *categorical* brain states, probabilities of transitioning between states and probabilities of producing words/elements in each state. CFGs, in contrast,

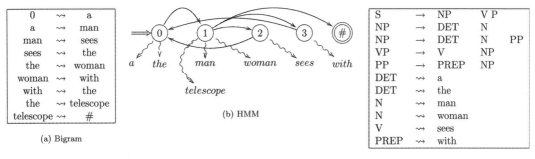

Figure 12.5
Three models for the production of a sentence (probabilities omitted for simplicity).

Step	State	Sound	Step	State	Sound	Step	State	Sound
1	0	a	1	0	a	1	S	-
2	a	man	2	1	man	2	NP VP	-
3	man	sees	3	2	sees	3	DET N VP	-
4	sees	the	4	0	the	4	N VP	a
5	the	woman	5	1	woman	5	VP	man
6	woman	with	6	3	with	6	V NP	-
7	with	the	7	0	the	7	NP	sees
8	the	telescope	8	1	telescope	8	DET N PP	-
9	telescope	-	9	#	-	9	N PP	the
10	#	-				10	PP	woman
				(b) HMM		11	PREP NP	-
	(a) Bigram					12	NP	with
						13	DET N	-
						14	N	the
						15	#	telescope

(c) Context-free grammar

Figure 12.6
Three corresponding derivation sequences in the production of a sentence.

allow for an *unbounded* number of brain states, which are characterized by a hierarchical and recursive representation. That is, by adopting CFGs we assume language users maintain a hierarchical representation in their brains while producing a sequence of elements/words; transitions between states are understood as additions to that representation, and the probabilities of producing elements/words are dependent on that representation. In Figure 12.5c we give an example of a context-free grammar model, used in Figure 12.6c to produce the same sentence as before. The production starts with the symbol "S," and then proceeds by repeatedly replacing a symbol on the left-hand side of one of the rules in Figure 12.5c with the symbols on its right-hand side. One can check that the states in Figure 12.6c are characterized by a sequence of varying (and unbounded) length.

Conclusions and Outlook

The approaches and studies discussed above have demonstrated that neither birdsong nor human language consists of a random jumble of notes; they have a clear structure. They have also demonstrated that there is structure at different levels, ranging from the sequencing of elements to that of songs, and from phones to sentences. In different bird species, we have seen that they may differ substantially in how their songs are organized and overall generalizations are hard to make. Standardization in delineating song units, level of analysis (sequencing of elements, syllables, or songs), and methods may clarify interspecific differences and similarities in song structure and may help to address the evolution of the complexity in song structure, at least for certain groups of birds. Another area that may benefit from a more sophisticated approach to studying song structure is that of the neural mechanism of birdsong production, which has, among other things, shown that certain brain areas seem particularly relevant for sequencing syllables and elements (Long & Fee, 2008).

We hope we have shown that the relation between popular models for birdsong and human language syntax can be best understood with reference to a class of formal models known as hidden Markov models. HMMs correspond to a very intuitive way of describing sequences of elements that obey particular rules. Formalizing descriptions of song structure in these terms can be useful for two reasons. First, identifying HMMs helps us understand the relation with other descriptions of grammatical structure—in particular with n-grams, which are weaker, with context-free grammars, which are more powerful, and with finite-state machines, which are equivalent. Second, identifying particular descriptions as equivalent to HMMs makes available a range of advanced techniques from statistics and learning theory. These techniques allow us to start answering fundamental questions like the following: What is the probability of producing a particular song? What is the most likely grammar underlying a particular repertoire of songs?

The formalism also allows for a straightforward extension to link various levels of analysis, such as the levels of element sequences and song sequences. So, the analyses by Todt (1970) of the complex ways in which blackbirds sequence elements as well as songs might not only be phrased in terms of an HMM, but this description may include another HMM that describes the rules by which songs are constructed from elements. This would result in a so-called *hierarchical* HMM, which is in turn a restricted version of (probabilistic) context-free grammars. Details of such formalisms go beyond the scope of this chapter, but their existence underlines the generality of the HMM perspective on birdsong.

So, how should we proceed with using the HMM approach? The extensive existing work on HMMs in statistical learning theory suggests a number of straightforward steps. HMMs define probability distributions over sequences. For any given HMM, it is almost trivial to calculate the probability of each of the possible sequences it can produce (this is *P(sequence|model)*, known as "data likelihood"). Given the true model, we can thus calculate how likely it is that a bird sings a particular song.

The most interesting questions, however, force us to draw inferences in the opposite direction. Given observed data on songs and utterances, what is the probability that a particular model underlies that data? This is *P(model |sequence)*, and we usually want to know which is the most probable model m (written as $m =$ argmax$_m$P(m|sequence)*, where "argmax . . ." reads as "the argument that maximizes . . .").

An increasingly popular approach in statistical learning theory connects data likelihood to the most probable model using Bayes's rule:

P(model|sequence) = P(sequence|model) * P(model)/P(sequence)

The most probable model is most easily found if we a priori assume a restricted set of possible models, and furthermore assume that all possible models are equally likely before we have seen any data. In that case, the Bayesian solution equals the so-called Maximum Likelihood solution.

These considerations give rise to a research program for describing song syntax and comparing syntaxes within and between species, including with humans. In this program, we have to go through the following steps in every analysis:

1. Consider the evidence for what the elementary building blocks of the song are.
2. Consider the evidence for what the relevant "brain states" are.
3. Compile a database with songs, and annotate them with as much of the available information as possible on elements and states.
4. From this, we can derive a set of possible models that might explain the observations—each model assigns a "likelihood" to empirical data.
5. Consider the supporting evidence for what the possible models are, and formulate this as an a priori probability of every model.
6. Apply Bayes's rule and a relevant optimization technique to find the most probable model underlying the data.
7. Use the model to make new testable predictions.

Interestingly, such a program is robust to uncertainty in steps 1-3: the controversy between lumpers and splitters need not be resolved. In the sketched approach, such uncertainties would lead to larger sets of possible models in step 4. Moreover, the analysis in step 6 would feed back to the controversy about the building blocks: some ways of lumping or splitting elements lead to better descriptions of the song syntax than others and might therefore be preferred.

Although a lot of related work has already been carried out for written human language in the field of computational linguistics, the sketched program would in most new species involve a considerable amount of work to be successful. Success depends strongly on the amount of data available in step 3—the more the better— and on supportive evidence in steps 1, 2 and 5 to constrain the set of models that need to be considered. Finally, if sufficient amounts of data are indeed available, nontrivial computational challenges might arise in step 6, for which we then would need efficient and accurate approximations to be worked out. So, while implementing the approach provides a challenge, it will bring the advantage of uniformity and of an analysis that includes all the different levels of song organization currently analyzed separately. Returning to our starting point, if we take seriously the examination of parallels between birdsong and language, we need such approaches to clarify whether, as seems the case so far, songs of all bird species can be satisfactory described with such models or whether there are species whose songs, like language, require more complex grammars.

Acknowledgments

Preparation of this chapter was supported by grant 051–12–047 of the NWO Evolution and Behavior program to CtC and RL, grant 051–07–002 of the NWO Cognition program to CtC and WZ, and a Lorentz Fellowship of the NIAS to CtC. We thank Moira Yip for useful comments.

References

Borensztajn, G., Zuidema, W., & Bod, R. (2008). Children's grammars grow more abstract with age: Evidence from an automatic procedure for identifying the productive units of language. *Topics in Cognitive Science*, *1*, 175–188.

Chatfield, C., & Lemon, R. E. (1970). Analysing sequences of behavioural events. *Journal of Theoretical Biology*, *29*, 427–445.

Chomsky, N. (1957). *Syntactic structures*. The Hague: Mouton.

Chomsky, N., & Halle, M. (1968). *The sound pattern of English*. New York: Harper & Row.

Clark, C. W., Marler, P., & Beeman, K. (1987). Quantitative analysis of animal vocal phonology: An application to *Swamp Sparrow* song. *Ethology*, *76*, 101–115.

Cynx, J. (1990). Experimental determination of a unit of song production in the zebra finch. *Journal of Comparative Psychology*, *104*, 3–10.

Franz, M., & Goller, F. (2002). Respiratory units of motor production and song imitation in the zebra finch. *Journal of Neurobiology*, *51*, 129–141.

Harnad, S. (Ed.). (1987). *Categorical perception: The groundwork of cognition*. Cambridge: Cambridge University Press.

Hauser, M. D., Chomsky, N., & Fitch, W. T. (2002). The faculty of language: What is it, who has it, and how did it evolve? *Science*, *298*, 1569–1579.

Hultsch, H., Mundry, R., & Todt, D. (1999). Learning, representation and retrieval of rule-related knowledge in the song system of birds. In A. D. Friederici & R. Menzel (Eds.), *Learning: Rule extraction and representation* (pp. 89–115). Berlin: De Gruyter.

Isaac, D., & Marler, P. (1963). Ordering of sequences of singing behaviour of mistle thrushes in relationship to timing. *Animal Behaviour*, *11*, 179–188.

Kogan, J. A., & Margoliash, D. (1998). Automated recognition of bird song elements from continuous recordings using dynamic time warping and hidden Markov models: A comparative study. *Journal of the Acoustical Society of America*, *103*, 2185–2196.

Lachlan, R. F., Peters, S., Verhagen, L., & ten Cate, C. (2010). Are there species-universal categories in bird song phonology and syntax? *Journal of Comparative Psychology*, *124*, 92–108.

Leitao, A., & Riebel, K. (2003). Are good ornaments bad armaments? Male chaffinch perception of songs with varying flourish length. *Animal Behaviour, 66*, 161–167.

Leitao, A., ten Cate, C., & Riebel, K. (2006). Within-song complexity in a songbird is meaningful to both male and female receivers. *Animal Behaviour, 71*, 1289–1296.

Lemon, R. E., & Chatfield, C. (1971). Organization of song in cardinals. *Animal Behaviour, 19*, 1–17.

Lemon, R. E., & Chatfield, C. (1973). Organization of song in rose-breasted grosbeaks. *Animal Behaviour, 21*, 28–44.

Levelt, W., & Wheeldon, L. (1994). Do speakers have access to a mental syllabary? *Cognition, 50*, 239–269.

Liberman, A., Cooper, F., Shankweiler, D., & Studdert-Kennedy, M. (1967). Perception of the speech code. *Psychological Review, 74*, 431–461.

Long, M. A., & Fee, M. S. (2008). Using temperature to analyse temporal dynamics in the songbird motor pathway. *Nature, 456*, 189–194.

Marler, P., & Pickert, R. (1984). Species-universal microstructure in the learned song of the swamp sparrow (melospiza-georgiana). *Animal Behaviour, 32*, 673–689.

Okanoya, K. (2004). Song syntax in Bengalese finches: Proximate and ultimate analyses. *Advances in the Study of Behavior, 34*, 297–346.

Podos, J., Peters, S., Rudnicky, T., Marler, P., & Nowicki, S. (1992). The organization of song repertoires in song sparrows: Themes and variations. *Ethology, 90*, 89–106.

Prather, J., Nowicki, S., Anderson, R., Peters, S., & Mooney, R. (2009). Neural correlates of categorical perception in learned vocal communication. *Nature Neuroscience, 12*, 221–228.

Shannon, C. E. (1948). A mathematical theory of communication. *Bell Systems Technical Journal, 27*, 379–423, 623–656.

Sturdy, C. B., Phillmore, L. S., & Weisman, R. G. (1999). Note types, harmonic structure, and note order in the songs of zebra finches (*Taeniopygia guttata*). *Journal of Comparative Psychology, 113*, 194–203.

Tan, P.-N., Steinback, M., & Kumar, V. (2006). *Introduction to data mining*. Boston: Pearson Addison Wesley.

Todt, D. J. (1968). Zur Steuerung unregelmässiger Verhaltensabläufe: Ergebnisse einer Analyse des Gesanges der Amsel (*Turdus merula*). In H. Mittelstaedt (Ed.), *Kybernetik* (pp. 465–485). Munich: Verlag Oldenbourg.

Todt, D. (1970). Gesang und gesangliche Korrespondenz der Amsel. *Naturwissenschaften, 57*, 61–66.

Todt, D. (1975). Short term inhibition of outputs occurring in the vocal behaviour of blackbirds (*Turdus merula* m.L.). *Journal of Comparative Physiology, 98*, 289–306.

Todt, D. (1977). Zur infradianen Rhythmik im Verhalten von Vertebraten—Ergebnisse aus Analysen des variablen Verhaltens von Singvögeln. *Nova Acta Leopoldina, 46*, 607–619.

Todt, D., & Wolffgramm, J. (1975). Ueberprüfung von Steuerungssystemen zur Strophenwahl der Amsel durch digitale Simulierung. *Biological Cybernetics, 17*, 109–127.

Zann, R. (1993). Structure, sequence and evolution of song elements in wild Australian zebra finches. *Auk, 110*, 702–715.

13 Phonological Awareness in Grey Parrots: Creation of New Labels from Existing Vocalizations

Irene M. Pepperberg

Although grey parrots (*Psittacus erithacus*) use elements of English speech referentially (Pepperberg, 1999),[1] these birds are still often regarded as mindless mimics. One reason for this belief is that only limited evidence exists to show that parrots—or any animal taught a human communication code—segment the human code; that is, recombine existing labels intentionally to describe novel situations or request novel items—rather than, for example, produce several labels that may simply apply to the situation. In the latter case, nonhuman subjects' productions have been regarded as descriptors of the entire situation, not as specific combinations to denote one element (e.g., apes' "water bird" for a swan, "cry hurt food" for a radish, Fouts & Rigby, 1977; dolphins' "ring-ball" during simultaneous play with two items, Reiss & McCowan, 1993; note Savage-Rumbaugh et al., 1993). In contrast, children in even early stages of normal language acquisition demonstrate intentional creativity (de Boysson-Bardies, 1999; Greenfield, 1991; Marschark, Everhart, Martin, & West, 1987; Tomasello, 2003). Another form of segmentation, intentional recombination of existing phonemes (parts of words) to produce targeted labels or their approximations (Greenfield, 1991; Peperkamp, 2003), has not previously been reported in animals; such phonological awareness has not only been considered basic to human language development (Carroll, Snowling, Hulme, & Stevenson, 2003), but also a uniquely human trait (Pepperberg, 2007a).

Notably, phonological awareness and segmentation require understanding that words consist of a finite number of sounds that can be recombined into an almost infinite number of patterns (limited only by constraints of a given language), and thus differ from the ability simply to discriminate human phonemes (e.g., /b/ from /p/), which is a widespread vertebrate trait (e.g., Kuhl & Miller, 1975). Furthermore, phonological awareness and segmentation develop over time. Children start by recognizing and producing words holistically (simple imitation; Studdert-Kennedy, 2002), then shift to recognizing words as being constructed via a rule-based phonology at about 3 years old (Carroll et al., 2003; Vihman, 1996) and apparently need training on sound-letter associations to focus on word phonology and not solely on word meaning (Carroll et al., 2003; Mann & Foy, 2003). To sound out—rather than mimic—a novel label, a child must segment the sound stream into discrete elements, recognize a match between those elements and bits (or close approximations) that exist in its own repertoire, and then recombine these elements in an appropriate sequence (see Gathercole & Baddeley, 1990; Treiman, 1995); such manipulation of individual parts of words is presumed to require development of an internal representation of phonological structure (Byrne & Liberman, 1999; note Edwards, Munson, & Beckman, 2011). Most animals, lacking speech, are never exposed to, nor trained or tested on, phonological awareness, nor are they expected to have internal representations of phonemes (Pepperberg, 2007a, 2007b, 2009).[2]

New evidence demonstrates, however, that vocal segmentation and phonological awareness (*sensu* Anthony & Francis, 2005) are not uniquely human (Pepperberg, 2007a, 2007b, 2009): my oldest speech-trained subject, a grey parrot named Alex, understood that his labels were made of individual phonological units that could be recombined in novel ways to create novel vocalizations. Thus parrots not only may use English labels referentially, but also understand how such labels are created from independent sound patterns. Alex developed this behavior through his considerable experience with English speech and sound-letter training. My younger birds, lacking such training, do not demonstrate this ability.

The study that demonstrated the behavioral differences that arise not only from training but also from exposure to human speech patterns (Pepperberg, 2007a) involved two grey parrots. Alex, then 27 years old, had had 26 years of intense training in interspecies communication: he had learned to identify, request, refuse, categorize, and quantify a large number (>100) of objects, colors, and shapes referentially, using English speech sounds (Pepperberg, 1999); he used English labels to answer questions concerning, for example, concepts of number, category, absence, relative size, and same-different (e.g., Pepperberg, 2006). He had also been trained to associate the letters B, CH, I, K, N, OR, S, SH, T with their corresponding appropriate phonological sounds (e.g., /bi/ for BI), the plastic or wooden symbols being his initial reward (later, he would ask for nuts, tickles, etc., after being told he was correct); his accuracy was well above chance ($1/9$, $p < .01$). He had not, however, received any training involving the combination of these letters. Arthur, 3½ years old, had had the equivalent of about a year of comparable interspecies communication training, but no training on phonemes; he had acquired four referential labels (Pepperberg & Wilkes, 2004). The birds lived with another grey parrot, Griffin, in a laboratory setting (for housing and day-to-day care, see Pepperberg & Wilkes, 2004).

The training technique—the Model/Rival procedure—has been described many times (e.g., Pepperberg, 1981, 1999) and was used for both Arthur and Alex on the label "spool." Arthur was trained first. After Arthur's training, Alex began to show

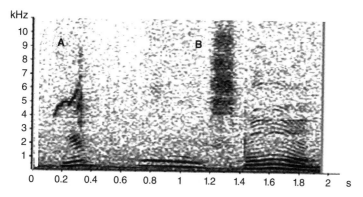

Figure 13.1
(a) Arthur's "spool" compared to (b) Pepperberg's "spool" (from Pepperberg, 2007a)

interest in the object, which he had previously ignored. We therefore initiated training on the object for Alex.

Interestingly, Arthur's, but not Alex's, pattern of acquisition followed the usual stages for birds in my lab (Pepperberg, 2007a, 2009). Their labels usually appear first as a vocal contour, then include vowels, and finally consonants (Patterson & Pepperberg, 1994, 1998). Arthur began with /u/ ("ooo"), added /l/, and then, because production of human /p/ is troublesome without lips, devised a novel solution. Unlike Alex, who learned to produce /p/ apparently via esophageal speech (Patterson & Pepperberg, 1998), Arthur produced a whistled, not plosive, /p/ in /sp/ (Figure 13.1A; Pepperberg, 2007a).

Arthur's /p/ was similar to what Lieberman (1984, p. 156) had predicted for parrot "speech" (Pepperberg, 2009). Specifically, Lieberman (1984) argued that birds cannot reliably produce humanlike formant structures, but produce whistles that, via interference patterns that create energy at defined frequencies, are translated by the human ear into speechlike sounds. However, only Arthur's /sp/ was whistled; /u/ (which could easily have been whistled) and /l/ resembled human speech (Pepperberg, 2007a), demonstrating basically the same formant structure that previous research (Patterson & Pepperberg, 1994, 1998) had revealed for Alex's vowels and stop consonants /p,b,d,g,k,t/ (see Beckers, Nelson, & Suthers, 2004, for discussion of tongue placement and formation of true formants in parrot vocalizations).

Alex, unlike Arthur and his usual pattern of acquisition, began by using a combination of existing phonemes and labels to identify the object: /s/ (trained independently in conjunction with the letter S) and *wool*, to form "s" (pause) "wool" ("s-wool"; /s-pause-wUl/; Figure 13.2; Pepperberg, 2007a). The pause seemingly provided space for the absent (and difficult) /p/ (possibly as a filler phoneme, preserving the targeted vocalization's syllable number or prosodic rhythm; see below).

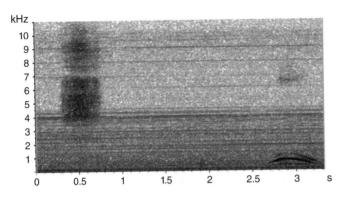

Figure 13.2
Alex's "s-wool" (/s-pause-wUl/) (from Pepperberg, 2007a)

Notably, no labels in his repertoire contained /sp/, nor did he know "pool," "pull," or any label that included /Ul/; he knew "paper," "peach," "parrot," "pick," etc.,[3] and "shape" and "sich" (*six*); thus, technically, /p/, /sh/, and /s/ but not /sp/ were available (Pepperberg, 2007a). He knew /u/ from labels such as "two" and "blue" (Pepperberg, 1999). As noted above, Alex had not had any training to combine the phonemes he had been taught to sound out, or to combine such phonemes and existing labels. Thus his combinatorial behavior was spontaneous.

Alex steadfastly retained "s-(pause)-wool" despite almost a year of instruction (Pepperberg, 2007a, 2007b), although normally only about 25 training sessions (at most, several weeks) are sufficient for learning a new label (Pepperberg, 1999). Griffin, a grey parrot who heard the other birds' sessions but was uninterested in spools and thus did not receive any "spool" training himself, was just beginning phoneme work; he did not exhibit any spool-related vocal behavior (Pepperberg, 2007a).

By 2004, Alex spontaneously uttered a human-sounding "spool" (/spul/; Figure 13.3, Pepperberg, 2007a) when Arthur was rewarded for labeling the object. Alex added the sound—which we hear, sonographically view, and transcribe, as—/p/ and shifted from /U/ to /u/. (*Note*: Both Alex's and my /u/'s are diphthongs, differing slightly from standard American English productions; Patterson & Pepperberg, 1994.) We never heard any intermediary form between "spool" and "s-(pause)-wool," preventing any statistical or other analysis of the developmental process (Pepperberg, 2007a).[4]

Alex's and Arthur's productions differed significantly, auditorily and sonographically (see Figures 13.1A, 13.3). Arthur incorporated an avian whistlelike /sp/; Alex's "spool" was distinctly human. Alex's vocal pattern closely resembles mine (Figure 13.1B), although students did 90% of the training. I had, however, been the principal

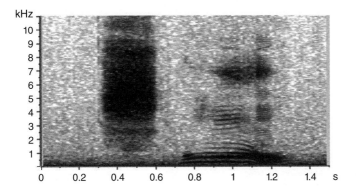

Figure 13.3
Alex's "spool" (/spul/) (from Pepperberg, 2007a)

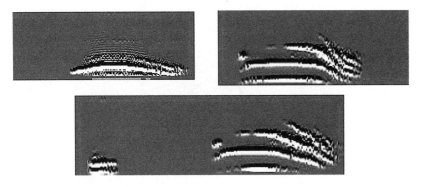

Figure 13.4
(a) Alex's /U/, (b) Alex's /u/, and (c) part of Pepperberg's /spu/ (from Pepperberg, 2007a)

trainer on *wool* about 20 years earlier (Pepperberg, 1999). Alex's "spool" shows true formant structures that closely approach, but are not identical to, mine (Figures 13.1B, 13.3; see Patterson & Pepperberg, 1994, 1998, for detailed comparisons of Alex's and my speech; identity is impossible because of the differences in vocal-tract sizes and Alex's lack of lips).

Figure 13.4 (Pepperberg, 2007a, 2007b) highlights Alex's vowel change, from /U/ to /u/. How he made the shift is unknown (Pepperberg, 2009). We could no longer eavesdrop on Alex's *solitary* practice (Pepperberg, Brese, & Harris, 1991) because the three parrots were now together 24/7. A gradual shift was unlikely had Alex maintained his previous pattern of private vocalizing, which involved significant portions of what would be considered stable end -rhyming in humans (e.g., "green, cheen, bean"; "mail, banail"; Pepperberg et al., 1991). An abrupt shift could indicate

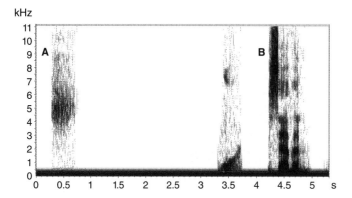

Figure 13.5
(a) Alex's "s-one" [/s-pause-wən/] followed by (b) Pepperberg's "seven" [/sEvIn/] (from Pepperberg, 2009)

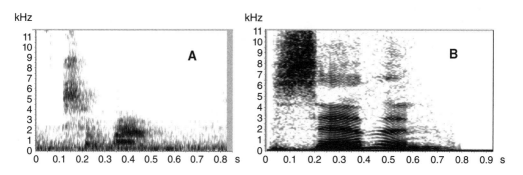

Figure 13.6
(a) Alex's "seben" [/sEbIn/] compared to (b) Pepperberg's "seven" [/sEvIn/] (from Pepperberg, 2009)

self-monitoring or some additional awareness that the appropriate vowel for "spool" derived from yet another label such as "two" (/tu/); such information was unavailable to Arthur (Pepperberg, 2009).

This acquisition pattern was not unique to "spool." Alex showed similar behavior while acquiring "seven" (Pepperberg, 2009), first in reference to the Arabic numeral, then for a set of objects. His first attempt at "seven" was "s . . . n," a bracketing using the phonemes /s/ and /n/. He quickly progressed to "s-pause-one" (Figure 13.5; Pepperberg, 2009; /s/-pause-/wən/), which differed considerably from my "seven," but followed the form of "s-pause-wool."

Eventually, he replaced "s-pause-one" with something like "seben," much closer to my "seven" (Figure 13.6; Pepperberg, 2009; sonograph expanded for reference).

These data provide evidence for a form of phonological awareness heretofore unseen in a nonhuman subject: Alex acoustically represented labels as do humans with respect to phonetic categories and understood that his labels were made of individual elements that could be recombined in various ways to produce new ones (Pepperberg, 2007a, 2007b). His previously reported vocal behavior patterns showed only that such awareness was possible, not that it existed. Discussion of these patterns demonstrates both their limitations and how they may have formed the basis for subsequent segmentation and phonological awareness.

Two patterns relate to ease of acquisition. First, my parrots' usual initial production of vowels while acquiring labels may reflect the relative ease of producing tonal sounds compared to those requiring, for example, plosive qualities for a subject lacking lips. Second, my birds' occasional production of acoustically perfect new labels after minimal or no training and without overt practice (Pepperberg, 1983, 1999)—for example, Alex's production of "carrot" the day after asking us what we were eating, or of the novel label "banerry" for an apple—may have involved segmentation, but not necessarily with intention (Pepperberg, 2007a). These labels did contain sounds already in his repertoire (e.g., for "carrot," /k/ from "key," the remainder from "parrot"; "banerry" derived from "banana" and "cherry"), but we cannot claim that Alex deliberately parsed labels in his repertoire to match a targeted utterance or to form novel vocalizations. Possibly "carrot" arose from acquired agility in manipulating his vocal tract to produce such sounds, or the new labels were simply created from phonotactically probable sequences involving beginnings and ends of existing labels (Storkel, 2001), or, for "banerry," from semantic relations (Pepperberg, 2007a).

Other possibilities could account for Alex's referential production of labels involving minimal pairs (Patterson & Pepperberg, 1998). His requests "Want corn" versus "Want cork," or "Want tea" versus "Want pea" (and refusal of alternatives), suggest the ability to segment phonemes from the speech stream (somewhat like nonhuman primates; Newport et al., 2004) and to recognize small phonetic differences ("tea" versus "pea") as meaningful, but he may not have deliberately parsed these labels when learning to produce them (Pepperberg, 2007a). Similarly, data (Patterson & Pepperberg, 1994, 1998) demonstrating that he (1) produced initial phonemes differently depending on subsequent ones (/k/ in "key" versus "cork") and (2) consistently recombined parts of labels according to their order in existing labels (i.e., combined beginnings of one label with the ends of others[5]) implied but did not prove that he engaged in such top-down processing (Ladefoged, 1982).

Two other closely related behavior patterns involved sound play and suggested a form of label parsing (Pepperberg, 1990; Pepperberg et al., 1991) but differ from true segmentation and phonological awareness (Pepperberg, 2007a). First, during private practice, Alex produced strings such as *mail chail benail* before producing the targeted, trained label *nail* (Pepperberg et al., 1991). Here his phoneme combinations seemed less a deliberate attempt to create a new label from specific sound patterns resembling the target than deliberate play with related, existing patterns in an attempt to hit on a configuration matching some remembered template. That is, he seemed to understand the combinatory nature of his utterances, but not necessarily how to split the novel targeted vocalization exactly, then match its components to those in his repertoire to create specific labels (Pepperberg, 2007a), both of which are required for phonological awareness and intentional segmentation. Second, in his trainers' presence, Alex babbled strings such as *grape, grain, chain, cane* in the absence of any objects. Our immediate mapping of these labels onto physical referents (Pepperberg, 1990) showed him the worth of such behavior (e.g., allowing him to request new interesting items), but we had no reason to believe his productions were intentional, other than to gain his trainers' attention. (*Note*: Only *grape* would have previously been used by trainers.) The rhyme awareness demonstrated in these behavior patterns—separate from phoneme awareness—is closely aligned to children's language skills (see Mann & Foy, 2003), but provides only supporting data that Alex viewed his labels as being constructed from individual sound patterns (Pepperberg, 2007a).

Only one other event suggested that Alex, previous to his actions with respect to "spool" and "seven," might have begun to understand label segmentation. The incident occurred at the MIT Media Lab; we were demonstrating how Alex could sound out his set of letters in front of a number of CEOs of various U.S. and international corporations, whose schedule allowed only about five to seven minutes with us. Alex was correctly responding to standard queries such as "What color /sh/?," requesting nuts as his reward. So that the short time available for the demonstration would not be spent watching Alex eat, after each response he was told to wait, that he could have his reward after a few more trials. Such a comment from trainers was unusual, in that he was always rewarded immediately after being told he was correct; only after an error, when he heard "no," were his requests denied. He became more and more agitated, asking for nuts with more emphasis after each correct response: His vocalizations became louder, with more emphasis on the label "nut." Finally, he looked at me and said, "Wanna nut . . . N-U-T!!," stating the individual sounds "nnn," "uhh," and "t" (/n/, /ə/, /t/, respectively). Again, he had never been trained to perform this kind of task, and although N and T had been trained individually, U had not. Somehow, on his own, he had deduced that it could be a separate sound. Because we were trying to complete other studies at the time, we dismissed this behavior as an anecdote and did not follow up with any type of testing. And, unfortunately, Alex died before additional studies could be designed.

The more recent data, however, when combined with previous evidence, suggests that Alex, much like a child, applied a phonological rule derived from knowledge of his repertoire (Pepperberg, 2007a): that sounds such as "pat" and "turn" can be recombined into a label for identifying a separate distinct item—*pattern*—having no referential correlation to the original utterances. Alex appeared to form the closest match based on segmentation and onset + nucleus + rhyme (Storkel, 2002). Arguably, the data could be considered stronger if Alex had known *pull* or *pool* and initially produced either "s-pause-pull" or "s-pause-pool." Given that /p/ is particularly difficult for a parrot, lacking lips, to produce (Patterson & Pepperberg, 1998), I believe production of "s-pause-wool" is actually more important, because, lacking exact matches, he took the closest, readily available sounds in his repertoire (i.e., "wool" is the only object label of ~50 documented in his repertoire resembling "spool") to form initial attempts at a novel vocalization, and by so doing, made the process transparent to his human trainers (Pepperberg, 2007a).

Alex may also have initially formed "s-pause-wool" so that two known utterances provided the overall structure and the pause was a place filler, somewhat like young children behave, until he could insert /p/ and adapt the vowel (Pepperberg, 2007a). Specifically, Peters (2001) suggests that children use fillers (a "holding tank") to preserve syllable number or prosodic rhythm of a target vocalization until the standard form is learned (note Leonard, 2001). Although Peters and colleagues primarily refer to earlier grammatical forms, not labels (but see Lleó, 2001), and Alex used a pause, not another phoneme, his behavior suggests (but cannot prove) an awareness of the need for something additional and somewhat different to complete the vocalization (Pepperberg, 2007a). Simply omitting or closing the gap—and responding on the basis of sound similarity—would have produced /swUl/ ("swull"), not /swul/ ("swooool")

Arguably, Alex might have applied a phonological rule for combining utterances (i.e., engaged in some metalinguistic analysis about beginnings and ends of labels) without truly understanding its basis (the nature of phonemes; Pepperberg, 2007a, 2009). Such an interpretation might explain labels he produced in sound play in the absence of referents (Pepperberg, 1990), but the specificity and consistent use of both the "s-pause-wool" and "s-pause-one" combinations argues against such an alternative explanation, as well as against "babble-luck" (a fortuitously correct but accidental combination; Thorndike, 1943). Instead, Alex had to have discriminated and extracted the appropriate speech sounds of the barely related target labels "spool" and "seven," generalized these to the closest related items in his repertoire, fit the existing sounds together—including a pause to maintain spacing for two separate absent sounds—in a particular serial order, and additionally had to link each novel phonology referentially with a specific item (Pepperberg, 2007a).

Do such data support a parrot's understanding of a phonetic "grammar" (e.g., Fitch & Hauser, 2004)? Alex did generate novel meaningful labels from a finite element set, but his rule system was relatively limited (Pepperberg, 2007a).

Nevertheless, the data presented here add another parallel between Alex's and young children's early label acquisition (Pepperberg, 1999, 2007a). Children's manipulation of individual word parts implies the existence of internal representations of words as divisible units, and normal children proceed in a fairly standard manner from babbling to full language. Alex never reached the level of any young child, nor would he likely ever have grammatically advanced beyond the use of simple sentence frames (e.g., "I want X," "I wanna go Y," X and Y being appropriate object or location labels). But any strides a bird makes toward language-like ability—such as, for example, comprehending recursive conjunctive sentences[6] or demonstrating the vocal segmentation described here—helps delineate similarities and differences between humans and nonhumans (Pepperberg, 2007b). Do these data affect the notion that language evolution must be based on phylogeny? Must anything language-like in nonhumans be a function of a common ancestry with humans (Savage-Rumbaugh et al., 1993)? Can only nonhuman primates exhibit human language precursors (cf. Gentner, Fenn, Margoliash, & Nusbaum, 2006)?

Alex's abilities were clearly not isomorphic with human language, but his data (and that of other studies; see Pepperberg & Shive, 2001) demonstrate that elements of linguistic behavior are not limited to primates, nor are neurological systems underlying such behavior. Avian neuroanatomy and its relation to the mammalian line is slowly being understood enough to determine specific parallels among oscine, psittacine, and mammalian structures, particularly with respect to vocal learning (e.g., Jarvis et al., 2005). Given the evolutionary distance between parrots and primates, the search for and arguments concerning responsible neural substrates and common behavior should be approached with care, but maybe not restricted to the primate line. The data presented here, plus our knowledge of avian vocal learning, of how social interaction affects such learning, and of birds' advanced cognition (Clayton, Dally, Gilbert, & Dickinson, 2005; Emery & Clayton, 2009; Gentner et al., 2006; Kenward et al., 2005; Kroodsma & Miller, 1996; Pepperberg, 1999, 2007a, 2007b, 2009; Weir, Chappell, & Kacelnik, 2002), all suggest *Aves* as an important model for determining the evolutionary pressures responsible for—and in developing testable theories about—complex communication systems, particularly those involving vocal learning (Pepperberg, 2007a).

Acknowledgments

I thank Diana Reiss (Hunter College, New York) and Donald Kroodsma (University of Massachusetts–Amherst) for creating the sonograms on Raven and Ofer Tchernichovski (CCNY) for the sonogram using Sound Analysis Pro. Research was supported by donors to *The Alex Foundation*. The writing of the chapter was supported in part by the Marc Haas Foundation.

Notes

1. No claim is made that Alex's speech was isomorphic with human language, only that the *elements* he produced were documented as being used referentially. Labels were both understood and used appropriately in contexts differing from and extending beyond training conditions.

2. Nonhuman primates have been trained and tested on their ability to segment human speech sounds (e.g., Newport, Hauser, Spaepen, & Aslin, 2004), but not on sound-letter associations or on productive recombination of speech elements.

3. Voice onset times for both Alex's and my /p/ fall solidly into the voiceless category and are distinct from the voiced /b/ (Patterson & Pepperberg, 1998).

4. For Wav. forms of both productions, contact impepper@media.mit.edu.

5. In over 22,000 vocalizations, we never observed backwards combinations such as "percup" instead of "cupper/copper" (Pepperberg et al., 1991).

6. For example, given various trays each holding seven objects of several colors, shapes, and materials, Alex responded to "What object/material is color-A and shape-B?" vs. "What shape is color-A and object/material-C?" vs. "What color is shape-B and object/material-C?" (Pepperberg, 1992).

References

Anthony, J. L., & Francis, D. (2005). Development of phonological awareness. *Current Directions in Psychological Science, 14,* 255–259.

Beckers, G. J. L., Nelson, B. S., & Suthers, R. A. (2004). Vocal-tract filtering by lingual articulation in a parrot. *Current Biology, 14,* 1592–1597.

Byrne, B., & Liberman, A. M. (1999). Meaninglessness, productivity and reading: Some observations about the relation between the alphabet and speech. In J. Oakhill & R. Beard (Eds.), *Reading development and the teaching of reading* (pp. 157–174). Oxford: Blackwell.

Carroll, J. M., Snowling, M. J., Hulme, C., & Stevenson, J. (2003). The development of phonological awareness in preschool children. *Developmental Psychology, 39,* 913–923.

Clayton, N. S., Dally, J., Gilbert, J., & Dickinson, A. (2005). Food caching by western scrub-jays (*Aphelocoma californica*) is sensitive to the conditions at recovery. *Journal of Experimental Psychology: Animal Behavior Processes, 31,* 115–124.

de Boysson-Bardies, B. (1999). *How language comes to children.* Cambridge, MA: MIT Press.

Edwards, J., Munson, B., & Beckman, M. E. (2011). Lexicon-phonology relationships and dynamics of early language development. *Journal of Child Language, 38,* 35–40.

Emery, N. J., & Clayton, N. S. (2009). Comparative social cognition. *Annual Review of Psychology, 60,* 87–113.

Fitch, W. T., & Hauser, M. D. (2004). Computational constraints on syntactic processing in a nonhuman primate. *Science, 303,* 377–380.

Fouts, R., & Rigby, R. (1977). Man-chimpanzee communication. In T. Sebeok (Ed.), *How animals communicate* (pp. 1034–1054). Bloomington: Indiana University Press.

Gathercole, S. E., & Baddeley, A. D. (1990). The role of phonological memory in vocabulary acquisition: A study of young children learning new names. *British Journal of Psychology, 81*, 439–454.

Gentner, T. Q., Fenn, K. J., Margoliash, D., & Nusbaum, H. C. (2006). Recursive syntactic pattern learning by songbirds. *Nature, 440*, 1204–1207.

Greenfield, P. (1991). Language, tools and brain: The ontogeny and phylogeny of hierarchically organized sequential behavior. *Behavioral and Brain Sciences, 14*, 531–595.

Jarvis, E. D., Güntürkün, O., Bruce, L., Csillag, A., Karten, A., Kuenzel, W., et al. (2005). Avian brains and a new understanding of vertebrate brain evolution. *Nature Reviews Neuroscience, 6*, 151–159.

Kenward, B., Weir, A. A. S., Rutz, C., & Kacelnik, A. (2005). Tool manufacture by naive juvenile crows. *Nature, 433*, 121.

Kroodsma, D. E., & Miller, E. H. (Eds.). (1996). *Ecology and evolution of acoustic communication in birds*. Ithaca, NY: Cornell University Press.

Kuhl, P. K., & Miller, J. D. (1975). Speech perception by the chinchilla: Voiced-voiceless distinction in alveolar plosive consonants. *Science, 190*, 69–72.

Ladefoged, P. (1982). *A course in phonetics*. San Diego: Harcourt Brace Jovanovich.

Leonard, L. B. (2001). Fillers across languages and language abilities. *Journal of Child Language, 28*, 257–261.

Lieberman, P. (1984). *The biology and evolution of language*. Cambridge, MA: Harvard University Press.

Lleó, C. (2001). Early fillers: Undoubtedly more than phonological stuffing. *Journal of Child Language, 28*, 262–265.

Mann, V. A., & Foy, J. G. (2003). Phonological awareness, speech development, and letter knowledge in preschool children. *Annals of Dyslexia, 53*, 149–173.

Marschark, M., Everhart, V. S., Martin, J., & West, S. A. (1987). Identifying linguistic creativity in deaf and hearing children. *Metaphor and Symbolic Activity, 2*, 281–306.

Newport, E. L., Hauser, M. D., Spaepen, G., & Aslin, R. N. (2004). Learning at a distance II. Statistical learning of non-adjacent dependencies in a non-human primate. *Cognitive Psychology, 49*, 85–117.

Patterson, D. K., & Pepperberg, I. M. (1994). A comparative study of human and parrot phonation: acoustic and articulatory correlates of vowels. *Journal of the Acoustical Society of America, 96*, 634–648.

Patterson, D. K., & Pepperberg, I. M. (1998). Acoustic and articulatory correlates of stop consonants in a parrot and a human subject. *Journal of the Acoustical Society of America, 103*, 2197–2215.

Peperkamp, S. (2003). Phonological acquisition: Recent attainments and new challenges. *Language and Speech, 46*, 87–113.

Pepperberg, I. M. (1981). Functional vocalizations by an African Grey parrot (*Psittacus erithacus*). *Zeitschrift für Tierpsychologie, 55*, 139–160.

Pepperberg, I. M. (1983). Cognition in the African grey parrot: Preliminary evidence for auditory/vocal comprehension of the class concept. *Animal Learning & Behavior, 11*, 179–185.

Pepperberg, I. M. (1990). Referential mapping: A technique for attaching functional significance to the innovative utterances of an African grey parrot (*Psittacus erithacus*). *Applied Psycholinguistics*, *11*, 23–44.

Pepperberg, I. M. (1992). Proficient performance of a conjunctive, recursive task by an African grey parrot (*Psittacus erithacus*). *Journal of Comparative Psychology*, *106*, 295–305.

Pepperberg, I. M. (1999). *The Alex studies*. Cambridge, MA: Harvard University Press.

Pepperberg, I. M. (2006). Emergence of linguistic communication: An avian perspective. In C. Lyon, C. L. Nehaniv, & A. Cangelosi (Eds.), *Emergence and evolution of linguistic communication* (pp. 355–386). London: Springer-Verlag.

Pepperberg, I. M. (2007a). Grey parrots do not always "parrot": Roles of imitation and phonological awareness in the creation of new labels from existing vocalizations. *Language Sciences*, *29*, 1–13.

Pepperberg, I. M. (2007b). When training engenders a failure to imitate in grey parrots (*Psittacus erithacus*). In M. Lopes (Ed.), *Proceedings of the AISB '07 Fourth International Symposium on Imitation in Animals and Artifacts*. Newcastle-upon-Tyne, UK: University of Newcastle.

Pepperberg, I. M. (2009). Grey parrot (*Psittacus erithacus*) vocal learning: Creation of new labels from existing vocalizations and issues of imitation. In P. Sutcliffe, L. Stanford, & A. Lommel (Eds.), *LACUS Forum 34: Speech and beyond* (pp. 21–30). Houston, TX: Linguistic Association of Canada and the United States, http://www.lacus.org/volumes/34.

Pepperberg, I. M., Brese, K. J., & Harris, B. J. (1991). Solitary sound play during acquisition of English vocalizations by an African grey parrot (*Psittacus erithacus*): Possible parallels with children's monologue speech. *Applied Psycholinguistics*, *12*, 151–178.

Pepperberg, I. M., & Shive, H. R. (2001). Simultaneous development of vocal and physical object combinations by a grey parrot (*Psittacus erithacus*): Bottle caps, lids, and labels. *Journal of Comparative Psychology*, *115*, 376–384.

Pepperberg, I. M., & Wilkes, S. R. (2004). Lack of referential vocal learning from LCD video by grey parrots *(Psittacus erithacus)*. *Interaction Studies: Social Behaviour and Communication in Biological and Artificial Systems*, *5*, 75–97.

Peters, A. M. (2001). Filler syllables: What is their status in emerging grammar? *Journal of Child Language*, *28*, 229–242.

Reiss, D., & McCowan, B. (1993). Spontaneous vocal mimicry and production by bottlenose dolphins (*Tursiops truncatus*): Evidence for vocal learning. *Journal of Comparative Psychology*, *107*, 301–312.

Savage-Rumbaugh, S., Murphy, J., Sevcik, R. A., Brakke, K. E., Williams, S. L., & Rumbaugh, D. M. (1993). Language comprehension in ape and child. *Monographs of the Society for Research in Child Development*, *233*, 1–258.

Storkel, H. (2001). Learning new words II: Phonotactic probability in verb learning. *Journal of Speech, Language, and Hearing Research: JSLHR*, *44*, 1312–1323.

Storkel, H. (2002). Restructuring of similarity neighbourhoods in the developing mental lexicon. *Journal of Child Language*, *29*, 251–274.

Studdert-Kennedy, M. (2002). Mirror neurons, vocal imitation, and the evolution of particulate speech. In M. I. Stamenov & V. Gallese (Eds.), *Mirror neurons and the evolution of brain and language* (pp. 207–227). Amsterdam: John Benjamins.

Thorndike, E. L. (1943). *Man and his works*. Cambridge, MA: Harvard University Press.

Tomasello, M. (2003). *Constructing a language*. Cambridge, MA: Harvard University Press.

Treiman, R. (1995). Errors in short-term memory for speech: A developmental study. *Journal of Experimental Psychology: Learning, Memory, and Cognition, 21*, 1197–1208.

Vihman, M. (1996). *Phonological development: The origins of language in the child*. Malden, MA: Blackwell.

Weir, A. A. S., Chappell, J., & Kacelnik, A. (2002). Shaping of hooks in New Caledonian crows. *Science, 297*, 981.

IV NEUROBIOLOGY OF SONG AND SPEECH

14 The Neural Basis of Links and Dissociations between Speech Perception and Production

Sophie K. Scott, Carolyn McGettigan, and Frank Eisner

Evidence for the links between speech perception and speech production can be seen in many ways. The language we acquire in childhood has a strong influence on the kinds of phonetic contrasts to which we are sensitive as adults (Goto, 1971), and we continually modulate the sound of our spoken voice in line with our acoustic environments (Lombard, 1911), to the extent that we even converge on pronunciations with cospeakers (Pardo, 2006). In development, a lack of auditory sensation (even a relatively moderate hearing loss) has a detrimental effect on speech production (Mogford, 1988), and deaf children can struggle with spoken language as well as other tasks such as reading that depend on speech perception (e.g., Furth, 1966). Some specific cognitive processes, such as verbal working memory, have been hypothesized to reflect the interplay of speech-perception and speech-production systems (Jacquemot & Scott, 2006). Semantic processes can be primed by speech production as well as perception: silently mouthing a word primes later responses to that word in an auditory lexical decision task, but not a visual lexical decision task (Monsell, 1987). Some models of speech perception specify that motor representations of a speaker's articulations form the "objects" of perception—in other words, that speech perception requires the representations used in speech production (Liberman, Delattre, & Cooper, 1952; Liberman, Cooper, Shankweiler, & Studdert-Kennedy, 1967; Liberman & Mattingly, 1985; Fowler, 1986).

However, speech perception and production also show some dissociations. In neuropsychology, patients with speech-production deficits following anterior brain damage can still comprehend spoken language (e.g., Blank, Bird, Turkheimer, & Wise, 2003; Crinion, Warburton, Lambon-Ralph, Howard, & Wise, 2006). In contrast, patients with speech-perception deficits following posterior brain damage do not have problems with the fluency of their speech, though the content of their speech production can be nonsensical (McCarthy & Warrington, 1990). This suggests that damage to the motor control of speech production can leave speech comprehension intact.

In development, speech-perception skills have been identified as driving language acquisition (Werker & Yeung, 2005). However, this does not necessarily link directly to speech production: variability in speech-perception skills measured at age 21 months does not correlate with variability in speech-production skills, and both are predicted by different behavioral tasks (Alcock & Krawczyk, 2010). Though a lack of auditory input is detrimental to the development of speech production (Mogford, 1988), this is not a bidirectional mechanism: the development of speech perception does not require speech-production skills. Developmental disorders that lead to severe speech-production difficulties (such as those that can be seen in cerebral palsy) do not necessarily affect speech perception and comprehension (Bishop, 1988), and there are accounts of adults who have grown up with severe dysarthria, yet who can produce fluent typed text shortly after being provided with a foot typewriter (Fourcin, 1975). This suggests that auditory input is necessary for normal speech-production skills to develop, but that speech-production skills are not essential for speech perception to develop.

Anatomically, links between speech perception and production can be seen in motor areas, which can be activated during speech perception, and in auditory areas, which can show both activation and suppression during speech production. In this chapter we review some of the anatomical bases for the links and dissociations between speech perception and speech production, using the organization of primate auditory neuroanatomy as an anatomical framework for speech perception. We also consider links between perception and production in higher-order linguistic areas.

Functional Organization of Speech Perception

Speech perception is associated with the dorsolateral temporal lobes: this is largely because speech is an acoustic signal, and the dorsolateral temporal lobes contain auditory cortical fields in primates. In primates, these auditory areas are associated with properties of both hierarchical and parallel processing (Rauschecker, 1998). The hierarchical processing is seen anatomically in the arrangement of the core (primary) auditory cortex, which is surrounded by the belt and parabelt fields (Rauschecker, 1998): core areas project to belt areas, and belt areas project on to parabelt regions (Kaas & Hackett, 1999). Hierarchical processing can also be seen neurophysiologically, with responses to progressively more complex sounds seen in recordings moving laterally from core to belt and parabelt regions (Rauschecker, 1998). The parallel properties of the auditory cortex are also seen in its anatomy. The pattern of connections between core, belt, parabelt, and beyond preserve the rostral-caudal organization of the core area. Rostral core areas project to mid- and rostral belt areas, rostral belt projects to mid- and rostral parabelt, and both rostral belt and parabelt project to anterior STG and frontal areas (Kaas & Hackett, 1999).

In contrast, caudal core auditory areas project to mid- and caudal belt areas, caudal belt areas project to mid- and caudal parabelt areas, and both caudal belt and parabelt regions project to frontal areas, which are adjacent but nonoverlapping with frontal areas that receive their projections from rostral auditory fields (Kaas & Hackett, 1999). Together, these rostral and caudal patterns of connections have been described as auditory "streams" of processing, analogous to those seen in the visual system. Also, as in the visual system, these parallel streams are associated with different kinds of auditory information: in nonhuman primates, rostral auditory areas show a higher sensitivity to different kinds of conspecific vocalizations (a "what" pathway), and caudal auditory areas show a greater sensitivity to the spatial location of sounds than their identity (a "where" pathway) (Tian, Reser, Durham, Kustov, & Rauschecker, 2001). Furthermore, medial caudal auditory fields are sensitive to somatosensory input as well as auditory (Fu et al., 2003; Smiley et al., 2007). This has been identified as part of an auditory "how" pathway that is probably neither anatomically nor functionally distinct from the "where" pathway (Rauschecker & Scott, 2009).

Patterns of hierarchical and parallel processing are also seen in human auditory areas. Functionally, hierarchical processing can be seen in auditory areas, where primary auditory cortex (PAC) is less sensitive to auditory structure than anterolateral fields. Thus, while PAC responds to any sound, contrasting the neural response to frequency-modulated tones with that to unmodulated tones reveals activation in regions lateral to the primary auditory cortex, running anterior and posterior into auditory association cortex (Hall et al., 2002). Posterior auditory fields in humans also show greater responses to spatial aspects of sounds than more anterior auditory regions (Alain, Arnott, Hevenor, Graham, & Grady, 2001).

In terms of speech perception, human primary auditory fields are not selectively responsive to speech (Mummery, Ashburner, Scott, & Wise, 1999). However, sensitivity to the acoustic structure of one's native language can be seen in early auditory areas lying just lateral to the primary auditory cortex (Jacquemot, Pallier, Le Bihan, Dehaene, & Dupoux, 2003; Scott, Rosen, Lang, & Wise, 2006), in fields that are responsive to aspects of acoustic structure such as amplitude modulation and harmonic structure (Scott & Johnsrude, 2003). Running lateral and anterior to PAC, the neural responses seen become progressively more sensitive to the linguistic information in speech, and less sensitive to the acoustic structure (Scott, Blank, Rosen, & Wise, 2000; Scott et al., 2006). Thus, responses can be seen in the left anterior superior temporal sulcus (STS) to speech, if the speech itself can be understood. In contrast, the same left anterior STS region is relatively insensitive to whether, for example, the speech sounds like a real person, or is noise-vocoded to sound like a harsh whisper. The anterior left STS has been suggested to be the location of auditory word forms (Cohen, Jobert, Le Bihan, & Dehaene, 2004), and is

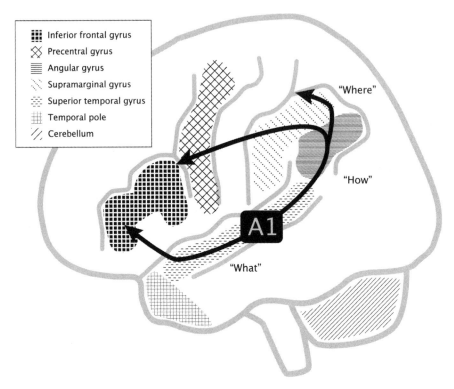

Figure 14.1
Lateral surface of the left hemisphere, showing important anatomical areas in the temporal lobe, inferior parietal lobe, and frontal cortex. Also shown are the "streams" of processing that have been suggested to underlie different aspects of auditory processing—for identification ("what"), sensorimotor processing ("how"), and spatial processing ("where").

certainly well placed to link to cortical areas associated with higher-order language functions, such as semantic representations and processes in the temporal pole (Patterson et al., 2007) as well as the frontal and parietal lobes (Obleser, Wise, Dresner, & Scott, 2007) (Figure 14.1).

In contrast to the intelligibility responses in more anterior temporal areas, posterior-medial auditory areas have been shown to respond during speech production, even if the speech produced is simply mouthed (i.e., there is no sound made) (Wise, Scott, Blank, Mummery, & Warburton, 2001; Hickok et al., 2000). This posterior-medial auditory area has been linked to working memory processes, for the rehearsal of both speech and nonspeech sounds (Hickok et al., 2003). This posterior-medial auditory area has also been hypothesized to process and represent how "doable" sounds are—is this a sound that one *could* make with one's articulators? (Warren, Wise, & Warren, 2005). This rostral-caudal distinction of functional organization in speech processing has been related to the putative streams of processing in the primate auditory cortex (Scott & Johnsrude, 2003; Rauschecker & Scott, 2009). The processing of linguistic meaning in speech is associated with a left-lateralized stream of processing, running lateral and anterior to PAC, toward the anterior STS—a "what" stream of processing (Figure 14.1). The sensorimotor responses in the posterior-medial auditory cortex have been postulated to form part of a "how" stream of processing (Figure 14.1), which may involve the same kind of cross-modal transformations involved in spatial aspects of sound processing (i.e., the posterior "how" and "where" pathways are in fact part of the same stream) (Rauschecker & Scott, 2009).

Speech Production

Speaking is a highly complex motor act, and thus involves extensive areas in the premotor and motor cortex, in the supplementary motor area, as well as in the basal ganglia and cerebellum. The motor cortex activation is associated with the direct control of individual muscles of articulation, while the premotor and supplementary motor areas are important in the coordination of muscle groups (e.g., Wise, Greene, Büchel, & Scott, 1999), and thus in more complex aspects of speech control.

In terms of the control of speech production, Broca's area has long been considered a critical region. Broca's original patient, Tan, could only speak the word *tan* (and some swearing): postmortem analyses revealed that he had a tumor in the posterior third of his left inferior frontal gyrus. However, while subsequent work has confirmed that Broca's area is important in speech production, more recent postmortem analyses have revealed that lesions only damaging Broca's area do not lead to the full range of problems associated with Broca's aphasia: instead, transient mutism is seen (Mohr et al., 1978). To see Broca's aphasia, with slow, difficult speech production and speech-sound errors, more widespread damage is needed, affecting the underlying white matter tracts. Furthermore, functional imaging of speech production has revealed that Broca's area is not necessarily activated during speech production. Instead the left anterior insula is activated during articulation (Wise et al., 1999), and the left anterior insula has also been identified as the common site of brain damage in speech apraxia (Dronkers, 1996) (Figure 14.2B). For Broca's area to be involved, the speech produced needs to be somewhat more complex: more activation is seen in the pars opercularis (part of Broca's area) for counting aloud than for single-word repetition, more activation for reciting a simple nursery rhyme than for counting, and yet more activation for propositional speech when contrasted with reciting nursery rhymes (Blank, Scott, Murphy, Warburton, & Wise, 2002). Thus, while the anterior insula is central to articulation itself, Broca's area is likely involved

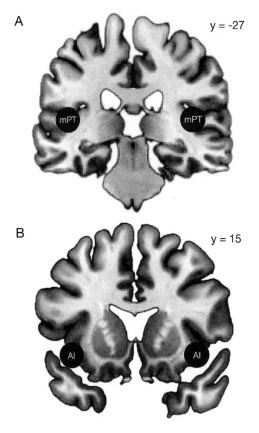

Figure 14.2
Two coronal slices of a structural MRI image of the brain. Panel A shows the location of the posterior medial planum temporale (mPT), and panel B shows the location of the anterior insula (AI).

in more high-order aspects of the planning or articulation, and has been argued to be involved in a range of nonverbal aspects of behavior (e.g., Schnur et al., 2009).

Motor Control in the Auditory Cortex?

In addition to these motor, subcortical, and cerebellar regions, the posterior-medial auditory area, discussed as part of the "how" pathway, is strongly activated during speech production, where it has been linked to sensorimotor interactions—where the sensation could be auditory and/or somatosensory—in the control of speech output (Wise et al., 2001; Warren et al., 2005) (Figure 14.2A). These interactions may represent auditory feedback projections important in monitoring the sounds of speech. However, a recent study of speech production and silent jaw and tongue movements found that the supratemporal plane was equally activated by speech and by silent jaw and tongue movements, while lateral STG areas were activated only by overt speech production (Dhanjal, Handunnetthi, Patel, & Wise, 2008). This implicates the posterior-medial planum in somatosensory as well as auditory processing. The posterior-medial auditory areas seem to perform cross-modal sensorimotor interactions during speech perception and production, and thus may form a central link between these two processes. Indeed, a specific role of somatosensory feedback has recently been posited in speech-production tasks: cochlear implant users alter their articulations to overcome physical disruptions of their speech, and this adaptation can occur even if the implant is switched off (and there is thus no acoustic feedback) (Nasir & Ostry, 2008).

While the involvement of the posterior-medial planum in silent articulation or silent repeated movements of the articulators is strong evidence that it has a role in speech production, it will be important to rule out some other process as leading to this activation—for example, some automatic auditory imagery—caused by the movement of the articulators. Wise et al. (2001) used silent rehearsal as a control condition for the articulation conditions, which revealed posterior-medial planum activation. This suggests that imagery alone activates this region less than articulation. However, the evidence that this region is driven solely by speech is weaker: it has been implicated in the rehearsal of music information (Hickok et al., 2003), and a direct comparison of speech production with non-speech-sound production, such as blowing a raspberry or whistling, reveals very similar activation in this region (Chang et al., 2009).

In the study of overt speech production by Blank et al. (2002), counting, nursery rhymes, and propositional speech were all contrasted with silence. All three conditions significantly activated the posterior-medial planum; however, there was significantly greater activation in this region for the recitation of familiar rhymes. This pattern was not seen in the anterior insula or in Broca's area, which were more generally involved in speech production across all three conditions (though a similar profile was seen in at the junction of the anterior insula and the frontal operculum). Since the speech produced in the nursery-rhyme condition followed specific rhythmic patterns very well, it is possible that the sensorimotor processes in posterior-medial planum are particularly relevant in the control of rhythmic structure in speech output. While human speech is not isochronously timed (unlike footsteps), there are underlying rhythmic characteristics that can be exploited by listeners and may be of special relevance in dialog (Scott, McGettigan, & Eisner, 2009). This may relate to its apparent nonspecificity to linguistic information, since we can produce rhythms with any kind of vocalization and other kinds of sound output (e.g., in music).

Suppression during Speech Production

In contrast to the posterior-medial activation during speech production, lateral auditory association areas have been shown to be suppressed during speech production
in humans (Houde, Nagarajan, Sekihara, & Merzenich, 2002; Wise et al., 1999) and
in nonhuman primates (Eliades & Wang, 2003, 2005). In nonhuman primates, the
suppression begins before the onset of voicing, suggesting that it represents a direct
input from the initiation of the vocal motor act (Eliades & Wang, 2003). Furthermore, the neurons that show these patterns of suppression are also those that
increase in activity if the frequency of the vocal feedback is altered, implicating this
suppression in the detection of alterations in speech production (Eliades & Wang,
2008). In humans, it has been suggested that this suppression is a simple mechanism
for establishing that incoming speech is self-generated, as can be found in other
sensory systems (e.g., Blakemore, Wolpert, & Frith, 1998). However, altering the
sounds of speech while people are talking (by spectrally alerting the speech in real
time or by introducing a delay) does lead to increased activation in medial and
lateral auditory areas (Tourville, Reilly, & Guenther, 2008; Takaso, Eisner, Wise, &
Scott, 2010) within the posterior temporal lobes. This suggests that some of these
responses may, as in the nonhuman primate work, be implicated in the detection of,
and correction for, real-time, online changes in the sounds of produced speech. Of
course, auditory stimulation is not the only kind of sensory consequence of speech
production, and movement of the articulators results in considerable activation of
the secondary somatosensory cortex. This secondary somatosensory activity is
reduced in simple speech production (counting) relative to silent tongue or jaw
movements, and is actually suppressed relative to rest during propositional speech
(Dhanjal et al., 2008). Thus the suppression of activation in lateral auditory areas is
also seen in somatosensory areas during speech production. It would be interesting
to know how this pattern of activation is altered when motoric aspects of speech
production are disrupted, leading to changes in articulation (Nasir & Ostry, 2008).

Speech Perception and Production: Motor Cortex

As outlined earlier, speech production makes heavy demands on the premotor and
primary motor cortex, and lesions to these motor speech areas lead to severe problems in speech production. Several recent functional imaging studies have reported
that speech perception also activates these speech-production areas, in addition to
the extensive dorsolateral temporal activation typically found with heard speech
(discussed above). This activation has been linked with both the motor theory of
speech perception—that a talker's gestures and intended gestures form the objects
of perception when listening to speech. A motor involvement in speech perception

also supports models of action perception that posit a central role for "mirror neurons"—that is, neural systems involved in both the production of an action and its observation—in perception. Thus, an fMRI study of syllable perception and production found peaks of activation for production that were also activated (to a smaller degree) by the perception of the same kinds of syllables (Wilson, Saygin, Sereno, & Iacoboni, 2004). Follow-up work from the same group showed that these motor areas are more activated by phonemes from outside the participants' linguistic experience than by phonemes from their own language (Wilson & Iacoboni, 2006). However, while there are clearly important responses in motor areas to sound (Scott et al., 2009), it is somewhat harder to link these motor responses to the linguistic context of speech. On a practical level, very few studies revealing premotor activation to speech sounds use any kind of comparison with a complex acoustic control condition, unlike the studies investigating the neural responses to speech in the temporal lobes. Conversely, very few functional imaging studies employing an adequate acoustic control condition reveal motor activation. This means that it can be hard to specify what, in the speech signal, is driving the motor responses to heard speech that have been detected (Scott et al., 2009). Recent reviews have suggested that the kinds of motor responses seen in the primate mirror system are too crude to characterize the kinds of representations argued for by the motor theory of speech perception (Lotto, Hickok, & Holt, 2009), and that the patterns of inference and construction seen in human communication are not well characterized by the simple responses seen in the motor system responses to observed actions (Toni, De Lange, Noordzij, & Hagoort, 2008). Recent papers have identified the motor cortex as a brain region recruited in metalinguistic aspects of language processing—that is, processes that are not driven by basic speech perception, but more by task-related aspects of the ways that subjects are required to engage with the speech stimuli (Davis, Johnsrude, Hervais-Adelman, & Rogers, 2008).

There is also evidence that premotor areas are more strongly engaged by some nonverbal vocalizations. In contrast to the relatively weak responses in the premotor cortex to speech sounds relative to silence, robust activations are seen in the premotor cortex to positive vocal expressions of emotion, such as laughter and cheering, relative to an acoustic control (Warren et al., 2006). We interpreted this motor activation as due to the behavior promoted by these positive, social emotions (e.g., smiling), rather than having to do with their comprehension (since the study also included other well-recognized, intense emotions such as disgust) or basic perceptual properties (since we included an acoustic control). We extended this role for the motor cortex in the mediation of aspects of social behavior in a recent review, where we hypothesized that motor responses seen to heard speech do not reflect their direct involvement in perception. Instead they appear to be a glimpse of the role of these areas when people are speaking and listening to speech in dialog—that is, when the motor system is central to the control of interspeaker behavior, such as the coordination of gestures and the accurate timing of turn taking (Scott et al., 2009).

Lateral Premotor vs. Anterior Insula in Speech Perception

The anterior insula, which has consistently been implicated in the production of speech (Dronkers, 1996; Wise et al., 1999), has also been implicated in the perception of speech (Wise et al., 1999) and in the unconscious repair of speech sounds in perception (Shahin et al., 2009). The anterior insula is the premotor cortex in humans, and it is possible that perceptual processes recruit the anterior insula in a more speech-specific manner than more dorsal and lateral motor areas.

Higher-Order Speech Comprehension and Production

Higher-order aspects of speech perception and production do seem to share a common neural basis. In a study of the cortical fields recruited when the predictability of a sentence (high or low) modulates speech intelligibility, activity was seen in the inferior frontal gyrus, the angular gyrus (Figure 14.1), the medial prefrontal cortex (Figure 14.3) and posterior cingulate to predictable sentences, contrasted with unpredictable sentences (Obleser et al., 2007). Very similar activations, plus activation in the left anterior temporal lobe, were seen in a study contrasting overt semantic processing of read words, suggesting that these regions form part of a distributed semantic network recruited in speech/language perception (Scott, Leff, & Wise, 2003). Within this network, activation in the medial prefrontal cortex (Figure 14.3) was specifically linked to the amount of time spent processing semantic information. The production of narrative speech—as opposed to simple repetition, counting, or reciting nursery rhymes—also showed strong activations in these regions (Blank et al., 2002; see also Menenti, Gierhan, Segaert, & Hagoort, 2011). A study directly comparing the perception and production of narrative speech reported common activation (contrasted with simple counting) in the left and right anterior lateral temporal lobes, extending into the temporal poles, medial prefrontal cortex, and temporal-occipital-parietal junction (Awad, Warren, Scott, Turkheimer, & Wise, 2007). These findings suggest strong links between the linguistic resources recruited for the comprehension and production of speech that is relatively complex in content. Thus, there are clearer links between the perception and production of speech in higher-order linguistic representations and processes than in the lower-level processes and representations that govern auditory and motoric factors in speech. It is possible, therefore, that these higher-order language areas underlie the kinds of semantic priming of lexical decision by silent

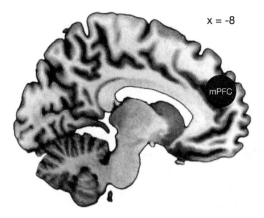

x = -8

Figure 14.3
Sagittal slice of a structural MRI of the brain, with the location of medial prefrontal cortex (mPFC) indicated.

mouthing (Monsell, 1987), although we would still need to account for the reason why silent mouthing primes later auditory lexical decision, but not visual lexical decision.

Discussion: From Perception to Production, and Vice Versa

The motor theory of speech perception places motor-control systems at the crux of speech-perception and speech-production links, a position that has been supported by uncritical interpretations of motor cortex activation by heard speech (e.g., Galantucci, Fowler, & Turvey, 2005; Wilson et al., 2004). However the picture is more complex than this: it is hard to find clear evidence that motor responses to heard speech result from a central role for the motor cortex in the perceptual processing of speech, as opposed to a general response to many other categories of sound, like environmental noises and emotional vocalizations. In contrast, posterior auditory areas seem to be very important in controlling speech production, and brain regions associated with linguistic information in speech perception and somatosensory processing are suppressed during speech production. Thus it is possible that the main link between speech perception and production lies in the posterior auditory cortex, rather than in the motor cortex, with the proviso that (as discussed above) it is not clear that the involvement of this area in movements of the articulators is speech specific, or has a more general auditory function.

As discussed briefly in the introduction, auditory perception is needed in development for speech production to develop normally (Mogford, 1988), but this relationship is not bidirectional—severe speech motor problems in development do not necessarily compromise speech-perception skills (Bishop, 1988). In the course of development, auditory information both about one's linguistic environment, and perhaps the sound of one's voice, is needed to develop speech-production skills, while speech perception does not require good speech-production skills to develop. In normal development, where hearing is not compromised, different factors appear to contribute to individual differences in speech perception and production during language acquisition (Alcock & Krawczyk, 2010). This pattern is broadly mirrored in the cortical bases of speech perception and production. While we need intact dorsolateral temporal lobes to understand speech, we do not need intact motor cortices—patients with anterior brain damage and expressive aphasias can perform normally on tests of speech comprehension (e.g., Crinion et al., 2006; Blank et al., 2003).

In functional imaging, speech-perception studies occasionally report motor activation, and the determining factor for seeing such activation has been linked to methodological factors (Scott et al., 2009). In contrast, speech-production studies commonly report activation of posterior-medial auditory areas, even if the speech is mouthed (i.e., is silent): the activation thus is not a simple result of hearing one's own voice when speaking. This suggests that auditory and somatosensory processing during speech production represent important elements of the control of spoken language, perhaps as part of a fast feed-forward network predicting the sensory consequences of intended actions (Rauschecker & Scott, 2009), or as part of a system driving speech production by representations of the acoustic possibilities of the articulators (Warren et al., 2005). Importantly, these auditory (and somatosensory) activations are dorsal, medial, and posterior to the neural responses to intelligibility in speech, suggesting that a different auditory pathway is used for mapping from perception to comprehension (the "what" pathway) than is used for coordinating perception and production mechanisms during speech output (the "how" pathway). Likewise, this posterior "how" pathway is much less sensitive to intelligibility than the anterior "what" pathway, and is activated by the rehearsal and production of a range of different sounds, including both mouth-related sound production (Chang et al., 2009) and the rehearsal of musical information (Hickok et al., 2003).

This anterior-posterior dissociation in the relationship between auditory areas and speech production might also explain why speech-perception skills do not necessarily correlate with speech-production skills in development. If different auditory networks are involved in the development of speech comprehension and the control of speech production, then they could follow different developmental trajectories. This dissociation might also account for the findings that, while Japanese listeners find it difficult to distinguish (the nonnative) English /r/ and /l/ sounds, there is only a moderate correlation between the perception and identification of these nonnative phonemes, and the accuracy of their production (Hattori & Iverson, 2009; Bradlow, Pisoni, Akahane-Yamada, & Tohkura, 1997).

Conclusions

We can see links and dissociations between speech perception and production at a number of levels in the brain. Speech perception is associated with the bilateral dorsolateral temporal lobes, and speech production with the bilateral motor and premotor cortex, the left anterior insula and left posterior-medial auditory cortex, in addition to subcortical regions. Within these perception and production systems, speech perception can lead to activation of motor areas, and the silent mouthing of words can lead to the activation of the auditory cortex. The precise functions of these links vary—the motor response to heard speech, for example, does not seem to reflect a basic, linguistic, perceptual mechanism (Scott et al., 2009). The posterior-medial auditory areas appear to have an important role in the control of speech production, though this again does not appear to be a speech-specific role, and may be associated with more general links between doable sounds and actions (Warren et al., 2005). In contrast, the suppression of activity in lateral STG areas, usually important in the perceptual processing of speech, may reflect mechanisms distinguishing self-generated speech from that of others. Medial and lateral posterior auditory areas have been implicated in the detection and compensation for distortions introduced in speech output (Tourville et al., 2008; Takaso et al., 2010), which may link some of these neural responses to monitoring of produced output.

It thus seems that the connections between speech perception and production do not rely on simple shared networks in the human brain. There seems to be a greater effect of acoustic processing on speech production than there is of speech production on speech perception. Moreover, there are at least two different ways in which the streams of processing in speech perception interact with speech production—one concerned with sensorimotor factors in speech production, and one concerned with the auditory processing of self-produced speech. It also appears that joint cortical networks for speech perception and production are more easily identified in domain-general linguistic processes and representations than in more peripheral perceptual and production systems. This may reflect the involvement of these higher-order language areas seen in more central linguistic functions for both speech perception and production.

References

Alain, C., Arnott, S. R., Hevenor, S., Graham, S., & Grady, C. L. (2001). "What" and "where" in the human auditory system. *Proceedings of the National Academy of Sciences of the United States of America*, *98*, 12301–12306.

Alcock, K. J., & Krawczyk, K. (2010). Individual differences in language development: Relationship with motor skill at 21 months. *Developmental Science*, *13*, 677–691.

Awad, M., Warren, J. E., Scott, S. K., Turkheimer, F. E., & Wise, R. J. (2007). A common system for the comprehension and production of narrative speech. *Journal of Neuroscience*, *27*, 11455–11464.

Bishop, D. V. M. (1988). Language development in children with abnormal structure or function of the speech apparatus. In D. V. M. Bishop & K. Mogford (Eds.), *Language development in exceptional circumstances* (pp. 220–238). New York: Churchill Livingstone.

Blakemore, S. J., Wolpert, D. M., & Frith, C. D. (1998). Central cancellation of self-produced tickle sensation. *Nature Neuroscience*, *1*, 635–640.

Blank, S. C., Bird, H., Turkheimer, F., & Wise, R. J. (2003). Speech production after stroke: The role of the right pars opercularis. *Annals of Neurology*, *54*, 310–320.

Blank, S. C., Scott, S. K., Murphy, K., Warburton, E., & Wise, R. J. (2002). Speech production: Wernicke, Broca, and beyond. *Brain*, *125*, 1829–1838.

Bradlow, A. R., Pisoni, D. B., Akahane-Yamada, R., & Tohkura, Y. (1997). Training Japanese listeners to identify English /r/ and /l/: IV. Some effects of perceptual learning on speech production. *Journal of the Acoustical Society of America*, *101*, 2299–2310.

Chang, S. E., Kenney, M. K., Loucks, T. M., Poletto, C. J., & Ludlow, C. L. (2009). Common neural substrates support speech and non-speech vocal tract gestures. *NeuroImage*, *47*, 314–325.

Cohen, L., Jobert, A., Le Bihan, D., & Dehaene, S. (2004). Distinct unimodal and multimodal regions for word processing in the left temporal cortex. *NeuroImage*, *23*, 1256–1270.

Crinion, J. T., Warburton, E. A., Lambon-Ralph, M. A., Howard, D., & Wise, R. J. (2006). Listening to narrative speech after aphasic stroke: The role of the left anterior temporal lobe. *Cerebral Cortex*, *16*, 1116–1125.

Davis, M. H., Johnsrude, I. S., Hervais-Adelman, A. G., & Rogers, J. C. (2008). Motor regions contribute to speech perception: Awareness, adaptation and categorization. *Journal of the Acoustical Society of America*, *123*, 3580.

Dhanjal, N. S., Handunnetthi, L., Patel, M. C., & Wise, R. J. (2008). Perceptual systems controlling speech production. *Journal of Neuroscience*, *28*, 9969–9975.

Dronkers, N. F. (1996). A new brain region for coordinating speech articulation. *Nature*, *384*, 159–161.

Eliades, S. J., & Wang, X. (2003). Sensory-motor interaction in the primate auditory cortex during self-initiated vocalizations. *Journal of Neurophysiology*, *89*, 2194–2207.

Eliades, S. J., & Wang, X. (2005). Dynamics of auditory-vocal interaction in monkey auditory cortex. *Cerebral Cortex*, *15*, 1510–1523.

Eliades, S. J., & Wang, X. (2008). Neural substrates of vocalization feedback monitoring in primate auditory cortex. *Nature*, *453*, 1102–1106.

Fourcin, A. J. (1975). Language development in the absence of expressive speech. In E. Lenneberg & E. Lenneberg (Eds.), *Foundations of language development* (pp. 263–268). Waltham, MA: Academic Press.

Fowler, C. A. (1986). An event approach to the study of speech-perception from a direct realist perspective. *Journal of Phonetics*, *14*, 3–28.

Fu, K. G., Johnston, T. A., Shah, A. S., Arnold, L., Smiley, J., Hackett, T. A., et al. (2003). Auditory cortical neurons respond to somatosensory stimulation. *Journal of Neuroscience, 23,* 7510–7515.

Furth, H. (1966). A comparison of reading test norms of deaf and hearing children. *American Annals of the Deaf, 111,* 461–462.

Galantucci, B., Fowler, C. A., & Turvey, M. T. (2005). The motor theory of speech perception reviewed. *Psychonomic Bulletin & Review, 13,* 361–377.

Goto, H. (1971). Auditory perception by normal Japanese adults of the sounds "L" and "R." *Neuropsychologia, 9,* 317–323.

Hall, D., Johnsrude, I. S., Haggard, M. P., Palmer, A. R., Akeroyd, M. A., & Summerfield, A. Q. (2002). Spectral and temporal processing in human auditory cortex. *Cerebral Cortex, 12,* 140–149.

Hattori, K., & Iverson, P. (2009). English /r/-/l/ category assimilation by Japanese adults: individual differences and the link to identification accuracy. *Journal of the Acoustical Society of America, 125,* 469–479.

Hickok, G., Buchsbaum, B., Humphries, C., & Muftuler, T. (2003). Auditory-motor interaction revealed by fMRI: Speech, music, and working memory in area Spt. *Journal of Cognitive Neuroscience, 15,* 673–682.

Hickok, G., Erhard, P., Kassubek, J., Helms-Tillery, A. K., Naeve-Velguth, S., Strupp, J. P., et al. (2000). A functional magnetic resonance imaging study of the role of left posterior superior temporal gyrus in speech production: Implications for the explanation of conduction aphasia. *Neuroscience Letters, 287,* 156–160.

Houde, J. F., Nagarajan, S. S., Sekihara, K., & Merzenich, M. M. (2002). Modulation of the auditory cortex during speech: An MEG study. *Journal of Cognitive Neuroscience, 14,* 1125–1138.

Jacquemot, C., Pallier, C., Le Bihan, D., Dehaene, S., & Dupoux, E. (2003). Phonological grammar shapes the auditory cortex: A functional magnetic resonance imaging study. *Journal of Neuroscience, 23,* 9541–9546.

Jacquemot, C., & Scott, S. K. (2006). What is the relationship between phonological short-term memory and speech processing? *Trends in Cognitive Sciences, 10,* 480–486.

Kaas, J. H., & Hackett, T. A. (1999). "What" and "where" processing in auditory cortex. *Nature Neuroscience, 2,* 1045–1047.

Liberman, A. M., Cooper, F. S., Shankweiler, D. P., & Studdert-Kennedy, M. (1967). Perception of the speech code. *Psychological Review, 74,* 431–461.

Liberman, A. M., Delattre, P., & Cooper, F. S. (1952). The role of selected stimulus-variables in the perception of the unvoiced stop consonants. *American Journal of Psychology, 65,* 497–516.

Liberman, A. M., & Mattingly, I. G. (1985). The motor theory of speech-perception revised. *Cognition, 21,* 1–36.

Lombard, E. (1911). Le signe de l'élévation de la voix. *Annales des Maladies de l'Oreille, du Larynx, du Nez et du Pharynx, 37,* 101–119.

Lotto, A. J., Hickok, G. S., & Holt, L. L. (2009). Reflections on mirror neurons and speech perception. *Trends in Cognitive Sciences, 13,* 110–114.

McCarthy, R. A., & Warrington, E. K. (1990). *Cognitive neuropsychology: A clinical introduction*. London: Academic Press.

Menenti, L., Gierhan, S. M. E., Segaert, K., & Hagoort, P. (2011). Shared language: Overlap and segregation of the neuronal infrastructure for speaking and listening revealed by functional MRI. *Psychological Science, 22,* 1173–1182.

Mogford, K. (1988). Oral language acquisition in the prelinguistically deaf. In D. V. M. Bishop & K. Mogford (Eds.), *Language development in exceptional circumstances* (pp. 110–131). New York: Churchill Livingstone.

Mohr, J. P., Pessin, M. S., Finkelstein, S., Funkenstein, H. H., Duncan, G. W., & Davis, K. R. (1978). Broca aphasia: Pathologic and clinical. *Neurology, 28,* 311–324.

Monsell, S. (1987). On the relation between lexical input and output pathways for speech. In A. Allport, D. G. Mackay, W. Prinz, & E. Scheerer (Eds.), *Language perception and production: Relationships between listening, speaking, reading, and writing* (pp. 273–311). London: Academic Press.

Mummery, C. J., Ashburner, J., Scott, S. K., & Wise, R. J. S. (1999). Functional neuroimaging of speech perception in six normal and two aphasic patients. *Journal of the Acoustical Society of America, 106,* 449–457.

Nasir, S. M., & Ostry, D. J. (2008). Speech motor learning in profoundly deaf adults. *Nature Neuroscience, 11,* 1217–1222.

Obleser, J., Wise, R. J., Dresner, M., & Scott, S. K. (2007). Functional integration across brain regions improves speech perception under adverse listening conditions. *Journal of Neuroscience, 27,* 2283–2289.

Pardo, J. S. (2006). On phonetic convergence during conversational interaction. *Journal of the Acoustical Society of America, 119,* 2382–2393.

Patterson, K., Nestor, P. J., & Rogers, T. T. (2007). Where do you know what you know? The representation of semantic knowledge in the human brain. *Nature Reviews: Neuroscience, 8,* 976–987.

Rauschecker, J. P. (1998). Cortical processing of complex sounds. *Current Opinion in Neurobiology, 8,* 516–521.

Rauschecker, J. P., & Scott, S. K. (2009). Maps and streams in the auditory cortex: How work in non-human primates has contributed to our understanding of human speech processing. *Nature Neuroscience, 12,* 718–724.

Schnur, T. T., Schwartz, M. F., Kimberg, D. Y., Hirshorn, E., Coslett, H. B., & Thompson-Schill, S. L. (2009). Localizing interference during naming: Convergent neuroimaging and neuropsychological evidence for the function of Broca's area. *Proceedings of the National Academy of Sciences of the United States of America, 106,* 322–327.

Scott, S. K., Blank, S. C., Rosen, S., & Wise, R. J. S. (2000). Identification of a pathway for intelligible speech in the left temporal lobe. *Brain, 123,* 2400–2406.

Scott, S. K., & Johnsrude, I. S. (2003). The neuroanatomical and functional organization of speech perception. *Trends in Neurosciences, 26,* 100–107.

Scott, S. K., Leff, A. P., & Wise, R. J. (2003). Going beyond the information given: A neural system supporting semantic interpretation. *NeuroImage, 19* (3), 870–876.

Scott, S. K., McGettigan, C., & Eisner, F. (2009). OPINION A little more conversation, a little less action—candidate roles for the motor cortex in speech perception. *Nature Reviews: Neuroscience, 10*, 295–302.

Scott, S. K., Rosen, S., Lang, H., & Wise, R. J. S. (2006). Neural correlates of intelligibility in speech investigated with noise vocoded speech—a Positron Emission Tomography study. *Journal of the Acoustical Society of America, 120*, 1075–1083.

Shahin, A. J., Bishop, C. W., & Miller, L. M. (2009). Neural mechanisms for illusory filling-in of degraded speech. *NeuroImage, 44*, 1133–1143.

Smiley, J. F., Hackett, T. A., Ulbert, I., Karmas, G., Lakatos, P., Javitt, D. C., et al. (2007). Multisensory convergence in auditory cortex, I. Cortical connections of the caudal superior temporal plane in macaque monkeys. *Journal of Comparative Neurology, 502*, 894–923.

Takaso, H., Eisner, F., Wise, R. J. S., & Scott, S. K. (2010). The effect of delayed auditory feedback on activity in the temporal lobe while speaking: A PET study. *Journal of Speech Hearing and Language Research, 53*, 226–236.

Tian, B., Reser, D., Durham, A., Kustov, A., & Rauschecker, J. P. (2001). Functional specialization in rhesus monkey auditory cortex. *Science, 292*, 290–293.

Toni, I., de Lange, F. P., Noordzij, M. L., & Hagoort, P. (2008). Language beyond action. *Journal of Physiology, Paris, 102*, 71–79.

Tourville, J. A., Reilly, K. J., & Guenther, F. H. (2008). Neural mechanisms underlying auditory feedback control of speech. *NeuroImage, 39*, 1429–1443.

Warren, J. E., Sauter, D. A., Eisner, F., Wiland, J., Dresner, M. A., Wise, R. J., et al. (2006). Positive emotions preferentially engage an auditory-motor "mirror" system. *Journal of Neuroscience, 13*, 13067–13075.

Warren, J. E., Wise, R. J., & Warren, J. D. (2005). Sounds do-able: Auditory-motor transformations and the posterior temporal plane. *Trends in Neurosciences, 28*, 636–643.

Werker, J. F., & Yeung, H. H. (2005). Infant speech perception bootstraps word learning. *Trends in Cognitive Sciences, 9*, 519–527.

Wilson, S. M., & Iacoboni, M. (2006). Neural responses to non-native phonemes varying in producibility: Evidence for the sensorimotor nature of speech perception. *NeuroImage, 33*, 316–325.

Wilson, S. M., Saygin, A. P., Sereno, M. I., & Iacoboni, M. (2004). Listening to speech activates motor areas involved in speech production. *Nature Neuroscience, 7*, 701–702.

Wise, R. J. S., Greene, J., Büchel, C., & Scott, S. K. (1999). Brain systems for word perception and articulation. *Lancet, 353*, 1057–1061.

Wise, R. J. S., Scott, S. K., Blank, S. C., Mummery, C. J., & Warburton, E. (2001). Identifying separate neural sub-systems within "Wernicke's area." *Brain, 124*, 83–95.

15 Neural Mechanisms of Auditory Learning and Memory in Songbirds and Mammals

Sharon M. H. Gobes, Jonathan B. Fritz, and Johan J. Bolhuis

Given the remarkable behavioral parallels of auditory-vocal learning in songbirds and humans (Doupe & Kuhl, 1999; Bolhuis, Okanoya, & Scharff, 2010; Moorman & Bolhuis, chapter 5, this volume), the question arises whether similar parallels can be found at the neural level. Jarvis (chapter 4, this volume) has outlined the evolutionary relationship of brain pathways involved in auditory-vocal learning. Indeed, it has become apparent that there are striking homologies between the brains of birds and mammals (Reiner et al., 2004; Jarvis et al., 2005). The nomenclature of the avian brain has been completely revised, reflecting these similarities (Reiner et al., 2004). The new nomenclature emphasizes the fact that a large part of the avian telencephalon is homologous with the mammalian cortex (Reiner et al., 2004; Jarvis et al., 2005). These neuroanatomical similarities have led to comparative studies of functional circuits. In mammals as well as birds, vocalizations that are imitated also have to be memorized. Birds, showing such remarkable skills in vocal learning, may be excellent model organims to study auditory memory. In this chapter we will discuss the similarities between auditory learning in birds and speech learning in humans. In addition, we will discuss auditory memory in species that do not learn their vocalizations, to investigate the general principles underlying auditory memory, independent of whether the memorized sounds are related to vocal production.

Comparative Behavioral Studies of Auditory Memory for Nonvocalizable Sounds

The use of random acoustic waveforms to probe auditory memory was introduced by Guttman and Julesz (1963). In their paradigm, a given sample of white noise was "frozen" and then repeated identically several times. The task requires detection of repetition of a cyclically repeated white noise segment of long duration (300 ms to 20 s). Subjects' ability to discriminate repeated frozen noise from unmodulated random white noise required recognition of the repetition, and hence memory of the repeated waveform. Kaernbach and his colleagues have explored the sensory memory for periodic frozen white noise in humans and a variety of animals (Kaernback & Schulze, 2002; Frey, Kaernbach, & Konig, 2003; Kaernbach, 2004; Kretschmar, Kalenscher, Güntürkün, & Kaernbach, 2008). Auditory sensory memory (for period length of the random waveform) varied from 360 ms in the Mongolian gerbil, 500 ms in the cat, 2,560 ms in the pigeon, to 20 seconds in the human. Zokoll and colleagues (Zokoll, Klump, & Langemann, 2007, 2008; Zokoll, Naue, Herrmann, & Langemann, 2008) studied auditory recognition memory in European starlings (*Sturnus vulgaris*) and humans using a slightly different behavioral paradigm and acoustic stimuli (delayed-match-to-sample discrimination between tones, birdsongs, or rates of sinusoidal amplitude modulation of white noise) and found that the persistence of auditory short-term memory in the starling varied with the acoustic stimulus (2–13 seconds for the noise stimuli, 4–20 seconds for the tonal stimuli). Thus, auditory sensory memory for (nonvocalizable) random waveforms may be greater in birds (and other animals) capable of vocal learning than in non–vocal learners, but currently, this would require further investigation because the behavioral paradigms and animal subjects were different in the two sets of studies. A recent study has explored long- as well as short-term auditory memory utilizing similar periodic (frozen) or nonperiodic white noise stimuli (Agus, Thorpe, & Pressnitzer, 2010). They found that human listeners had a long-term memory of repetitions in noise samples that had been presented multiple times as compared with freshly made, unique noise stimuli. Thus the subjects had implicitly and rapidly learned the details of random acoustic waveforms and retained them in long-term auditory memory storage. It would be valuable to develop a parallel set of studies to investigate long-term auditory memory in birds and mammals in both vocal and non–vocal learners.

In Search of the Auditory Engram

Auditory Memory Systems in Nonhuman Mammals

There are multiple forms of auditory memory including rapid and short-lasting sensory memory; recognition memory, pitch, timbre, and loudness-specific memory; and short-term memory, working memory, and long-term memory (Demany & Semal, 2008). Given the importance of vocal recognition, birds and mammals often show remarkable long-term auditory memories for familiar voices (Godard, 1991; Miller, 1979; Insley, 2000; Charrier, Mathevon, & Jouventin, 2001; Sharp, McGowan, Wood, & Hatchwell, 2005). In primates, a critical substrate for auditory memory is the auditory association cortex in the temporal lobe, which consists of a complex network of brain regions in the secondary and tertiary auditory cortex (see Figure 15.1). These areas include the medial and lateral belt regions and parabelt regions (Figure 15.2b) that project to regions in the prefrontal cortex, which also encode acoustic information in short-term and working memory (Bodner, Kroger, & Fuster,

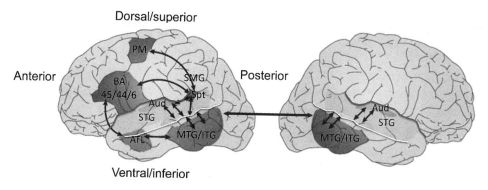

Dorsal/superior

Anterior   Posterior

Ventral/inferior

Figure 15.1
Brain regions involved in human language. Two broad processing streams are shown. The first is a ventral stream for speech recognition and comprehension that is largely bilaterally organized and flows into the temporal lobe—mapping sound into meaning. The phonological stages of spoken word recognition occur in this ventral stream in the superior temporal lobe (STG and STS) bilaterally. The second is a dorsal stream for sensorimotor integration that is left-hemisphere dominant and involves structures at the parietal-temporal junction (Spt) and frontal lobe (Broca's area). Spt supports auditory-motor integration for vocal-tract actions including speech and singing and is connected with frontal speech-production related areas. ATL plays a critical role in syntactic and/or semantic integration in sentence processing. Arrows indicate connectivity/flow of information. Abbreviations: ATL: anterior temporal lobe; Aud: early processing stages of auditory cortex (primary auditory core areas in the sylvian fissure on the superior temporal plane); BA: Broca's area (comprising Brodmann's areas 45/44/6); ITG: inferior temporal gyrus; MTG: middle temporal gyrus; PM: premotor cortex; SMG: supramarginal gyrus; Spt: sylvian parietal temporal region (left hemisphere only); STG, superior temporal gyrus. Black line: sylvian fissure; white line: superior temporal sulcus (STS). Modified and reproduced, with permission, from Hickok, 2009, copyright 2009 Elsevier. All rights reserved.

1996; Kaas & Hackett, 1999; Kaas, Hackett, & Tramo, 1999; Romanski, Bates, & Goldman-Rakic, 1999; Kaas & Hackett, 2000; Hackett, Preuss, & Kaas, 2001; Wise, 2003; Hackett, 2011).

Currently, there is insufficient evidence for one-to-one homologies between the avian and the mammalian brain (Jarvis et al., 2005). Within the forebrain Field L complex, Field L2 receives auditory connections from the thalamus and in turn projects onto Field L1 and Field L3 (Figure 15.2a). These two regions project to the caudomedial mesopallium (CMM) and the caudomedial nidopallium (NCM), respectively. Bolhuis and Gahr (2006) have argued that the Field L complex in birds may be homologous to the primary auditory cortex, in the primate superior temporal plane, which also consists of three "core" regions that receive inputs from the thalamus (Kaas et al., 1999; Wise, 2003) (Figure 15.2b). Thus, the projection regions of the Field L complex (the NCM and CMM) may be homologous to the belt and

A. Auditory and vocal pathways in the songbird brain

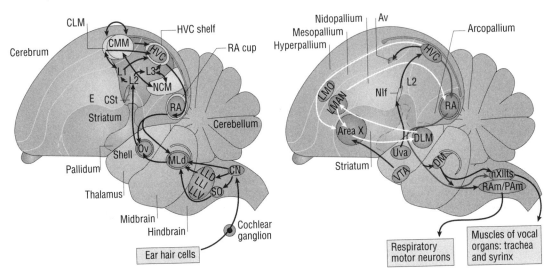

B. Speech-related regions in the human brain

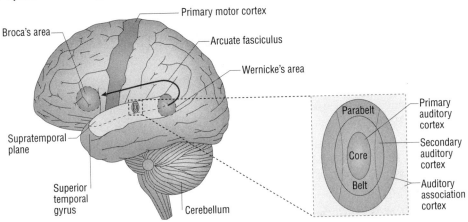

Figure 15.2
Schematic diagrams of composite views of parasagittal sections of the songbird brain and the human brain. (a) Diagram of a songbird brain giving approximate positions of nuclei and brain regions involved in auditory perception and memory (left, auditory pathways) and in vocal production and sensorimotor learning (right, vocal pathways). White areas represent brain regions that show increased neuronal activation when the bird hears song. The light-gray nuclei in the song system show increased neuronal activation when the bird is singing. (b) Schematic view of the left side of the human brain, with some of the regions involved in speech and language (see text and Figure 15.1 for details). Abbreviations: Area X, Area X of the striatum; Av, avalanche; CLM, caudal lateral mesopallium; CN, cochlear nucleus; CSt,

parabelt regions in the mammalian auditory association cortex (Bolhuis & Gahr, 2006).

There is evidence from a number of studies using different experimental methodologies that the primary auditory cortex and secondary auditory association cortex are involved in auditory recognition memory in monkeys and rats (Gottlieb, Vaadia, & Abeles, 1989; Colombo, Damato, Rodman, & Gross, 1990; Wan et al., 2001; Fritz, Mishkin, & Saunders, 2005; Weinberger, 2004). Learning and storage of emotional and fear-related auditory memories in rats also involves the auditory cortex (Sacco & Sacchetti, 2010; Letzkus et al., 2011). Species-specific vocalizations are encoded in areas of the auditory association cortex (including belt, parabelt, and insula) (Tian, Reser, Durham, Kustov, & Rauschecker, 2001; Petkov et al., 2008; Remedios, Logothetis, & Kayser, 2009; Perrodin, Kayser, Logothetis, & Petkov, 2011) and also in the ventrolateral prefrontal cortex (Russ, Ackelson. Baker, & Cohen, 2008; Romanski & Averbeck, 2009). Neuroanatomical and neuroimaging evidence suggests that the monkey's auditory cortex sends substantial projections to stimulus-quality processing areas in the ventral prefrontal cortex (the "what" pathway) and to spatial processing areas in the dorsal prefrontal cortex (the "where" pathway), and also to the monkey homologs of Broca's area in the premotor cortex (Petrides & Pandya, 2009). Broca's area in humans may have evolved from a brain region in our primate ancestor that served higher control of complex hierarchical sequences of gestural and vocal action, with the evolution of communication leading to human speech (Petrides & Pandya, 2009). The auditory-premotor connection, much weaker in monkeys, is mediated in humans by the dense projection pathway formed by the arcuate fasciculus interconnecting Wernicke's and Broca's areas (Figure 15.2b).

Localization of the Neural Substate for Auditory Memory in Birds

The "Template" of Song Learning

In the early days of songbird research, Konishi (1965) proposed a theory to describe the mechanism underlying song learning in birds. The concept of the "template,"

---

caudal striatum; DLM, dorsal lateral nucleus of the medial thalamus; DM, dorsal medial nucleus of the thalamus; E, entopallium; L1, L2, L3, subdivisions of Field L; LLD, lateral lemniscus, dorsal nucleus; LLI, lateral lemniscus, intermediate nucleus; LLV, lateral lemniscus, ventral nucleus; LMAN, lateral magnocellular nucleus of the anterior nidopallium; LMO, lateral oval nucleus of the mesopallium; MLd, dorsal lateral nucleus of the mesencephalon; NIf, interfacial nucleus of the nidopallium; nXIIts, tracheosyringeal portion of the nucleus hypoglossus (nucleus XII); Ov, ovoidalis; PAm, para-ambiguus; RA, robust nucleus of the arcopallium; RAm, retroambiguus; SO, superior olive; Uva, nucleus uvaeformis; VTA, ventral tegmental area. Modified and reproduced, with permission, from Bolhuis et al., 2010, copyright 2010 Nature Publishing Group. All rights reserved.

which is the internal representation of the tutor song, has guided research undertaken to understand the neural mechanisms for tutor song (auditory) memory ever since (Konishi, 1965, 2004). It has been suggested that songbirds are born with a "crude template" because in the absence of a suitable song model they will still develop some species-specific characteristics (Marler & Sherman, 1985; see Moorman & Bolhuis, chapter 5, this volume). By hearing conspecific — or "tutor" — song, the crude template is thought to be refined into a more precise representation of the tutor song. During the sensorimotor phase of song learning, the young bird starts to vocalize, and it is thought that it matches its song output with the template formed in the first phase because the song of the young bird is gradually modified and will increasingly resemble the tutor song. Eventually the bird will sing a crystallized song that, in the case of age-limited learners such as the zebra finch, does not change during adulthood. Attempts to localize the neural representation for the "template" have traditionally concentrated on the neural system for song production, the "song system."

Does the Song System Contain the Neural Representation of Tutor Song Memory?

The production of song in birds is controlled by an elaborate system of interconnected nuclei known as the "song system" (Nottebohm, Stokes, & Leonard, 1976) (Figure 15.2a). Some nuclei in the song system are larger in males than in females (Nottebohm & Arnold, 1976) and the song system is unique to birds that learn their song. The song system is subdivided into the song motor pathway (SMP), involved in song production (Bottjer, Miesner, & Arnold, 1984; Scharff & Nottebohm, 1991; Kao, Doupe, & Brainard, 2005; Mooney, 2009), and the anterior forebrain pathway (AFP, see Figure 15.2a), involved in vocal motor learning and auditory feedback processing (Bottjer et al., 1984; Scharff & Nottebohm, 1991; Kao et al., 2005; Brainard & Doupe, 2000, 2002). Lesions to nuclei in the AFP, including lMAN and Area X, do not affect adult song but cause abnormal song development in young zebra finches (Bottjer et al., 1984; Sohrabji, Nordeen, & Nordeen, 1990; Scharff & Nottebohm, 1991). Juvenile males with lesions in Area X do not develop crystallized songs, while juveniles with lesions in lMAN produce aberrant but stable songs. It has therefore been suggested that the AFP may contain the neural substrate for auditory memory (Basham, Nordeen, & Nordeen, 1996; Nordeen & Nordeen, 2004). It is difficult to estimate how much birds that received permanent lesions to the AFP have memorized from their tutor because they never develop normal adult song. The impaired strength of song learning found in these birds could thus be the result of deficits in auditory learning and/or sensorimotor learning. To circumvent this problem, the song system nucleus lMAN was temporarily and reversibly inactivated by injections with an NMDA (N-methyl-D-aspartic acid) receptor blocker on days that birds were exposed to tutor song in the sensorimotor phase of song learning, but not between sessions (Basham et al., 1996). Therefore, normal sensorimotor learning between sessions should have been unaltered while only auditory learning during tutoring session should have been affected. Song learning was significantly impaired in birds that received the experimental treatment, consistent with the suggestion that the AFP is involved in memorization of tutor song. However, in zebra finches, there is overlap between the memorization phase and the sensorimotor learning phase. In the study of Basham and colleagues (1996), lMAN was inactivated in the sensorimotor phase; these findings could be due to an effect on sensorimotor integration (Bolhuis & Gahr, 2006) during the tutoring sessions. This could have resulted in reduced song imitation without affecting auditory memory processes. In this interpretation it is assumed that when auditory and/or visual interaction is possible between the young bird and the tutor, this has a positive influence on the sensorimotor integration processes that are essential for vocal learning, independently of the formation of auditory memory (Nordeen & Nordeen, 2004). The exact role of the AFP in song learning and possibly memorization of tutor song thus remains unclear.

A recent study (Aronov, Andalman, & Fee, 2008) suggests that the production of subsong in the zebra finch requires RA (robust nucleus of the arcopallium) and LMAN (lateral magnocellular nucleus of the nidopallium), a forebrain nucleus involved in sensorimotor learning but not in adult song production (Bottjer et al., 1984; Scharff & Nottebohm, 1991). The circuit for adult song production consists of the HVC, RA, and brainstem motor nuclei. However, subsong does not require the HVC, a key premotor nucleus for singing in adult birds. Therefore, juvenile subsong is driven by a circuit different from the premotor circuit for song production in adults. It is not known whether this is also true of human babbling.

Song Memory beyond the Song System

Electrophysiological responsiveness of neurons is commonly measured to investigate the neural response to specific stimuli. Alternatively, the expression of immediate early genes (IEGs) can also be used as a marker for neuronal activation, because they respond rapidly to stimulation by neuronal depolarization (Sagar, Sharp, & Curran, 1988; Moorman, Mello, & Bolhuis, 2011). The detectable levels of mRNA and protein decrease back to baseline shortly after the stimulus is removed. This technique (as compared to electrophysiology) makes it possible to investigate the neuronal response during a certain time frame of several brain regions simultaneously. Expression of the IEG *ZENK* (an acronym of *zif-268*, *egr-1*, NGF-1A, and *krox-24*; "Zenk" is used to indicate the protein product of this gene) has revealed that nuclei in the song system are activated when the bird is singing (Jarvis & Nottebohm, 1997). The motor act of singing is sufficient to induce *ZENK* in the song system, because neuronal activation in the song system has been demonstrated in deafened birds that do not hear themselves sing (Jarvis & Nottebohm, 1997). In contrast, regions outside the song system, such as the NCM and the CMM (Figure 15.2a), are activated when the bird hears song (Mello, Vicario, & Clayton, 1992; Mello & Clayton, 1994). Neuronal activation in these regions occurs irrespective of the motor act of singing. There is expression of *ZENK* in birds that are exposed to songs played from tapes as well as in birds that hear themselves sing but not in deafened birds (Mello et al., 1992; Mello & Clayton, 1994; Jarvis & Nottebohm, 1997). The expression of *ZENK* decreases after repeated exposure to the same stimulus but can be induced again when the animal is exposed to a novel stimulus (Mello, Nottebohm, & Clayton, 1995). Habituation of the *ZENK* response—a decline in the expression levels after repeated exposure that is persistent over at least 24 hours—could be an indication that the NCM is involved in processes related to auditory memory (Mello et al., 1995).

In zebra finch males, it was found that expression of Zenk was induced in the NCM and CMM when adult male zebra finches were reexposed to tutor song (Bolhuis, Zijlstra, Den Boer-Visser, & Van der Zee, 2000). The expression of Zenk in the NCM was positively correlated with the fidelity of song imitation only after re-exposure to the tutor song (Bolhuis et al., 2000; Bolhuis, Hetebrij, Den Boer-Visser, De Groot, & Zijlstra, 2001; Terpstra, Bolhuis, & Den Boer-Visser, 2004). That is, Zenk expression in birds that had copied many elements of their father's song was greater than in birds that had copied few elements. These results led to the suggestion that the NCM contains the neural substrate for the representation of tutor song memory.

This correlation between strength of song learning and Zenk expression was present in birds that were raised by means of tape tutoring (Bolhuis et al., 2000) as well as in socially reared zebra finches (Bolhuis et al., 2001). In good learners (birds that copied many elements from their song tutor) the BOS resembles the tutor song closely. It was therefore suggested that this correlation could be due to reactivation of BOS memory and not tutor song memory (Marler & Doupe, 2000). Terpstra et al. (2004) addressed this issue and showed that Zenk expression in the NCM was related to the strength of song learning only in birds that were reexposed to their tutor's song, but not in birds that were exposed to their own song. In this study, another group of birds was exposed to a novel song to investigate whether this correlation could be due to processes related to attention (i.e., good learners always pay more attention when they listen to songs), which results in higher levels of Zenk after song exposure (Marler & Doupe, 2000). Paying more attention to songs might be the reason why they learned the tutor song better during the memorization phase. In the group exposed to novel song, there was no correlation with the strength of song learning in the NCM (Terpstra et al., 2004), which makes it unlikely that the reported correlation in the tutor song group is due to attention.

Alternatively, the Zenk response in the NCM might be dependent on stimulus salience. "Stimulus salience" refers to the characteristics of the stimulus that make the stimulus stand out from other stimuli, and the salience of a stimulus could facilitate attention. Exposure to songs with a greater salience, such as long song bouts in starlings (Gentner, Hulse, Duffy, & Ball, 2001) or more complex songs in budgerigars (*Melopsittacus undulatus*) (Eda-Fujiwara, Satoh, Bolhuis, & Kimura, 2003), leads to more Zenk induction in the NCM as compared to less salient songs. The songs from the tutors of good learners could have had a greater salience, resulting in the reported correlation. However, the length of the tutor song did not correlate significantly with IEG expression in NCM (Bolhuis et al., 2001). This renders it unlikely that the salience of the tutor songs explains the reported results, although we cannot exclude that an unknown characteristic related to song salience was different between the used tutor songs.

A recent electrophysiological study, however, showed that neurons in the NCM of adult zebra finch males showed steeper rates of habituation to novel song than to tutor song (Phan et al., 2006), indicating that neurons in the NCM treat tutor song like other conspecific songs that they are previously familiarized with (Chew et al., 1995). Habituation rates that are different from those of novel song can be interpreted as a neural representation of memory for song. In addition, these authors found that the "familiarity index" (a measure that indicates how much a song behaves like a familiar song with respect to the habituation rates to novel songs) of NCM neurons correlated significantly and positively with the strength of song learning (Phan, Pytte, & Vicario, 2006). It is unlikely that in the NCM the correlations with the strength of song learning and IEG expression as well as with the familiarity index of habituation rates are both caused only by stimulus salience and independent of a representation of memory. Thus, evidence from IEG as well as electrophysiological studies indicates that the NCM might be (part of) the neural substrate for tutor song memory.

Recently, we have shown that the NCM is necessary for recognition of the tutor song, but that its integrity is not required for song production or sound discrimination in adult male zebra finches (Gobes & Bolhuis, 2007; Figure 15.3). These findings suggest that the NCM contains (part of) the neural representation of tutor song memory, and that access to this representation is not necessary for song production. In this study, the subjects were adults and had crystallized songs. It is possible that song acquisition in juveniles that are learning their song requires an intact NCM because the neural representation of tutor song is formed during the memorization phase (Gobes, Zandbergen, & Bolhuis, 2010; Bolhuis et al., 2010). In addition, during the sensorimotor phase, when the young bird's song output is thought to be matched with the stored representation of tutor song (Konishi, 1965), development of the bird's own song may involve access to this representation (London & Clayton,

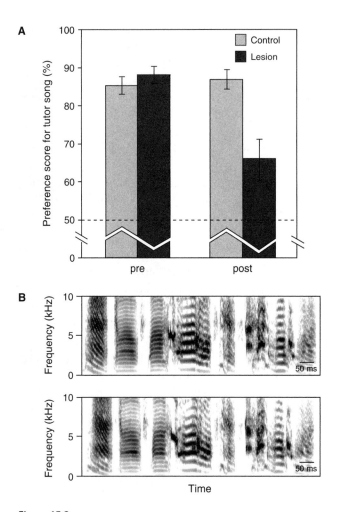

Figure 15.3
Neural dissociation between birdsong recognition and production in the zebra finch. **A.**
Lesions to the NCM impair song recognition in zebra finch males. Song preference scores
(expressed as a percentage) were measured by calculating the amount of time spent near a
speaker that broadcast the song of the bird's tutor compared to a speaker that broadcast a
novel zebra finch song. Before surgery ("pre"), birds in both groups showed a strong preference for the song of the tutor over a novel song. After surgery ("post"), birds in the sham-
operated group had maintained their preference, while this was significantly impaired in the
group that received lesions to the NCM. **B.** Representative spectrograms of the song of a
zebra finch male before (top) and after (bottom) surgery show that song production was not
altered by lesions to the NCM. Adapted, with permission, from Gobes & Bolhuis (2007), ©
2007 Cell Press.

2008; see below). In juvenile zebra finches that are in the middle of the sensorimotor learning period, IEG expression is greater after exposure to tutor song than after exposure to novel song (Gobes et al., 2010). The juveniles' songs in this study already resembled the tutor songs to some degree, and thus this result indicates that the NCM is involved in tutor song recognition memory. Interestingly, during sleep at night, IEG expression correlated with the strength of song learning only in juveniles not reexposed to their tutor's song during the previous day (Gobes et al., 2010).

Is the neural representation of the tutor song that we presume to be located in the NCM necessary for vocal learning? If that hypothesis is correct, we predicted (Bolhuis et al., 2000; Gobes & Bolhuis, 2007) that lesions to the NCM of juvenile zebra finches should impair not only the formation of tutor song recognition memory but also song development. So far, this hypothesis has been tested only indirectly, in an elegant study by London and Clayton (2008). To investigate whether IEG expression in the NCM is necessary for song learning, London and Clayton (2008) infused an inhibitor of the activation of extracellular signal regulated kinase (ERK) in the NCM of juvenile zebra finches during tutoring sessions in which they were exposed to a song tutor. The transcription of *ZENK*, as well as other immediate early genes such as *c-fos* and *Arc*, is regulated by ERK (Velho, Pinaud, Rodrigues, & Mello, 2005). Juveniles that received this treatment during tutoring sessions developed poor imitations of the tutor song, whereas normal discrimination of songs was unaffected by infusion with this inhibitor (London & Clayton, 2008). Birds that received an inactive compound structurally similar to the inhibitor developed songs that resembled the tutor song. These findings suggest that a molecular response (regulated by ERK) in the NCM is necessary for normal song learning to occur. As such, these results support the hypothesis (Bolhuis & Gahr, 2006; Bolhuis, 2008) that the NCM contains a neural representation of the tutor song that is important for vocal learning, while allowing for the possibility that simultaneous activity in other regions in the songbird brain is equally important for song acquisition (Bolhuis et al., 2000; Bolhuis & Gahr, 2006; Gobes & Ölveczky, 2011).

Female zebra finches do not learn to produce a song, but develop a preference for the song of their father. Investigating the neural mechanisms of song memory is thus not complicated by interference with similarity to their own song, as in males. There was increased IEG expression in the CMM of female zebra finches when they were exposed to their father's song, compared to novel song (Terpstra, Bolhuis, Riebel, Van der Burg, & Den Boer-Visser, 2006). In zebra finch females, lesions to the CMM, but not to the song system nucleus HVC, impaired a preference for conspecific over heterospecific song (MacDougall-Shackleton, Hulse, & Ball, 1998). In contrast, lesions to song system nuclei in songbirds impair normal production of song, without affecting recognition of previously learned song examples in an operant discrimination task (Nottebohm et al., 1976; Gentner, Hulse, Bentley, & Ball, 2000). Taken together with our recent findings (Gobes & Bolhuis, 2007), showing that lesions to the NCM do not affect song production but impair song recognition (Figure 15.3), these results reveal a complete double dissociation of the effects of lesions to rostral and caudal brain regions on song in adult zebra finches, with rostral regions involved in song production, and caudal regions involved in auditory memory for song.

Auditory Memory for Speech, Music, and Environmental Sounds in Humans

In humans, the neural substrates of speech perception and memory include the auditory association cortex in the superior temporal gyrus (Scott & Wise, 2004; Viceic et al., 2006), while motor representations of speech involve regions in the frontal cortex centered around Broca's area (Hickok & Poeppel, 2000; Hickok, Buchsbaum, Humphries, & Muftuler, 2003; Demonet et al., 2005; Hickok, 2009) (Fig. 15.1). fMRI studies have shown that regions in the superior temporal gyrus (STG) are involved in speech perception and memory, working memory for voices or FM tones and also for encoding of environmental sounds (Hickok & Poeppel, 2000; Hickok et al., 2003; Lewis et al., 2004; Rama et al., 2004; Viceic et al., 2006; Brechmann et al., 2007; Kraut et al., 2006). In the study of auditory working memory for voices (Rama et al., 2004), broader activation of the prefrontal and parietal cortices was also observed. The neural basis for auditory memory in humans has been explored for musical themes or sound sequences. The work of Halpern and Zatorre (1999) suggests that retrieval of familiar melodies from musical semantic memory is mediated by structures in the right frontal lobe, in conjunction with right superior temporal regions (auditory association cortex) and SMA (supplementary motor area). A recent study of anticipatory imagery of familiar music (Leaver et al., 2009) found marked activation of the rostral prefrontal and premotor cortex, which may have to do with encoding of predictable sequences. Similarly, when subjects heard a familiar piece of music that they did not know how to play, brain activation was observed in the auditory areas of the STG, but when subjects listened to equally familiar pieces of music that they had learned to play, activation was found bilaterally in a frontoparietal motor-related circuit associated with the human mirror neuron system and including Broca's area, the premotor region, the intraparietal sulcus, and the interior parietal region (Lahav et al., 2007). Related findings show activation of the mirror network system for identification of sounds made by tools, but not for identification of animal sounds (Lewis, Brefczynski, Phinney, Janik, & DeYoe, 2005). Thus, there may be differential encoding of sound sequences that are in the individual's motor repertoire. Perfect or absolute pitch in humans is not common (about 1 in 10,000 people have it). These individuals can label up to 75+

musical frequencies. Intriguingly, some people with absolute pitch can only label notes produced by one kind of instrument that they play—for example, absolute piano (Levitan & Rogers, 2005).

Interactions between Auditory Memory and Vocal Production Regions

Recent evidence shows functional interactions between the temporal and frontal cortex in human speech (Ojemann, 1991; Hickok & Poeppel, 2000; Hickok et al., 2003; Dehaene-Lambertz et al., 2006; Okada & Hickok, 2006). Beginning early in life, in humans, temporal and frontal regions strongly interact. In preverbal human infants, the superior-temporal cortex (or auditory association cortex) and the frontal cortex are both active during speech perception and memory tasks (Dehaene-Lambertz et al., 2006; Imada et al., 2006). In newborn infants (5 days old) there is activation (measured by magnetoencephalography) in the superior temporal cortex (Wernicke's area), but not in the inferior frontal cortex (Broca's area) when they are exposed to human speech (Imada et al., 2006). In older (3-month-old) infants that were exposed to sentences, there was activation (measured with fMRI) in the superior-temporal cortex as well as in Broca's area (Dehaene-Lambertz et al., 2006). The activation in Broca's area was specific for speech repetition, indicating a role for Broca's area in memory in preverbal infants. Six- and twelve-month-old infants also exhibit activation in the superior-temporal cortex as well as in the inferior-frontal cortex when exposed to speech (Imada et al., 2006). Taken together, these findings suggest that the human superior-temporal cortex is (part of) the neural substrate for speech perception in neonates and that Broca's area in the frontal cortex becomes active later in development, around the age that infants start to produce sounds themselves. The interaction between these areas might be of importance for speech-production learning. In adults, Broca's area has also been shown to play multiple roles—and is involved in both production and comprehension of syntax (Friederici, Meyer, & Von Cramon, 2000).

There is a parallel in the organization of the avian brain, where the primary premotor nucleus for the production of song, the HVC, exhibits preferential neuronal responsiveness (measured electrophysiologically) to the song of the tutor early in development (35–69 dph) (Nick & Konishi, 2005) (cf. activity after exposure to speech in the frontal cortex of human infants when they start babbling). Later in development (in zebra finches of 70–90 dph and in adults), neurons in HVC preferentially respond to the bird's own song (BOS) (Margoliash & Konishi, 1985; Margoliash, 1986; Volman, 1993; Nick & Konishi, 2005), indicative of a (motor) representation of the bird's own song in this nucleus. In adults, neuronal activation in HVC is also induced by singing (Jarvis & Nottebohm, 1997; Kimpo & Doupe,

1997). In songbirds, only recently a functional connection between the neural substrate for the representation of learned auditory sounds (the NCM/CMM complex) and the vocal control system that can produce these sounds (in the form of the song system nucleus HVC) has been demonstrated (Bauer et al., 2008; see Figure 15.2a). In humans as well as birds, regions that are involved in the perception, memorization, and production of vocalizations may interact throughout vocal learning (Bolhuis et al., 2010).

Evolution and Mechanisms of Birdsong, Speech, and Language

What are the implications of our review of the neural mechanisms of auditory learning and memory for the evolution of birdsong, speech, and language? When considering the evidence discussed in this book on the mechanisms underlying auditory-vocal behavior in songbirds and mammals, an evolutionary scenario emerges where three factors are important (see also Beckers, Bolhuis, Okanoya, & Berwick, 2012). First, there is increasing evidence for neural and genetic homology (Jarvis et al., 2005; Bolhuis et al., 2010; Fisher & Scharff, 2009; Jarvis, chapter 4, this volume; see also part V of the present volume), where similar genes and brain regions are involved in auditory learning and vocal production, not only in songbirds and humans, but also in apes, monkeys, and mice. Second, there is evolutionary convergence with regard to the mechanisms of auditory-vocal learning, which proceeds in essentially the same way in songbirds and human infants, but not in nonhuman primates or mice (Doupe & Kuhl, 1999; Bolhuis et al., 2010; Kikusui et al., 2011; Moorman & Bolhuis, chapter 5, this volume). Third, analyses (Berwick, Okanoya, Beckers, & Bolhuis, 2011; Beckers et al., 2012) have shown that claims for context-free grammar abilities in songbirds (Gentner et al., 2006; Abe & Watanabe, 2011; see also Okanoya, chapter 11, this volume; Moorman & Bolhuis, chapter 5, this volume) are premature, and that there is no evidence to suggest that nonhuman animals possess the combinatorial complexity of human language syntax. The neural and genetic homologies suggest that both songbirds and mammals may be excellent animal models for the study of the mechanisms of auditory learning and vocal production and their evolution. The evolutionary convergence of auditory-vocal learning implies that songbirds are better models for the study of the evolution and mechanisms of speech than nonhuman primates or mice. On the basis of their critical review, Beckers et al. (2012) concluded that presently there is no credible animal model for the study of the neural substrate of human language. However, it may be that the neural and genetic mechanisms that evolved from a common ancestor, combined with the auditory-vocal learning ability that evolved in both humans and songbirds, enabled the emergence of language uniquely in the human lineage.

References

Abe, K., & Watanabe, D. (2011). Songbirds possess the spontaneous ability to discriminate syntactic rules. *Nature Neuroscience, 14*, 1067–1074.

Agus, T. R., Thorpe, S. J., & Pressnitzer, D. (2010). Rapid formation of robust auditory memories: Insights from noise. *Neuron, 66*, 610–618.

Aronov, D., Andalman, A. S., & Fee, M. S. (2008). A specialized forebrain circuit for vocal babbling in the juvenile songbird. *Science, 320*, 630–634.

Basham, M. E., Nordeen, E. J., & Nordeen, K. W. (1996). Blockade of NMDA receptors in the anterior forebrain impairs sensory acquisition in the zebra finch (*Poephila guttata*). *Neurobiology of Learning and Memory, 66*, 295–304.

Bauer, E. E., Coleman, M. J., Roberts, T. F., Roy, A., Prather, J. F., & Mooney, R. (2008). A synaptic basis for auditory-vocal integration in the songbird. *Journal of Neuroscience, 28*, 1509–1522.

Beckers, G. J. L., Bolhuis, J. J., Okanoya, K., & Berwick, R. C. (2012). Birdsong neurolinguistics: Songbird context-free grammar claim is premature. *NeuroReport, 23*, 139–145.

Berwick, R. C., Okanoya, K., Beckers, G. J. L., & Bolhuis, J. J. (2011). Songs to syntax: The linguistics of birdsong. *Trends in Cognitive Sciences, 15*, 113–121.

Bodner, M., Kroger, J., & Fuster, J. M. (1996). Auditory memory cells in dorsolateral prefrontal cortex. *NeuroReport, 7*, 1905–1908.

Bolhuis, J. J. (2008). Chasin' the trace: The neural substrate of bird song memory. In H. P. Zeigler & P. Marler (Eds.), *Neuroscience of Birdsong* (pp. 269–279). Cambridge: Cambridge University Press.

Bolhuis, J. J., & Gahr, M. (2006). Neural mechanisms of birdsong memory. *Nature Reviews: Neuroscience, 7*, 347–357.

Bolhuis, J. J., Hetebrij, E., den Boer-Visser, A. M., de Groot, J. H., & Zijlstra, G. G. O. (2001). Localized immediate early gene expression related to the strength of song learning in socially reared zebra finches. *European Journal of Neuroscience, 13*, 2165–2170.

Bolhuis, J. J., Okanoya, K., & Scharff, C. (2010). Twitter evolution: Converging mechanisms in birdsong and human speech. *Nature Reviews: Neuroscience, 11*, 747–759.

Bolhuis, J. J., Zijlstra, G. G. O., den Boer-Visser, A. M., & van der Zee, E. A. (2000). Localized neuronal activation in the zebra finch brain is related to the strength of song learning. *Proceedings of the National Academy of Sciences of the United States of America, 97*, 2282–2285.

Bottjer, S., Miesner, E., & Arnold, A. (1984). Forebrain lesions disrupt development but not maintenance of song in passerine birds. *Science, 224*, 901–903.

Brainard, M. S., & Doupe, A. J. (2000). Auditory feedback in learning and maintenance of vocal behaviour. *Nature Reviews: Neuroscience, 1*, 31–40.

Brainard, M. S., & Doupe, A. J. (2002). Interruption of a basal ganglia-forebrain circuit prevents plasticity of learned vocalizations. *Nature, 404*, 762–766.

Brechmann, A., Gaschler-Markefski, B., Sohr, M., Yoneda, K., Kaulisch, T., & Scheich, H. (2007). Working memory-specific activity in auditory cortex: Potential correlates of sequential processing and maintenance. *Cerebral Cortex, 17*, 2544–2552.

Charrier, I., Mathevon, N., & Jouventin, P. (2001). Mother's voice recognition by seal pups. *Nature, 412*, 873.

Chew, S. J., Mello, C., Nottebohm, F., Jarvis, E., & Vicario, D. S. (1995). Decrements in auditory responses to a repeated conspecific song are long-lasting and require 2 periods of protein-synthesis in the songbird forebrain. *Proceedings of the National Academy of Sciences of the United States of America, 92*, 3406–3410.

Colombo, M., Damato, M. R., Rodman, H. R., & Gross, C. G. (1990). Auditory association cortex lesions impair auditory short-term-memory in monkeys. *Science, 247*, 336–338.

Dehaene-Lambertz, G., Hertz-Pannier, L., Dubois, J., Meriaux, S., Roche, A., Sigman, M., et al. (2006). Functional organization of perisylvian activation during presentation of sentences in preverbal infants. *Proceedings of the National Academy of Sciences of the United States of America, 103*, 14240–14245.

Demany, L., & Semal, C. (2008). The role of memory in auditory perception. In W. A. Yost, A. N. Popper, & R. R. Fay (Eds.), *Auditory perception of sound sources* (pp. 77–114). New York: Springer-Verlag.

Demonet, J. F., Thierry, G., & Cardebat, D. (2005). Renewal of the neurophysiology of language: Functional neuroimaging. *Physiological Reviews, 85*, 49–95.

Doupe, A. J., & Kuhl, P. K. (1999). Birdsong and human speech: Common themes and mechanisms. *Annual Review of Neuroscience, 22*, 567–631.

Eda-Fujiwara, H., Satoh, R., Bolhuis, J. J., & Kimura, T. (2003). Neuronal activation in female budgerigars is localized and related to male song complexity. *European Journal of Neuroscience, 17*, 149–154.

Fisher, S. E., & Scharff, C. (2009). FOXP2 as a molecular window into speech and language. *Trends in Genetics, 25*, 166–177.

Frey, H. P., Kaernbach, C., & Konig, P. (2003). Cats can detect repeated noise stimuli. *Neuroscience Letters, 346*, 45–48.

Friederici, A., Meyer, M., & von Cramon, D. Y. (2000). Auditory language comprehension: An event-related fMRI study on the processing of syntactic and lexical information. *Brain and Language, 74*, 289–300.

Fritz, J., Mishkin, M., & Saunders, R. C. (2005). In search of an auditory engram. *Proceedings of the National Academy of Sciences of the United States of America, 102*, 9359–9364.

Gentner, T. Q., Fenn, K. M., Margoliash, D., & Nusbaum, H. C. (2006). Recursive syntactic pattern learning by songbirds. *Nature, 440*, 1204–1207.

Gentner, T. Q., Hulse, S. H., Bentley, G. E., & Ball, G. F. (2000). Individual vocal recognition and the effect of partial lesions to HVc on discrimination, learning, and categorization of conspecific song in adult songbirds. *Journal of Neurobiology, 42*, 117–133.

Gentner, T. Q., Hulse, S. H., Duffy, D., & Ball, G. F. (2001). Response biases in auditory forebrain regions of female songbirds following exposure to sexually relevant variation in male song. *Journal of Neurobiology, 46*, 48–58.

Gobes, S. M. H., & Bolhuis, J. J. (2007). Birdsong memory: A neural dissociation between song recognition and production. *Current Biology, 17*, 789–793.

Gobes, S. M. H., & Ölveczky, B. P. (2011). The sensorimotor nucleus NIf is required for normal acquisition and production of birdsong. *Program No. 303.15. 2011 Neuroscience Meeting Planner*. Washington, DC: Society for Neuroscience. Online.

Gobes, S. M. H., Zandbergen, M. A., & Bolhuis, J. J. (2010). Memory in the making: Localized brain activation related to song learning in young songbirds. *Proceedings of the Royal Society B: Biological Sciences, 277,* 3343–3351.

Godard, R. (1991). Long-term memory of individual neighbors in a migratory songbird. *Nature, 350,* 228–229.

Gottlieb, Y., Vaadia, E., & Abeles, M. (1989). Single unit activity in the auditory cortex of a monkey performing a short term memory task. *Experimental Brain Research, 774,* 139–148.

Guttman, N., & Julesz, B. (1963). Lower limits of auditory analysis. *Journal of the Acoustical Society of America, 35,* 610.

Hackett, T. A. (2011). Information flow in the auditory cortical network. *Hearing Research, 271,* 133–146.

Hackett, T. A., Preuss, T. M., & Kaas, J. H. (2001). Architectonic identification of the core region in the auditory cortex of macaques, chimpanzees and humans. *Journal of Comparative Neurology, 441,* 197–222.

Halpern, A. R., & Zatorre, R. J. (1999). When that tune runs through your head: A PET investigation of auditory imagery for familiar melodies. *Cerebral Cortex, 9,* 697–704.

Hickok, G. (2009). The functional neuroanatomy of language. *Physics of Life Reviews, 6,* 121–143.

Hickok, G., Buchsbaum, B., Humphries, C., & Muftuler, T. (2003). Auditory-motor interaction revealed by fMRI: Speech, music, and working memory in area Spt. *Journal of Cognitive Neuroscience, 15,* 673–682.

Hickok, G., & Poeppel, D. (2000). Towards a functional neuroanatomy of speech perception. *Trends in Cognitive Sciences, 4,* 131–138.

Imada, T., Zhang, Y., Cheour, M., Taulu, S., Ahonen, A., & Kuhl, P. K. (2006). Infant speech perception activates Broca's area: A developmental magnetoencephalography study. *NeuroReport, 17,* 957–962.

Insley, S. J. (2000). Long-term vocal recognition in the northern fur seal. *Nature, 406,* 404–405.

Jarvis, E., Gunturkun, O., Bruce, L., Csillag, A., Karten, H., Kuenzel, W., et al. (2005). Avian brains and a new understanding of vertebrate brain evolution. *Nature Reviews: Neuroscience, 6,* 151–159.

Jarvis, E., & Nottebohm, F. (1997). Motor-driven gene expression. *Proceedings of the National Academy of Sciences of the United States of America, 94,* 4097–4102.

Kaas, J. H., & Hackett, T. A. (1999). "What" and "where" processing in auditory cortex. *Nature Neuroscience, 2,* 1045–1047.

Kaas, J. H., & Hackett, T. A. (2000). Subdivisions of auditory cortex and processing streams in primates. *Proceedings of the National Academy of Sciences of the United States of America, 97,* 11793–11799.

Kaas, J. H., Hackett, T. A., & Tramo, M. J. (1999). Auditory processing in primate cerebral cortex. *Current Opinion in Neurobiology, 9,* 164–170.

Kaernbach, C. (2004). The memory of noise. *Experimental Psychology, 51,* 240–248.

Kaernbach, C., & Shulze, H. (2002). Auditory sensory memory for random waveforms in the Mongolian gerbil. *Neuroscience Letters, 329,* 37–40.

Kao, M. H., Doupe, A. J., & Brainard, M. S. (2005). Contributions of an avian basal ganglia-forebrain circuit to real-time modulation of song. *Nature, 433,* 638–643.

Kikusui, T., Nakanishi, K., Nakagawa, R., Nagasawa, M., Mogi, K., & Okanoya, K. (2011). Cross fostering experiments suggest that mice songs are innate. *PLoS ONE, 6,* e17721.

Kimpo, R. R., & Doupe, A. J. (1997). FOS is induced by singing in distinct neuronal populations in a motor network. *Neuron, 18,* 315–325.

Konishi, M. (1965). The role of auditory feedback in the control of vocalization in the white-crowned sparrow. *Zeitschrift für Tierpsychologie, 22,* 770–783.

Konishi, M. (2004). The role of auditory feedback in birdsong. In H. P. Zeigler & P. Marler (Eds.), *Behavioral neurobiology of birdsong* (pp. 463–475). New York: New York Academy of Sciences.

Kraut, M. A., Pitcock, J. A., Calhoun, V., Li, J., Freeman, T., & Hart, J., Jr. (2006). Neuroanatomic organization of sound memory in humans. *Journal of Cognitive Neuroscience, 18,* 1877–1888.

Lahav, A., Saltzman, E., & Schlaug, G. (2007). Action representation of sound: Audiomotor recognition network while listening to newly acquired actions. *Journal of Neuroscience, 27,* 308–314.

Leaver, A. M., van Lare, J., Zielinski, B., Halpern, A. R., & Rauschecker, J. P. (2009). Brain activation during anticipation of sound sequences. *Journal of Neuroscience, 29,* 2477–2485.

Letzkus, J. J., Wolff, S. B., Meyer, E. M., Tovote, P., Courtin, J., Herry, C., et al. (2011). A disinhibitory microcircuit for associative fear learning in the auditory cortex. *Nature, 480,* 331–335.

Levitan, D. J., & Rogers, S. E. (2005). Absolute pitch: perception, coding and controversies. *Trends in Cognitive Sciences, 9,* 26–33.

Lewis, J. W., Brefczynski, J. A., Phinney, R. E., Janik, J. J., & DeYoe, E. A. (2005). Distinct cortical pathways for processing tool versus animal sounds. *Journal of Neuroscience, 25,* 5148–5158.

Lewis, J. W., Wightman, F. L., Brefczynski, J. A., Phinney, R. E., Binder, J. R., & DeYoe, E. A. (2004). Human brain regions involved in recognizing environmental sounds. *Cerebral Cortex, 14,* 1008–1021.

London, S. E., & Clayton, D. F. (2008). Functional identification of sensory mechanisms required for developmental song learning. *Nature Neuroscience, 11,* 579–586.

MacDougall-Shackleton, S. A., Hulse, S. H., & Ball, G. F. (1998). Neural bases of song preferences in female zebra finches (*Taeniopygia guttata*). *NeuroReport, 9,* 3047–3052.

Margoliash, D. (1986). Preference for autogenous song by auditory neurons in a song system nucleus of the white-crowned sparrow. *Journal of Neuroscience, 6,* 1643–1661.

Margoliash, D., & Konishi, M. (1985). Auditory representation of autogenous song in the song system of white-crowned sparrows. *Proceedings of the National Academy of Sciences of the United States of America, 82,* 5997–6000.

Marler, P., & Doupe, A. J. (2000). Singing in the brain. *Proceedings of the National Academy of Sciences of the United States of America, 97,* 2965–2967.

Marler, P., & Sherman, V. (1985). Innate differences in singing behavior of sparrows reared in isolation from adult conspecific song. *Animal Behaviour, 33*, 57–71.

Mello, C. V., & Clayton, D. F. (1994). Song-induced Zenk gene-expression in auditory pathways of songbird brain and its relation to the song control-system. *Journal of Neuroscience, 14*, 6652–6666.

Mello, C. V., Nottebohm, F., & Clayton, D. F. (1995). Repeated exposure to one song leads to a rapid and persistent decline in an immediate early gene's response to that song in zebra finch telencephalon. *Journal of Neuroscience, 15*, 6919–6925.

Mello, C. V., Vicario, D. S., & Clayton, D. F. (1992). Song presentation induces gene-expression in the songbird forebrain. *Proceedings of the National Academy of Sciences of the United States of America, 89*, 6818–6822.

Miller, D. B. (1979). Long-term recognition of father's song by female zebra finches. *Nature, 280*, 389–391.

Mooney, R. (2009). Neural mechanisms for learned birdsong. *Learning & Memory (Cold Spring Harbor, NY), 16*, 655–669.

Moorman, S., Mello, C. V., & Bolhuis, J. J. (2011). From songs to synapses: Molecular mechanisms of birdsong memory. *BioEssays, 33*, 377–385.

Nick, T. A., & Konishi, M. (2005). Neural song preference during vocal learning in the zebra finch depends on age and state. *Journal of Neurobiology, 62*, 231–242.

Nordeen, K. W., & Nordeen, E. J. (2004). Synaptic and molecular mechanisms regulating plasticity during early learning. *Annals of the New York Academy of Sciences, 1016*, 416–437.

Nottebohm, F., & Arnold, A. P. (1976). Sexual dimorphism in vocal control areas of the songbird brain. *Science, 194*, 211–213.

Nottebohm, F., Stokes, T. M., & Leonard, C. M. (1976). Central control of song in the canary, *Serinus canarius. Journal of Comparative Neurology, 165*, 457–486.

Ojemann, G. A. (1991). Cortical organization of language. *Journal of Neuroscience, 11*, 2281–2287.

Okada, K., & Hickok, G. (2006). Left posterior auditory-related cortices participate both in speech perception and speech production: Neural overlap revealed by fMRI. *Brain and Language, 98*, 112–117.

Perrodin, C., Kayser, C., Logothetis, N. K., & Petkov, C. I. (2011). Voice cells in the primate temporal lobe. *Current Biology, 21*, 1408–1415.

Petkov, C. I., Kayser, C., Steudel, T., Whittingstall, K., Augath, M., & Logothetis, N. K. (2008). A voice region in the monkey brain. *Nature Neuroscience, 11*, 367–374.

Petrides, M., & Pandya, D. N. (2009). Distinct parietal and temporal pathways to the homologues of Broca's area in the monkey. *PLoS Biology, 7*, e1000170.

Phan, M. L., Pytte, C. L., & Vicario, D. S. (2006). Early auditory experience generates long-lasting memories that may subserve vocal learning in songbirds. *Proceedings of the National Academy of Sciences of the United States of America, 103*, 1088–1093.

Rama, P., Poremba, A., Sala, J. B., Yee, L., Malloy, M., Mishkin, M., et al. (2004). Dissociable functional cortical topographies for working memory maintenance of voice identity and location. *Cerebral Cortex, 14*, 768–780.

Reiner, A., Perkel, D. J., Bruce, L. L., Butler, A. B., Csillag, A., Kuenzel, W., et al. (2004). Revised nomenclature for avian telencephalon and some related brainstem nuclei. *Journal of Comparative Neurology*, *473*, 377–414.

Remedios, R., Logothetis, N. K., & Kayser, C. (2009). An auditory region in the primate insular cortex responding preferentially to vocal communication sounds. *Journal of Neuroscience*, *29*, 1034–1045.

Romanski, L. M., & Averbeck, B. B. (2009). The primate cortical auditory system and neural representation of conspecific vocalizations. *Annual Review of Neuroscience*, *32*, 315–346.

Romanski, L. M., Bates, J. F., & Goldman-Rakic, P. S. (1999). Auditory belt and parabelt projections to the prefrontal cortex in the rhesus monkey. *Journal of Comparative Neurology*, *403*, 141–157.

Russ, B. E., Ackelson, A. L., Baker, A. E., & Cohen, Y. E. (2008). Coding of auditory-stimulus identity in the auditory non-spatial processing stream. *Journal of Neurophysiology*, *99*, 87–95.

Sacco, T., & Sacchetti, B. (2010). Role of secondary sensory cortices in emotional memory storage and retrieval in rats. *Science*, *329*, 649–656.

Sagar, S. M., Sharp, F. R., & Curran, T. (1988). Expression of c-fos protein in brain-metabolic mapping at the cellular-level. *Science*, *240*, 1328–1331.

Scharff, C., & Nottebohm, F. (1991). A comparative-study of the behavioral deficits following lesions of various parts of the zebra finch song system—Implications for vocal learning. *Journal of Neuroscience*, *11*, 2896–2913.

Scott, S. K., & Wise, R. J. S. (2004). The functional neuroanatomy of the prelexical processing of speech. *Cognition*, *92*, 13–45.

Sharp, S. P., McGowan, A., Wood, M. J., & Hatchwell, B. J. (2005). Learned kin recognition cues in a social bird. *Nature*, *434*, 1127–1130.

Sohrabji, F., Nordeen, E. J., & Nordeen, K. W. (1990). Selective impairment of song learning following lesions of a forebrain nucleus in the juvenile zebra finch. *Behavioral and Neural Biology*, *53*, 51–63.

Terpstra, N. J., Bolhuis, J. J., & den Boer-Visser, A. M. (2004). An analysis of the neural representation of birdsong memory. *Journal of Neuroscience*, *24*, 4971–4977.

Terpstra, N. J., Bolhuis, J. J., Riebel, K., van der Burg, J. M. M., & den Boer-Visser, A. M. (2006). Localized brain activation specific to auditory memory in a female songbird. *Journal of Comparative Neurology*, *494*, 784–791.

Tian, B., Reser, D., Durham, A., Kustov, A., & Rauschecker, J. P. (2001). Functional specialization in rhesus monkey auditory cortex. *Science*, *292*, 290–293.

Velho, T. A. F., Pinaud, R., Rodrigues, P. V., & Mello, C. V. (2005). Co-induction of activity-dependent genes in songbirds. *European Journal of Neuroscience*, *22*, 1667–1678.

Viceic, D., Fornari, E., Thiran, J. P., Maeder, P. P., Meuli, R., Adriani, M., et al. (2006). Human auditory belt areas specialized in sound recognition: A functional magnetic resonance imaging study. *NeuroReport*, *17*, 1659–1662.

Volman, S. F. (1993). Development of neural selectivity for birdsong during vocal learning. *Journal of Neuroscience*, *13*, 4737–4747.

Wan, H., Warburton, E. C., Kusmierek, P., Aggleton, J. P., Kowalska, D. M., & Brown, M. W. (2001). Fos imaging reveals differential neuronal activation of areas of rat temporal cortex by novel and familiar sounds. *European Journal of Neuroscience*, *14*, 118–124.

Weinberger, N. M. (2004). Long-term memory traces in primary auditory cortex. *Nature Reviews: Neuroscience*, *5*, 279–290.

Wise, R. J. S. (2003). Language systems in normal and aphasic human subjects: Functional imaging studies and inferences from animal studies. *British Medical Bulletin*, *65*, 95–119.

Zokoll, M. A., Klump, G. M., & Langemann, U. (2007). Auditory short-term memory persistence for tonal signals in a songbird. *Journal of the Acoustical Society of America*, *121*, 2842–2851.

Zokoll, M. A., Klump, G. M., & Langemann, U. (2008). Auditory memory for temporal characteristics of sound. *Journal of Comparative Physiology A: Neuroethology, Sensory, Neural, and Behavioral Physiology*, *194*, 457–467.

Zokoll, M. A., Naue, N., Herrmann, C. S., & Langemann, U. (2008). Auditory memory: A comparison between humans and starlings. *Brain Research*, *1220*, 33–46.

16 Age Effects in Language Acquisition and Attrition

Christophe Pallier

The notion that children are especially gifted at learning languages, compared to adults, is certainly not revolutionary. For example, the French philosopher Michel de Montaigne (1533–1592) reported that, when he was a child, his father only hired servants who could speak Latin and gave them strict orders to always speak this language to him or in his presence. The aim, of course, was to maximize the chances of making the future philosopher become fluent in Latin. The notion that "the younger, the better" concerning language learning is widespread in the general public and has been invoked to justify the introduction of foreign language education in elementary schools in many countries. Interestingly, research has shown that adults actually outperform young children in the first stages of second language learning (Krashen, Long, & Scarcella, 1982), and that the benefits of early exposure to foreign language in the classroom are far from obvious (Burstall, 1975; Singleton & Ryan, 2004). Yet, it remains undeniable that the age of acquisition of a language, at least in naturalistic situations if not in the classroom, is clearly negatively correlated with eventual proficiency, especially for phonological and grammatical skills (for reviews see Birdsong, 2005; Hyltenstam & Abrahamson, 2003; DeKeyser, 2000).

Understanding the basis of the age effect on language acquisition is important not only for theoretical but also for practical reasons. One popular explanation is that the brain of young children is especially "plastic" and that, under the influence of maturational factors, this plasticity progressively diminishes, resulting in essentially stable language circuits. This notion found a staunch advocate in the Canadian neurosurgeon Wilder Penfield, who claimed that "for the purpose of learning languages, the human brain becomes progressively stiff and rigid after the age of nine" (Penfield & Roberts, 1959, p. 236). Going further in his book *Biological Foundations of Language*, Lenneberg (1967) developed the theory that language acquisition in humans was subject to a critical period. More precisely, he proposed that the human brain was equipped with specialized mechanisms to acquire language that functioned only during a certain time window. According to Lenneberg, these mechanisms start working around 2 years of age and "after puberty, automatic acquisition from mere exposure to a given language seems to disappear, and foreign languages have to be taught and learned through a conscious and labored effort" (p. 176.) Sensitivity to language input would have the shape displayed in Figure 16.1A. Steven Pinker (1994, p. 294) expressed a similar view and reasoned that "once a language is acquired, the neural machinery for language acquisition can be dismantled as keeping it would incur unnecessary metabolic costs."

It must be noted that, in the literature, one often encounters upper age limits for the critical period that are lower than those advanced by Lenneberg or Penfield. For example, 6 years is often mentioned as the upper age limit for the acquisition of an accent-free second language (e.g., Long, 1990). Pinker (1994, p. 293) stated that "acquisition of a normal language is guaranteed for children up to the age of six, is steadily compromised from then until shortly after puberty, and is rare thereafter" (the age function would have the shape depicted in Figure 16.1B). There is actually evidence that even starting to learn a second language as early as 4 to 6 years of age does not necessarily ensure nativelike levels in speech production or perception (Flege, Munro, & MacKay, 1995; Pallier, Bosch, & Sebastian-Gallés, 1997), nor even in grammatical processing (Weber-Fox & Neville, 1996). Effects of age are therefore present even before 6 years. Concerning the closure of the critical period, empirical data on second language (L2) acquisition provide little evidence for a discontinuity at puberty. Birdsong (2005) convincingly argued that the age effects on L2 extend after puberty (and maybe across the whole life span; see Hakuta, Bialystok, & Wiley, 2003) and essentially decrease in a monotonous fashion, as shown in Figure 16.1D.

If the age effect on L2 acquisition is really due to an irreversible loss of neural plasticity under the influence of maturational factors (i.e., a decline in neuronal or synaptic density with age), then the conclusion is clear: it is critical to expose children to new languages as soon as possible. The research reviewed in this chapter will show that the reality is more complex.

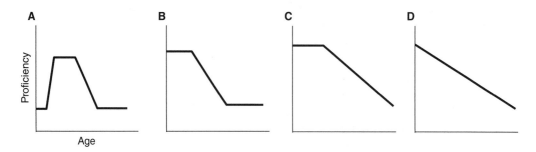

Figure 16.1
Various potential relationships between age of exposure and ultimate attainment in a language (adapted from Birdsong, 2005)

Early Sensitivity to Language

In 1967, Lenneberg proposed that the onset of the critical period for language acquisition was around 20 months of age. At that time, only the speech-production behavior of children was easily accessible to investigation. In the ensuing years, research looking at the perceptual capabilities of children, initiated by the discovery of Eimas et al. (1971) that 1-month-old infants could discriminate phonetic contrasts, established that infants are sensitive to language much earlier and that learning starts from birth and even in the womb. For example, it was shown that neonates (1 to 4 days old) prefer to listen to their mother's voice (DeCasper & Fifer, 1980; Mehler, Bertoncini, Barrière, & Jassik-Gerschenfeld, 1978) and to their maternal language (Mehler et al., 1988; Moon, Cooper, & Fifer, 1993). During the first year of life, infants become attuned to the phonology of the ambient language(s), learning the phonemic repertoire (Werker & Tees, 1984; Kuhl, Williams, Lacerda, Stevens, & Lindblom, 1992), the phonotactics (Jusczyk, Friederici, et al., 1993), and the prosodic characteristics (Jusczyk, Cutler, et al., 1993). At the babbling stage, starting around 8 to 10 month of age, the productions of babies are already influenced by the language spoken in their surroundings (De Boysson-Bardies, Sagart, & Durand, 1984). At the end of the first year, they start to associate words and meanings (Hallé & De Boysson-Bardies, 1994). The profile of sensitivity depicted in Figure 16.1A, with an onset at 2 years, is no longer tenable: there is no evidence that language acquisition is, as it were, "switched on" at a given time point. Yet, one important question remains concerning whether these early abilities reflect language-specific or general learning mechanisms; the debate has not been completely resolved (see, e.g., Elman et al., 1997). At the very least, the behavioral studies suggest that human infants are innately attracted to speech signals (Colombo & Bundy, 1983; Jusczyk & Bertoncini, 1988), which makes sense from an evolutionary perspective.

The definite answer to the question of whether the human brain is hardwired to process speech will ultimately come from brain studies. For the moment, very few functional brain imaging studies on infants exist (Dehaene-Lambertz, Dehaene, & Hertz-Pannier, 2002; Dehaene-Lambertz et al., 2006; Peña et al., 2003). Their main conclusion is that the infant's brain does not respond diffusely to speech but relies on the same perisylvian network of areas as in adults, with some left-hemisphere dominance already apparent.

Peña et al. (2003) and Dehaene-Lambertz et al. (2002) presented recordings to young infants in their maternal language. The first study used functional magnetic resonance imaging (fMRI) with 3-month-old babies, while the second used near-infrared spectroscopy (NIRS) with neonates. The same utterances were also presented backward, yielding sounds that are as complex as speech but cannot be produced by a vocal tract and violate universal prosodic rules. The NIRS captors showed stronger activity over the left hemisphere than the right when neonates listened to forward speech compared to backward speech. The anatomically more precise magnetic resonance technique showed activations in the left temporal (superior temporal gyrus) and parietal (angular gyrus) regions. Moreover, the latter region reacted more strongly to forward speech than to backward speech, as did a right prefrontal region in the awakened children (some of the infants were asleep in the scanner, while others were awake).

Using fMRI again, Dehaene-Lambertz et al. (2006) presented short sentences to 3-month-old infants. As in the previous experiment, activations were detected in the temporal lobes bilaterally (with more activation in the left hemisphere than in the right). Active clusters were also found in the right and left insula and in the left inferior frontal gyrus, as typically observed in adults. The stimuli were played at a slow pace that allowed the authors to examine the temporal delays of the brain responses. The analysis of these delays revealed an adultlike structure: the fastest responses were recorded in the vicinity of Heschl's gyrus, whereas responses became increasingly slower toward the posterior part of the superior temporal gyrus and toward the temporal poles and inferior frontal regions (Broca's area).

In brief, the cerebral activations in very young infants listening to speech are remarkably similar to those observed in adults; they are not limited to unimodal auditory regions and extend to remote frontal regions, which used to be considered barely functional at this age. This observation refutes the theory of progressive lateralization of language advanced by Lenneberg (1967). According to him, both hemispheres were responsive to speech at the onset of the critical period and the maturational process made the left become progressively dominant over the right. Functional brain imaging studies of language comprehension show bilateral activations, both in children and in adults, with a relative left dominance. It must be stressed, however, that data currently available on infants remain quite scarce. Despite the results with backward speech, it is too early to categorically claim that there are brain areas that respond specifically to speech and not to other types of sounds of similar complexity.

Effect of Delay on First Language Acquisition

Most humans are typically exposed to language(s) from infancy and learn it (them) without difficulty. Only a few reports exist on children who grew up in extreme social isolation and received very little language input until puberty. Itard (1964) described the case of Victor, the "wild boy of Aveyron," who was discovered running naked in woods in the south of France when he was about 12. With the help of his instructor, he acquired a rudimentary vocabulary and learned to respond to simple written expressions, yet he never learned to articulate speech properly. (Because he had a throat wound, it is not clear if this reflected a cognitive limitation.) More details are known about the case of Genie (Fromkin, Krashen, Curtiss, Rigler, & Rigler, 1974; Curtiss, 1977). Genie was found at the age of 13 after having suffered extreme social deprivation. She became able to understand and produce speech, acquired a fair amount of vocabulary, and was able to build sentences. Yet she failed to achieved normal linguistic competence after many years of training. For example, she continued to form negative sentences by putting *no* at the beginning of sentences. Her case suggests that limited language acquisition is possible after puberty (for another case of relatively late L1 acquisition (after 9 years), see also Vargha-Khadem et al., 1997). However, it must be acknowledged that in Genie's case as in Victor's, little is known about their early experiences with language, which makes it difficult to draw strong conclusions.

More extensive data come from studies on deaf adults who learn sign language at different ages (Newport, 1990; Mayberry & Eichen, 1991; Mayberry & Fischer, 1989). Many deaf children are born to hearing parents and, for the most part, are not exposed to a full-fledged language until they enter schools for the deaf and learn sign language. This line of research has shown that, even if sign language can be learned at any age, there are clear effects of age of first exposure on the ultimate proficiency: the earlier deaf students were exposed to language, the better they perform in various language tasks (memory for sentence and story, shadowing, sentence and story comprehension, and grammatical judgment tasks). According to Mayberry (1998, p. 8), delayed acquisition of sign language affects the processing of "both simple and complex syntactic structures and impacts all levels of linguistic structure, namely, phonology, morphology, the lexicon, syntax, and semantics."

This age effect on ultimate proficiency in L1 begins quite early. Newport (1990) found that children who began learning sign language at age 4 did not perform as well as those exposed to sign language from birth. Studies of auditory and language development in congenitally deaf children who receive cochlear implants (an auditory prosthesis that stimulates the auditory nerve in order to transmit acoustic information to the central auditory system) reveal that behavioral benefits of early implantation can be observed even in the range of 1 to 3 years of age (McConkey Robbins, Burton Koch, Osberger, Zimmerman-Philips, & Kishon-Rabin, 2004; Tomblin, Barker, Spencer, Zhang, & Gantz, 2005). Measuring auditory evoked potentials, Gilley, Sharma, and Dorman (2008) found that their topography was influenced by the age of implantation, suggesting a cortical reorganization with age.

Are endogenous maturational factors the only cause of the age effect on L1 acquisition? This hypothesis can be rejected. A study by Mayberry, Lock, and Kazmi (2001) clearly established the major role of *deprivation*—that is, the fact that the deaf children who did not learn American Sign Language (ASL) in infancy were deprived of normal linguistic input in their first years of life. The authors compared two groups of deaf adults who had both learned ASL relatively late, between 9 and 15 years of age. ASL was the L1 for the participants of the first group, who were congenitally deaf. Participants in the second group were born with normal hearing and had started to acquire English before they became deaf. Therefore, ASL was their L2 (note that the grammar of ASL differs markedly from the grammar of English; Klima & Bellugi, 1979). Mayberry et al. (2001) found that the second group largely outperformed the first in ASL. If only maturational factors were at play then the proficiency in ASL should only depend on age of acquisition of ASL. Mayberry et al.'s results demonstrate the crucial role of early experience with language.

The studies on the effect of delay on the acquisition of an L1 demonstrate that linguistic deprivation has rapid detrimental effects on ultimate proficiency. Because deprivation is the usual test applied to assess critical periods in animals, it appears undeniable that there is a critical period for normal first language acquisition in humans. This may be an instance of *experience-expectant* plasticity as defined by Greenough et al. (1987): the immature brain expects "language" in the environment. In the absence of linguistic stimulation, the brain areas that normally subserve language processing may either deteriorate or be recruited for other functions. This last interpretation is supported by data from Lee et al. (2001) showing that the benefits of cochlear implantation are inversely related to the amount of metabolism in the temporal lobes. In other words, the deaf who have "abnormally" low metabolism in the temporal region (before receiving cochlear implants) profit more from implants than those who have higher metabolism, presumably because in the latter case, these areas are recruited for extralinguistic functions (see also Neville & Bavelier, 2002).

Effect of Delay on Second Language Acquisition

As mentioned in the introduction, research has confirmed the view that the age of acquisition of a second language is a potent factor for ultimate attainment (for reviews see Long, 1990; Birdsong, 1999; DeKeyser & Larson-Hall, 2005). Unlike the situation with L1, the age effect on L2 cannot be explained by a lack of language input in the first years of life. Incidentally, Mayberry (1998) noted that the effects of age are less dramatic for L2 than for L1 acquisition, suggesting that different mechanisms may be at play.

Some versions of the critical period hypothesis claim that, after puberty, a second language is acquired in a fundamentally different way than the first because the brain circuits for language acquisition are no longer operational. This predicts, then, that L1 and L2 should be supported by (at least partially) different brain areas.

This view is not supported by brain imaging studies on the cortical representations of L1 and L2 in bilinguals (for reviews see Perani & Abutalebi, 2005; Pallier & Argenti, 2003). In studies using fMRI or positron-emission tomography (PET), the patterns of activation elicited by the use of the first or the second language are quite similar for the two languages in highly proficient bilinguals, and this is little affected by their age of acquisition. In less proficient bilinguals, L2 sometimes provokes stronger or more diffuse activations than L1 in classic language areas but there is no clear evidence for L2-specific brain areas. Considering data from bilingual aphasia as well, Paradis (2004) concluded that a bilingual's language subsystems are represented in the same cerebral areas at the macroanatomical level. Given the current resolution of brain imaging techniques (about 2 mm), it is quite possible that L1 and L2 are differentiated at the microanatomical level, but the main fact remains that a second language, even acquired late, does not rely on fundamentally distinct brain systems from the first language. This result refutes a version of the critical period hypothesis, according to which the circuits that support L1 have lost plasticity and L2 must be supported by different circuits.

Language Loss

The hypothesis that the neural circuits subserving language lose plasticity predicted that the effects of learning a language in the first years of life should be irreversible. In the same way that one never forgets how to ride a bicycle, one should therefore never forget one's maternal language.

International adoption provides a way to assess this idea. The overwhelming majority of foreign children adopted in new families stop using their maternal language (Maury, 1995; Isurin, 2000). Pallier et al. (2003) contacted organizations in charge of international adoption and managed to recruit a small sample of young adults born in Korea who had been adopted by French-speaking families. They came to France when they were between 3 and 8 years old and had not been exposed to Korean since then. All claimed to have completely forgotten Korean (though some had memories of their life in Korea).

Three behavioral experiments were designed to assess their residual knowledge of the Korean language. The adoptees' performances was compared to that of a control group of native French speakers who had never been exposed to Korean, nor to any Asian language. The Korean sentence identification experiment involved recognizing sentences in Korean among recordings in different languages. In the word recognition experiment, subjects heard two Korean words and had to choose which was the translation of a given French word. Lastly, in the speech segment detection experiment, the task was to decide if speech fragments were present in sentences in various languages, including Korean. The results show similar patterns of performance for the adoptees and for the control group of native French speakers, validating the adoptees' claim that they have largely forgotten their first language.

While the subjects performed the speech segment detection task, their brain activity was monitored using functional magnetic resonance. The analyses of fMRI data showed, for each of the adoptees, no detectable difference in brain activity when comparing the cerebral responses to Korean sentences versus Japanese or Polish sentences, two languages to which the adoptees had never been exposed. Thus, brain imaging data and behavioral data converge in the conclusion that years of exposure to a language in childhood are not sufficient to maintain a solid knowledge of this language.

This result can be interpreted in two different ways. First, the Korean language may have been "erased" from the brain of the adoptees. This would constitute strong evidence against versions of the critical period hypothesis that state that some "neural connections" become fixed in the early years of life, as a result of learning and/or because of maturational factors. These hypotheses predicted that the adoptees (at least those arriving at older ages) should have displayed a considerable sensitivity to Korean. It must be noted, however, that because the subjects arrived in France before the age of 10, we cannot exclude the possibility that irreversible changes occur at puberty.

A second possible interpretation is that the paradigms used in Pallier et al., (2003) lacked sensitivity and that further testing may uncover effects of the early exposure to Korean. With Valerie Ventureyra, I ran a series of behavioral experiments to more thoroughly test the remnants of Korean in the adoptees (Ventureyra, 2005). In a nutshell, we found virtually no significant difference between the adoptees and native French speakers. For example, the adoptees were not better at perceiving the differences between Korean plain, tense, and aspirated stop consonants, a phonemic contrast in Korean (Ventureyra, Pallier, & Yoo, 2004).

One important question is whether the adoptees could relearn their native language faster or better than people who have never been exposed to Korean. This would provide evidence for remnant traces of early exposure to Korean. From an anecdotal point of view, the adoptees who visited Korea for short stays (from a few days to a few months) did not miraculously "recover" the ability to speak or comprehend the language, nor did the few of them who attended Korean lectures.

There is some evidence that early exposure to a language leads to an advantage when one relearns it later (Tees & Werker, 1984; Oh, Jun, Knightly, & Au, 2003; Au, Knightly, Jun, & Oh, 2002; Knightly, Jun, Oh, & Au, 2003; Au, Oh, Knightly, Jun, & Romo, 2008). For example, Oh et al. (2003) evaluated the perception and production of Korean consonants by three groups enrolled in Korean language classes: one group had spoken Korean regularly for a few years during childhood, another group had heard Korean regularly during childhood but had spoken Korean minimally, and the last group consisted of novice learners. The first two groups performed better than the novice learners, demonstrating long-term benefits of early childhood experience with Korean.

Au et al. (2008) tested adult learners of Spanish who had spoken Spanish as their native language before age 7 and only minimally, if at all, thereafter until they began to relearn Spanish around age 14 years. They spoke Spanish with a more nativelike accent than typical late L2 learners. On grammar measures, although far from reliably nativelike, they also outperformed typical late L2 learners. These results suggest that while simply overhearing a language during childhood could help adult learners speak it with a more nativelike phonology, speaking a language regularly during childhood could help relearners use it with more nativelike grammar as well as phonology.

As mentioned above, it would be highly desirable to know whether the adoptees also have "dormant traces" of the language they have been exposed to in their childhood. In the relearning studies cite above, the subjects were not completely severed from the language of interest. For example, in the Oh et al. (2003) study the nonnovice subjects overheard Korean on average 4 hours per week. Therefore, their situation was quite different from that of adoptees who have not been exposed at all to Korean since adoption. Whether the adoptees would relearn their first language faster than novice learners remains an unsolved empirical question. Nevertheless, the studies on adoptees that show the ability to comprehend a language can be lost suggest that the "plasticity" of the language-learning system is considerable up to the age of 10 years.

Conclusion

I started from a seemingly simple idea: that the brain is especially "plastic" in very young children and that, under the influence of maturational factors, this plasticity is progressively lost, resulting in an essentially stable adult brain. Instead, the research reviewed in the chapter suggests that

• When children are not exposed to a first language in the early years of life, their language acquisition is compromised: they are not going to master a language like native users. However, this effect is not simply a maturational effect but is a consequence of an "abnormal" experience: linguistic deprivation in the early environment (Mayberry et al., 2001). One putative explanation is that the brain circuits for language are reused for other functions (Lee et al., 2001).

• There are indisputable age effects on ultimate proficiency in the second language (L2). However, the shape of the age effect on L2 is more or less linear and does not show a clear discontinuity (Hakuta et al., 2003; Birdsong, 2005). It certainly does not have the same origin as the age effect on L1 because L2 learners have not been deprived of any language input and their brain has, presumably, developed in a "normal way."

• Studies on internationally adopted children suggest that it is possible to lose understanding of a first language, even after 10 years of exposure. There is therefore still considerable plasticity in the language circuits until that age. An interesting observation it that studies on language loss in adult immigrants show much less dramatic forgetting (Köpke & Schmid, 2004; Köpke, 2004), maybe reflecting changes in brain plasticity around puberty.

• In babies, the same brain areas are activated by language as in adults, undermining the notion of progressive lateralization put forward by Lenneberg in one version of the critical period hypothesis.

• Brain imaging studies (PET or fMRI) of bilinguals found that they rely on the same macroanatomical brain areas to process L1 and L2 even when L2 has been acquired after L1, as long as proficiency in L2 is high (Perani & Abutalebi, 2005). This refutes a simple version of the critical period hypothesis, according to which the brain circuits underlying L1 have lost plasticity and L2 must be learned by different circuits.

Data on first language acquisition demonstrate that there is indeed a critical period for language acquisition in humans in the sense that a lack of language stimulation in the early years has irreversible consequences. This critical period for a first language does not explain the effect of age on second language acquisition, inasmuch as second language learners have not suffered from linguistic deprivation in childhood. The effects of age on second language learning, which begin early, are unlikely to involve simple maturational loss of plasticity, because plasticity is still considerable at 10 years of age, as studies on adoptees show.

The reality is therefore considerably more complex than entailed by a simplistic notion of maturational loss of plasticity. Yet, one must recognize that the critical period hypothesis for language acquisition has generated, and is still generating, a lot of research that has improved our understanding of the mechanisms of language acquisition.

References

Au, T. K.-F., Knightly, L. M., Jun, S.-A., & Oh, J. S. (2002). Overhearing a language during childhood. *Psychological Science, 13*, 238–243.

Au, T. K.-F., Oh, J. S., Knightly, L. M., Jun, S.-A., & Romo, L. F. (2008). Salvaging a childhood language. *Journal of Memory and Language, 58*, 998–1011.

Birdsong, D. (Ed.). (1999). *Second language acquisition and the critical period hypothesis.* Mahwah, NJ: LEA.

Birdsong, D. (2005). Interpreting age effects in second language acquisition. In J. F. Kroll & A. M. B. de Groot (Eds.), *Handbook of bilingualism: Psycholinguistic approaches* (pp. 109–127). Oxford: Oxford University Press.

de Boysson-Bardies, B., Sagart, L., & Durand, C. (1984). Discernible differences in the babbling of infants according to target language. *Journal of Child Language, 11*, 1–15.

Burstall, C. (1975). Primary French in the balance. *Educational Research, 17*, 193–197.

Colombo, J., & Bundy, R. (1983). Infant response to auditory familiarity and novelty. *Infant Behavior and Development, 6*, 305–311.

Curtiss, S. (1977). *Genie: A psycholinguistic study of a modern day "Wild Child."* New York: Academic Press.

DeCasper, A. J., & Fifer, W. P. (1980). Of human bonding: Newborns prefer their mother's voices. *Science, 208*, 1174–1176.

Dehaene-Lambertz, G., Dehaene, S., & Hertz-Pannier, L. (2002). Functional neuroimaging of speech perception in infants. *Science, 298*, 2013–2015.

Dehaene-Lambertz, G., Hertz-Pannier, L., Dubois, J., Mériaux, S., Roche, A., Sigman, M., et al. (2006). Functional organization of perisylvian activation during presentation of sentences in preverbal infants. *Proceedings of the National Academy of Sciences of the United States of America, 103*, 14240–14245.

DeKeyser, R. M. (2000). The robustness of critical period effects in second language acquisition. *Studies in Second Language Acquisition, 22*, 499–533.

DeKeyser, R. M., & Larson-Hall, J. (2005). What does the critical period really mean? In J. F. Kroll & A. M. B. de Groot (Eds.), *Handbook of bilingualism: Psycholinguistic approaches* (pp. 88–108). Oxford: Oxford University Press.

Eimas, P. D., Siqueland, E. R., Jusczyk, P. W., & Vigorito, J. (1971). Speech perception in infants. *Science, 171*, 303–306.

Elman, J. L., Bates, E. A., Johnson, M. H., Karmiloff-Smith, A., Parisi, D., & Plunkett, K. (1997). *Rethinking innateness: A connectionist perspective on development.* Cambridge, MA: MIT Press.

Flege, J. E., Munro, M. J., & MacKay, I. R. A. (1995). Factors affecting strength of perceived foreign accent in a second language. *Journal of the Acoustical Society of America, 97*, 3125–3134.

Fromkin, V., Krashen, S., Curtiss, S., Rigler, D., & Rigler, M. (1974). The development of language in genie: A case of language acquisition beyond the "critical period." *Brain and Language, 1*, 83–107.

Gilley, P. M., Sharma, A., & Dorman, M. F. (2008). Cortical reorganization in children with cochlear implants. *Brain Research, 1239*, 56–65.

Greenough, W. T., Black, J. E., & Wallace, C. S. (1987). Experience and brain development. *Child Development, 58*, 539–559.

Hakuta, K., Bialystok, E., & Wiley, E. (2003). Critical evidence: A test of the critical-period hypothesis for second-language acquisition. *Psychological Science, 14*, 31–38.

Hallé, P., & de Boysson-Bardies, B. (1994). Emergence of an early receptive lexicon: Infants' recognition of words. *Infant Behavior and Development, 17*, 119–129.

Hyltenstam, K., & Abrahamson, N. (2003). Maturational constraints in SLA. In C. J. Doughty & M. J. Long (Eds.), *The handbook of second language acquisition* (pp. 539–588). Oxford: Blackwell.

Isurin, L. (2000). Deserted island or a child's first language forgetting. *Bilingualism: Language and Cognition, 3*, 151–166.

Itard, J. (1964). Mémoire et rapport sur Victor de l'Aveyron. In L. Malson (Ed.), *Les enfants sauvages: Mythes et réalité* (pp. 119–246). Paris: Union Général d'éditions 10/18.

Jusczyk, P., & Bertoncini, J. (1988). Viewing the development of speech perception as an innately guided learning process. *Language and Speech, 31*, 217–238.

Jusczyk, P., Cutler, A., & Redanz, N. J. (1993). Infants' preference for the predominant stress patterns of English words. *Child Development, 64*, 675–687.

Jusczyk, P., Friederici, A., Wessels, J., Svenkerud, V., & Jusczyk, A. (1993). Infants' sensitivity to the sound patterns of native language words. *Journal of Memory and Language, 32*, 402–420.

Klima, E., & Bellugi, U. (1979). *The signs of language*. Cambridge, MA: Harvard University Press.

Knightly, L. M., Jun, S.-A., Oh, J. S., & Au, T. K.-F. (2003). Production benefits of childhood overhearing. *Journal of the Acoustical Society of America, 114*, 465–474.

Köpke, B. (2004). Neurolinguistic aspects of attrition. *Journal of Neurolinguistics, 17*, 3–30.

Köpke, B., & Schmid, M. S. (2004). Language attrition: The next phase. In M. S. Schmid, B. Köpke, M. Keijzer, & L. Weilemar (Eds.), *First language attrition: Interdisciplinary perspectives on methodological issues* (pp. 1–46). Amsterdam: John Benjamins.

Krashen, S., Long, M., & Scarcella, R. (Eds.). (1982). *Child-adult differences in second language acquisition*. Rowley, MA: Newbury House.

Kuhl, P., Williams, K., Lacerda, F., Stevens, K., & Lindblom, B. (1992). Linguistic experience alters phonetic perception in infants by 6 months of age. *Science, 255*, 606–608.

Lee, D. S., Lee, J. S., Oh, S. H., Kim, S.-K., Kim, J.-W., Chung, J.-K., et al. (2001). Cross-modal plasticity and cochlear implants. *Nature, 409*, 149–150.

Lenneberg, E. H. (1967). *Biological foundations of language*. New York: Wiley.

Long, M. (1990). Maturational constraints on language development. *Studies in Second Language Acquisition, 12*, 251–285.

Maury, F. (1995). *Les mécanismes intrapsychiques de l'adoption internationale et interraciale: L'adoption des enfants coréens en France*. Doctoral dissertation, Université de Paris VIII.

Mayberry, R. (1998). The critical period for language acquisition and the deaf child's language comprehension: A psycholinguistic approach. *Bulletin d'Audiophonologie: Annales Scientifiques de L'Université de Franche-Comté, 15*, 349–358.

Mayberry, R. I., & Eichen, E. B. (1991). The long-lasting advantage of learning sign language in childhood: Another look at the critical period for language acquisition. *Journal of Memory and Language, 30*, 486–512.

Mayberry, R. I., & Fischer, S. D. (1989). Looking through phonological shape to lexical meaning: The bottleneck of non-native sign language processing. *Memory & Cognition, 17*, 740–754.

Mayberry, R. I., Lock, E., & Kazmi, H. (2001). Linguistic ability and early language exposure. *Nature*, *417*, 38.

McConkey Robbins, A., Burton Koch, D., Osberger, M. J., Zimmerman-Philips, S., & Kishon-Rabin, L. (2004). Effect of age at cochlear implantation on auditory skill development in infants and toddlers. *Archives of Otolaryngology: Head & Neck Surgery*, *130*, 570–574.

Mehler, J., Bertoncini, J., Barrière, M., & Jassik-Gerschenfeld, D. (1978). Infant recognition of mother's voice. *Perception*, *7*, 491–497.

Mehler, J., Jusczyk, P., Lambertz, G., Halsted, N., Bertoncini, J., & Amiel-Tison, C. (1988). A precursor of language acquisition in young infants. *Cognition*, *29*, 143–178.

Moon, C., Cooper, R., & Fifer, W. (1993). Two-day-olds prefer their native language. *Infant Behavior and Development*, *16*, 495–500.

Neville, H., & Bavelier, D. (2002). Human brain plasticity: Evidence from sensory deprivation and altered language experience. *Progress in Brain Research*, *138*, 177–188.

Newport, E. L. (1990). Maturational constraints on language learning. *Cognitive Science*, *14*, 11–28.

Oh, J. S., Jun, S.-A., Knightly, L. M., & Au, T. K.-F. (2003). Holding on to childhood language memory. *Cognition*, *86*, B53–B64.

Pallier, C., & Argenti, A.-M. (2003). Imagerie cérébrale du bilinguisme. In O. Etard & N. Tzourio-Mazoyer (Eds.), *Cerveau et langage: Traité de sciences cognitives* (pp. 183–198). Paris: Hermès Science.

Pallier, C., Bosch, L., & Sebastian-Gallés, N. (1997). A limit on behavioral plasticity in speech perception. *Cognition*, *64*, B9–B17.

Pallier, C., Dehaene, S., Poline, J.-B., LeBihan, D., Argenti, A.-M., Dupoux, E., et al. (2003). Brain imaging of language plasticity in adopted adults: Can a second language replace the first? *Cerebral Cortex*, *13*, 155–161.

Paradis, M. (2004). *A neurolinguistic theory of bilingualism*. Amsterdam: John Benjamins.

Peña, M., Maki, A., Kovačić, D., Dehaene-Lambertz, G., Koizumi, H., Bouquet, F., et al. (2003). Sounds and silence: An optical topography study of language recognition at birth. *Proceedings of the National Academy of Sciences of the United States of America*, *100*, 11702–11705.

Penfield, W., & Roberts, L. (1959). *Speech and brain mechanisms*. Princeton, NJ: Princeton University Press.

Perani, D., & Abutalebi, J. (2005). The neural basis of first and second language processing. *Current Opinion in Neurobiology*, *15*, 202–206.

Pinker, S. (1994). *The language instinct*. New York: Morrow.

Singleton, D., & Ryan, L. (2004). *Language acquisition: The age factor* `(2nd ed.). Clevedon, UK: Multilingual Matters.

Tees, R. C., & Werker, J. F. (1984). Perceptual flexibility: Maintenance or recovery of the ability to discriminate non-native speech sounds. *Canadian Journal of Psychology*, *38*, 579–590.

Tomblin, J., Barker, B., Spencer, L., Zhang, X., & Gantz, B. (2005). The effect of age at cochlear implant initial stimulation on expressive language growth in infants in toddlers. *Journal of Speech, Language, and Hearing Research: JSLHR*, *48*, 853–867.

Vargha-Khadem, F., Carr, L., Isaacs, E., Brett, E., Adams, C., & Mishkin, M. (1997). Onset of speech after left hemispherectomy in a nine-year-old boy. *Brain*, *120*, 159–182.

Ventureyra, V. (2005). *À la recherche de la langue perdue: Etude psycholinguistique de l'attrition de la première langue chez des Coréens adoptés en France*. Doctoral dissertation, Ecole des Hautes Etudes en Sciences Sociales, Paris.

Ventureyra, V., Pallier, C., & Yoo, H.-Y. (2004). The loss of first language phonetic perception in adopted Koreans. *Journal of Neurolinguistics*, *17*, 79–91.

Weber-Fox, C. M., & Neville, H. J. (1996). Maturational constraints on functional specializations for language processing: ERP and behavioral evidence in bilingual speakers. *Journal of Cognitive Neuroscience*, *8*, 231–256.

Werker, J. F., & Tees, R. C. (1984). Cross-language speech perception: Evidence for perceptual reorganization during the first year of life. *Infant Behavior and Development*, *7*, 49–63.

17 A "Birdsong Perspective" on Human Speech Production

Hermann Ackermann and Wolfram Ziegler

Acoustic communication in songbirds and humans is characterized by a variety of "common themes" (Doupe & Kuhl, 1999). Most importantly, the about 4,500+ avian species (Williams, 2004) pertaining to the suborder Passeri (oscines) of the order Passeriformes ("perching birds") rely—besides utilizing innate calls—on a repertoire of songs, learned or at least refined by imitation of the vocalizations of conspecific tutors (Kroodsma, 2005; Catchpole & Slater, 2008). Hearing-dependent vocal learning has otherwise emerged only in parrots and hummingbirds, some bat species, cetaceans (i.e., whales, dolphins)—and humans. As a consequence, the electrophysiological and molecular-biological underpinnings of these remarkable capabilities cannot be adequately studied in monkeys, and the oscine song system should provide a more appropriate model system. The network of distinct brain nuclei, subserving the acquisition and production of stereotyped crystallized songs, roughly divides into two circuits: the *vocal motor pathway* (VMP) and the *anterior forebrain pathway* (AFP). Despite extensive diversity at the behavioral level such as time course of vocal learning, repertoire size, and so on, the "song network" of the oscine brain shows, by and large, an identical organization across species (Brenowitz & Beecher, 2005). Most noteworthy, AFP has been found to be highly preserved across vertebrate taxa and displays striking similarities to the basal ganglia loops of mammals, including humans (Doupe, Perkel, Reiner, & Stern, 2005). Furthermore, VMP and the corticobulbar tracts of our species share some organizational principles such as monosynaptic projections to brainstem centers, engaged in the control of the peripheral vocal apparatus (Nottebohm, Stokes, & Leonard, 1976; for recent reviews see Jarvis, 2004a, 2004b; Jarvis, chapter 4, this volume; Bolhuis, Okanoya, & Scharff, 2010). Against this background, the present chapter aims at a comparison of the oscine "song brain" with the cerebral network of human speech production—as defined on the basis of clinical and, more recently, functional imaging data.

Acoustic Communication in Songbirds and Humans: Common Themes

The spoken language of our species engages a variety of knowledge systems and computational capabilities (e.g., Whitney, 1998; Jackendoff, 2003):

• A finite repertoire of sound categories (phonemes) and the respective combinatorial constraints (phonology)
• A vocabulary of word forms, which may form composite entities (morphology)
• A set of rules allowing for the construction of phrases and sentences from words (syntax)
• A language-specific rhythmic and melodic organization of words and phrases (linguistic prosody)
• The assignment of meaning to words and sentences (semantics)
• The modulation of the meaning of verbal utterances during conversation by a speaker's intentions or by social context (pragmatics)

Among the various subcomponents of spoken language, "the phonology (sound structure), the rules of ordering sounds, and perhaps the prosody (in the sense that it involves control of frequency, timing, and amplitude) are the levels at which birdsong can be most usefully compared with language, and more specifically with spoken speech" (Doupe & Kuhl, 1999, p. 573). Thus, any extrapolation from the "song circuitry" of oscine brains—relatively well characterized at the neuroanatomical, molecular-genetic, and electrophysiological level—to the less accessible cerebral mechanisms of spoken language should be, by and large, restricted to phonological aspects of acoustic communication.

Avian song and human speech show, nevertheless, crucial differences in design features even at the level of sound structure: spoken languages are based on a digital code, characterized by a set of discrete meaningless units—that is, speech-sound categories (phonemes)—and a set of combinatorial (phonotactic) constraints (Burling, 2005; Figure 17.1). Principally, the vocabularies derived from this knowledge base enable us to refer to anything "we can think about," whether present or displaced, whether real or imagined. A further set of combinatorial rules (syntax) allows for the construction of sentences from lexical items, portraying propositions about actual or possible states of the world. These two levels of sequential organization ("duality of patterning") are considered a unique trait of our species, distinct from any system of animal communication (Hockett, 1966). By contrast, the extant data on the communicative effects of oscine vocalizations point at "the rather simple hypothesis that song has the two main functions of mate attraction and territorial defence" (Catchpole & Slater, 2008, p. 235). As a consequence, oscine vocal signals appear to be tightly and exclusively bound to motivational states—for example, reflecting the activity level of circulating steroid hormones (Bottjer & Johnson, 1997).

The display of sociobiological traits, affective states, and emotional attitudes plays a crucial role during social interactions in our species as well. A speaker's mood and intentions are conveyed not only by words and phrases, such as "I'm feeling really fine today," but also by the "music" accompanying these words (i.e., the prosody and vocal signature of spoken language; Kreiman & Sidtis, 2011). Variation in the pitch, loudness, rate, and rhythm of stressed and unstressed segments provides information about the feelings of a speaker and her or his attitudes toward the conversation partner. Furthermore, voice quality signals physical fitness and attractiveness (Hughes, Harrison, & Gallup, 2002). Hence human speech not only portrays referential and propositional information, but also encompasses speaker-related, sociobiological, motivational, and affective markers. Unlike propositional language, based on a digital linguistic code, the "paralinguistic" modulation of verbal utterances has a graded (i.e., analog) signal structure (Burling, 2005). As concerns both signal structure and semiotic content, oscine song compares to the sociobiological and affective-prosodic aspects of spoken language—the analog channel of human acoustic communication—rather than the digital-phonological domain (i.e., the segmental-phonemic architecture of speech). Variability in repertoire size or sequential complexity of oscine songs appears to provide graded sound source–related information on, for instance, a bird's physical fitness (Catchpole & Slater, 2008; but see Kroodsma, 2005, pp. 127–128).

More recently, the term *syntax* has been applied to the vocalizations of the Bengalese finch since the sequential organization of the songs of this species can be described in terms of a finite-state model of note-to-note transitions (Okanoya, 2004; Okanoya, chapter 11, this volume). The functional role of vocal sequencing in Bengalese finches nevertheless appears restricted to the domains of territorial defense and mate attraction. Reduced linearity of the songs—as determined on the basis of a second-order Markovian model—has been reported, for example, to induce higher serum estradiol levels in female listeners (Okanoya, 2004; Berwick, Okanoya, Beckers, & Bolhuis, 2011; see Beckers, Bolhuis, Okanoya, & Berwick, 2012, for a recent discussion).

Despite structural and semiotic discrepancies, human speech-sound inventories and oscine song repertoires display interesting developmental similarities: In both instances, the acquisition of mature vocal behavior

- Requires exposure to the acoustic signals of conspecific adults
- Is influenced by innate predispositions of the central auditory system
- Occurs—within some limits—more easily during distinct critical (sensitive) periods of the life cycle

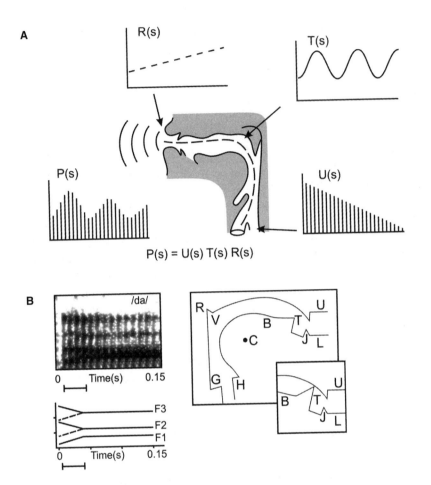

A

R(s)

T(s)

P(s)

U(s)

P(s) = U(s) T(s) R(s)

B

/da/

0   Time(s)   0.15

F3
F2
F1

0   Time(s)   0.15

R   V   B   T   U
         •C   J   L
G   H

B   T   U
         J   L

Figure 17.1
A discrete set of articulatory organs subserves the acoustic encoding of the phonological structure of verbal utterances. (A) Besides a few exceptions, modulation of expiratory airflow at the level of the vocal folds (larynx) and several supralaryngeal structures (pharynx, velum, tongue, and lips) gives rise to the various speech sounds of human languages. Oscillations of the vocal folds generate a laryngeal source signal U(s) of a harmonic structure (fundamental frequency plus "overtones"). Subsequently, U(s) is filtered by the resonance characteristics of the supralaryngeal cavities T(s) and the radiation function of the vocal tract R(s). As a consequence, a vowel sound—transmitted from the mouth—encompasses at the acoustic level a fairly distinct pattern of "peaks and troughs" (P(s) = formant structure) of its spectral energy distribution ("source-filter model of speech production"; e.g., Ladefoged, 2005). (B) As a rule, constrictions of the vocal tract at distinct locations are associated with the various consonants of a language system—for example, occlusion of the oral cavity via contact of the tip of the tongue with the alveolar ridge of the upper jaw (behind the incisors) in case of /d/- and /t/-sounds (right panel with insert; T = tip of the tongue, B = tongue body, U = upper lips, L = lower lips, J = lower jaw with teeth). Such maneuvers have a specific impact on the

- Shows strong interaction with social context (Doupe & Kuhl, 1999; Kuhl, 2003; Bolhuis et al., 2010)

These phenomena thus warrant a comparative analysis of the cerebral organization of acoustic communication in oscine species and humans and can be expected to further elucidate the functional-neuroanatomical basis of speech motor learning in humans. First, however, the inborn calls of birds and primates, including the nonverbal vocalizations of our species, will be briefly addressed — serving as a background for the discussion of learned behavior within the acoustic domain.

Brain Networks Subserving Innate Vocalizations in Songbirds and Primates

A repertoire of calls, varying in size, can be found in all avian species (Marler, 2004). Unlike songs, these vocalizations, as a rule, are characterized by shorter duration, a "monosyllabic" structure, less tonal virtuosity, if any, and a genetically specified sound pattern, not shaped by vocal-learning processes. At the functional level, however, "The calls of birds actually range far more widely than songs, approaching in diversity and complexity the calls of primates" (Marler, 2004, p. 132). These inherited communicative acoustic signals seem to be mediated predominantly by the *dorsomedial subregion* (DM) of *nucleus intercollicularis* (ICo) and the central mesencephalic grey (Dubbeldam & Den Boer-Visser, 2002). For example, low-threshold microstimulation of ICo has been found to elicit simple call-like vocalizations in adult male zebra finches and canaries (Vicario & Simpson, 1995).

Human acoustic communication also encompasses nonverbal affective vocalizations such as laughter or cries, also characterized as well by a largely inherited acoustic structure and an inherent sound-to-meaning relationship (Wild, Rodden, Grodd, & Ruch, 2003). Clinical observations suggest that the motor-control mechanisms involved in these intrinsic vocalizations are separate from speech production. For example, patients with bilateral damage to the corticobulbar tracts (i.e., the efferent projections of the primary motor cortex to the caudal cranial nerve nuclei of the brainstem) may show preserved affective vocalizations such as laughter or crying, despite compromised voluntary vocal-tract activities — in its extreme a complete inability to speak or to sing ("automatic-voluntary movement dissociation";

---

temporal course of spectral energy distribution. Thus, syllables comprising a voiced stop consonant followed by a vowel are characterized by distinct formant transients, or up- and downward shifts of formant structure across time (left panels show formant transients of / da/: upper-left panel = spectrogram, lower-left panel = schematic display, dashed lines = formant transients of syllable /ba/; figures adapted from Kent & Read, 2002; for more information see http://www.phys.unsw.edu.au/jw/voice.html).

Mao, Coull, Golper, & Rau, 1989; for a review see Ackermann & Ziegler, 2010). In nonhuman primates, projections from the rostral part of the mesial wall of the frontal lobes via midbrain periaqueductal grey and adjacent tegmentum (i.e., the mammalian analogs of DM and ICo) to a pontine vocal-pattern generator and, finally, to the brainstem cranial nerve nuclei, subserve the control of motor aspects of vocal behavior—bypassing the lower sensorimotor and adjacent premotor cortex (Jürgens, 2002; Hage & Jürgens, 2006). This circuit represents the most plausible functional-neuroanatomical substratum of nonverbal vocal behavior in our species as well. Across avian and primate taxa, production of genetically specified acoustic communicative signals thus seems to depend on similar brain networks.

Cerebral Correlates of Mature (Crystallized) Song Production in Oscine Species

Organization of the Vocal Motor Pathway (VMP)

The production of learned vocalizations in songbirds critically depends on the VMP circuitry, extending from HVC (proper name; for more on oscine brain nomenclature see Bolhuis et al., 2010; Jarvis, chapter 4, this volume; Gobes, Fritz, & Bolhuis, chapter 15, this volume) via the *robust nucleus of the arcopallium* (RA) to both the tracheosyringeal portion of the hypoglossal nucleus and respiration-related areas of the brainstem (Farries & Perkel, 2008). At any level, damage to VMP significantly compromises song production (Nottebohm et al., 1976). Bilateral HVC lesions even may result in "silent song"—that is, a complete inability to sing concomitant with seemingly preserved motivation for vocal behavior. Furthermore, the connections of RA—the oscine analog to the human motor cortex—to its brainstem targets show a monosynaptic and topographic (myotopic) organization (Wild, 2008). Besides lesion data, electrophysiological studies (e.g., microstimulation of VMP structures) provide further evidence for a critical contribution of this circuit to motor aspects of song production (Vu, Mazurek, & Kuo, 1994; Vicario & Simpson, 1995).

By contrast to the mammalian larynx (see below), the oscine syrinx includes two "voice boxes" with exclusively ipsilateral innervation via the respective brainstem centers.[1] During song production, the activities of the two sound sources must be precisely adjusted to each other (Suthers & Zollinger, 2008). Among other things, vocal motor control involves syringeal gating processes, giving rise to lateralization effects at the peripheral level (for a review see Suthers, 1997). By contrast, the oscine "song circuit" lacks any morphological asymmetries and any commissural connections above the hypoglossal nucleus. An interhemispheric network at the brainstem level, therefore, is assumed to subserve the coordination of the vocal apparatus (see, e.g., Figure 11 in Vu et al., 1994, or Figure 1 in Ashmore, Wild, & Schmidt, 2005), and nonauditory motor feedback loops might synchronize neural activity at the level of HVC and RA during singing (Vu et al., 1994).

The distinct contribution of the various VMP components to the specification of learned motor routines is still a matter of debate. Hierarchical models assume HVC and RA to sequence syllabic and subsyllabic units, respectively (Yu & Margoliash, 1996; but see Fee, Kozhevnikov, & Hahnloser, 2004, for an alternative concept). More specifically, HVC might participate in the sequencing of syllables and in the control of song tempo (Vu et al., 1994), whereas RA appears to engage in the scaling of acoustic parameters such as pitch (Sober, Wohlgemuth, & Brainard, 2008).

Participation of the Anterior Forebrain Pathway (AFP) in Crystallized Song Production

By contrast to VMP, the AFP circuitry provides an indirect connection between HVC and RA (Farries & Perkel, 2008). This loop arises from a separate subpopulation of HVC neurons—as compared to the neurons directly targeting RA—and projects via Area X (a specialized region of the anteromedial striatum) and the medial part of the *dorsolateral anterior thalamic nucleus* (DLM) to the *lateral magnocellular nucleus of the anterior nidopallium* (LMAN). Area X and LMAN are considered the oscine equivalents of the mammalian basal ganglia (striatum and pallidum) and their cortical target areas, respectively (Brainard & Doupe, 2002; Doupe et al., 2005). AFP not only makes an essential contribution to hearing-dependent vocal learning (see below), but also has a significant—though more subtle—influence on mature song production. For example, bilateral cochlear removal (deafening) in adult male canaries and zebra finches gives rise to slow, but substantial deterioration ("decrystallization") of the temporal and/or spectral structure of the learned syllable sequences (Nordeen & Nordeen, 1992; Roy & Mooney, 2007). Bilateral damage to LMAN, however, may prevent song decrystallization due to bilateral deafening (Brainard & Doupe, 2000, 2001). Such observations, concomitant with electrophysiological data (e.g., Roy & Mooney, 2007), strongly suggest AFP to be engaged in auditory-perceptual—vocal-motor interactions relevant for the maintenance of crystallized songs.

The contribution of AFP to vocal plasticity in adult oscine birds apparently is not restricted to the evaluation of reafferent auditory input. As a rule, the syllables or motifs of memorized adult song repertoires are characterized by a rather stereotyped acoustic structure across successive renditions, even over periods of months or years. Nevertheless, crystallized vocalizations may display some variation with respect to temporal organization, fundamental frequency contour, or the sequential arrangement of song units—depending, for example, on social context. In particular, the absence (undirected song) or presence of a female (directed or female-directed song) has a significant impact on vocal behavior, including the structure of the acoustic signal. For example, male zebra finches show enhanced song tempo in the presence of females (Cooper & Goller, 2006). Since faster singing was found to be correlated with autonomic responses, context-dependent modification of tempo appears to be mediated by shifts in motivational states. Similarly, female-directed song of male Bengalese finches exhibits significantly less F0 variability across renditions of syllables of the same type than undirected vocal behavior (Sakata, Hampton, & Brainard, 2008). The impact of social signals and their associated motivational states on vocal behavior seems to depend on the AFP circuitry since, for example, bilateral electrolytic damage to LMAN eliminates prelesional differences of F0 variability between undirected and directed song in male zebra finches (Kao & Brainard, 2006).

Most noteworthy, human speech production shows some similarities to these trade-offs between invariance and plasticity of vocal output in oscine species. On the one hand, the phonological pattern of a language must demonstrate a high degree of stability and auditory-motor entrenchment since otherwise the speech code could not subserve acoustic communication. On the other hand, there is "space" for extensive phonetic variability during speech production, and speakers are known to exploit these resources—in terms of a flexible adjustment of articulatory processes to a conversation partner (e.g., Pardo, 2006). Such high-level adaptive mechanisms result, eventually, in long-term diachronic changes of the sound structure of a language (Delvaux & Soquet, 2007) and depend, presumably, on auditory-motor crosstalks along temporo-parieto-frontal cortical connections (Duffau, 2008), Later in this chapter, lower-level adaptive mechanisms will be briefly addressed that appear to be engaged in the implicit regulation of vocal loudness during conversation. This mechanism has been assumed to rely, in analogy to the zebra finch model, on basal ganglia loops.

Cerebral Correlates of Motor Aspects of Spoken Language in Humans

Cortical Representation of Vocal-Tract Muscles: Electrophysiological Data

Electrical stimulation experiments during brain surgery found the primary motor cortex to encompass a comprehensive topographic representation of the human body (see Woolsey, Erickson, & Gilson, 1979, for a review of the earlier work). More recent studies based on noninvasive stimulation techniques were able to corroborate this concept of a "motor homunculus" (e.g., Rödel, Laskawi, & Markus, 2003). Systematic intraoperative "speech-area mapping" studies revealed electrical stimulation of precentral and—less frequently—postcentral "orofacial areas" at either side of the brain to elicit vocalizations in terms of a "sustained or interrupted vowel cry, which at times may have a consonant component" (Penfield & Roberts, 1959, pp. 119–121). Besides the lateral surface of the hemispheres, this procedure elicits involuntary vocalizations, dysfluencies, or slowing of spoken language at the level of the supplementary motor area (SMA) within the medial wall of the frontal lobes (Penfield & Roberts, 1959; Fried et al., 1991; Figure 17.2). These effects are predominantly, but not exclusively, bound to the language-dominant hemisphere. Apart from the

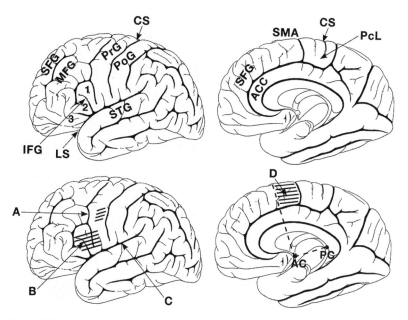

Figure 17.2

Cortical areas engaged in speech motor control. *Upper row*. Major gyri and sulci of the lateral (left) and medial aspect (right) of the left/right hemisphere: IFG = inferior frontal gyrus, segregating into an opercular (1), a triangular (2), and an orbital part (3), MFG/SFG = middle/superior frontal gyrus, PrG/PoG = pre-/postcentral gyrus, STG = superior temporal gyrus, LS = lateral sulcus (sylvian fissure), CS = central sulcus (Rolandic sulcus). The posterior part of SFG houses at its medial side the so-called supplementary motor area (SMA), the dashed line perpendicular to a plane through anterior (AC) and posterior commissure (PC) roughly corresponds to the anterior SMA limit (= SMA proper), PcL = paracentral lobule, ACC = anterior cingulate cortex.

Lower row. The shaded areas (horizontal lines) refer to the cortical regions engaged in speech motor control: bilateral primary motor cortex (A), opercular part of left IFG and lower-left PrG (B), left SMA proper (D). In addition, the rostral part of the intrasylvian cortex (anterior insula) in the depth of the lateral sulcus (LS) is assumed to contribute to speech motor control (C; see Ackermann & Ziegler, 2010, for more details).

dorsal bank of the cingulate sulcus, electrical stimulation of the anterior cingulate cortex (ACC)—located ventral and rostral to SMA—fails to evoke involuntary vocal behavior in humans (Penfield & Roberts, 1959). By contrast, simple speech tasks, like the production of prespecified verbal responses to an auditory stimulus, have been reported to give rise to hemodynamic activation of distinct ACC subregions (Paus, Petrides, Evans, & Meyer, 1993). As compared to precentral areas, SMA responses may show a more complex acoustic pattern—for example, in terms of loudness and pitch fluctuations. Production of "intelligible words," however, was never observed during electrical cortical stimulation. Since, furthermore, the observed involuntary "cries," by and large, do not display the perceptual qualities of (well-articulated) speech sounds, a direct impact on the motor execution apparatus rather than higher-order phonological operations must be assumed.

Apart from vocalizations resembling simple speech utterances, electrical surface stimulation may cause "an inability to vocalize spontaneously" ("speech arrest," Penfield & Roberts, 1959; see, e.g., Pascual-Leone, Gates, & Dhuna, 1991, for a more recent approach, based on transcranial magnetic stimulation). At the level of the right hemisphere, elicitation of speech arrest is exclusively restricted to vocalization-related areas. By contrast, the respective susceptible zone of the language-dominant side extends to the *inferior frontal gyrus* (IFG; Ojemann, 1994; Figure 17.2). Principally, speech arrest could reflect disruption of either speech motor control mechanisms ("motor speech arrest") or preceding higher-order processes of language formulation such as the generation of the sound structure of verbal utterances. Apart from pre- and postcentral areas, systematic exploration of the perisylvian cortex found motor speech arrest to be limited to a small segment of the posterior IFG. At that location, electrical stimuli were found to interrupt all verbal utterances, irrespective of task condition, and to compromise even the ability to mimic single orofacial gestures, indicating this area serves as a "final motor pathway" of the cerebral circuitry engaged in speech production. Taken together, stimulation studies revealed at least three distinct cortical areas to be engaged in motor aspects of speech production: the mesiofrontal SMA, the border zone of IFG and adjacent precentral gyrus, as well as the primary motor cortex (Figure 17.2).

Speech Motor Deficits Subsequent to Acquired Brain Lesions

Corticobulbar System

Bilateral damage to the lower primary motor cortex and/or the respective efferent projections to caudal cranial nerve nuclei may give rise to the syndrome of spastic (paretic) dysarthria, characterized by slow speaking rate, reduced range of orofacial movements, velar insufficiency, and hyperadduction of the vocal folds (Duffy, 2005; Ackermann, Hertrich, & Ziegler, 2010). In its extreme, complete speechlessness (anarthria/aphonia) may develop. Emotional vocal behavior remains intact as long as the efferent bulbar projections of limbic structures and mesiofrontal motor areas are spared (see above). Unilateral dysfunctions of the upper motor neuron, as a rule, give rise to mild and transient speech motor deficits only. With the exception of, by and large, lower face muscles, the caudal brainstem nuclei subserving the innervation of the vocal-tract apparatus receive input from both cerebral hemispheres. This organizational pattern presumably allows for the recovery of phonatory as well as articulatory functions within a typical time interval of several days to a few weeks. Electrophysiological and functional imaging data indicate preexisting uncrossed motor pathways of the respective intact hemisphere to participate in the compensation of the initial speech motor deficits (Riecker, Wildgruber, Grodd, & Ackermann, 2002).

Geschwind (1969) assumed left-hemisphere dominance of speech production to extend, beyond linguistic aspects of language processing, to the cortical motor areas "steering" vocal-tract movements. The neural control mechanisms of speech production thereby avoid lateral competition for the innervation of midline vocal-tract muscles, preventing the emergence of asynchronous input to the relevant brainstem nuclei. However, clinical investigations found speech motor deficits in a substantial proportion of patients with right-hemisphere corticobulbar dysfunctions (e.g., Urban et al., 2006). Electrophysiological investigations revealed, furthermore, the muscle fibers of either side of the tongue to receive the same information from each hemisphere in terms of inhibitory and facilitatory control signals (Muellbacher, Boroojerdi, Ziemann, & Hallett, 2001). Rather than reflecting strictly lateralized control mechanisms, spoken language thus appears to engage a bilateral corticobulbar network, conveying identical signals to the caudal brainstem nuclei at either side.

Ventrolateral-Frontal and Intrasylvian Cortex

The syndrome of acquired apraxia of speech (AOS) encompasses phoneme errors, distortions of consonants as well as vowels, and a nonfluent "speech stream," exhibiting trial-and-error groping behavior (Ziegler, 2008). Unlike the dysarthrias, basic motor functions such as muscle tone and force generation are spared. As a consequence, higher-order deficits of speech motor control are assumed in this disorder. More specifically, psycholinguistic studies point at an impaired capability of AOS patients to "plan" speech movements at the level of syllable-sized or even larger linguistic units. Speech motor control is thus often assumed to encompass two hierarchically organized processing stages: the planning of a sequence of phonetic gestures followed by the specification and execution of single gestures. AOS has most often been observed to arise from ischemic infarctions of the left medial cerebral artery and, hence, must be considered a syndrome of the language-dominant hemisphere. The ventrolateral-frontal premotor cortex (posterior IFG or opercular precentral areas) and/or the anterior insula in the depth of the lateral sulcus are considered the most likely neuroanatomical substrates of this constellation.

Medial Wall of the Frontal Lobes

Unilateral—predominantly left-sided—SMA lesions sometimes give rise to reduced spontaneous verbal behavior, in the absence of any central-motor disorders of vocal-tract muscles or any deterioration of higher-order language functions. Furthermore, dysfluent (i.e., stuttering-like) speech utterances in terms of sound prolongations and syllable repetitions have been observed in subjects with mesiofrontal lesions of the dominant hemisphere (see Ackermann & Ziegler, 2010, for a review). In contrast to unilateral disorders, damage to the medial wall of both hemispheres, encroaching on SMA and its projections to ACC, may elicit the syndrome of akinetic mutism, characterized by a lack of self-initiated motor activities, including speech production. In view of these clinical data, the mesiofrontal cortex appears to mediate motivational aspects of verbal motor behavior, and SMA, more specifically, has been assumed to operate as a "starting mechanism of speech," required for the initiation and maintenance of fluent verbal utterances (Botez & Barbeau, 1971).

Corticosubcortical Motor Loops

Early stages of idiopathic Parkinson's disease have been considered the most feasible paradigm of a striatal dysfunction, although the underlying neurodegenerative process emerges outside the basal ganglia. The characteristic motor signs (i.e., akinesia, bradykinesia, hypokinesia, and rigidity) reflect presynaptic dopamine depletion within striatal target structures. Basically the same pathomechanisms are assumed to act on the vocal-tract muscles (Duffy, 2005). As a consequence, the speech motor deficits of Parkinson's disease are usually called hypokinetic/rigid dysarthria. Monopitch, reduced stress, monoloudness, and imprecise consonants represent the most salient perceptual signs of this disorder (Duffy, 2005, p. 196). Damage to the thalamic "relay-stations" of the basal ganglia loops or disruption of midbrain dopaminergic pathways may give rise to a similar constellation (Ackermann & Ziegler, 2010). In accordance with the auditory-perceptual features of Parkinsonian dysarthria, acoustic measurements revealed lower overall speech volume during reading and conversation tasks as well as reduced variation of fundamental frequency and speech intensity during sentence productions (e.g., Ho, Iansek, & Bradshaw, 1999). Thus, the loss of prosodic variability of spoken language represents a core feature of the observed profile of speech motor deficits. As a consequence, these patients may—because of a flat and soft voice—appear to suffer from depression, in spite of uncompromised mood.

Perceptual deficits do not pertain to the classic catalog of the signs and symptoms of Parkinson's disease. There is, however, some preliminary evidence for impaired higher-order auditory functions. Whereas these patients converse with a softer voice than healthy people, they are, nevertheless, able to elevate their vocal intensity to normal levels when consciously focusing on speech production. By contrast, automatic gain adjustment to implicit cues has been found significantly compromised (Ho et al., 1999). Furthermore, investigations of volume perception revealed PD patients to overestimate the loudness of their own speech, both during reading and conversation tasks (Ho, Bradshaw, & Iansek, 2000). These data can be interpreted to reflect compromised evaluation of reafferent speech-related auditory input: subjects overestimate the loudness of their own voice and, as a consequence, underscale, in the absence of explicit cues, the volume of speech output.[2]

Cerebellar dysfunctions in humans give rise to distinct speech motor deficits, called ataxic dysarthria (Ackermann, 2008). Unlike spoken language, there is so far no evidence for a significant contribution of this organ to birdsong production, conceivably due to a different mode of vocal output sequencing. In oscine species, the syllables of a phrase (i.e., the most plausible motor control units) are separated by minibreaths to replenish air supply (Suthers & Zollinger, 2008). By contrast, articulatory units of verbal utterances are not organized like beads on a string, but overlap in a context-dependent manner (coarticulation). The cerebellum, subserving the online sequencing of syllables into fast, smooth, and rhythmically organized larger utterances such as words and phrases, presumably participates in the control of coarticulation effects (Ackermann, 2008).

Functional Imaging Studies

Positron emission tomography (PET) or functional magnetic resonance imaging (fMRI) have been exploited for the investigation of the cerebral networks underlying speech production. Using lexical and nonlexical mono- or polysyllabic items as test materials, these studies have found the brain network of motor aspects of speech production to encompass the mesiofrontal cortex, opercular parts of the precentral gyrus and posterior IFG (Broca's area), the anterior insula at the floor of the lateral sulcus, the "mouth region" of the primary sensorimotor cortex, the basal ganglia, the thalamus, and the cerebellar hemispheres (for a review see Ackermann & Riecker, 2010). By and large, clinical and functional imaging data converge on the same brain circuitry of motor aspects of speech production (Figure 17.3).

Brain Structures Engaged in the Acquisition of Learned Vocal Behavior in Songbirds and Humans

It is well established that AFP displays striking morphological and physiological similarities to mammalian basal ganglia loops, including our species (Doupe et al., 2005). Lesion studies revealed AFP to represent an essential prerequisite to

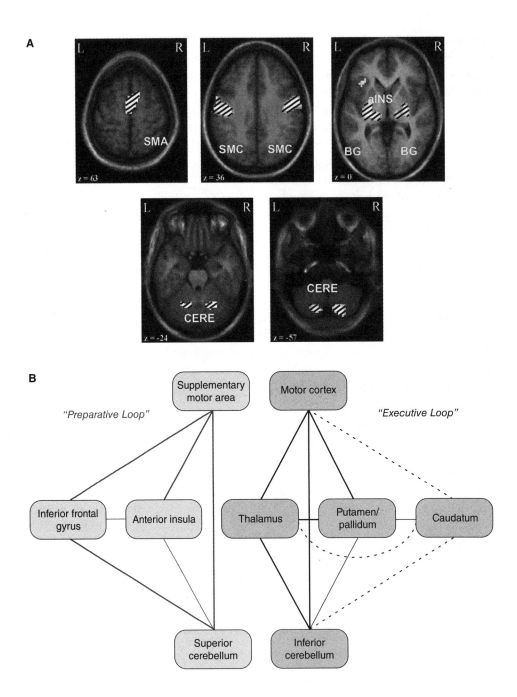

hearing-dependent vocal learning. For example, damage to at least two major AFP components (i.e., LMAN and Area X) during early stages of vocal development in juvenile passerine birds disrupts vocal learning (e.g., Scharff & Nottebohm, 1991). These results would suggest that in humans damage to the basal ganglia soon after birth, at least prior to the onset of babbling, hampers acquisition of the speech-sound inventory of an infant's mother tongue. However, acute brain lesions in full-term neonates around delivery or during early infancy—for example, subsequent to infectious diseases or asphyxia (hypoxia)—often also disrupt the corticospinal and corticobulbar tracts. In the presence of a tetraplegia concomitant with spastic dysarthria, the specific contribution of the basal ganglia to the observed articulatory/phonatory deficits cannot easily be determined. Nevertheless, there is a distinct—but rare—variant of acquired perinatal brain dysfunctions that predominantly affects subcortical brain nuclei. Under specific conditions, such as uterine rupture or umbilical cord prolapse, compromised oxygen supply in an infant prior to or during delivery (birth asphyxia) may give rise to predominant damage to the basal ganglia and the thalamus (and eventually brainstem), concomitant with relative preservation of the cerebral cortex and underlying white matter. So far, only a single clinical study has addressed the speech-production capabilities of children with bilateral lesions of the basal ganglia and thalamus subsequent to birth asphyxia (Krägeloh-Mann et al., 2002). Obviously, 9 subjects out of a group of 17 patients were completely unable to produce articulate speech at age 2 to 9 years. Six of the remaining eight individuals were reported to have suffered from "dysarthria" at clinical examination (3–17 years). Unfortunately, a more detailed evaluation of cranial nerve functions, verbal and nonverbal vocal behavior, and so on, was not conducted. It thus remains to be established whether the "mute" subjects had been unable to learn the phonological code of their mother tongue—in spite of, by and large, uncompromised corticobulbar tracts. Nevertheless, however, functional imaging studies have provided some preliminary evidence for a specific involvement of the left-hemisphere

◄ Figure 17.3

"Minimal brain network" of speech motor control. (A) The displays show the hemodynamic main effects (functional magnetic resonance imaging, fMRI) during repetitions of the syllable /ta/, displayed on transverse sections of the averaged anatomical reference images (z = distance to the intercommissural plane; L = left, R = right hemisphere). Significant responses emerge within SMA, bilateral sensorimotor cortex (SMC), bilateral basal ganglia (BG), left anterior insula (aINS), left inferior frontal gyrus (not shown), and both cerebellar hemispheres (CERE). (B) Quantitative functional connectivity analyses point at a segregation of the "minimal brain network" of speech motor control into two subsystems, engaged in preparation (left side) and execution (right) of vocal-tract movement sequences (bold lines = correlation coefficient > 0.9, thin lines = 0.75–0.9, low and intermediate correlations not depicted; for further details see Ackermann & Riecker, 2010).

putamen in motor aspects of second language learning (Klein, Zatorre, Milner, Meyer, & Evans, 1994; Frenck-Mestre, Anton, Roth, Vaid, & Viallet, 2005).

Conclusions

Humans share with other primates a limbic-mesencephalic channel of acoustic communication, subserving the control of nonverbal vocalizations, and the innate calls of avian taxa appear to depend on analogous pathways. A second channel of acoustic communication in our species, encompassing the SMA, primary motor areas, ventrolateral-frontal and intrasylvian cortex, as well as corticosubcortical loops, supports learned spoken language. In spite of considerable differences in signal structure (analog versus digital code) and semiotic content (motivational versus referential information), the cerebral network of human speech motor control shows several parallels to the brain circuits involved in the acquisition/production of learned songs in oscine species. Such commonalities loom even larger against the background of the entirely different cerebral organization of inherited calls in birds and primates, including humans.

Most importantly, the brain networks underlying both human speech and oscine song production encompass three presumably hierarchical processing stages associated with the control of ongoing vocal behavior: (i) premotor-frontal areas / HVC, (ii) primary motor cortex / RA, and (iii) brainstem centers. Primary motor regions and RA display a topographic organization of muscle groups and project in a monosynaptic fashion to their brainstem targets—apart from respiratory motor neurons. Furthermore, premotor structures (i.e., human ventrolateral-frontal / intrasylvian cortex and oscine HVC) seem to code vocal motor actions of higher "grain-size" such as rhythmic entities like syllables or even larger units, although strict lateralization effects at this control level are restricted to our species. The topographically organized sensorimotor cortex, connected via monosynaptic projections to lower motor neurons, has been assumed to subserve the "fractionation of movements" (Brooks, 1986). Movement fractionation probably represents a prerequisite for the construction of new movement sequences by the imitation—within the auditory domain—of a tutor.

The specific contribution of the basal ganglia to the acquisition of spoken language remains to be further elucidated, in particular, future investigations of infants with perinatal basal ganglia lesions should include more detailed analyses of their speech motor deficits. Besides requiring vocal learning, the maintenance of a mature song repertoire in oscine species depends on the continuous evaluation of a bird's own vocalizations. There is preliminary evidence that in humans the basal ganglia also participate in speech-related processing of reafferent auditory information. Most remarkably, both LMAN lesions, the target structure of oscine AFP, and striatal dysfunctions in PD patients cause a loss of acoustic variability in vocal behavior. In oscine species, residual variability of skilled (vocal) behavior might help a bird to explore motor space and, as a consequence, to support continuous optimization of performance ("adaptive plasticity"; Turner & Brainard, 2007). According to this model, the basal ganglia actively "inject" behavioral variation into the oscine vocal motor system, paving the way for adaptive behavioral changes. Adaptive plasticity based on incidental variation might also represent an important mechanism of human vocal interactions since it decreases the social distance between communication partners (Meltzoff, 2002) and may—in the long run—contribute to diachronic changes in the sound repertoire of a language community (Wedel, 2006). So far, however, there is no evidence available that the basal ganglia of our species—like those of passerine birds—mediate adaptive plasticity by actively "injecting behavioral variation" into the speech motor system. Furthermore, auditory-motor integration and adaptive changes within the domain of human speech production depend crucially—over and above any potential contribution of the basal ganglia—on a superordinate cortical network, including the sensorimotor and auditory association areas.

Notes

1. Apart from the cerebral control mechanisms, songbirds and humans differ at the peripheral level of acoustic communication. The vocal-tract mechanisms of sound generation in oscine species will not be further addressed here (for more details see Jarvis, chapter 4, this volume; Fee & Long, chapter 18, this volume; Gobes et al., chapter 15, this volume). Regarding our species, Figure 17.1 provides a brief sketch of the phonatory and articulatory operations engaged in human speech production.

2. Besides Parkinson's disease, Huntington's chorea, an autosomal-dominant hereditary disease, represents a further paradigm of striatal disorders. However, the speech motor deficits of this disease have rarely been analyzed and will not be considered further (see Ackermann & Ziegler, 2010, for more details).

References

Ackermann, H. (2008). Cerebellar contributions to speech production and speech perception: Psycholinguistic and neurobiological perspectives. *Trends in Neurosciences*, *31*, 265–272.

Ackermann, H., Hertrich, I., & Ziegler, W. (2010). Dysarthria. In J. S. Damico, N. Müller, & M. J. Ball (Eds.), *The handbook of language and speech disorders* (pp. 362–390). Malden, MA: Wiley-Blackwell.

Ackermann, H., & Riecker, A. (2010). Cerebral control of motor aspects of speech production: Neurophysiological and functional imaging data. In B. Maassen & P. van Lieshout (Eds.), *Speech motor control: New developments in basic and applied research* (pp. 117–134). Oxford: Oxford University Press.

Ackermann, H., & Ziegler, W. (2010). Brain mechanisms underlying speech motor control. In W. J. Hardcastle, J. Laver, & F. E. Gibbon (Eds.), *The handbook of phonetic sciences* (2nd ed., pp. 202–250). Malden, MA: Wiley-Blackwell.

Ashmore, R. C., Wild, J. M., & Schmidt, M. F. (2005). Brainstem and forebrain contributions to the generation of learned motor behaviors for song. *Journal of Neuroscience, 25,* 8543–8554.

Beckers, G. J. L., Bolhuis, J. J., Okanoya, K., & Berwick, R. C. (2012). Birdsong neurolinguistics: Songbird context-free grammar claim is premature. *NeuroReport, 23,* 139–145.

Berwick, R. C., Okanoya, K., Beckers, G. J. L., & Bolhuis, J. J. (2011). Songs to syntax: The linguistics of birdsong. *Trends in Cognitive Sciences, 15,* 113–121.

Bolhuis, J. J., Okanoya, K., & Scharff, C. (2010). Twitter evolution: Converging mechanisms in birdsong and human speech. *Nature Reviews: Neuroscience, 11,* 747–759.

Botez, M. I., & Barbeau, A. (1971). Role of subcortical structures, and particularly of the thalamus, in the mechanisms of speech and language: A review. *International Journal of Neurology, 8,* 300–320.

Bottjer, S. W., & Johnson, F. (1997). Circuits, hormones, and learning: Vocal behavior in songbirds. *Journal of Neurobiology, 33,* 602–618.

Brainard, M. S., & Doupe, A. J. (2000). Interruption of a basal ganglia—forebrain circuit prevents plasticity of learned vocalizations. *Nature, 404,* 762–766.

Brainard, M. S., & Doupe, A. J. (2001). Postlearning consolidation of birdsong: Stabilizing effects of age and anterior forebrain lesions. *Journal of Neuroscience, 21,* 2501–2517.

Brainard, M. S., & Doupe, A. J. (2002). What songbirds teach us about learning. *Nature, 417,* 351–358.

Brenowitz, E. A., & Beecher, M. D. (2005). Song learning in birds: Diversity and plasticity, opportunities and challenges. *Trends in Neurosciences, 28,* 127–132.

Brooks, V. B. (1986). *The neural basis of motor control.* New York: Oxford University Press.

Burling, R. (2005). *The talking ape: How language evolved.* Oxford: Oxford University Press.

Catchpole, C. K., & Slater, P. J. B. (2008). *Bird song: Biological themes and variations* (2nd ed.). Cambridge: Cambridge University Press.

Cooper, B. G., & Goller, F. (2006). Physiological insights into the social-context-dependent changes in the rhythm of the song motor program. *Journal of Neurophysiology, 95,* 3798–3809.

Delvaux, V., & Soquet, A. (2007). The influence of ambient speech on adult speech productions through unintentional imitation. *Phonetica, 64,* 145–173.

Doupe, A. J., & Kuhl, P. K. (1999). Birdsong and human speech: Common themes and mechanisms. *Annual Review of Neuroscience, 22,* 567–631.

Doupe, A. J., Perkel, D. J., Reiner, A., & Stern, E. A. (2005). Birdbrains could teach basal ganglia research a new song. *Trends in Neurosciences, 28,* 353–363.

Dubbeldam, J. L., & den Boer-Visser, A. M. (2002). The central mesencephalic grey in birds: Nucleus intercollicularis and substantia grisea centralis. *Brain Research Bulletin, 57,* 349–352.

Duffau, H. (2008). The anatomo-functional connectivity of language revisited: New insights provided by electrostimulation and tractography. *Neuropsychologia, 46,* 927–934.

Duffy, J. R. (2005). *Motor speech disorders: Substrates, differential diagnosis, and management* (2nd ed.). St. Louis, MO: Elsevier Mosby.

Farries, M. A., & Perkel, D. J. (2008). The songbird brain in comparative perspective. In H. P. Zeigler & P. Marler (Eds.), *Neuroscience of birdsong* (pp. 63–71). Cambridge: Cambridge University Press.

Fee, M. S., Kozhevnikov, A. A., & Hahnloser, R. H. R. (2004). Neural mechanisms of vocal sequence generation in the songbird. *Annals of the New York Academy of Sciences, 1016,* 153–170.

Frenck-Mestre, C., Anton, J. L., Roth, M., Vaid, J., & Viallet, F. (2005). Articulation in early and late bilinguals' two languages: Evidence from functional magnetic resonance imaging. *NeuroReport, 16,* 761–765.

Fried, I., Katz, A., McCarthy, G., Sass, K. J., Williamson, P., Spencer, S. S., et al. (1991). Functional organization of human supplementary motor cortex studied by electrical stimulation. *Journal of Neuroscience, 11,* 3656–3666.

Geschwind, N. (1969). Problems in the anatomical understanding of the aphasias. In A. L. Benton (Ed.), *Contributions to clinical neuropsychology* (pp. 107–128). Chicago: Aldine.

Hage, S. R., & Jürgens, U. (2006). On the role of the pontine brainstem in vocal pattern generation: A telemetric single-unit recording study in the squirrel monkey. *Journal of Neuroscience, 26,* 7105–7115.

Ho, A. K., Bradshaw, J. L., & Iansek, R. (2000). Volume perception in Parkinsonian speech. *Movement Disorders, 15,* 1125–1131.

Ho, A. K., Iansek, R., & Bradshaw, J. L. (1999). Regulation of Parkinsonian speech volume: The effect of interlocuter distance. *Journal of Neurology, Neurosurgery, and Psychiatry, 67,* 199–202.

Hockett, C. F. (1966). The problem of universals in language. In J. H. Greenberg (Ed.), *Universals of language* (2nd ed., pp. 1–22). Cambridge, MA: MIT Press.

Hughes, S. M., Harrison, M. A., & Gallup, G. G. (2002). The sound of symmetry: Voice as a marker of developmental instability. *Evolution and Human Behavior, 23,* 173–180.

Jackendoff, R. (2003). *Foundations of language: Brain, meaning, grammar, evolution.* Oxford: Oxford University Press.

Jarvis, E. D. (2004a). Brains and birdsong. In P. Marler & H. Slabbekoorn (Eds.), *Nature's music: The science of birdsong* (pp. 226–271). Amsterdam: Elsevier.

Jarvis, E. D. (2004b). Learned birdsong and the neurobiology of human language. In H. P. Zeigler & P. Marler (Eds.), *Behavioral neurobiology of birdsong* (pp. 749–777). New York: New York Academy of Sciences (*Annals of the New York Academy of Sciences*, vol. 1016).

Jürgens, U. (2002). Neural pathways underlying vocal control. *Neuroscience and Biobehavioral Reviews, 26,* 235–258.

Kao, M. H., & Brainard, M. S. (2006). Lesions of an avian basal ganglia circuit prevent context-dependent changes to song variability. *Journal of Neurophysiology, 96,* 1441–1455.

Kent, R. D., & Read, C. (2002). *The acoustic analysis of speech* (2nd ed.). Albany, NY: Singular / Thomson Learning.

Klein, D., Zatorre, R. J., Milner, B., Meyer, E., & Evans, A. C. (1994). Left putaminal activation when speaking a second language: Evidence form PET. *NeuroReport, 5,* 2295–2297.

Krägeloh-Mann, I., Helber, A., Mader, I., Staudt, M., Wolff, M., Groenendaal, F., et al. (2002). Bilateral lesions of thalamus and basal ganglia: Origin and outcome. *Developmental Medicine and Child Neurology, 44,* 477–484.

Kreiman, J., & Sidtis, D. (2011). *Foundations of voice studies: An interdisciplinary approach to voice production and perception.* Malden, MA: Wiley-Blackwell.

Kroodsma, D. (2005). *The singing life of birds: The art and science of listening to birdsong.* Boston: Houghton Mifflin.

Kuhl, P. K. (2003). Human speech and birdsong: Communication and the social brain. *Proceedings of the National Academy of Sciences of the United States of America, 100,* 9645–9646.

Ladefoged, P. (2005). *Vowels and consonants: An introduction to the sounds of languages* (2nd ed.). Malden, MA: Blackwell.

Mao, C. C., Coull, B. M., Golper, L. A., & Rau, M. T. (1989). Anterior operculum syndrome. *Neurology, 39,* 1169–1172.

Marler, P. (2004). Bird calls: A cornucopia for communication. In P. Marler & H. Slabbekoorn (Eds.), *Nature's music: The science of birdsong* (pp. 132–177). Amsterdam: Elsevier.

Meltzoff, A. N. (2002). Imitation as a mechanism of social cognition: Origins of empathy, theory of mind, and the representation of action. In U. Goswami (Ed.), *Blackwell handbook of childhood cognitive development* (pp. 6–25). Oxford: Blackwell.

Muellbacher, W., Boroojerdi, B., Ziemann, U., & Hallett, M. (2001). Analogous corticocortical inhibition and facilitation in ipsilateral and contralateral human motor cortex representations of the tongue. *Journal of Clinical Neurophysiology, 18,* 550–558.

Nordeen, K. W., & Nordeen, E. J. (1992). Auditory feedback is necessary for the maintenance of stereotyped song in adult zebra finches. *Behavioral and Neural Biology, 57,* 58–66.

Nottebohm, F., Stokes, T. M., & Leonard, C. M. (1976). Central control of song in the canary, *Serinus canarius. Journal of Comparative Neurology, 165,* 457–486.

Ojemann, G. A. (1994). Cortical stimulation and recording in language. In A. Kertesz (Ed.), *Localization and neuroimaging in neuropsychology* (pp. 35–55). San Diego: Academic Press.

Okanoya, K. (2004). The Bengalese finch: A window on the behavioral neurobiology of birdsong syntax. *Annals of the New York Academy of Sciences, 1016,* 724–735.

Pardo, J. S. (2006). On phonetic convergence during conversational interaction. *Journal of the Acoustical Society of America, 119,* 2382–2393.

Pascual-Leone, A., Gates, J. R., & Dhuna, A. (1991). Induction of speech arrest and counting errors with rapid-rate transcranial magnetic stimulation. *Neurology, 41,* 697–702.

Paus, T., Petrides, M., Evans, A. C., & Meyer, E. (1993). Role of the human anterior cingulate cortex in the control of oculomotor, manual, and speech responses: A positron emission tomography study. *Journal of Neurophysiology, 70,* 453–469.

Penfield, W., & Roberts, L. (1959). *Speech and brain-mechanisms*. Princeton, NJ: Princeton University Press.

Riecker, A., Wildgruber, D., Grodd, W., & Ackermann, H. (2002). Reorganization of speech production at the motor cortex and cerebellum following capsular infarction: A follow-up fMRI study. *Neurocase, 8,* 417–423.

Rödel, R. M. W., Laskawi, R., & Markus, H. (2003). Tongue representation in the lateral cortical motor region of the human brain as assessed by transcranial magnetic stimulation. *Annals of Otology, Rhinology, and Laryngology, 112,* 71–76.

Roy, A., & Mooney, R. (2007). Auditory plasticity in a basal ganglia—forebrain pathway during decrystallization of adult birdsond. *Journal of Neuroscience, 27,* 6374–6387.

Sakata, J. T., Hampton, C. M., & Brainard, M. S. (2008). Social modulation of sequence and syllable variability in adult birdsong. *Journal of Neurophysiology, 99,* 1700–1711.

Scharff, C., & Nottebohm, F. (1991). A comparative study of the behavioral deficits following lesions of various parts of the zebra finch song system: Implications for vocal learning. *Journal of Neuroscience, 11,* 2896–2913.

Sober, S. J., Wohlgemuth, M. J., & Brainard, M. S. (2008). Central contributions to acoustic variation in birdsong. *Journal of Neuroscience, 28,* 10370–10379.

Suthers, R. A. (1997). Peripheral control and lateralization of birdsong. *Journal of Neurobiology, 33,* 632–652.

Suthers, R. A., & Zollinger, S. A. (2008). From brain to song: The vocal organ and vocal tract. In H. P. Zeigler & P. Marler (Eds.), *Neuroscience of birdsong* (pp. 78–98). Cambridge: Cambridge University Press.

Turner, E. C., & Brainard, M. S. (2007). Performance variability enables adaptive plasticity of "crystallized" adult birdsong. *Nature, 450,* 1240–1244.

Urban, P. P., Rolke, R., Wicht, S., Keilmann, A., Stoeter, P., Hopf, H. C., et al. (2006). Left-hemispheric dominance for articulation: A prospective study on acute ischaemic dysarthria at different localizations. *Brain, 129,* 767–777.

Vicario, D. S., & Simpson, H. B. (1995). Electrical stimulation in forebrain nuclei elicits learned vocal patterns in songbirds. *Journal of Neurophysiology, 73,* 2602–2607.

Vu, E. T., Mazurek, M. E., & Kuo, Y. C. (1994). Identification of a forebrain motor programming network for the learned song of zebra finches. *Journal of Neuroscience, 14,* 6924–6934.

Wedel, A. B. (2006). Exemplar models, evolution and language change. *Linguistic Review, 23,* 247–274.

Whitney, P. (1998). *The psychology of language*. Boston: Houghton Mifflin.

Wild, B., Rodden, F. A., Grodd, W., & Ruch, W. (2003). Neural correlates of laughter and humour. *Brain, 126,* 2121–2138.

Wild, J. M. (2008). Birdsong: Anatomical foundations and central mechanisms of sensorimotor integration. In H. P. Zeigler & P. Marler (Eds.), *Neuroscience of birdsong* (pp. 136–151). Cambridge: Cambridge University Press.

Williams, H. (2004). Birdsong and singing behavior. *Annals of the New York Academy of Sciences, 1016,* 1–30.

Woolsey, C. N., Erickson, T. C., & Gilson, W. E. (1979). Localization in somatic sensory and motor areas of human cerebral cortex as determined by direct recording of evoked potentials and electrical stimulation. *Journal of Neurosurgery*, *51*, 476–506.

Yu, A. C., & Margoliash, D. (1996). Temporal hierarchical control of singing in birds. *Science*, *273*, 1871–1875.

Ziegler, W. (2008). Apraxia of speech. In G. Goldenberg & B. Miller (Eds.), *Handbook of clinical neurology* (Vol. 88, 3rd series, pp. 269–285). London: Elsevier.

18 Neural Mechanisms Underlying the Generation of Birdsong: A Modular Sequential Behavior

Michale S. Fee and Michael A. Long

The songbird affords a unique opportunity to understand the physiological basis of complex sequential behaviors, such as speech and language. Our knowledge now extends from the detailed biophysics of the avian vocal organ to the basic functional principles of the central neural circuitry that controls vocalization. The brain areas involved in song production have been identified and the connections between them well characterized. In addition, the firing patterns of many of the neuronal cell types within these areas have been characterized during singing. Finally, by manipulating these circuits and observing the effects on vocal output it has been possible to directly test specific, circuit-level hypotheses about how these circuits generate the spatially and temporally patterned neural activity that underlies learned vocalizations in birds. Together, these studies yield a coherent and testable picture of how songbirds produce their songs, and perhaps more generally, how the vertebrate brain implements complex sequential learned behaviors such as speech.

Modularity and Temporal Hierarchy in Birdsong and Human Speech

Songbirds and humans generate complex sequences of vocal gestures that carry behavioral significance (Pinker, Bloom, Barkow, Cosmides, & Tooby, 1992). One of the most striking similarities between birdsong and human speech is the complex hierarchical and modular organization of these behaviors (Doupe & Kuhl, 1999). Speech has organization at the level of phonemes, syllables, words, phrases, and so on (Bock & Levelt, 1995). Similarly, while the songs of different avian species appear quite diverse (Greene, 1999), a relatively common feature of birdsong is a hierarchical structure of notes, syllables, phrases (Konishi, 1985), and even higher-level song "categories" (Todt & Hultsch, 1998). Perhaps the fundamental unit in the hierarchy of birdsong is the song syllable (Cynx, 1990), which is usually operationally defined as a burst of sound bounded by silence. Adult birds singing fully learned (crystallized) songs typically have a repertoire of distinct syllables, each with a stereotyped duration and acoustic pattern. For example, zebra finch song is composed of a small

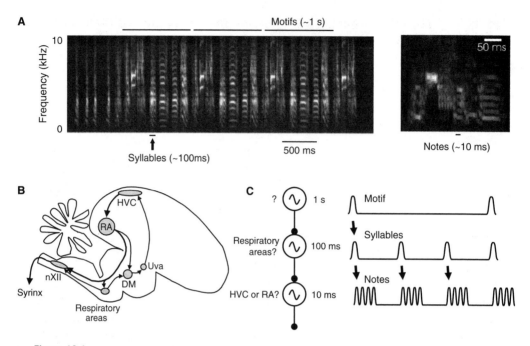

Figure 18.1
Timescales of zebra finch song. (A) A spectrogram showing a typical bout of zebra finch song. Light colors indicate high power. A song bout lasts about a second and is composed of repeated song motifs that are made of strings of syllables separated by silent gaps. *Right*, an expanded view of one song syllable exhibiting a complex sequence of brief notes. (B) Schematic diagram of the major brain areas involved in song production. HVC and RA send descending projections to vocal motor neurons (nXIIts). Also shown are descending projections from RA to respiratory areas, which project back to HVC through the thalamic nucleus Uva. (C) In principle, the different timescales of song could be controlled by different oscillatory frequencies that may arise in different regions of the song control circuitry.

number (two to eight) of identifiable syllables (Zann, 1996), each having a duration on the order of 100 ms (Figure 18.1A). Syllables may be acoustically simple, nothing more than an unmodulated harmonic stack or whistle, or they may be very complex, with an uninterrupted sequence of broadband noise, clicks, harmonic stacks, and whistles (Price, 1979). In many cases, the song can modulate between these different notes on a timescale of 10 ms or less (Figure 18.1A, right) (Fee, Shraiman, Pesaran, & Mitra, 1998).

Within the hierarchical organization of birdsong, syllables are often organized into larger groupings in a species-specific manner. In zebra finches, syllables are sung in a fairly linear and stereotyped sequence to produce a song motif with a duration of about 0.5 to 1 second, which is then repeated a variable number of times in a bout of singing (Price, 1979; Zann, 1996). In other birds, such as the canary, syllables are organized into phrases in which a single syllable is repeated multiple times for about a second, regardless of the duration of the syllable (Gardner, Naef, & Nottebohm, 2005; Mundinger, 1995). A bout of canary song typically consists of several tens of phrases, each containing a different syllable. In yet another form of syntactic hierarchy, the songs of Bengalese finches contain seemingly improvised strings of syllables that are sequenced nondeterministically, in what has been described as a Markov process (Jin & Kozhevnikov, 2011; Okanoya, 2004). There is as yet little evidence that different combinations of these elements have semantic content, as is the case for speech (but see Gentner, Fenn, Margoliash, & Nusbaum, 2006). However, at the level of a motor act, birdsong exhibits a degree of flexibility and structural complexity often thought unique to human speech.

A fundamental open question about the neural basis of vocal behavior, indeed of any complex sequential behavior, is this: What brain mechanisms underlie the timing of the elements of the behavior? Several studies have emphasized the timescales inherent in speech (Drullman, Festen, & Plomp, 1994; Greenberg & Arai, 2004) and have pointed to the possible involvement of different brain circuits or mechanisms in the processing of different timescales (Ghitza & Greenberg, 2009). Birdsong shares this range of temporal structure and rhythmicity (Saar & Mitra, 2008). Is the timing of notes, syllables, and motifs controlled by separate hierarchically organized circuits that operate at different timescales (Figure 18.1C)?

Another fundamental question relates to the temporal ordering of the elements of a behavior (Lashley, 1951). How is the sequence of phonemes generated to make a word, or the sequence of notes to make a songbird syllable? How is a sequence of words generated to form a sentence, or a sequence of syllables to form a zebra finch song motif? Recent advances in understanding the brain mechanisms underlying learned vocalizations in the songbird have begun to reveal answers to these questions. Here we argue that each different syllable is a behavioral module, generated by a distinct synaptically-connected chain of neurons. The hierarchical organization of birdsong arises from the action of higher-order circuitry that can initiate these modules flexibly, allowing some birds to produce complex syntactic structure. We argue that the multiple timescales of birdsong arise not from distinct circuits operating at different timescales, but from the sequential execution of multiple behavioral modules. Finally, we argue that a similar modular neuronal organization may underlie other complex learned vertebrate behaviors, including human speech.

Song Control Nuclei in the Brain

The muscles of the avian vocal organ are controlled by a discrete set of brain areas that are exclusively devoted to singing behavior (Figure 18.1B) (Nottebohm, Stokes,

& Leonard, 1976). One of these areas is nucleus RA (robust nucleus of the arcopallium), which resides in the avian forebrain. Functionally, RA is homologous to the primary motor cortex (Jarvis et al, 2005): descending control of the vocal behavior comes from roughly 8,000 neurons in RA that project directly to brainstem motor neurons that innervate syringeal muscles and respiratory motor areas in the brainstem (Sturdy, Wild, & Mooney, 2003). RA receives excitatory projections from two other forebrain areas. One is an anterior forebrain nucleus necessary for song learning in juvenile birds, but not for singing in adult birds (Bottjer, Miesner, & Arnold, 1984; Scharff & Nottebohm, 1991; Sohrabji, Nordeen, & Nordeen, 1990). The other projection comes from nucleus HVC (proper name, formerly high vocal center), which contains roughly 40,000 neurons that project to RA (Wang et al, 2002). HVC in turn receives excitatory inputs from a number of brain areas, including an anterior thalamic region called Uva (nucleus uvaeformis) (Nottebohm, Kelley, & Paton, 1982). Uva, HVC, and RA form the "song motor pathway" and are necessary for normal adult song (Aronov, Andalman, & Fee, 2008; Coleman & Vu, 2005; Nottebohm, Kelley, & Paton, 1982; Simpson & Vicario, 1990).

The role of these song control nuclei in song production has been the subject of intense investigation in the past three decades. RA neurons generate complex sequences of bursts during singing (Figure 18.2A) (Leonardo & Fee, 2005; Margoliash & Yu, 1996). During the song motif, each RA neuron generates about 10 high-frequency bursts of spikes (median burst duration of 8 ms); between bursts the neurons are silent. The pattern of spikes generated by these neurons is extremely stereotyped and is precisely timed with respect to the ongoing vocalization (Chi & Margoliash, 2001).

In contrast to RA neurons, which burst multiple times per song motif, HVC neurons that project to RA ($HVC_{(RA)}$ neurons) burst extremely sparsely—at most once per song motif (Hahnloser, Kozhevnikov, & Fee, 2002; Kozhevnikov & Fee, 2007). The bursts last roughly 6 ms, corresponding to only 1% of the duration of a typical song motif, and are locked to the ongoing vocalization with submillisecond precision (Figure 18.2B). However, across a population of $HVC_{(RA)}$ neurons, there is no clear relation between the acoustic structure of the song and the timing of bursts, which appear to be distributed, perhaps uniformly, at all times within the song[1] (Kozhevnikov & Fee, 2007). It is important to note that the bursts in HVC are short compared to the duration of a syllable. Thus, a single syllable is associated with a sequence of 5-30 non-overlapping populations of HVC neurons. Because each $HVC_{(RA)}$ neuron is active at only one moment in the song, and each moment in time may be associated with activity in a unique group of neurons, it has been hypothesized that this sparse sequence of bursts codes for time, or temporal order, throughout the entire song (Fee, Kozhevnikov, & Hahnloser, 2004).

A Simple Model of Song Generation

This view of the firing patterns in HVC led to a specific hypothesis for how the entire song is generated (Figure 18.2C,D) (Fee et al., 2004; Leonardo & Fee, 2005). First, $HVC_{(RA)}$ neurons, as a population, burst sequentially throughout the song, each neuron generating a single burst at one unique time in the song. Second, the group of $HVC_{(RA)}$ neurons active at each moment drive, with short latency, a brief burst of activity in a subset of RA neurons. Conversely, the pattern of activity in RA at each moment in the song is driven by the population of $HVC_{(RA)}$ neurons active at that moment. Finally, the complex burst sequences in RA activate downstream respiratory and syringeal motor neurons, which then produce the vocalization.

How might the patterns of bursts in RA be translated into vocal output? One possibility is that the activity of the motor neuron pool associated with one syringeal muscle simply results from the converging synaptic input of RA neurons (Leonardo & Fee, 2005). There is evidence for a myotopic organization within RA, so each motor neuron pool should receive input from roughly 1,000 RA neurons (Vicario, 1991; Vicario & Nottebohm, 1988). Thus, even though RA neurons generate relatively sparse and binary spike trains, the convergent input from so many RA neurons would result in an effective excitatory drive that has high analog resolution.[2] Little is known about whether modulations in syringeal muscle tension are encoded in nXIIts motor neuron pools by temporal summation, by recruitment, or by more complex nonlinear phenomena.

This model emphasizes the idea that dynamics in the motor pathway downstream of HVC are fast—essentially that the song is generated in thin slices in time that interact minimally with each other (Fiete, Hahnloser, Fee, & Seung, 2004). This view is consistent with the short timescale of bursts in HVC and RA, and is supported by experiments showing that brief electrical stimulation in RA can elicit perturbations in the vocal output that are brief (~10–20 ms) and have short latency (~10 ms) (Fee et al., 2004) (but see Ashmore, Wild, & Schmidt, 2005; Vu, Mazurek, & Kuo, 1994). Further support for this view comes from studies of temporal variation in song that are consistent with fine-grained temporal coding in the motor pathway (Glaze & Troyer, 2007). However, this aspect of the model should be subjected to further experimental tests. An ideal approach would be to locally and transiently inactivate synaptic transmission at different levels of the motor pathway during singing using optogenetic silencing (Chow et al., 2010). We would predict that brief silencing (on a 10 ms timescale) of the input from HVC to ventral RA, or the input from RA to nXIIth, would produce only a similarly brief perturbation of the vocal output, consistent with the idea that the motor pathway produces the song in thin (~10 ms), independent "slices" in time.

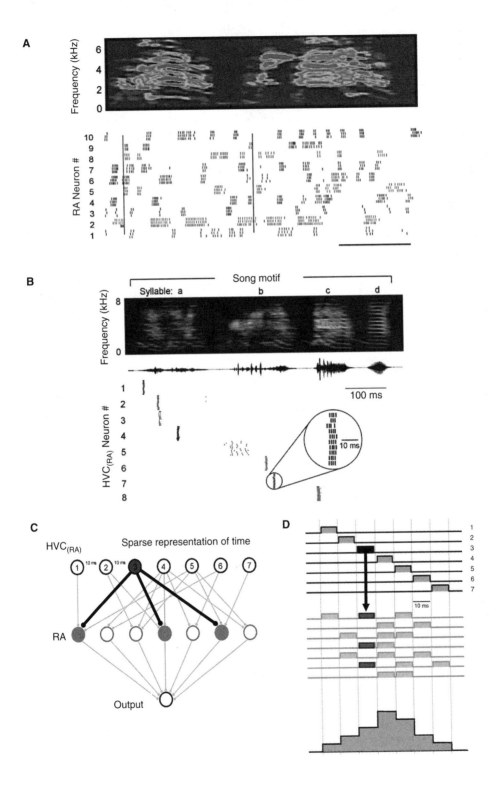

A

Frequency (kHz)

RA Neuron #

B

Song motif

Syllable: a   b   c   d

Frequency (kHz)

100 ms

HVC(RA) Neuron #

10 ms

C

HVC(RA)   Sparse representation of time

1 10 ms 2 10 ms 3 4 5 6 7

RA

Output

D

1
2
3
4
5
6
7

10 ms

Where Are the Dynamics?

We have described the firing of HVC neurons as coding for time or temporal order throughout the song. At each time step $HVC_{(RA)}$ neurons drive bursts in a subset of RA neurons, which then produce the appropriate vocal output for that time step. Missing from these descriptions is a localization of the dynamics that step the song forward in time. What initiates and controls the timing of the sparse sequence in HVC? Among the several models in which circuitry in HVC could give rise to the observed activity, perhaps the simplest is the proposal that the $HVC_{(RA)}$ network could have a chainlike synaptic organization (Figure 18.3A), such that the group of $HVC_{(RA)}$ neurons active at each moment drives the next group of neurons, and so on—like a chain of dominoes (Jin, Ramazanoğlu, & Seung, 2007; Li & Greenside, 2006; Long, Jin, & Fee, 2010). An alternative possibility is that timing in HVC is imposed by upstream structures, such as Uva.[3] For example, Uva could produce a fast oscillation that controls the timing of the sequence in HVC (like a metronome). One could also imagine other possibilities—for instance, Uva could exhibit sparse sequential activation, such that the timing and sequence of HVC bursts is directly controlled by Uva input (Figure 18.3B).

How can these two possibilities be experimentally distinguished—that song timing is controlled entirely by dynamics intrinsic to HVC or by dynamics upstream of HVC? One approach recently adopted is to directly manipulate dynamics in HVC and observe changes in singing behavior (Fee & Long, 2011). The two possibilities described above yield very different predictions for the consequence of slowing down of dynamics in HVC. In the intrinsic dynamics model, the speed at which the HVC chain (and thus the song) progresses is determined by the time it takes for one group of HVC neurons to activate the next group. Slowing the dynamics in HVC would cause every step in this chain to occur more slowly. These delays would accumulate at each link, causing the entire chain, as well as the song, to proceed more

◄ **Figure 18.2**
Spiking activity of neurons in the vocal motor pathway during singing. (A) Raster plot of song-related spiking patterns of 10 RA neurons recorded in the same bird. Each tick mark indicates the time of a single spike. For each neuron, spiking activity is shown for three repetitions of the song motif. Spikes for different neurons are aligned to each other using the syllable onsets as time-alignment points (vertical black lines). (B) Raster plot of spiking of 8 identified HVC neurons that project to RA. For each neuron, spikes are shown for 10 repetitions of the song motif. Note that each $HVC_{(RA)}$ neuron bursts at a single moment of the song. (C) and (D) A simple model of sequence generation in the songbird motor pathway. $HVC_{(RA)}$ neurons burst sequentially throughout the song motif in small ensembles of 200 coactive neurons. At each moment the active $HVC_{(RA)}$ neurons activate an ensemble of RA neurons, which converge onto motor neurons to activate muscles in the vocal organ.

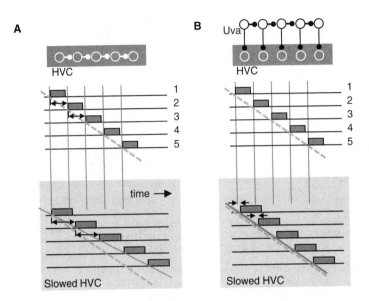

Figure 18.3
Two simple models for the sparse sequential activation of HVC$_{(RA)}$ neurons can be distinguished by slowing HVC dynamics. (A) In one model, circuit dynamics that control song timing reside within HVC. For example, the HVC network could have a chainlike synaptic connectivity such that one group of active HVC neurons activates the next group of neurons. Activity propagates through the chain at a speed determined by the latency from one group to the next (black arrows, top). In this model, slowing biophysical dynamics (action potential propagation, synaptic transmission, membrane time constant, etc.) in HVC would slow the propagation speed through the network and cause the song to slow down. (B) An alternative model in which circuit dynamics that control song timing reside upstream of HVC, for example in Uva. In this example, the HVC sequence is directly driven by input from Uva. Slowing biophysical dynamics in HVC might increase the latency from Uva input to the HVC burst, but this would be the same for each step in the sequence. In this model, the song would not be slowed down.

slowly (Figure 18.3A, bottom). In contrast, if HVC timing were controlled by an independent, autonomous clock upstream of HVC (i.e., Uva), we would *not* expect the song to slow down. Slowing dynamics in HVC might slow the synaptic input from Uva to HVC, but this increased latency would be the same for every element in the HVC sequence. Thus, the delays would not accumulate and the speed of the HVC sequence and the song would be unchanged (Figure 18.3B, bottom).

Altering Song Dynamics through Focal Cooling

Nearly every aspect of neuronal dynamics strongly varies with temperature. The speed of axonal propagation, the release of synaptic vesicles, as well as the latency and time course of postsynaptic currents all occur more slowly at lower temperatures (Thompson, Masukawa, & Prince, 1985; Volgushev, Vidyasagar, Chistiakova, & Eysel, 2000). Thus small, localized temperature changes can be used to examine the dynamic impact of groups of neurons on behavior (Andalman, Foerster, & Fee, 2011; Aronov, Veit, Goldberg, & Fee, 2011; Long & Fee, 2008; Pires & Hoy, 1992; Yamaguchi, Gooler, Herrold, Patel, & Pong, 2008), as well as to inactivate them (Ferster, Chung, & Wheat, 1996). We have developed several chronically implanted devices suitable for cooling superficial or deep brain structures in small animals (Andalman et al., 2011; Aronov & Fee, 2011; Long & Fee, 2008). These were built using commercially available Peltier-based thermoelectric heat-pump components. For cooling HVC (Figure 18.4A), the device has two gold cooling pads placed symmetrically over the left and right HVC (which are located within 500 μm of the brain surface). The temperature of HVC could be controlled continuously down to 30°C by passing current through the device in one direction, or up to 44°C by passing current in the other direction. In addition, the temperature change was localized to HVC: the maximal temperature change in RA resulting from the cooling device placed over HVC was less than 1°C. Birds continued to produce directed song over this entire range of HVC temperatures,[4] although singing became more difficult to elicit at the colder extremes.

HVC cooling produced a dramatic slowing of the song, in some cases by as much as 45%. Despite this, the acoustic structure of the song was barely affected by the cooling (Figure 18.4B). The fractional increase in the duration (dilation) of song motifs, measured from the beginning of the first syllable to the end of the last syllable, was found to increase nearly linearly over a range of applied cooling current from 0 to 1A (corresponding to a temperature change of 0 °C to -6.5 °C, Figure 18.4D). The slope of this relation was found to be on average -3%/°C (Figure 18.4E) (Long & Fee, 2008).

To test the models outlined above, we needed to specifically analyze the effect of HVC cooling on subsyllabic acoustic structure, not just on the duration of the entire song motif. This was done in two ways—using a dynamic time-warping algorithm[5] (Chi, Wu, Haga, Hatsopoulos, & Margoliash, 2007), and by measuring the duration of the syllables—both of which yielded similar results. The average temperature-dependent dilation of subsyllabic acoustic structure was found to be −2.88±0.12%/°C (Long & Fee, 2008). This finding was predicted by the model in which song timing is governed by neuronal dynamics within HVC (Figure 18.3A), and is completely incompatible with the predictions of a model in which the timing of subsyllabic song structure is controlled by an autonomous clock upstream of HVC, such as in Uva (Figure 18.3B).

While this result does not support a model in which dynamics in RA play a role in song timing, such a view cannot be completely ruled out from the HVC cooling experiments. It is possible to imagine a model in which oscillatory dynamics in RA

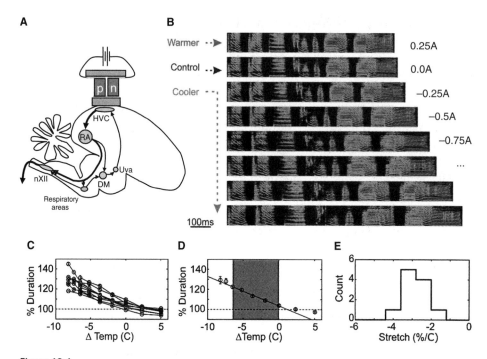

Figure 18.4

Cooling HVC causes slowing of the song. (A) Bilateral cooling of HVC is achieved using a miniature solid-state heat pump. Temperature change is a monotonic function of the electrical current applied to the Peltier device. (B) Spectrograms of the song of one bird at different levels of current. The song is compressed at warmer temperatures and stretched at colder temperatures. (C) Fractional change in the motif duration as a function of HVC temperature for 10 birds with the example in B indicated by the arrow. (D) Average change in motif duration is roughly linear over a range from 0 to $-6.5°C$. The slope of this relation gives a quantitative measure of the effect of temperature change. (E) Distribution of song stretch for all 10 birds. The average stretch is $-3.01\%/°C$. Roughly the same stretch is observed for all song timescales: the fast acoustic structure within syllables, gaps between syllables, the intervals between syllable onsets, and the intervals between motif onsets (not shown).

control song timing (Margoliash & Yu, 1996; Mooney, 1992) and that the descending drive from HVC modulates the speed of the RA dynamics. In this case, cooling HVC could reduce this drive and slow the song. A test of this possibility was carried out by direct bilateral cooling of RA. Temperature changes in RA were found to have no observable effect on song speed at any timescale (Long & Fee, 2008). Thus, biophysical dynamics in RA were found not to play a significant role in the control of song timing.

It has also been possible to test hypotheses about the control of syllable timing. Syllable patterns are tightly linked to respiration. Generally speaking, syllables are generated during expiration, while inspiration primarily occurs during the silent gaps between syllables (Goller & Cooper, 2004; Hartley & Suthers, 1989). It has recently been shown that brainstem respiratory areas project to Uva, thus providing a mechanism by which these areas could influence song timing (Ashmore, Renk, & Schmidt, 2008), or even serve as a clock that controls the timing of syllable onsets (Schmidt & Ashmore, 2008). The HVC cooling experiments provide a direct test of this hypothesis. If syllable onsets are controlled by an autonomous clock outside of HVC, the cooling HVC should have no effect on the interval between syllable onsets. It can be seen from the spectrograms of the songs with cooled HVC that the intervals between the onsets of neighboring syllables were clearly dilated by HVC cooling (Figure 18.4B). A quantitative analysis showed that syllable-onset intervals were stretched by $-3.05\pm0.11\%/°C$ during HVC cooling, suggesting that syllable timing is not controlled by an autonomous clock in respiratory brain areas, or in any area outside of HVC. Similarly, it was found that HVC cooling caused the interval between motif onsets within a bout of singing to stretch by $-3.19\pm0.24\%/°C$, suggesting that motif timing is likewise not controlled by an autonomous clock in any area outside of HVC.

A Chain Model of Timing Control

If the timing of syllable onsets and motif onsets is not controlled by neural clocks outside HVC, then how is the timing of these elements controlled? One possibility is that song timing at every scale is controlled by a separate biophysical process within HVC. There could be one oscillator in HVC that controls motif onsets, another that controls syllable onsets, and yet another process that controls the timing of the sparsely firing neurons that project to RA. If all of these processes reside within HVC, then cooling HVC would slow all of them. Of course, the cooling experiments cannot rule out the possibility of an "oscillator"-based network. Another possibility, described above, is that song timing is controlled by a traveling wave of activity that propagates through a "chain"-like HVC network (Jin et al., 2007; Li & Greenside, 2006) (Figure 18.3A).

It is possible to experimentally distinguish between an "oscillator" mechanism and a "chain" mechanism because these two models predict that neurons would receive very different inputs. For example, in an oscillator model, neurons should receive oscillatory synaptic input, and the time at which the neuron bursts would be determined by when this input reaches the threshold for spiking. In contrast, in a chain model, neurons would get little input until they received a large synaptic drive from the previously active set of neurons. Recent technological advances have permitted us to record intracellularly from HVC neurons in a singing bird (Long et al., 2010). Intracellular membrane potentials in HVC during singing show no evidence of oscillatory activity in the HVC network. Instead, $HVC_{(RA)}$ neurons receive a strong excitatory input only within the 5–10 ms period before they burst, consistent with the chain model. Of course, these experiments do not prove that sequential activity in HVC derives from a chainlike network connectivity. Further tests of this idea will be required, perhaps involving detailed reconstruction of the HVC network by serial electron microscopy (Briggman, Helmstaedter, & Denk, 2011; Denk & Horstmann, 2004; Seung, 2009).

Is song timing entirely controlled by dynamics within HVC? In the context of the chain model, we could ask whether $HVC_{(RA)}$ neurons are activated in a chain by synaptic connections entirely within HVC, or might this process involve connections outside HVC? Indeed, there are well-established feedback pathways from RA to the midbrain/brainstem, and back to HVC through the thalamic nucleus Uva (Schmidt, Ashmore, & Vu, 2004; Striedter & Vu, 1998). These projections, in addition to playing a role in interhemispheric synchronization, could be involved in the maintenance or propagation of activity in HVC. One might even imagine that each burst in HVC could be driven by a preceding burst in HVC via these feedback pathways, around which bursts of activity could rapidly cycle (Figure 18.5A). Of course, this model would predict that cooling RA should also slow the song, since this would also slow the propagation of activity around the feedback loop. Inconsistent with this prediction, bilateral cooling of RA was shown to have no effect on song timing (Long & Fee, 2008). Thus, rapid circulation of activity on a 10 ms timescale through feedback pathways is therefore not likely involved in the generation of sequences in HVC.

The circulation of activity through the thalamic feedback pathway may be important on a longer timescale, however, perhaps at the level of the song motif. For example, HVC might contain a long chain of synaptically connected $HVC_{(RA)}$ neurons that generates a single song motif. The neurons at the end of the motif chain could be connected through the thalamic feedback pathway to the HVC neurons at the beginning of the chain (Figure 18.5B), forming a continuous loop. In this case, the beginning of the next song motif could be initiated by the end of the previous motif, thus explaining how HVC cooling could stretch the interval between motif

A HVC

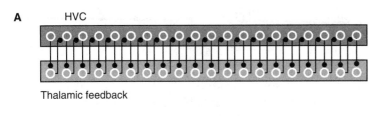

Thalamic feedback

B HVC

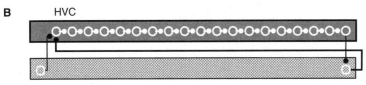

Thalamic feedback

C HVC

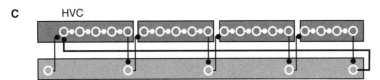

Thalamic feedback

Figure 18.5
Three different models of the role of feedback connections from RA back to HVC through the brainstem respiratory areas and thalamic nucleus Uva. (A) A model in which activity cycles rapidly around a brainstem,/thalamic feedback loop. (B) A different model in which sequential activity in HVC results from a synaptically connected chain of neurons within HVC. Feedback acts to restart the next motif at the end of the previous motif. (C) A model in which short syllable-length chains in HVC form "modules." In this model the brainstem/thalamic feedback loop acts to detect the end of one module and immediately activate the next one. Models A and B are inconsistent with various cooling experiments (see text), whereas model C is consistent with a number of experiments that support a syllable-level modular organization in HVC.

onsets by the same fractional amount as structure at the finest timescale. In this view, we could think of the HVC circuitry as representing, or enacting, a single behavioral module that generates a song motif.

Another possibility is that this modularity operates at a timescale shorter than the song motif. For example, HVC might contain multiple ~100 ms chains, each of which generates a syllable-length module. The neurons at the end of one chain could be connected through thalamic feedback circuitry to the neurons at the beginning of the next (Figure 18.5C). In this view, the chain for each song element could be activated by thalamic feedback at the end of the previous syllable, explaining how

HVC cooling could stretch the interval between syllable onsets by the same fractional amount as structure at all other timescales.

The idea that song is modular at the level of syllables has been suggested previously (Glaze & Troyer, 2006) and is supported by several lines of evidence. Flashes of light cause the interruption of song selectively at the ends of syllables or at acoustic transitions within complex multinote syllables (Cynx, 1990). In addition, a detailed analysis of song timing reveals that the durations of silent gaps between song syllables are more variable than the durations of syllables (Glaze & Troyer, 2006). Thus, the links between song syllables appear to be more flexible in their timing and more susceptible to external influences, suggesting that the links between song syllables are mediated by a different mechanism than the structure within syllables. The model shown in Figure 18.5C naturally captures this feature: the song syllable is generated by a rigid synaptically connected chain of neurons in HVC, whereas the links between syllables are mediated by a more flexible midbrain/thalamic feedback connection.

Additional Evidence for HVC Modules: Hemispheric Interactions

We have been discussing the possible modular organization of HVC in the context of midbrain/thalamic projections to HVC because these projections would be required to activate the hypothesized modules in HVC. However, these feedback connections are also the most probable means of interhemispheric interactions in the songbird (Schmidt, Ashmore, & Vu, 2004; Wild, Williams, & Suthers, 2000). Thus, perhaps the strongest clues as to any modular organization of HVC may come from examining the role of the midbrain/thalamic feedback connections in the context of their essential role in synchronizing the two hemispheres of the brain.

How does the vocal motor system maintain synchrony throughout the song? To address this question, Schmidt (2003) performed simultaneous multiunit recordings (thought to be dominated by global interneuronal activity) in HVC of both hemispheres and found that left and right HVC were continuously active at all points in the song. He also found that left and right HVCs exhibited a brief episode of correlated activity prior to the onset of each syllable, and also at some acoustic transitions within long complex syllables. These observations led to the suggestion that the two HVCs are bilaterally synchronized by thalamic inputs prior to each syllable onset and possibly at some subsyllabic transitions (Schmidt, 2003).

How can we determine if synchronization occurs only at the beginning of motifs, or at multiple time points in the motif as suggested by these findings (depicted in Figures 18.5B and 18.5C, respectively)? Unilateral cooling of HVC provides a unique way to examine the dynamics of hemispheric synchronization. By cooling one hemisphere, we should be able to cause the cooled HVC to run significantly (~30%)

slower than the uncooled HVC. Thus, if the two HVCs are synchronized only at the beginning of the motif, after which the two HVC chains run independently, cooling one HVC should cause a gradual desynchronization of the HVCs by as much as several hundred milliseconds by the end of the motif (~30% of the motif duration), producing a dramatic and progressive distortion of the acoustic signal during the motif (Figure 18.6B). In contrast, if the two HVCs are resynchronized at multiple time points during the motif (e.g., at syllable onsets; Figure 18.6C), then the chains in the two hemispheres will never be desynchronized by more than a few tens of milliseconds (~30% of a syllable duration), which might not even be visible in the song. In fact, unilateral HVC cooling did not produce a progressive distortion of the song, but rather produced normal-sounding song motifs with an intermediate amount of stretching (Figure 18.6D,E). This finding is not consistent with a model in which HVC synchronization occurs only at motif onset, but is consistent with the idea that HVC contains multiple syllable-related modules and that hemispheric coordination occurs by a bilaterally synchronized activation of these modules in HVC.

We have suggested that feedback circuitry may act to detect the end of one syllable and rapidly and bilaterally initiate the next syllable, thus simultaneously continuing the song sequence and resynchronizing the two HVCs (Andalman et al., 2011; Long & Fee, 2008). This model poses an interesting problem: If there are two HVC clocks running simultaneously, which one determines when the next syllable is triggered? We can imagine a hypothetical situation in which the next syllable is always triggered by one dominant HVC. In this case, cooling the dominant HVC would cause the syllable-onset intervals to stretch, whereas cooling the nondominant HVC would have a negligible effect. In contrast to this prediction, the unilateral cooling experiments showed that cooling either side produced a stretch of the song, inconsistent with the idea that one dominant HVC consistently initiates the next syllable.

Let us consider another hypothesis: Imagine that initiation of the next syllable was driven by whichever HVC finished the previous syllable first. In this case, cooling the left HVC would slow down the dynamics on the left side, allowing the right (uncooled) HVC to complete each syllable first. The result would be that the syllable onsets would occur at the normal intervals, as though there were no cooling. Conversely, if the next syllable were triggered by whichever HVC finished last, then unilateral cooling should produce the same stretch of syllable-onset intervals as bilateral cooling.

So how do we explain the surprising result that unilateral HVC cooling produces an intermediate degree of song stretching? One possible solution is suggested by the fact that unilateral cooling results in a highly nonuniform stretch of the song. That is, left HVC cooling causes some song elements to stretch and leaves others

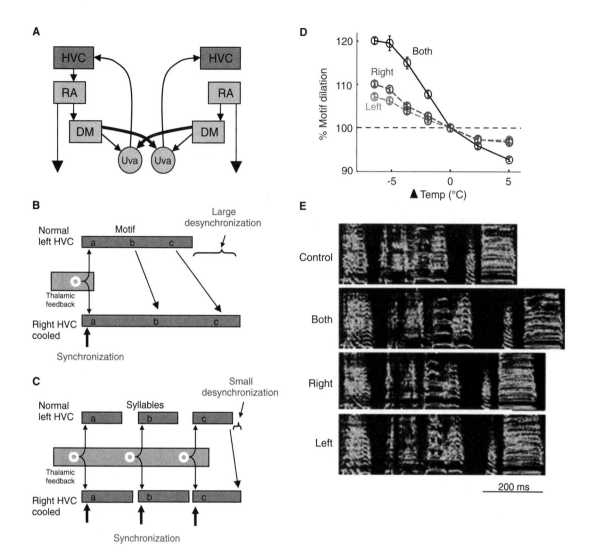

Figure 18.6

The role of the brainstem/thalamic feedback connections can be examined by unilateral cooling of HVC. (A) The bilateral projection from respiratory areas to Uva is a major site of bilateral interaction in the song motor system and is thought to be essential for the bilateral synchronization of the two HVCs (Ashmore, Renk, & Schmidt, 2008). (B) A model in which synchronized initiation of the two HVCs occurs only at motif onsets, after which the two HVCs operate independently. Cooling only one HVC should create a large desynchronization of the two sides, resulting in degradation of the song at the end of the motif. (C) In contrast, if the two HVCs are synchronized at every syllable onset, unilateral cooling of HVC should create only a small desynchronization at the end of each syllable. The effect would likely not be observable in the song. (D, E) Unilateral cooling produces a song stretch intermediate to that seen for bilateral cooling and does not cause song degradation. This result is inconsistent with the model of bilateral synchronization only at motif onsets (panel B).

unstretched. Remarkably, the effects of right HVC cooling are often complementary (anticorrelated) to that of left HVC cooling (Figure 18.7A). The results of the unilateral cooling experiments suggest a model in which responsibility for triggering the next HVC module alternates between the left and right HVC. In the circuit shown in Figure 18.7B, the onset of module *b* is initiated by the end of module *a* in the left HVC, and the onset of module *c* is initiated by the end of module *b* in the right HVC. Such a rapid switching of timing control would have two consequences. First, left HVC cooling would stretch the onset intervals between some song elements and not others (Figure 18.7C) and right HVC cooling would have a complementary effect. Second, unilateral cooling would result in an intermediate degree of song stretch compared to bilateral cooling.

A model like this might also help explain another recent finding. While bilateral electrical stimulation in HVC typically causes the song to terminate (Vu et al., 1994), unilateral stimulation produces a complex alternating pattern of effectiveness in terminating the song (Wang, Herbst, Keller, & Hahnloser, 2008). For example, stimulation in left HVC terminates the song primarily during one set of song elements, while right HVC stimulation terminates the song primarily during the complementary set of song elements. If we consider the possible effect of HVC stimulation in the model described above (Figure 18.7B), we would predict a similar pattern. If we assume that stimulation in left HVC interrupts the ongoing chain in left HVC, then only stimulation applied during element *a*, *c*, or *e* should terminate the song because the left HVC is responsible for triggering the next HVC module specifically at the end of these elements. Similarly, stimulation in right HVC should terminate the song only if applied during elements *b*, *d*, or *f*. It would be interesting to carry out unilateral HVC cooling and unilateral HVC stimulation in the same bird to determine if the rapid hemispheric switching revealed by these two techniques correspond in time.

Summary and Future Directions

We have synthesized findings from a wide range of experiments to propose a model for the temporal control of birdsong. Our model suggests that the multiple timescales apparent in song arise not from different brain mechanisms operating at different timescales (and produced in different brain circuits), but instead result from the sequential execution of behavioral modules. We imagine that each syllable is generated by different synaptically connected chains of neurons in HVC. Each chain forms a ~100 ms behavioral module that can be activated by the thalamic nucleus Uva. We also suggest that these behavioral modules are initiated at the completion of the previous module by feedback circuitry through the midbrain and thalamus, thus synchronizing the two HVCs and permitting the flexible assembly of complex

A

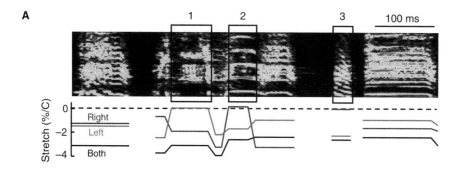

B

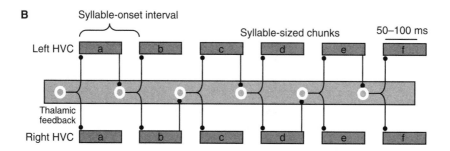

C

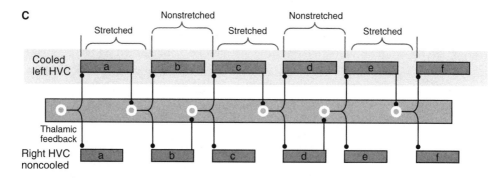

Figure 18.7
A model of switching bilateral interactions. (A) Unilateral cooling of HVC produces nonuniform stretching of song elements. Right HVC cooling stretches element 1 but not element 2
or 3, while left HVC cooling stretches elements 2 and 3 but not element 1. (B) A model of
the role of brainstem-thalamic feedback in which responsibility for synchronizing and initiat-
ing the next module alternates between the two HVCs. For example, the module for song
element *b* is initiated by the end of the element *a* in the left HVC, while song element *c* is
initiated by the end of element *b* in the right hemisphere, and so on. (C) This model predicts
the nonuniform stretch of song elements. Cooling the left HVC will increase the onset interval
between elements *a* and *b*, but will have no effect on the interval between elements *b* and *c*.

sequences of syllable modules. The basic outlines of the model presented here for vocal sequence generation in the songbird are consistent with the existing neurophysiological and behavioral data. However, many aspects of this set of hypotheses remain to be fully tested.

Another important issue is the correspondence between the hypothesized modules in HVC and song syllables. The unilateral cooling and HVC stimulation experiments in zebra finches both suggest that the rapid switching of control from one hemisphere to the other is not tightly linked to the onsets or offsets of syllables, and often occurs in the middle of many syllables (see Figure 18.7A). It is possible that this irregularity is a consequence of the highly stereotyped temporal ordering of syllables in zebra finch song. It would be interesting to see if the greater flexibility in syllable ordering exhibited by other songbird species, such as Bengalese finches, is associated with a tighter correspondence between the hypothesized HVC modules and song syllables.

Greater song flexibility is likely mediated by brain structures other than the ones we have been discussing in the zebra finch. Indeed, in Bengalese finches, the thalamic feedback connections within the song circuit involve another brain structure, nucleus interface (NIf), that projects to HVC and receives a projection from Uva (Akutagawa & Konishi, 2010; Nottebohm et al., 1982). Importantly, lesions of NIf in Bengalese finches reduce song complexity, resulting in a simple linear syllable syntax (Hosino & Okanoya, 2000). What role do Uva and NIf play in selecting the next syllable? One could imagine that Uva or NIf inputs to HVC directly select which syllable is generated next. Or alternatively, Uva inputs could simply act as a trigger that initiates the next syllable, while intrinsic circuitry in HVC (perhaps biased by input from NIf) selects which module comes next (Hosino & Okanoya, 2000; Jin, 2009).

The answer to these questions could have a direct impact on our understanding of how the vertebrate brain generates complex hierarchical behaviors, of which speech and language are arguably the most elaborate examples. It is not difficult to imagine that synaptically connected chains of neurons form behavioral modules that underlie words at the phonological level, or perhaps common word phrases, or perhaps even concepts at the highest level. The songbird provides an unprecedented model system in which to experimentally explore, at a detailed biophysical and

---

(Note that cooling the left HVC would cause some slowing of the neuronal sequence coding element b, but not of the interval between b and c, because this interval is controlled by the right HVC). Right HVC cooling would have the opposite effect, increasing interval between b and c, but not between a and b. This model also predicts the intermediate average stretch to the entire song produced by unilateral cooling, because cooling each side alone stretches only half of the intervals.

circuit level, the neural hardware that creates these modules and allows the brain to flexibly select and combine them to perform complex motor and cognitive behaviors.

Acknowledgments

The authors thank Jesse Goldberg for comments on the manuscript. This work is supported by funding from the National Institutes of Health to M.S.F. (MH067105 and DC009183) and M.A.L. (DC009280).

Notes

1. Only half of the recorded $HVC_{(RA)}$ neurons produced a burst during singing. Of the half that were not active during singing, 50% produced a single burst during learned calls and the remainder were not observed to spike at all. Thus, of the 40,000 $HVC_{(RA)}$ neurons in each hemisphere, only 20,000 are song related. If each song-related neuron is active for ~1% of the motif duration, and we assume that $HVC_{(RA)}$ activity is uniformly distributed during the song, then at each moment in time we expect a population of ~200 coactive $HVC_{(RA)}$ neurons.

2. Even though, on average, only 10% of RA neurons are active at any time, one might guess that during periods of intense activation, the fraction of RA neurons converging to an individual muscle might approach 100%. Thus, with a simple binary coding scheme in which every RA neuron has the same downstream effect, RA could code for 1,000 levels of activity, corresponding to 10 bits of analog resolution for each muscle.

3. There is a projection to HVC from another brain area, NIf (nucleus interface), which may be involved in song syntax in birds with more complex song. However, a role for this nucleus in zebra finch song production can be ruled out because complete bilateral lesions of NIf have no effect on song in these birds (Cardin, Raksin, & Schmidt, 2005).

4. Singing was elicited by presenting a female zebra finch to the experimental bird after applying the current to the Peltier device for at least two minutes, long enough for the temperature in HVC to reach equilibrium.

5. This technique is based on a two-dimensional matrix of cross correlations between a vector of acoustic features calculated at all pairs of time points in samples of the cooled and control songs. Because the cooled and control songs were so acoustically similar, this matrix has a ridge, or line, of high correlation that runs near the diagonal, and the local slope of this line indicates the local stretching of the cooled song.

References

Akutagawa, E., & Konishi, M. (2010). New brain pathways found in the vocal control system of a songbird. *Journal of Comparative Neurology, 518*, 3086–3100.

Andalman, A. S., Foerster, J. N., & Fee, M. S. (2011). Control of vocal and respiratory patterns in birdsong: Dissection of forebrain and brainstem mechanisms using temperature. *PLoS ONE, 6*, e25461.

Aronov, D., Andalman, A. S., & Fee, M. S. (2008). A specialized forebrain circuit for vocal babbling in the juvenile songbird. *Science, 320,* 630–634.

Aronov, D., & Fee, M. S. (2011). Analyzing the dynamics of brain circuits with temperature: Design and implementation of a miniature thermoelectric device. *Journal of Neuroscience Methods, 197,* 32–47.

Aronov, D., Veit, L., Goldberg, J. H., & Fee, M. S. (2011). Two distinct modes of forebrain circuit dynamics underlie temporal patterning in the vocalizations of young songbirds. *Journal of Neuroscience, 31,* 16353–16368.

Ashmore, R. C., Renk, J. A., & Schmidt, M. F. (2008). Bottom-Up activation of the vocal motor forebrain by the respiratory brainstem. *Journal of Neuroscience, 28,* 2613–2623.

Ashmore, R. C., Wild, J. M., & Schmidt, M. F. (2005). Brainstem and forebrain contributions to the generation of learned motor behaviors for song. *Journal of Neuroscience, 25,* 8543–8554.

Bock, K., & Levelt, W. (1995). Language production: Grammatical encoding. In M. Gernsbacher (Ed.), *Handbook of psycholinguistics* (pp. 945–984). New York: Academic Press.

Bottjer, S. W., Miesner, E. A., & Arnold, A. P. (1984). Forebrain lesions disrupt development but not maintenance of song in passerine birds. *Science, 224,* 901–903.

Briggman, K. L., Helmstaedter, M., & Denk, W. (2011). Wiring specificity in the direction-selectivity circuit of the retina. *Nature, 471,* 183–188.

Cardin, J. A., Raksin, J. N., & Schmidt, M. F. (2005). Sensorimotor nucleus Nlf is necessary for auditory processing but not vocal motor output in the avian song system. *Journal of Neurophysiology, 93,* 2157–2166.

Chi, Z., & Margoliash, D. (2001). Temporal precision and temporal drift in brain and behavior of zebra finch song. *Neuron, 32,* 899–910.

Chi, Z., Wu, W., Haga, Z., Hatsopoulos, N. G., & Margoliash, D. (2007). Template-based spike pattern identification with linear convolution and dynamic time warping. *Journal of Neurophysiology, 97,* 1221–1235.

Chow, B. Y., Han, X., Dobry, A. S., Qian, X., Chuong, A. S., Li, M., et al. (2010). High-Performance genetically targetable optical neural silencing by light-driven proton pumps. *Nature, 463,* 98–102.

Coleman, M. J., & Vu, E. T. (2005). Recovery of impaired songs following unilateral but not bilateral lesions of nucleus uvaeformis of adult zebra finches. *Journal of Neurobiology, 63,* 70–89.

Cynx, J. (1990). Experimental determination of a unit of song production in the zebra finch (*Taeniopygia guttata*). *Journal of Comparative Psychology, 104,* 3–10.

Denk, W., & Horstmann, H. (2004). Serial block-face scanning electron microscopy to reconstruct three-dimensional tissue nanostructure. *PLoS Biology, 2,* e329.

Doupe, A. J., & Kuhl, P. K. (1999). Birdsong and human speech: Common themes and mechanisms. *Annual Review of Neuroscience, 22,* 567–631.

Drullman, R., Festen, J. M., & Plomp, R. (1994). Effect of reducing slow temporal modulations on speech reception. *Journal of the Acoustical Society of America, 95,* 2670–2680.

Fee, M. S., Kozhevnikov, A. A., & Hahnloser, R. H. R. (2004). Neural mechanisms of vocal sequence generation in the songbird. *Annals of the New York Academy of Sciences, 1016,* 153–170.

Fee, M. S., & Long, M. A. (2011). New methods for localizing and manipulating neuronal dynamics in behaving animals. *Current Opinion in Neurobiology, 21,* 693–700.

Fee, M. S., Shraiman, B., Pesaran, B., & Mitra, P. P. (1998). The role of nonlinear dynamics of the syrinx in the vocalizations of a songbird. *Nature, 395,* 67–71.

Ferster, D., Chung, S., & Wheat, H. (1996). Orientation selectivity of thalamic input to simple cells of cat visual cortex. *Nature, 380,* 249–252.

Fiete, I. R., Hahnloser, R. H. R., Fee, M. S., & Seung, H. S. (2004). Temporal sparseness of the premotor drive is important for rapid learning in a neural network model of birdsong. *Journal of Neurophysiology, 92,* 2274–2282.

Gardner, T. J., Naef, F., & Nottebohm, F. (2005). Freedom and rules: The acquisition and reprogramming of a bird's learned song. *Science, 308,* 1046–1049.

Gentner, T. Q., Fenn, K. M., Margoliash, D., & Nusbaum, H. C. (2006). Recursive syntactic pattern learning by songbirds. *Nature, 440,* 1204–1207.

Ghitza, O., & Greenberg, S. (2009). On the possible role of brain rhythms in speech perception: Intelligibility of time-compressed speech with periodic and aperiodic insertions of silence. *Phonetica, 66,* 113–126.

Glaze, C. M., & Troyer, T. W. (2006). Temporal structure in zebra finch song: Implications for motor coding. *Journal of Neuroscience, 26,* 991–1005.

Glaze, C. M., & Troyer, T. W. (2007). Behavioral measurements of a temporally precise motor code for birdsong. *Journal of Neuroscience, 27,* 7631–7639.

Goller, F., & Cooper, B. G. (2004). Peripheral motor dynamics of song production in the zebra finch. *Annals of the New York Academy of Sciences, 1016,* 130–152.

Greenberg, S. & Arai, T. (2004). What are the essential cues for understanding spoken language? *IEICE Transactions on Information and Systems E, Series D, 87,* 1059–1070.

Greene, E. (1999). Toward an evolutionary understanding of song diversity in oscines. *Auk, 116,* 299–301.

Hahnloser, R. H. R., Kozhevnikov, A. A., & Fee, M. S. (2002). An ultra-sparse code underlies the generation of neural sequences in a songbird. *Nature, 419,* 65–70.

Hartley, R. S., & Suthers, R. A. (1989). Airflow and pressure during canary song: Direct evidence for mini-breaths. *Journal of Comparative Physiology A: Neuroethology, Sensory, Neural, and Behavioral Physiology, 165,* 15–26.

Hosino, T., & Okanoya, K. (2000). Lesion of a higher-order song nucleus disrupts phrase level complexity in Bengalese finches. *NeuroReport, 11,* 2091–2095.

Jarvis, E. D., Güntürkün, O., Bruce, L., Csillag, A., Karten, H., Kuenzel, W., et al. (2011). Opinion: Avian brains and a new understanding of vertebrate brain evolution. *Nature Reviews Neuroscience.6,* 151–159.

Jin, D. Z. (2009). Generating variable birdsong syllable sequences with branching chain networks in avian premotor nucleus HVC. *Physical Review E: Statistical, Nonlinear, and Soft Matter Physics, 80,* 051902.

Jin, D. Z., & Kozhevnikov, A. A. (2011). A compact statistical model of the song syntax in Bengalese finch. *PLoS Computational Biology*, *7*, e1001108.

Jin, D. Z., Ramazanoğlu, F. M., & Seung, H. S. (2007). Intrinsic bursting enhances the robustness of a neural network model of sequence generation by avian brain area HVC. *Journal of Computational Neuroscience*, *23*, 283–299.

Konishi, M. (1985). Birdsong: From behavior to neuron. *Annual Review of Neuroscience*, *8*, 125–170.

Kozhevnikov, A. A., & Fee, M. S. (2007). Singing-related activity of identified HVC neurons in the zebra finch. *Journal of Neurophysiology*, *97* (6), 4271–4283.

Lashley, K. (1951). The problem of serial order in behavior. In L. A. Jeffress (Ed.), *Cerebral mechanisms in behavior* (pp. 112–131). New York: Wiley.

Leonardo, A., & Fee, M. S. (2005). Ensemble coding of vocal control in birdsong. *Journal of Neuroscience*, *25*, 652–661.

Li, M., & Greenside, H. (2006). Stable propagation of a burst through a one-dimensional homogeneous excitatory chain model of songbird nucleus HVC. *Physical Review E: Statistical, Nonlinear, and Soft Matter Physics*, *74*, 911–918.

Long, M. A., & Fee, M. S. (2008). Using temperature to analyse temporal dynamics in the songbird motor pathway. *Nature*, *456*, 189–194.

Long, M. A., Jin, D. Z., & Fee, M. S. (2010). Support for a synaptic chain model of neuronal sequence generation. *Nature*, *468*, 394–399.

Margoliash, D., & Yu, A. C. (1996). Temporal hierarchical control of singing in birds. *Science*, *273*, 1871–1875.

Mooney, R. (1992). Synaptic basis for developmental plasticity in a birdsong nucleus. *Journal of Neuroscience*, *12*, 2464–2477.

Mundinger, P. C. (1995). Behaviour-genetic analysis of canary song: Inter-strain differences in sensory learning, and epigenetic rules. *Animal Behaviour*, *50*, 1491–1511.

Nottebohm, F., Kelley, D. B., & Paton, J. A. (1982). Connections of vocal control nuclei in the canary telencephalon. *Journal of Comparative Neurology*, *207*, 344–357.

Nottebohm, F., Stokes, T. M., & Leonard, C. M. (1976). Central control of song in the canary, serinus canarius. *Journal of Comparative Neurology*, *165*, 457–486.

Okanoya, K. (2004). The Bengalese finch: A window on the behavioral neurobiology of birdsong syntax. *Annals of the New York Academy of Sciences*, *1016*, 724–735.

Pinker, S., Bloom, P., Barkow, J., Cosmides, L., & Tooby, J. (1992). Natural language and natural selection. In J. H. Barklow, L. Cosmides, & J. Tooby (Eds.), *The adapted mind: Evolutionary psychology and the generation of culture* (pp. 451–493). New York: Oxford University Press.

Pires, A., & Hoy, R. R. (1992). Temperature coupling in cricket acoustic communication. II. Localization of temperature effects on song production and recognition networks in gryllus firmus. *Journal of Comparative Physiology A: Neuroethology, Sensory, Neural, and Behavioral Physiology*, *171*, 79–92.

Price, P. H. (1979). Developmental determinants of structure in zebra finch song. *Journal of Comparative and Physiological Psychology*, *93*, 260.

Saar, S., & Mitra, P. P. (2008). A technique for characterizing the development of rhythms in bird song. *PLoS ONE, 3,* e1461.

Scharff, C., & Nottebohm, F. (1991). A comparative study of the behavioral deficits following lesions of various parts of the zebra finch song system: Implications for vocal learning. *Journal of Neuroscience, 11,* 2896–2913.

Schmidt, M. F. (2003). Pattern of interhemispheric synchronization in hvc during singing correlates with key transitions in the song pattern. *Journal of Neurophysiology, 90,* 3931–3949.

Schmidt, M. F., & Ashmore, R. C. (2008). Integrating breathing and singing: Forebrain and brainstem mechanisms. In H. P. Zeigler & P. Marler (Eds.), *Neuroscience of Birdsong* (pp. 285–322). Cambridge: Cambridge University Press.

Schmidt, M. F., Ashmore, R. C., & Vu, E. T. (2004). Bilateral control and interhemispheric coordination in the avian song motor system. *Annals of the New York Academy of Sciences, 1016,* 171–186.

Seung, H. S. (2009). Reading the book of memory: Sparse sampling versus dense mapping of connectomes. *Neuron, 62,* 17–29.

Simpson, H. B., & Vicario, D. S. (1990). Brain pathways for learned and unlearned vocalizations differ in zebra finches. *Journal of Neuroscience, 10,* 1541–1556.

Sohrabji, F., Nordeen, E. J., & Nordeen, K. W. (1990). Selective impairment of song learning following lesions of a forebrain nucleus in the juvenile zebra finch. *Behavioral and Neural Biology, 53,* 51–63.

Striedter, G. F., & Vu, E. T. (1998). Bilateral feedback projections to the forebrain in the premotor network for singing in zebra finches. *Journal of Neurobiology, 34,* 27–40.

Sturdy, C. B., Wild, J. M., & Mooney, R. (2003). Respiratory and telencephalic modulation of vocal motor neurons in the zebra finch. *Journal of Neuroscience, 23,* 1072–1086.

Thompson, S. M., Masukawa, L. M., & Prince, D. A. (1985). Temperature dependence of intrinsic membrane properties and synaptic potentials in hippocampal CA1 neurons in vitro. *Journal of Neuroscience, 5,* 817–824.

Todt, D., & Hultsch, H. (1998). How songbirds deal with large amounts of serial information: Retrieval rules suggest a hierarchical song memory. *Biological Cybernetics, 79,* 487–500.

Vicario, D. S. (1991). Organization of the zebra finch song control system: Functional organization of outputs from nucleus robustus archistriatalis. *Journal of Comparative Neurology, 309,* 486–494.

Vicario, D. S., & Nottebohm, F. (1988). Organization of the zebra finch song control system: I. Representation of syringeal muscles in the hypoglossal nucleus. *Journal of Comparative Neurology, 271,* 346–354.

Volgushev, M., Vidyasagar, T. R., Chistiakova, M., & Eysel, U. T. (2000). Synaptic transmission in the neocortex during reversible cooling. *Neuroscience, 98,* 9–22.

Vu, E. T., Mazurek, M. E., & Kuo, Y. C. (1994). Identification of a forebrain motor programming network for the learned song of zebra finches. *Journal of Neuroscience, 14,* 6924–6934.

Wang, C. Z. H., Herbst, J. A., Keller, G. B., & Hahnloser, R. H. R. (2008). Rapid interhemispheric switching during vocal production in a songbird. *PLoS Biology, 6,* e250.

Wang, N., Hurley, P., Pytte, C., Kirn, J. R. (2002) Vocal control neuron incorporation decreases with age in the adult zebra finch. *Journal of Neuroscience*, *22*, 10864–10870.

Wild, J. M., Williams, M. N., & Suthers, R. A. (2000). Neural pathways for bilateral vocal control in songbirds. *Journal of Comparative Neurology*, *423*, 413–426.

Yamaguchi, A., Gooler, D., Herrold, A., Patel, S., & Pong, W. W. (2008). Temperature-dependent regulation of vocal pattern generator. *Journal of Neurophysiology*, *100*, 3134–3143.

Zann, R. A. (1996). *The zebra finch: A synthesis of field and laboratory studies*. Oxford: Oxford University Press.

19   Auditory-Vocal Mirror Neurons for Learned Vocal Communication

Jonathan F. Prather and Richard Mooney

Why Search for Mirror Neurons in Songbirds?

Imitation is an essential engine for propagating human culture, enabling people to transmit art, music, speech, and language from one generation to the next. The young child's ability to vocally imitate the speech of parents and peers is arguably one of the most essential forms of learning for human societies, because it provides the foundation for spoken language (Locke, 1993). The sensorimotor interactions that underlie human speech learning and communication remain poorly understood. An emerging idea is that sensorimotor neurons selectively active during both the execution and observation of specific gestures (i.e., mirror neurons) could play an important role in the learning, perception, and production of speech and language (Iacoboni et al., 2005; Kohler et al., 2002; Rizzolatti, 2005; Rizzolatti & Arbib, 1999, 1998; Rizzolatti & Craighero, 2004). Explicitly testing this idea is impractical, and consequently whether mirror neurons are important to human speech and language remains a matter of substantial debate. In this context, an important goal is to develop a suitable animal model in which to search for auditory-vocal mirror neurons and explore how they function to enable learned vocal communication.

Songbirds afford two great advantages in attaining this goal. First, they are one of the few nonhuman animals to communicate using learned vocalizations. Indeed, despite the fundamental importance of speech learning to human societies, and the widespread use of vocal communication by other animals, vocal learning in nonhuman species is quite rare. Importantly, studies in nonhuman primates have failed to uncover evidence of vocal learning. Oscine songbirds (order: Passeriformes) culturally transmit their courtship songs from one generation to the next, providing an experimentally tractable system in which to study mechanisms of vocal imitation and learned vocal communication (Doupe & Kuhl, 1999; Marler, 1970; Marler & Tamura, 1964). Moreover, both songbirds and humans learn to produce a complex and temporally precise sequence of vocal gestures using auditory signals originating from tonotopically organized hair cells of the inner ear. Therefore, even though vocal learning in birds and humans evolved independently, the brains of juvenile songbirds and humans must accomplish highly similar and challenging sensorimotor transformations. The second great advantage of using songbirds to search for auditory-vocal mirror neurons is that the neuronal circuitry for singing and song learning, known as the song system, is well described and amenable to cellular- and synaptic-level analysis (Dutar, Vu, & Perkel, 1998; Farries & Perkel, 2000; Mooney, 2000, 1992; Mooney & Prather, 2005; Nottebohm, Kelley, & Paton, 1982; Nottebohm, Stokes, & Leonard, 1976). Importantly, the advent of miniaturized recording technologies has enabled experimentalists to analyze the behavior of anatomically identified song system neurons in singing and listening birds (Fee & Leonardo, 2001; Hahnloser, Kozhevnikov, & Fee, 2002; Leonardo & Fee, 2005; Yu & Margoliash, 1996). These features render the songbird an exceptional organism in which to search for auditory-vocal mirror neurons, to explore their involvement in learned vocal communication, and to analyze the synaptic and circuit mechanisms that give rise to their complex sensorimotor properties.

Such an undertaking can be informed by the realization that mirror neurons for learned vocal communication are predicted to display three key features. First, individual mirror neurons should display a systematic auditory-vocal correspondence. Second, their auditory properties should be tightly linked to vocal perception. Third, they should be strategically located to influence learned vocal communication. Following a brief introduction to song learning and the song system, this chapter discusses recent advances that identify song system neurons displaying all three of these features. Consideration is then given to how the activity of auditory-vocal mirror neurons is likely to be harnessed for vocal learning and communication and how synaptic and experiential mechanisms give rise to this auditory vocal correspondence.

Song Learning and Song as a Communication Signal

Songbirds learn to sing during a juvenile sensitive period comprising two distinct phases—sensory learning and sensorimotor learning—both of which depend on auditory experience (Mooney, Prather, & Roberts, 2008). During sensory learning, a young bird listens to and memorizes one or more tutor songs, usually those of the male parent or a nearby adult male of the same species (Immelmann, 1969; Marler & Peters, 1982b, 1982c, 1987, 1988). One consequence of this auditory imprinting process is that geographically separate populations of songbirds of the same species display regional dialects, similar to the different regional dialects of human speech (Marler & Tamura, 1964; Thorpe, 1958). During the ensuing phase of sensorimotor learning, the pupil relies on auditory feedback to match its song to the memorized model, which is often referred to as the song "template." Sensorimotor learning begins with "subsong," which resembles infant babbling in its rambling and poorly structured quality, and advances to "plastic" song, which though more structured than plastic song, is still highly variable from one song bout to the next (Immelmann, 1969; Konishi, 1965; Marler & Peters, 1982a, 1982c; Marler & Waser, 1977; Price, 1979). During the plastic song phase of sensorimotor learning, song also exhibits slower adaptive changes, rendering it increasingly similar to the memorized tutor song. Sensorimotor learning terminates in song crystallization, a developmental process wherein the song becomes highly stereotyped and usually much less dependent on auditory feedback (Konishi, 1965; Lombardino & Nottebohm, 2000).

The adult male's crystallized song is a highly effective communication signal that serves to attract mates and defend territory from other males. Both of these functions engage the adult songbird's acute auditory perceptual abilities, which psychophysical studies suggest are on par with those of humans (Dooling, 1978). Sensitive perceptual abilities enable a breeding male to detect subtle acoustical features distinguishing songs of familiar neighbors from those of intruders, thus enabling him to more efficiently defend his territory. When a breeding male hears an intruder's song, one way he responds is by singing (Hyman, 2003). During this antiphonal behavior, known as countersinging, the male's role rapidly switches from receiver to sender. Notably, such a dual role also is required of humans when they engage in vocal dialog, necessitating rapid switching between sensory and motor representations of the vocalization. It has been widely hypothesized that this dual role could be facilitated by mirror neurons that display a systematic sensorimotor correspondence, although auditory-vocal mirror neurons have not been described in primates or other mammals. Countersinging in birds thus affords a highly relevant communication context in which to search for auditory-vocal mirror neurons.

The Song System

Like human speech, and in contrast to most other vertebrate vocalizations, birdsong reflects the executive influence of the telencephalon on vocal and respiratory activity (for a review of brainstem and peripheral song mechanisms, see Wild, 2004; Mooney et al., 2008). The songbird's brain is distinguished by a network of nuclei, referred to collectively as the song system, that controls singing through the muscles of the syrinx (i.e., the bird's vocal organ) and respiration (Nottebohm et al., 1982, 1976). The song system comprises two major pathways—a song motor pathway and an anterior forebrain pathway.

The song motor pathway (SMP) generates precise motor signals necessary for song production and includes the telencephalic nuclei HVC and RA and the brainstem nucleus XIIts. Specifically, individual HVCRA neurons burst in a temporally sparse manner during singing and function as a population to generate a precise timing signal integrated via convergent and divergent synaptic connections HVCRA axons make with RA neurons, which then transmit activity to the brainstem vocal network (Hahnloser et al., 2002; Leonardo & Fee, 2005). The anterior forebrain pathway (AFP) is necessary to acute song variability and slower forms of vocal plasticity, and comprises an indirect pathway from HVC to RA that includes Area X, DLM, and LMAN (Bottjer et al., 1984; Nottebohm et al., 1982; Okuhata & Saito, 1987; Olveczky, Andalman, & Fee, 2005; Scharff & Nottebohm, 1991). Our current understanding is that song variability depends on the activity of LMAN synapses on RA song premotor neurons. These synapses, which lie adjacent to those from HVCRA neurons, evoke NMDA receptor mediated synaptic currents that are thought to induce variability in the timing signals emanating from HVC (Canady, Burd, DeVoogd, & Nottebohm, 1988; Kao, Doupe, & Brainard, 2005; Kao, Wright, & Doupe, 2008; Mooney, 1992; Olveczky et al., 2005). Notably, LMAN activity is necessary for auditory feedback perturbations, such as deafening or exposure to distorted auditory feedback (DAF), to trigger increased song variability and plasticity (Brainard & Doupe, 2000; Williams & Mehta, 1999). Thus, it is likely that the AFP receives feedback-related information, even though the singing-related activity of LMAN neurons appears to be insensitive to acute feedback perturbations (Hessler & Doupe, 1999; Leonardo, 2004).

A noteworthy organizational feature of the song system is that the SMP and AFP receive song-related input from two different populations of projection neurons located in the telencephalic song nucleus HVC (these two different cell types are referred to as HVCRA and HVCX cells, based on their projections; HVC also contains several different classes of interneurons) (Hahnloser et al., 2002; Kozhevnikov & Fee, 2007; Prather, Peters, Nowicki, & Mooney, 2008). Although HVC receives input from other brain areas, including the telencephalic nucleus NIf and the thalamic nucleus Uva, HVC appears to be the highest site in the song system containing an explicit song motor representation. Notably, NIf neurons resemble HVC neurons in that both display time-locked activity during singing (McCasland, 1987). However, unlike HVC lesions, which permanently block singing, NIf lesions only transiently disrupt song (Cardin, Raksin, & Schmidt, 2005). In contrast, although Uva lesions can permanently disrupt song, Uva neurons do not display activity locked to song features (Coleman & Vu, 2005; Williams & Vicario, 1993).

The essential role for hearing in song learning indicates that auditory information must influence the song motor network. Indeed, HVC has emerged as the earliest site where auditory information is integrated with explicit song motor commands. Auditory presentation of the bird's own song (BOS) strongly excites neurons in both the SMP and the AFP, and these auditory responses depend on input from HVC (Doupe & Konishi, 1991; Roy & Mooney, 2009; Vicario & Yohay, 1993). On the one hand, studies in the zebra finch, a semidomesticated songbird widely used for songbird neurobiology, reveal that these responses are most reliably detected when the bird is either sleeping or lightly anesthetized (Cardin & Schmidt, 2003; Dave, Yu, &

Margoliash, 1998; Schmidt & Konishi, 1998). However, recordings made in several other freely behaving songbird species show that auditory responses can be expressed in HVC during periods of quiet wakefulness, consistent with their serving a role in learned vocal communication (McCasland & Konishi, 1981; Prather, Nowicki, Anderson, Peters, & Mooney, 2009; Prather et al., 2008). Moreover, auditory responses in HVC, as well as other parts of the song system, are highly selective for the bird's own song (i.e., the BOS) (Doupe & Konishi, 1991; Margoliash, 1983, 1986; Theunissen et al., 2004; Theunissen & Doupe, 1998). Selectivity for the BOS develops in parallel with sensorimotor learning, indicating an effect of auditory feedback and hinting at a functional linkage by which auditory information could influence vocal learning (Doupe, 1997; Volman, 1993). Nonetheless, efforts to detect real-time feedback signals in either HVC or the AFP have been largely fruitless (Hessler & Doupe, 1999; Kozhevnikov & Fee, 2007; Leonardo, 2004; Prather et al., 2008) (for a possible exception, see Sakata, Hampton, & Brainard, 2008; this issue also is treated in more detail in the following section). Beyond a potential role in vocal learning, auditory responsive neurons in the song system are likely to serve a role in song perception, because lesions made either in HVC or the AFP can impair the bird's ability to recognize conspecific songs (Brenowitz, 1991; Gentner, Hulse, Bentley, & Ball, 2000; Scharff, Nottebohm, & Cynx, 1998). Taken together, these findings advance HVC as a fruitful site to explore how auditory and vocal motor information is integrated to enable learned vocal communication.

Coda: Parallels between Songbird and Human Brains

For the uninitiated, song system anatomy can appear relatively arcane and challenging to relate to mammalian brain architecture. In this light, it may be useful to draw several anatomical parallels, even though vocal learning evolved independently in birds and humans. First, executive influence of the cortex on human speech is exerted by Broca's area and the regions of the lateral motor cortex that indirectly control the various muscles important to phonation, including those of the larynx, tongue, orofacial region, and respiratory system (Burns & Fahy, 2010; Simonyan & Horwitz, 2011). In songbirds, this executive influence is exerted by projections from the telencephalic nucleus HVC to the caudal telencephalic nucleus RA, and thence to vocal motor neurons and respiratory premotor neurons in the medulla. With reference to human cortical organization, RA projection neurons can be analogized to layer V pyramidal neurons in the face motor cortex, while HVC can be viewed as an analog of the supragranular layers of the face motor cortex or Broca's area.

A second parallel is that the songbird's AFP displays organizational features strongly similar to mammalian cortical-basal ganglia pathways: Area X contains local neurons with properties highly similar to medium spiny neurons in the mammalian striatum and output neurons that make inhibitory synapses onto thalamic neurons, much like mammalian pallidal neurons (Doupe, Perkel, Reiner, & Stern, 2005; Farries, 2001). Although the exact role of cortical-basal ganglia pathways in human vocal communication is not well understood, both hypo- and hyperkinetic basal ganglia disorders can disrupt speech (Martnez-Sanchez, 2010; Velasco Garcia, Cobeta, Martin, Alonso-Navarro, & Jimenez-Jimenez, 2011). Furthermore, mutations of a forkhead transcription factor (FoxP2) that is highly expressed in the human striatum leads to orofacial dyspraxias and impaired speech learning (Lai, Fisher, Hurst, Vargha-Khadem, & Monaco, 2001; Lai, Gerrelli, Monaco, Fisher, & Copp, 2003; MacDermot et al., 2005). Interestingly, expression of the avian homolog of FoxP2 is enriched in Area X of songbirds (Teramitsu, Kudo, London, Geschwind, & White, 2004), and knockdown of its expression in Area X in juvenile male zebra finches degrades the quality of song imitation (Haesler et al., 2007; White, Fisher, Geschwind, Scharff, & Holy, 2006).

A third parallel is that speech and birdsong demand exquisite interactions between auditory and vocal systems. In humans, clinical evidence points to connections between the tertiary auditory cortex (i.e., Wernicke's area) and speech motor areas (i.e., Broca's area) as a fundamental substrate for these auditory-vocal interactions (Catani & Mesulam, 2008; Geschwind, 1970). In songbirds, emerging functional and anatomical evidence strongly suggests that connections from secondary regions of the auditory telencephalon (i.e., CM) to HVC are the substrate for these interactions (Bauer et al., 2008; Roy & Mooney, 2009).

Although these parallels should not be taken too literally, they do raise the possibility that vocal learning in birds and humans depends on similar brain mechanisms. Furthermore, they reinforce the notion that auditory-vocal integration necessary to learned vocal communication can be explored at the cellular and synaptic level in the songbird's HVC. Finally, from a practical standpoint, the segregated projections from different HVC projection neuron types to RA and Area X afford experimentalists the possibility of distinguishing vocal motor signals from related signals conveyed to basal ganglia pathways (Hahnloser et al., 2002; Kozhevnikov & Fee, 2007; Mooney, 2000; Prather et al., 2008). This distinction may be harder to make when recording from cortical neurons in vocalizing mammals.

Auditory-Vocal Mirror Neurons in Songbirds

Armed with this perspective, and in collaboration with Stephen Nowicki and Susan Peters, two experts in songbird behavior, we explored the auditory and vocal motor representations of identified HVC neurons. This effort built on a longstanding collaboration between our labs to explore neural representations of song in the swamp sparrow, and the results of this research provided the first evidence of auditory-vocal mirror neurons for learned vocal communication (Prather et al., 2009, 2008).

Neural Recordings in Countersinging Swamp Sparrows

Swamp sparrows exhibit two traits conducive to systematically examining the auditory and vocal properties of song system neurons. First, a captive male swamp sparrow will, at least occasionally, sing in response to hearing its own songs or the songs of other swamp sparrows played through an audio speaker. In the laboratory, countersinging in response to a bird's own songs enables a neurophysiologist to rapidly assess the auditory and singing-related representations of one and the same behavior by individual neurons. Second, male swamp sparrows typically sing several different song types, each of which consists of a highly stereotyped multinote syllable repeated 10 to 20 times in a continuous trill. Thus, recording a neuron's activity during a single bout of countersinging can be sufficient to characterize its behavior during the sensory presentation and motor performance of many iterations of a vocal gesture. Moreover, because an individual has several song types and because countersinging can be triggered by playback of the bird's own songs, a bird that hears one of its song types will sometimes sing the same song type (i.e., matched countersinging) and at other times sing another (nonmatched countersinging). These symmetrical and asymmetrical forms of countersinging allow the extent of any sensorimotor correspondence to be more fully probed.

These facets of swamp sparrow singing behavior are very informative, but recording associated neural activity is especially challenging, because a countersinging bird responding to a perceived intruder is in a highly aroused state. A miniature motorized microdrive developed by Michale Fee provided two significant advantages in this regard. First, it is sufficiently lightweight (~1.25 g) for a swamp sparrow to easily carry on its head as it hops and flies around its cage (Fee & Leonardo, 2001). Second, because this drive permits several extracellular electrodes to be precisely positioned under remote control, individual neurons can be isolated without handling the bird. Together, these features enable microdrive recordings of individual neurons as unperturbed birds engage in singing and other naturalistic behaviors.

Individual HVCX Cells Display an Auditory-Vocal Correspondence

Using this lightweight drive to record from antidromically identified HVC cells in adult male swamp sparrows, we found that HVCX cells display a precise auditory-vocal correspondence (Prather et al., 2008). When the sparrow passively listened to songs played through a speaker, an individual HVCX cell responded to only one song type in the bird's repertoire, with different HVCX cells responding to different song types (interestingly, HVCRA neurons in the awake sparrow were unresponsive to song playback, even though they respond to BOS playback in anesthetized sparrows (Mooney et al., 2001)). The highly phasic responses of HVCX neurons occurred reliably at a precise time in each syllable and depended on a specific sequence of notes in the effective syllable (Figure 19.1A). Neural recordings

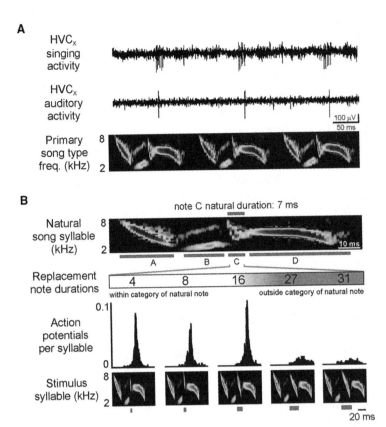

Figure 19.1
HVCX neurons in the adult swamp sparrow display a precise auditory-vocal correspondence, and their auditory properties are tightly linked to song perception (Prather et al. 2008, 2009). (a) HVCX cells are phasically active during the presentation or production of one song type in the bird's repertoire, and the syllable-locked timing of that activity is nearly identical when the bird sings (top panel) or listens to playback (middle panel) of that song type (bottom spectrogram). (b) The same cell was also tested for its responses to song playbacks in which the duration of one note in each syllable had been changed systematically (note C in the top spectrogram; replacement note durations indicated in second panel). The neuron responded strongly to songs with replacement notes shorter than 20 ms, but responded very weakly to songs with longer replacement notes (histograms in third row). Field experiments confirmed that the neural response boundary predicts a categorical perceptual boundary for note duration (Prather et al., 2009). Neural data in each panel were collected from the same neuron in a freely behaving adult male swamp sparrow.

made during matched and nonmatched bouts of countersinging also revealed that individual HVCX neurons displayed the most robust singing-related activity for the song type that evoked an auditory response in the playback condition. Even more remarkably, these neurons displayed almost identical patterns of activity when the bird sang that song type, firing at exactly the same time in the effective syllable as when the bird was quietly listening to the song played through a speaker.

Additional experiments by others and us have also confirmed the presence of an HVCX sensorimotor correspondence in the Bengalese finch (Fujimoto, Hasegawa, & Watanabe, 2011; Prather et al., 2008). This conservation across distantly related species suggests that colocalization of sensory and motor activity in HVCX cells may play an important role in shaping how vocal signals are perceived and performed. Specifically, HVCX neurons in adult male Bengalese finches are active in association with not only individual song syllables but also the specific transitions between consecutive syllables (Fujimoto et al., 2011). Taking advantage of the natural variance of Bengalese finch song sequence (Okanoya, 2004), those experiments reveal that HVCX activity encodes specific behavioral sequences from among many possible trajectories. The demonstration of an auditory-vocal correspondence in the same neurons that encode specific features of vocal sequence provides a potential mechanism through which auditory perception of vocal patterns may guide the generation of motor commands to imitate those patterns (Fujimoto et al., 2011). An important future goal will be to record the activity of HVCX cells in young finches and sparrows to determine the extent to which the patterns observed in adults are also evident during juvenile imitative learning.

These recordings demonstrated that HVCX cells display similar activity when they listen to or sing the same vocal gesture, but they do not resolve whether the singing-related activity is auditory or motor in nature. Indeed, an initially intriguing idea was that singing-related activity of HVCX neurons was a real-time auditory feedback signal. However, experiments employing DAF revealed that the singing-related activity in HVCX cells was unaffected by acutely disrupting auditory feedback (Kozhevnikov & Fee, 2007; Prather et al., 2008). Thus singing activity in HVCX cells appears to be motor-related, which can account for the singing-related signals that can be detected downstream in the AFP even in deafened birds (Hessler & Doupe, 1999). The ultimate source of this motor-related signal is likely to be the HVCRA cell population that directly and indirectly connects to HVCX cells through HVC's local synaptic network (Mooney & Prather 2005). Evidently, HVCX neurons receive a corollary discharge of the song motor signal precisely delayed by the local synaptic network to mimic the auditory signal evoked by the associated vocalization. Thus, HVCX cells exhibit one of the hallmarks of a mirror neuron: they display a systematic sensorimotor correspondence.

Auditory Responses of HVCX Neurons Predict a Perceptual Boundary

One characteristic predicted of an auditory vocal mirror neuron is that its auditory properties should be tightly linked to the individual's vocal perception. As previously mentioned, lesions to HVC can impair a songbird's ability to distinguish conspecific songs, implicating auditory responsive HVC neurons in song perception. One hint that swamp sparrow HVCX neurons facilitate song perception is that they respond to other swamp sparrow songs containing note sequences similar to the requisite sequence in the effective song type from the bird's own repertoire (Prather et al., 2008). The ability of HVCX cells to respond to other birds' songs raises the possibility that they could facilitate song perception, and are not simply involved in the processing of self-generated vocalizations. However, establishing a tighter link to perception requires comparing neuronal and perceptual responses.

Fortunately, this comparison was simplified by an earlier field study that established that swamp sparrows perceive continuous changes in song note duration in an all-or-none, or categorical, manner (Nelson & Marler, 1989). This remarkable ability of the brain to group stimuli that vary in a continuous manner into discrete perceptual categories facilitates a wide range of communication behaviors, including human speech (Diehl et al., 2004). To begin to explore the link between HVC auditory-vocal mirror neurons and categorical perception, we measured how individual HVCX cells responded when the freely behaving sparrow heard variants of the effective song type in which the duration of a single note in each syllable of the trill had been systematically varied (Prather et al., 2009). Indeed, HVCX neurons respond categorically to changes in note duration, indicating that their activity is tightly linked to perception (Figure 19.1b).

One potential discrepancy was that this neuronal response boundary differed from the previously published perceptual boundary (Nelson & Marler, 1989). Notably, this perceptual boundary was measured in a New York sparrow population geographically distinct from the Pennsylvania population used in our neural recordings. Because different sparrow populations learn different song dialects (Balaban, 1988), one intriguing idea is that the perceptual boundary for note duration may be influenced by learning and thus may differ between these two populations. Indeed, a parallel set of field studies that we conducted confirmed that the Pennsylvania population's perceptual boundary differed from the New York population and agreed with the neural boundary we had measured in the lab setting (Prather et al., 2009). Thus, the perceptual boundary for note duration was accurately predicted by the auditory responses measured in HVCX neurons, including a subset from which it was possible to collect singing-related activity and document that they exhibited a systematic auditory-vocal correspondence. This tight link between auditory properties and song perception lends further support to the idea that HVCX neurons function as auditory-vocal mirror neurons.

Functions of Auditory-Vocal Mirror Neurons in Perception and Learning

As previously mentioned, one expectation is that auditory-vocal mirror neurons will populate brain regions where they can affect learned vocal communication. In the songbird, auditory vocal HVCX neurons occupy a pivotal position where they could influence receptive and expressive aspects of song communication. Their position in the sensorimotor hierarchy also could enable them to play an important role in song learning.

Functional Implications for Communication

A role for HVCX neurons in receptive aspects of vocal communication is strongly supported by the close parallel between their auditory response properties and the bird's categorical perceptual boundaries, as well as by the deleterious effects of HVC and AFP lesions on song recognition (Brenowitz, 1991; Gentner et al., 2000; Scharff et al., 1998). As theorized for mirror neurons generally, auditory-vocal mirror neurons could facilitate perception by enabling the listener to categorize the songs of other birds in reference to its own repertoire. Because a songbird learns its song repertoire, song perception mediated by HVC auditory-vocal mirror neurons should be strongly influenced by learning. Two features are consistent with this view. First, lesions in the AFP, to which HVC mirror neurons send their axons, disrupt a bird's ability to distinguish different conspecific songs, with the most substantial deficits for songs most like its own (Scharff et al., 1998). Second, auditory selectivity in HVC, as well in the AFP, is strongly influenced by the sensorimotor effects of singing (Doupe, 1997; Volman, 1993). The importance of self-experience is underscored by the finding that HVC and AFP neurons acquire BOS selectivity even in birds that are made to sing spectrally distorted songs by cutting the vocal nerve or partially blocking airflow through the syrinx (Roy & Mooney, 2007; Solis & Doupe, 1999). Although a perceptual mechanism dependent on self-experience may provide a highly narrow filter through which to recognize song, it may also enable a degree of sensitivity not readily achieved without explicit sensorimotor interactions. Additionally, some juvenile songbirds, including swamp sparrows, produce plastic songs from many (> 10) different tutors, but retain only a small subset of these songs in their crystallized repertoire (Marler & Peters, 1982a). One possibility is that over-production followed by attrition permanently broadens the range of songs that can be discriminated through a sensitive filter dependent on sensorimotor experience of self-generated vocalization. If this model is correct, then HVC should contain a permanent record of transiently learned songs.

In support of that idea, we found that the adult HVC contains a persistent representation of juvenile experience (Prather et al., 2010). In a set of hand-reared swamp sparrows collected from the wild only a few days after hatching, we presented them with 21 different songs throughout their lives. In addition, we regularly sampled the vocal output of each bird throughout development and into adulthood. This record of the birds' lifetime of auditory and vocal experience revealed a subset of songs that were imitated during development but eliminated from the adult repertoire. The electrophysiological representation of those eliminated song types revealed that neurons in the adult HVC can be even more responsive to songs from the bird's developmental past than to any song in its adult repertoire. In addition, a small number of cells responded to tutor songs the bird heard during early development but for which no evidence of motor imitation was ever detected. Responses to song types that were heard but not imitated would have been nearly impossible to detect without a comprehensive knowledge of each bird's life history, and the ability of some HVC neurons to respond to song types not present in the adult repertoire could enable HVC to play a broader role in song recognition than would be possible if its neurons simply encoded the bird's current repertoire.

Auditory-vocal mirror neurons also could facilitate expressive aspects of vocal communication. More specifically, auditory activation of these neurons, which are embedded in the song motor network, could guide subsequent vocalization. In adult swamp sparrows, this process could enable a breeding male to select the song from its repertoire that most closely matches the song of a neighbor, resulting in matched countersinging. Young adult chipping sparrows (Liu & Nottebohm, 2007) and white-crowned sparrows in the late stages of plastic song also selectively crystallize songs in their repertoire most like those of nearby breeding males (Nelson & Marler, 1994), a developmental process of auditory-guided vocal matching that could be facilitated as well by auditory-vocal mirror neurons.

Functional Implications for Song Learning

Beyond serving a perceptual role in adult birds, auditory-vocal mirror neurons could also facilitate sensorimotor learning. An intriguing observation in support of this idea is that the auditory activity of individual HVCX neurons fails to accurately represent specific song features in a manner reminiscent of inaccurate imitation of those features (Prather, Peters, Nowicki, & Mooney 2012). Young swamp sparrows tutored with a trill that has been artificially accelerated well beyond species-typical norms sometimes produce brief bursts of accurately imitated syllables and trill rate that are separated by gaps of silence (Podos, 1996; Podos, Nowicki, & Peters, 1999). This "broken syntax" represents a major departure from the swamp sparrow's typical pattern of a continuous trill. In recordings of HVC auditory-vocal mirror neurons in adult swamp sparrows raised hearing tutor songs with normal trill rates, the auditory responses of HVCX neurons fail to follow highly accelerated trills, providing a possible sensory correlate of broken syntax. Although an individual cell can respond to some of the individual syllables in a highly accelerated trill, those responses are separated by gaps of several syllables' duration in which the HVCX cell fails to respond (Prather, Peters, Nowicki, & Mooney, 2012). This parallel between features of HVCX auditory processing and imitative song learning suggests one way in which HVCX auditory-vocal mirror neurons could influence song learning. More precisely, if the auditory responses of HVCX neurons are used to guide motor learning, the failure of HVC neurons to faithfully encode accurate auditory representations of accelerated trills could result in the generation of broken syntax independent of any motor constraints.

Behavioral evidence suggests that sensorimotor learning depends on a neural comparator that detects differences between auditory feedback and the template, generating an error signal that adaptively modifies the song motor network to subsequently minimize these differences. Although the nature of any comparator circuit remains enigmatic, two features of plastic song—namely trial-and-error variations in performance and evaluation by auditory feedback—are reminiscent of reinforcement learning. A general feature of reinforcement learning algorithms is that they evaluate performance by comparing performance outcome to the predicted outcome. Realized in the context of sensorimotor learning, a neuron providing such a prediction might display motor-related singing activity that systematically corresponds to the auditory signal evoked by the associated vocal gesture.

The striking sensorimotor correspondence exhibited by HVCX neurons raises the possibility that the singing-related activity of HVCX cells provides a motor-based prediction of auditory feedback. In the context of a comparator circuit, combining this predictive signal with the actual feedback signal could be used to compute an error signal. Assuming that feedback insensitivity characterizes the entire HVCX cell population, these neurons could provide one of the inputs to the comparator. Based on current knowledge, this arrangement would localize the comparator to the AFP or to other HVC neurons, including interneurons and HVCRA cells (Mooney & Prather, 2005). Another possibility is that HVCX neurons are the sites of comparison, but singing-related corollary discharge overwhelms the feedback signal. This may be especially likely in the adult birds used as subjects in (Prather et al., 2009, 2008), because their crystallized songs are relatively insensitive to feedback perturbations. As noted previously, one important step will be to determine whether HVCX neurons display an auditory-vocal correspondence during sensorimotor learning. A second step will be to test whether juvenile HVCX neurons are sensitive to acute feedback perturbations, when song changes most rapidly in response to altered feedback signals.

Synaptic Mechanisms for Generating the Auditory-Vocal Correspondence

A major goal in systems neuroscience is to understand the mechanisms by which neuronal networks give rise to higher-order functions, including perception and complex behavior. In this regard, a distinct advantage afforded by the songbird is that the neural networks that give rise to singing and song perception can be analyzed with cellular and synaptic resolution. Specifically, studies in songbirds can begin to illuminate the synaptic and circuit mechanisms that produce the precise sensorimotor correspondence exhibited by HVCX neurons. Indeed, analysis of the HVC microcircuitry using both in vivo and in vitro intracellular methods already provides substantial insights into the synaptic mechanisms giving rise to the precise auditory-vocal correspondence in HVCX neurons.

Two sets of findings stemming from such analysis indicate that the systematic auditory-vocal correspondence in HVCX neurons is established by local circuit mechanisms in HVC. First, a synaptic substrate for conveying song-related motor activity from HVCRA cells to HVCX neurons has been identified using intracellular recordings from identified neurons in brain slices (Dutar, Petrozzino, Vu, Schmidt, & Perkel, 2000; Dutar et al., 1998; Mooney & Prather, 2005). Dual intracellular recordings reveal that HVCRA cells are linked to HVCX cells via direct monosynaptic connections and disynaptic feedforward inhibitory projections (Mooney & Prather, 2005). This feedforward inhibitory linkage provides a plausible means by which corollary discharge transmitted to HVCX cells could be delayed so that it matches the timing of associated auditory feedback signals. Second, several observations stemming from in vivo intracellular recordings in HVC made in anesthetized birds indicate that the precise spike timing exhibited by HVCX cells depends on local inhibition. Intracellular recordings made from HVCX cells in both zebra finches and swamp sparrows reveal that BOS playback evokes strong membrane hyperpolarizations punctuated by highly phasic action potential bursts (Mooney, 2000; Mooney et al., 2001). These epochs of membrane hyperpolarization correlate closely with BOS-evoked firing in interneurons (Mooney, 2000; Rosen & Mooney, 2006), which make monosynaptic inhibitory connections onto HVCX cells (Mooney & Prather, 2005). Moreover, intracellular blockade of inhibitory input onto individual HVCX neurons shows that this inhibition is critical for regulating precisely when HVCX neurons fire action potentials in response to BOS playback (Rosen & Mooney, 2003). Finally, recordings made in either NIf or CM, both of which provide auditory input to HVC, fail to detect neurons that display either highly phasic BOS-evoked responses or a precise sensorimotor correspondence (Bauer et al., 2008; Coleman & Mooney, 2004).

Taken together, these various observations support the notion that the temporally precise sensorimotor correspondence exhibited by HVCX neurons is a product of local circuit mechanisms. These observations also raise the possibility that this precise correspondence is the product of an experience-dependent process wherein song-related corollary discharge emanating from HVCRA to HVCX neurons is "trained" by an auditory feedback signal (Troyer & Doupe, 2000). Of course, this raises the obvious question of the source of the training signal. In this light, a recent report that putative HVC interneurons in the Bengalese finch may respond to acute feedback perturbations suggests a likely source (Sakata & Brainard, 2008). Indeed, an attractive idea is that feedback perturbations act to acutely modulate the activity of HVC interneurons, which then over a slower time course retrain the corollary discharge signal. If this process of retraining subsequently modulates AFP activity, it could enable auditory feedback perturbations to exert temporally specific effects on song performance.

References

Balaban, E. (1988). Cultural and genetic variation in swamp sparrows (*Melospiza georgiana*). *Behaviour*, *105*, 250–290.

Bauer, E. E., Coleman, M. J., Roberts, T. F., Roy, A., Prather, J. F., & Mooney, R. (2008). A synaptic basis for auditory-vocal integration in the songbird. *Journal of Neuroscience*, *28*, 1509–1522.

Bottjer, S. W., Miesner, E. A., & Arnold, A. P. (1984). Forebrain lesions disrupt development but not maintenance of song in passerine birds. *Science*, *224*, 901–903.

Brainard, M., & Doupe, A. (2000). Interruption of a forebrain-basal ganglia circuit prevents plasticity of learned vocalizations. *Nature*, *404*, 762–766.

Brenowitz, E. A. (1991). Altered perception of species-specific song by female birds after lesions of a forebrain nucleus. *Science*, *251*, 303–305.

Burns, M. S., & Fahy, J. (2010). Broca's area: Rethinking classical concepts from a neuroscience perspective. *Topics in Stroke Rehabilitation*, *17*, 401–410.

Canady, R. A., Burd, G. D., DeVoogd, T. J., & Nottebohm, F. (1988). Effect of testosterone on input received by an identified neuron type of the canary song system: A Golgi/electron microscopy/degeneration study. *Journal of Neuroscience*, *8*, 3770–3784.

Cardin, J. A., Raksin, J. N., & Schmidt, M. F. (2005). Sensorimotor nucleus NIf is necessary for auditory processing but not vocal motor output in the avian song system. *Journal of Neurophysiology*, *93*, 2157–2166.

Cardin, J. A., & Schmidt, M. F. (2003). Song system auditory responses are stable and highly tuned during sedation, rapidly modulated and unselective during wakefulness, and suppressed by arousal. *Journal of Neurophysiology*, *90*, 2884–2899.

Catani, M., & Mesulam, M. (2008). The arcuate fasciculus and the disconnection theme in language and aphasia: History and current state. *Cortex*, *44*, 953–961.

Coleman, M. J., & Mooney, R. (2004). Synaptic transformations underlying highly selective auditory representations of learned birdsong. *Journal of Neuroscience*, *24*, 7251–7265.

Coleman, M. J., & Vu, E. T. (2005). Recovery of impaired songs following unilateral but not bilateral lesions of nucleus uvaeformis of adult zebra finches. *Journal of Neurobiology*, *63*, 70–89.

Dave, A. S., Yu, A. C., & Margoliash, D. (1998). Behavioral state modulation of auditory activity in a vocal motor system. *Science*, *282*, 2250–2254.

Diehl, R. L., Lotto, A. J., & Holt, L. L. (2004). Speech perception. *Annual Review of Psychology*, *55*, 149–179.

Dooling, R. (1978). Behavior and psychophysics of hearing in birds. *Journal of the Acoustical Society of America, 64*(S1), S4.

Doupe, A., & Kuhl, P. (1999). Birdsong and human speech: common themes and mechanisms. *Annual Review of Neuroscience, 22,* 567–631.

Doupe, A. J. (1997). Song- and order-selective neurons in the songbird anterior forebrain and their emergence during vocal development. *Journal of Neuroscience, 17,* 1147–1167.

Doupe, A. J., & Konishi, M. (1991). Song-selective auditory circuits in the vocal control system of the zebra finch. *Proceedings of the National Academy of Sciences of the United States of America, 88,* 11339–11343.

Doupe, A. J., Perkel, D. J., Reiner, A., & Stern, E. A. (2005). Birdbrains could teach basal ganglia research a new song. *Trends in Neurosciences, 28,* 353–363.

Dutar, P., Petrozzino, J., Vu, H., Schmidt, M., & Perkel, D. (2000). Slow synaptic inhibition mediated by metabotropic glutamate receptor activation of GIRK channels. *Journal of Neurophysiology, 84,* 2284–2290.

Dutar, P., Vu, H. M., & Perkel, D. J. (1998). Multiple cell types distinguished by physiological, pharmacological, and anatomic properties in nucleus Hvc of the adult zebra finch. *Journal of Neurophysiology, 80,* 1828–1838.

Farries, M. A. (2001). The oscine song system considered in the context of the avian brain: Lessons learned from comparative neurobiology. *Brain, Behavior and Evolution, 58,* 80–100.

Farries, M. A., & Perkel, D. J. (2000). Electrophysiological properties of avian basal ganglia neurons recorded in vitro. *Journal of Neurophysiology, 84,* 2502–2513.

Fee, M. S., & Leonardo, A. (2001). Miniature motorized microdrive and commutator system for chronic neural recording in small animals. *Journal of Neuroscience Methods, 112,* 83–94.

Fujimoto, H., Hasegawa, T., & Watanabe, D. (2011). Neural coding of syntactic structure in learned vocalizations in the songbird. *Journal of Neuroscience, 31,* 10023–10033.

Gentner, T. Q., Hulse, S. H., Bentley, G. E., & Ball, G. F. (2000). Individual vocal recognition and the effect of partial lesions to HVc on discrimination, learning, and categorization of conspecific song in adult songbirds. *Journal of Neurobiology, 42,* 117–133.

Geschwind, N. (1970). The organization of language and the brain. *Science, 170,* 940–944.

Haesler, S., Rochefort, C., Georgi, B., Licznerski, P., Osten, P., & Scharff, C. (2007). Incomplete and inaccurate vocal imitation after knockdown of FoxP2 in songbird basal ganglia nucleus Area X. *PLoS Biology, 5,* e321.

Hahnloser, R. H., Kozhevnikov, A. A., & Fee, M. S. (2002). An ultra-sparse code underlies the generation of neural sequences in a songbird. *Nature, 419,* 65–70.

Hessler, N. A., & Doupe, A. J. (1999). Singing-related neural activity in a dorsal forebrain-basal ganglia circuit of adult zebra finches. *Journal of Neuroscience, 19,* 10461–10481.

Hyman, J. (2003). Countersinging as a signal of aggression in a territorial songbird. *Animal Behaviour, 65,* 1179–1185.

Iacoboni, M., Molnar-Szakacs, I., Gallese, V., Buccino, G., Mazziotta, J. C., & Rizzolatti, G. (2005). Grasping the intentions of others with one's own mirror neuron system. *PLoS Biology, 3,* e79.

Immelmann, K. (1969). Song development in zebra finch and other estrildid finches. In R. A. Hinde (Ed.), *Bird vocalisations* (pp. 61–74). London: Cambridge University Press.

Kao, M. H., Doupe, A. J., & Brainard, M. S. (2005). Contributions of an avian basal ganglia–forebrain circuit to real-time modulation of song. *Nature, 433,* 638–643.

Kao, M. H., Wright, B. D., & Doupe, A. J. (2008). Neurons in a forebrain nucleus required for vocal plasticity rapidly switch between precise firing and variable bursting depending on social context. *Journal of Neuroscience, 28,* 13232–13247.

Kohler, E., Keysers, C., Umilta, M. A., Fogassi, L., Gallese, V., & Rizzolatti, G. (2002). Hearing sounds, understanding actions: Action representation in mirror neurons. *Science, 297,* 846–848.

Konishi, M. (1965). The role of auditory feedback in the control of vocalization in the white-crowned sparrow. *Zeitschrift für Tierpsychologie, 22,* 770–783.

Kozhevnikov, A. A., & Fee, M. S. (2007). Singing-related activity of identified HVC neurons in the zebra finch. *Journal of Neurophysiology, 97,* 4271–4283.

Lai, C. S., Fisher, S. E., Hurst, J. A., Vargha-Khadem, F., & Monaco, A. P. (2001). A forkhead-domain gene is mutated in a severe speech and language disorder. *Nature, 413,* 519–523.

Lai, C. S., Gerrelli, D., Monaco, A. P., Fisher, S. E., & Copp, A. J. (2003). FOXP2 expression during brain development coincides with adult sites of pathology in a severe speech and language disorder. *Brain, 126,* 2455–2462.

Leonardo, A. (2004). Experimental test of the birdsong error-correction model. *Proceedings of the National Academy of Sciences of the United States of America, 101,* 16935–16940.

Leonardo, A., & Fee, M. S. (2005). Ensemble coding of vocal control in birdsong. *Journal of Neuroscience, 25,* 652–661.

Liu, W. C., & Nottebohm, F. (2007). A learning program that ensures prompt and versatile vocal imitation. *Proceedings of the National Academy of Sciences of the United States of America, 104,* 20398–20403.

Locke, J. L. (1993). *The child's path to spoken language.* Cambridge, MA: Harvard University Press.

Lombardino, A. J., & Nottebohm, F. (2000). Age at deafening affects the stability of learned song in adult male zebra finches. *Journal of Neuroscience, 20,* 5054–5064.

MacDermot, K. D., Bonora, E., Sykes, N., Coupe, A. M., Lai, C. S., Vernes, S. C., et al. (2005). Identification of FOXP2 truncation as a novel cause of developmental speech and language deficits. *American Journal of Human Genetics, 76,* 1074–1080.

Margoliash, D. (1983). Acoustic parameters underlying the responses of song-specific neurons in the white-crowned sparrow. *Journal of Neuroscience, 3,* 1039–1057.

Margoliash, D. (1986). Preference for autogenous song by auditory neurons in a song system nucleus of the white-crowned sparrow. *Journal of Neuroscience, 6,* 1643–1661.

Marler, P. (1970). Birdsong and speech development: Could there be parallels? *American Scientist, 58,* 669–673.

Marler, P., & Peters, S. (1982a). Developmental overproduction and selective attrition: New processes in the epigenesis of birdsong. *Developmental Psychobiology, 15,* 369–378.

Marler, P., & Peters, S. (1982b). Long-term storage of learned birdsongs prior to production. *Animal Behaviour*, *30*, 479–482.

Marler, P., & Peters, S. (1982c). Structural changes in song ontogeny in the swamp sparrow Melospiza georgiana. *Auk*, *99*, 446–458.

Marler, P., & Peters, S. (1987). A sensitive period for song acquisition in the song sparrow, Melospiza melodia: A case of age-limited learning. *Ethology*, *76*, 89–100.

Marler, P., & Peters, S. (1988). Sensitive periods for song acquisition from tape recordings and live tutors in the swamp sparrow, Melospiza georgiana. *Ethology*, formerly *Zeitschrift für Tierpsychologie*, *77*, 76–84.

Marler, P., & Tamura, M. (1964). Culturally transmitted patterns of vocal behaviour in sparrows. *Science*, *146*, 1483–1486.

Marler, P., & Waser, M. S. (1977). Role of auditory feedback in canary song development. *Journal of Comparative and Physiological Psychology*, *91*, 8–16.

Martnez-Sanchez, F. (2010). Speech and voice disorders in Parkinson's disease. *Revista de Neurologia*, *51*, 542–550.

McCasland, J. S. (1987). Neuronal control of birdsong production. *Journal of Neuroscience*, *7*, 23–39.

McCasland, J. S., & Konishi, M. (1981). Interaction between auditory and motor activities in an avian song control nucleus. *Proceedings of the National Academy of Sciences of the United States of America*, *78*, 7815–7819.

Mooney, R. (1992). Synaptic basis for developmental plasticity in a birdsong nucleus. *Journal of Neuroscience*, *12*, 2464–2477.

Mooney, R. (2000). Different subthreshold mechanisms underlie song-selectivity in identified HVc neurons of the zebra finch. *Journal of Neuroscience*, *20*, 5420–5436.

Mooney, R., Hoese, W., & Nowicki, S. (2001). Auditory representation of the vocal repertoire in a songbird with multiple song types. *Proceedings of the National Academy of Sciences of the United States of America*, *98*, 12778–12783.

Mooney, R., & Prather, J. (2005). The HVC microcircuit: The synaptic basis for interactions between song motor and vocal plasticity pathways. *Journal of Neuroscience*, *25*, 1952–1964.

Mooney, R., Prather, J., & Roberts, T. (2008). Neurophysiology of birdsong learning. In H. Eichenbaum (Ed.), *Memory systems* (pp. 441–474). Oxford: Elsevier.

Nelson, D. A., & Marler, P. (1989). Categorical perception of a natural stimulus continuum: Birdsong. *Science*, *244*, 976–978.

Nelson, D. A., & Marler, P. (1994). Selection-based learning in bird song development. *Proceedings of the National Academy of Sciences of the United States of America*, *91*, 10498–10501.

Nottebohm, F., Kelley, D. B., & Paton, J. A. (1982). Connections of vocal control nuclei in the canary telencephalon. *Journal of Comparative Neurology*, *207*, 344–357.

Nottebohm, F., Stokes, T. M., & Leonard, C. M. (1976). Central control of song in the canary, Serinus canarius. *Journal of Comparative Neurology*, *165*, 457–486.

Okanoya, K. (2004). The Bengalese finch: A window on the behavioral neurobiology of birdsong syntax. *Annals of the New York Academy of Sciences*, *1016*, 724–735.

Okuhata, S., & Saito, N. (1987). Synaptic connections of a forebrain nucleus involved with vocal learning in zebra finches. *Brain Research Bulletin*, *18*, 35–44.

Olveczky, B. P., Andalman, A. S., & Fee, M. S. (2005). Vocal experimentation in the juvenile songbird requires a basal ganglia circuit. *PLoS Biology*, *3*, e153.

Podos, J. (1996). Motor constraints on vocal development in a songbird. *Animal Behaviour*, *51*, 1061–1070.

Podos, J., Nowicki, S., & Peters, S. (1999). Permissiveness in the learning and development of song syntax in swamp sparrows. *Animal Behaviour*, *58*, 93–103.

Prather, J. F., Nowicki, S., Anderson, R. C., Peters, S., & Mooney, R. (2009). Neural correlates of categorical perception in learned vocal communication. *Nature Neuroscience*, *12*, 221–228.

Prather, J. F., Peters, S., Nowicki, S., & Mooney, R. (2008). Precise auditory-vocal mirroring in neurons for learned vocal communication. *Nature*, *451*, 305–310.

Prather, J. F., Peters, S., Nowicki, S., & Mooney, R. (2010). Persistent representation of juvenile experience in the adult songbird brain. *Journal of Neuroscience*, *30*, 10586–10598.

Prather, J. F., Peters, S., Nowicki, S., & Mooney, R. (2012). Sensory constraints on birdsong syntax: Neural responses to swamp sparrow songs with accelerated trill rates. *Animal Behaviour*, *83*, 1411–1420.

Price, R. (1979). Developmental determinants of structure in zebra finch song. *Journal of Comparative and Physiological Psychology*, *93*(2), 260–277.

Rizzolatti, G. (2005). The mirror neuron system and its function in humans. *Anatomy and Embryology*, *210*, 419–421.

Rizzolatti, G., & Arbib, M. A. (1998). Language within our grasp. *Trends in Neurosciences*, *21*, 188–194.

Rizzolatti, G., & Arbib, M. A. (1999). From grasping to speech: Imitation might provide a missing link [reply]. *Trends in Neurosciences*, *22*, 152.

Rizzolatti, G., & Craighero, L. (2004). The mirror-neuron system. *Annual Review of Neuroscience*, *27*, 169–192.

Rosen, M. J., & Mooney, R. (2003). Inhibitory and excitatory mechanisms underlying auditory responses to learned vocalizations in the songbird nucleus HVC. *Neuron*, *39*, 177–194.

Rosen, M. J., & Mooney, R. (2006). Synaptic interactions underlying song-selectivity in the avian nucleus HVC revealed by dual intracellular recordings. *Journal of Neurophysiology*, *95*, 1158–1175.

Roy, A., & Mooney, R. (2007). Auditory plasticity in a basal ganglia-forebrain pathway during decrystallization of adult birdsong. *Journal of Neuroscience*, *27*, 6374–6387.

Roy, A., & Mooney, R. (2009). The song nucleus NIf is unnecessary for song decrystallization triggered by distorted auditory feedback. *Journal of Neurophysiology*, *102* (2), 979–991.

Sakata, J. T., & Brainard, M. S. (2008). Online contributions of auditory feedback to neural activity in avian song control circuitry. *Journal of Neuroscience*, *28*, 11378–11390.

Sakata, J. T., Hampton, C. M., & Brainard, M. S. (2008). Social modulation of sequence and syllable variability in adult birdsong. *Journal of Neurophysiology*, *99*, 1700–1711.

Scharff, C., & Nottebohm, F. (1991). A comparative study of the behavioral deficits following lesions of various parts of the zebra finch song system: Implications for vocal learning. *Journal of Neuroscience, 11,* 2896–2913.

Scharff, C., Nottebohm, F., & Cynx, J. (1998). Conspecific and heterospecific song discrimination in male zebra finches with lesions in the anterior forebrain pathway. *Journal of Neurobiology, 36,* 81–90.

Schmidt, M. F., & Konishi, M. (1998). Gating of auditory responses in the vocal control system of awake songbirds. *Nature Neuroscience, 1,* 513–518.

Simonyan, K., & Horwitz, B. (2011). Laryngeal motor cortex and control of speech in humans. *Neuroscientist, 17,* 197–208.

Solis, M. M., & Doupe, A. J. (1999). Contributions of tutor and bird's own song experience to neural selectivity in the songbird anterior forebrain. *Journal of Neuroscience, 19,* 4559–4584.

Teramitsu, I., Kudo, L. C., London, S. E., Geschwind, D. H., & White, S. A. (2004). Parallel FoxP1 and FoxP2 expression in songbird and human brain predicts functional interaction. *Journal of Neuroscience, 24,* 3152–3163.

Theunissen, F. E., Amin, N., Shaevitz, S. S., Woolley, S. M., Fremouw, T., & Hauber, M. E. (2004). Song selectivity in the song system and in the auditory forebrain. *Annals of the New York Academy of Sciences, 1016,* 222–245.

Theunissen, F. E., & Doupe, A. J. (1998). Temporal and spectral sensitivity of complex auditory neurons in the nucleus HVc of male zebra finches. *Journal of Neuroscience, 18,* 3786–3802.

Thorpe, W. (1958). The learning of song patterns by birds, with especial reference to the song of the chaffinch. *Ibis, 100,* 535–570.

Troyer, T., & Doupe, A. J. (2000). An associational model of birdsong sensroimotor learning. I. Efference copy and the learning of song syllables. *Journal of Neurophysiology, 84,* 1204–1223.

Velasco Garcia, M. J., Cobeta, I., Martin, G., Alonso-Navarro, H., & Jimenez-Jimenez, F. J. (2011). Acoustic analysis of voice in Huntington's disease patients. *Journal of Voice, 25,* 208–217.

Vicario, D. S., & Yohay, K. H. (1993). Song-selective auditory input to a forebrain vocal control nucleus in the zebra finch. *Journal of Neurobiology, 24,* 488–505.

Volman, S. F. (1993). Development of neural selectivity for birdsong during vocal learning. *Journal of Neuroscience, 13,* 4737–4747.

White, S. A., Fisher, S. E., Geschwind, D. H., Scharff, C., & Holy, T. E. (2006). Singing mice, songbirds, and more: Models for FOXP2 function and dysfunction in human speech and language. *Journal of Neuroscience, 26,* 10376–10379.

Wild, J. M. (2004). Functional neuroanatomy of the sensorimotor control of singing. *Annals of the New York Academy of Sciences, 1016,* 438–462.

Williams, H., & Mehta, N. (1999). Changes in adult zebra finch song require a forebrain nucleus that is not necessary for song production. *Journal of Neurobiology, 39,* 14–28.

Williams, H., & Vicario, D. S. (1993). Temporal patterning of song production: Participation of nucleus uvaeformis of the thalamus. *Journal of Neurobiology, 24,* 903–912.

Yu, A. C., & Margoliash, D. (1996). Temporal hierarchical control of singing in birds. *Science, 273,* 1871–1875.

20 Peripheral Mechanisms of Vocalization in Birds: A Comparison with Human Speech

Gabriël J. L. Beckers

Due to their extreme diversity in acoustic structure and sometimes high degree of complexity, birdsongs are among the most prominent of all animal acoustic signals. How birds achieve their vocal performances has been a central research question for years (e.g., Hérisant, 1753; Cuvier & Duvernoy, 1846; Häcker, 1900; Rüppell, 1933; Greenwalt, 1968; Gaunt & Nowicki, 1998; Nowicki & Marler, 1988; Suthers & Goller, 1997; Brackenbury, 1982). Initially, such research was driven by intrinsic interest in the physiological mechanisms underlying the behavior, and not so much by an interest from a comparative perspective with human speech. Although bird vocalization has long been known to share a number of interesting characteristics with human speech (Doupe & Kuhl, 1999; Bolhuis, Okanoya, & Scharff, 2010; Moorman & Bolhuis, chapter 5, this volume), it was originally believed that the underlying peripheral production mechanisms of birds and humans were fundamentally different. This view was rooted in the fact that birds and humans do not share a homologous voice organ, as well as in the observation that the acoustic structure of canonically studied birdsongs does not resemble that of speech (Figure 20.1). In recent years, however, the burgeoning interest in bird vocal learning as a model system for speech acquisition (Zeigler & Marler, 2004) has fueled a cascade of new studies into bird vocal production mechanisms, with more emphasis on a comparative perspective. This body of work has expanded our view of how birds vocalize in sometimes surprising ways and suggests that some key mechanisms in speech that were thought to be uniquely human may in fact have had a long evolutionary history, or at least may have evolved multiple times independently. Here I provide a general overview of current knowledge on peripheral vocal production mechanisms in birds, with a focus on aspects particularly relevant in a comparison with human speech. For reviews of this topic that are not speech-focused, I refer the reader to Suthers and Zollinger (2008) as well as Mindlin and Laje (2005).

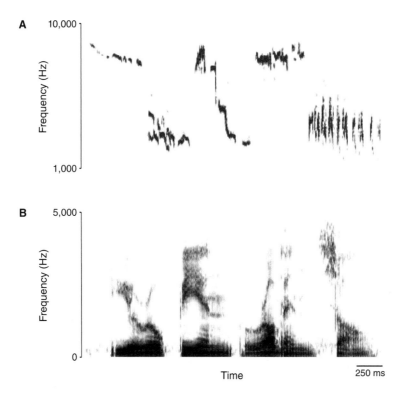

Figure 20.1

Spectrograms of (a) birdsong of the European robin, *Erithacus rubecula* (recording by courtesy of Claudia Ramenda), and (b) human speech, the author saying "European robin song." Spectrograms calculated with a short-time Fourier transform, Gaussian window of 13 ms, 40 dB dynamic range. Note that the range of the frequency axis differs between the subfigures.

Physical Principles

The biophysical fundamentals that underlie vocal production in all tetrapods, including birds and humans, are similar. The energy needed to produce sound originates from airflow through the respiratory tract that is induced by the action of respiratory muscles. Somewhere in this tract, the airflow is modulated in a voice organ by vibrating membranes or labia, a process that causes acoustic pressure waves. These vibrations are not caused directly by oscillatory muscle activity in the vocal organ, but arise passively from the elastic properties of the membranes that are induced to oscillate by airflow. Indeed, excised vocal organs in both birds (Rüppell, 1933; Fee, Shraiman, Pesaran, & Mitra, 1998) and mammals (e.g., Finnegan & Alipour, 2009) can be induced to phonate ex situ without neuromuscular control. The sound waves generated by the voice organ propagate through the respiratory tract to its distal end, and then radiate into the outside environment. The upper part of the respiratory tract is also called the "vocal tract" because it is often adapted to achieve specific acoustic resonance patterns for vocalization. The exact characteristics of the sound that is produced depend on the properties of each of these three fundamental components—the respiratory system, the voice organ, and the vocal tract—and on how they interact. Complex vocalizations with time-varying features may arise from the modulation of the physical properties of any or all these components during phonation, and require coordinated neural activity patterns from motor circuits in the brain (Fee and Long, chapter 18, this volume) as well as somatosensory feedback (Suthers, Goller, & Wild, 2002). However, complex vocalizations need not necessarily be generated by complex control systems (Gardner, Cecchi, Magnasco, Laje, & Mindlin, 2001; Fee et al., 1998).

It is appropriate to note that this review focuses on the production of communicative sounds in which a voice organ is involved. This does not include all communicative sounds. For example, in human speech the phonemes \a\ and \o\ in the words *pat* and *hot* originate from the voice, but the phonemes \p\, \h\, and \t\ do not (Ladefoged, 2005). The latter arise from fundamentally different sound sources in the upper vocal tract, and their role in communication is no less important than that of voiced sounds. The situation in birds, however, is less clear. Harmonic sounds, which are all sounds that have a periodic waveform (including "pure tones"), and nonlinear combinations thereof are likely produced by the voice organ. This is confirmed for cases in which this has been investigated directly (Goller & Larsen, 1997a; Jensen, Cooper, Larsen, & Goller, 2007; Larsen & Goller, 2002), but these represent only a minute fraction of the 9,000 extant bird species, almost all of which produce sounds for communication. Nonvoiced communication sounds in birds may include noisy ones, such as hissing in swans, but the topic has remained virtually unstudied so far. Vocalizations in birds may sometimes be hard to distinguish from sonations, which are sounds produced by external structures such as the bill, wings, tail, feet, and feathers (Darwin, 1871; Bostwick & Prum, 2003; Clark, Elias, & Prum, 2011).

The Respiratory System in Vocalization

Both humans and birds phonate mostly during the expiratory part of the respiratory cycle, although phonation during inspiration may sometimes occur (Goller & Daley, 2001; Gaunt, Gaunt, & Casey, 1982; Crystal, 2007). In humans, airflow through the respiratory tract is caused by air pressure variation in the lungs, while the situation is birds is much more complicated (McLelland, 1989; Fedde, 1998) and not completely understood. In addition to their lungs, birds possess an elaborate system of air sacs that are partly interconnected and connected to the lungs. The air sacs do not play a direct role in gas exchange with the circulatory system, but store air and act as bellows to ventilate the lungs and enable more efficient respiration.

Relevant to phonation is that the avian voice organ, the syrinx, is suspended in one of these air sacs, the clavicular air sac. The human vocal organ, the larynx, in contrast, is not enclosed in any air space. The clavicular air sac cavity may act as a resonator that somehow is involved in shaping the sound the bird eventually emits. This idea remains to be tested but is supported by the existence of very thin membranes in the syrinx of different groups of birds, which may allow for acoustic coupling between the air sac and the lumen of the syrinx. Indeed, in songbirds, suborder *Passeri*, such thin membranes (medial tympaniform membranes) do not seem to be directly involved as a primary sound source (Goller & Larsen, 1997a) although they may increase its efficiency (Fee, 2002). The syrinx in Eurasian collared doves, *Streptopelia decaocto*, as well as in other dove species in the same genus, has an additional very thin dorsal membrane (Ballintijn, ten Cate, Nuijens, & Berkhoudt, 1994) that is morphologically well separated from the primary sound generators (Goller & Larsen, 1997b) and that may specifically serve to couple the clavicular air sac acoustically to the syringeal lumen. Another consequence of the fact that the bird voice organ is situated inside an air sac is that the dramatic variation in air sac pressure during vocalization may directly modulate tension in the sound-generating structures in the syrinx, causing frequency modulation of the sound produced (Beckers, Suthers, & ten Cate, 2003a; Amador, Goller, & Mindlin, 2008).

Although birdsong often appears as a continuous flow of phonation, it is in fact often punctuated by brief silent episodes during which there are short inspirations known as minibreaths (Calder, 1970; Hartley & Suthers, 1989; Allan & Suthers, 1994; Goller & Suthers, 1996a, 1996b). These very fast inspirations during vocalization are not a passive phenomenon due to the elastic recoil forces of compressed air sacs after expiration, but arise from respiratory muscle activity (Wild, Goller, & Suthers, 1998).

The Vocal Source

The Voice Organ

Perhaps the most interesting difference in vocal production between birds and other tetrapods is the fact that birds have evolved a specialized voice organ, the syrinx, situated at the junction of the trachea and bronchi, deep inside the thoracic cavity. In all other tetrapods, including humans, voiced sounds are produced by the larynx, situated at the cranial end of the trachea. Birds also possess a larynx, but it has no known function in sound generation in this group. Because the syrinx is an evolutionary innovation whose single function is to produce sound, its design may be less constrained in this respect than that of the larynx, which also has important functions in respiration and feeding. This may explain why birds are such impressive vocal performers, as judged by many human listeners.

The syrinx comprises a cartilaginous or bony framework that contains vibratory membranes or labia, analogous to the laryngeal human vocal folds, and musculature that modulates the geometry configuration of the framework and tension of its vibratory components. The specifics of the morphology of the syrinx may differ between species, and even basic design may differ between higher-order taxa such as families (King, 1989; Myers, 1917; Miller, 1934; Ballintijn et al., 1994; Ames, 1971; Chamberlain, Gross, Cornwell, & Mosby, 1968). Indeed, syringeal morphology was one of the more important features for taxonomic classification of birds before molecular phylogenetic techniques became available (e.g., Beddard, 1898). The most complex syrinx is found in songbirds, which are also the most virtuous singers among birds. However, for birds in general a strict relationship between the complexity of syringeal morphology and a measure of complexity of the vocalizations produced remains elusive (Gaunt, 1983). There is also no clear relationship between syringeal morphology and vocal imitation learning: both parrots (order Psittaciformes) and songbirds may be excellent vocal learners, and even great imitators of human speech, but their syringes are fundamentally different.

The songbird syrinx (Figure 20.2) is unique in the animal kingdom in that it is a duplex voice organ. The extrapulmonary end of each bronchus has a pair of labia that act as a voice source, which can act independently during vocalization (Suthers, 1990), but can also interact mechanically or acoustically (Jensen, Cooper, Larsen, & Goller, 2007; Laje, Sciamarella, Zanella, & Mindlin, 2008; Nowicki & Capranica, 1986). The two-voice capabilities of the songbird syrinx are exploited in various ways across species to spectacularly enhance vocal performance.

In many songbirds, each voice specializes in its own vocal register. The left voice is normally used for lower notes and the right voice for higher ones, enabling a wide frequency range for the song as a whole (Suthers, Goller, & Pytte, 1999). Moreover, the use of this vocal strategy is not necessarily limited to sequential notes, because many songbirds will also sing independent notes simultaneously (Figure 20.3). This so-called two-voice phenomenon has long been known to birdsong aficionados and a duplex-voice system had been suspected (history discussed in Greenwalt, 1968). However, it was Suthers (1990) who recorded for the first time airflow in each syrinx-half in parallel and was able to provide direct proof that this explanation is indeed correct. Single notes can also be produced in a two-voiced way: Northern cardinals, *Cardinalis cardinalis*, sing seemingly continuous pure-tone notes that are modulated over an extremely wide frequency range, by starting phonation on one side of the syrinx and seamlessly switching to the other in the middle of the note (Suthers, 1997).

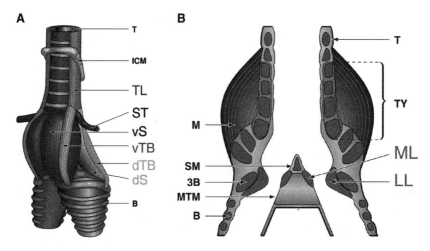

Figure 20.2

The tracheobronchial syrinx of songbirds is a complex bipartite structure situated where the trachea bifurcates into the bronchi. (a) Schematic ventrolateral external view showing the syringeal muscles. (b) Horizontal section through a songbird syrinx illustrating the two labial sound sources ML and LL. BC, bronchial cartilage; 3B, third bronchial cartilage; BL, bronchial lumen; dS, m. syringealis dorsalis; dTB, m. tracheobronchialis dorsalis; ICM, membrane of the interclavicular air sac; LL, lateral labium; M, syringeal muscle; ML, median labium; MTM, median tympaniform membrane; SM, semilunar membrane; ST, m. sternotrachealis; T, trachea; TL, m. tracheolateralis; TY, tympanum; vS, m. syringealis ventralis; vTB, m. tracheobronchialis ventralis. (Both subfigures from Larsen & Goller, 2002. Reproduced with permission.)

Another way the bipartite nature of the syrinx can be exploited in vocal performance is by enhancing respiration strategies during singing. Waterschlager canaries, *Serinus canaria*, have optimized this kind of respiration by using one side of the syrinx to take minibreaths between repeated notes while keeping the other side in phonatory position (Calder, 1970; Hartley & Suthers, 1989). This specialization increases the duration of the period in which fast syllable repetition can be sustained.

Of course, these and other specializations that are possible due to the duplex voice organ in songbirds do not occur in humans and other animals that only possess a single voice. It should be noted, however, that humans do use two simultaneous sound sources in speech. In voiced consonants such as the \z\ in *zebra*, for example, a harmonic sound is generated by the laryngeal voice, while at the same time a constriction formed between the tongue and teeth produces a noisy sound. Mammalian voices do sometimes appear to generate two simultaneous and independent voiced sounds, but these originate from the dynamics of a single voice in which the vocal folds do not vibrate synchronously (Wilden, Herzel, Peters, & Tembrock, 1998).

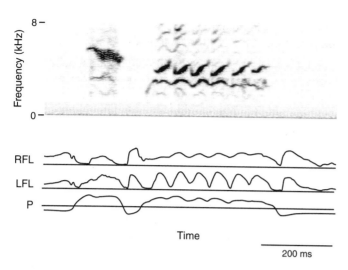

Figure 20.3

Two-voice phenomenon in the song of a Northern mockingbird (*Mimus polyglottos*), which has a bipartite syrinx as shown in Figure 20.2. The top panel shows a spectrogram of the vocalization, while the three plots at the bottom show concurrently recorded physiological parameters. Note that the lower-frequency trace in the second phonation is produced by the right side of the syrinx, while the simultaneous higher phonations are produced by the left side. RFL, airflow through the right bronchus; LFL, airflow through the left bronchus; P: air pressure in the cranial thoracic air sac. (Figure by courtesy of Sue Anne Zollinger.)

Mechanism of Voice Production

Exactly how sound is generated during voiced phonation in humans has been intensively studied and is relatively well known (Titze, 1994; Stevens, 1999). In the larynx, two oscillating vocal folds act as a pneumatic valve that modulates the expiratory airflow from the lungs. In part of the oscillation cycle, when the two vocal folds momentarily close off the airway, the airflow is interrupted. The result is a stream of brief air pulses that are released into the supralaryngeal airway and that generate acoustic pressure waves. The frequency of opening and closing of this valve-type source determines the frequency of the sound waves, and thus the pitch of the voice. Because the produced sound wave is periodic but not sinusoidal, its frequency representation consists of a fundamental frequency, typically around 100 Hz in male speech and 200 Hz in female speech (Ladefoged, 2005), and a series of harmonic overtones at integer multiples of this fundamental. This "harmonic stack" is visible as a series of parallel bands in spectrograms and is indicative of voiced phonation (i.e., of sounds that originate from vibrating vocal folds) in humans and other mammals.

In birds, in contrast, especially in song vocalization, the sound wave is often sinusoidal in nature, which is reflected in a single frequency component (a "pure tone") in spectrographic representation. This phenomenon has been considered difficult to explain on the basis of a humanlike pneumatic valve mechanism, and it has therefore been suggested that birds vocalize using a fundamentally different source mechanism (Nottebohm, 1976; Casey & Gaunt, 1985; Gaunt & Nowicki, 1998; Fletcher, 1989). Moreover, because typical birdsong sound resembles whistling in humans, it has been suggested that birds produce song using the same underlying sound-production principle, which is based on an aerodynamic mechanism. This hypothesis seems also to be favored by the general public: I know of at least six languages (Danish, Dutch, English, French, German, and Portuguese) in which it is said colloquially that birds "whistle" when they sing. However, despite the wide international agreement by the lay public on this issue, the hypothesis has not survived experimental scrutiny. Nowicki (1987) recorded songbirds singing in heliox gas and did not find a shift in fundamental frequency, as would be predicted on the basis of a whistle mechanism, or any other mechanism in which the fundamental frequency depends on the resonance characteristics of a source-coupled air cavity. A similar result was later obtained for pure-tonal song in the Eurasian collared dove, *Streptopelia decaocto* (Ballintijn & ten Cate, 1998). Furthermore, during spontaneous coo vocalizations of the same species as well as of a sister species, the ringdove, *Streptopelia risorea*, Beckers, Suthers, and ten Cate (2003b) recorded sound inside the trachea and clavicular air sac, close to the syringeal voice source. Their results showed that their pure-tonal coos are not produced as such by the syrinx, but rather as a human voicelike harmonic stack, the fundamental frequency of which corresponds to the sound that is eventually radiated from the bird (Figure 20.4). The harmonic overtones are filtered out by the vocal tract (Beckers et al, 2003b; Riede, Beckers, Blevins, & Suthers, 2004; Fletcher, Riede, Beckers, & Suthers, 2004). Thus, pure-tonal birdsong, in doves at least, can be explained on the basis of a humanlike voice mechanism.

The voice mechanisms underlying spectrally more complex bird vocalization has been observed directly with high-speed video and an angiofiberscope in zebra finches, *Taeniopygia guttata*, starlings, *Sturnus vulgaris*, and hooded crows, *Corvus corone cornix*, and is similar to that found in the human vocal-fry resister (Jensen, Cooper, Larsen, & Goller, 2007).

Taken together, the most straightforward explanation that emerges from the experiments and observations so far is that the syringeal voice source mechanism in bird vocalization, pure-tonal or not, is similar in principle to that of the laryngeal voice in humans, and is based on a pneumatic valve.

Modulation of Voice Features
Source phonation in birds and humans is usually modulated to generate time-varying frequency patterns. In birds there are two fundamentally different ways in

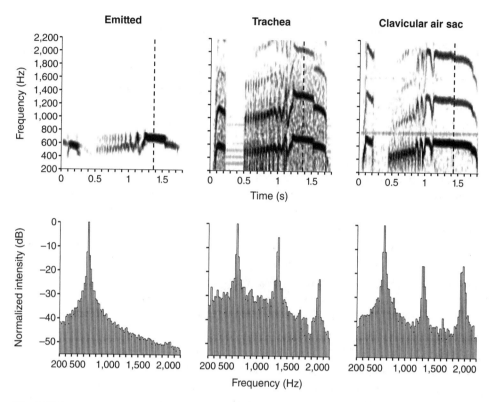

Figure 20.4

Pure-tonal vocalization in the ringdove, *Streptopelia risoria*, originates from syringeal voice sound that qualitatively resembles that produced by a laryngeal human voice. The emitted sound is pure-tonal, while the sound inside the tracheal lumen and in the clavicular air sac is multicomponent harmonic. Upper panels show spectrograms, while lower panels show the power spectrum at the time point indicated by the dashed line in the upper panels. (Adapted from Beckers et al., 2003b.)

which this is achieved. First, coordinated neuromuscular control of the respiratory and syringeal systems can directly modulate biophysical characteristics of the voice, so that spectral patterns and overall amplitude of the sound that is produced changes over time. This type of voice modulation is similar to that occurring in human speech. Louder speech sounds are achieved either through increasing airflow from the lungs or by increasing glottal resistance, both of which are controlled by muscle actions. Similarly, the pitch variation of speech sounds is mostly controlled by intrinsic and extrinsic laryngeal muscles that are able to change the resonance properties of the vocal folds (Stevens, 1999).

In songbirds, the fundamental frequency of phonation (pitch) is controlled by the action of intrinsic syringeal muscles (Goller & Suthers, 1996b; Elemans, Mead, Rome, & Goller., 2008), which presumably regulate the tension of the vibrating labia (Figure 20.5a,b). Many nonsongbirds, comprising more than half of all bird species, however, lack intrinsic syringeal musculature and are still able to modulate the fundamental frequency of their vocalizations. Beckers, Suthers, & ten Cate. (2003a) showed that in the coo vocalizations of ringdoves, which is a nonsongbird without intrinsic syringeal musculature, the fundamental frequency correlates closely with the air pressure in the clavicular air sac (Figure 20.5c). Presumably, the pressure variation modulates the sound-generating membranes directly. Frequency modulation during brief transient onsets and offsets of notes, which could not be explained by air pressure variation, was later shown by Elemans, Zaccarelli, & Herzel (2008); also Elemans, Spierts, Muller, Van Leeuwen, & Goller, (2004) to correlate with the activity of extrinsic syringeal muscles. In a different nonsongbird, the great kiskadee, *Pitangus sulfuratus*, the overall pattern of fundamental frequency variation is also highly correlated with air sac pressure (Amador et al., 2008). An experiment to specifically test the involvement of syringeal musculature—denervating the syrinx, and thereby abolishing coordinated muscle activity—did not have an effect on the strong frequency modulation patterns in this species (Amador et al., 2008). Thus, to conclude, the results so far suggest that frequency modulation in songbirds is largely caused by specialized syringeal musculature, while in nonsongbirds it is largely caused by air pressure variation in the air sac system. However, these results need to be verified in more species to confirm this generalization.

An interesting recent finding on the bird voice modulation is that superfast syringeal muscles actuate extremely fast frequency and amplitude variation, both in a songbird, the starling, *Sturnus vulgaris* (Elemans et al., 2008) and in a nonsongbird, the ringdove (Elemans et al., 2004). In the former, intrinsic syringeal muscles directly underlie modulation frequencies of over 200 Hz. The performance of these muscles thus ranks among the fastest known in vertebrates (Rome, 2006). Fast muscles also occur in the mammalian larynx (Hoh, 2005), but they are not nearly as fast as those found in songbirds.

A second and fundamentally different way in which voice characteristics may be modulated in birds is through nonlinear interactions. For example, in songbirds the two sides of the duplex syrinx may not only be used independently, but they may also interact acoustically or mechanically to create more complex voice patterns. In black-capped chickadees, *Parus atricapillus*, each syringeal side vibrates at its own frequency but they interact nonlinearly; the resulting sound contains the fundamental frequencies and harmonic overtones from both sides, but also their sum or difference (heterodyne) frequencies (Nowicki & Capranica, 1986). In such cases it is

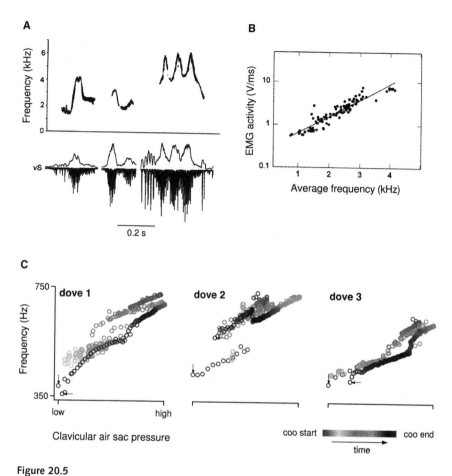

Figure 20.5
Frequency modulation of voice source phonation in (a) and (b) a songbird, the brown
thrasher, *Toxostoma rufum*, where EMG of ventral syringeal muscles is positively correlated
with song frequency, and (c) in a nonsongbird, the ringdove, *Streptopelia risoria*, where overall
song frequency is correlated with clavicular air sac pressure. vS = EMG of M. syringealis
ventrals. (Subfigures a and b from Goller and Suthers, 1996b, reproduced with permission;
subfigure c from Beckers et al., 2003a.)

perhaps misleading to speak of a two-voice system, because the two syringeal halves behave as one system to produce a more complex source sound. Most likely, at least the extremely rapid periodic modulations found in bird vocalizations, which sometimes exceed 500 Hz and are unlikely to be under direct muscle control, originate from such nonlinear interactions.

Fee et al. (1998) showed that the role of nonlinear dynamics in birdsong is even more extensive than this and also underlies other types of vocal complexity. In zebra finches, *Taeniopygia guttata*, seemingly qualitative changes in phonation, such as frequency jumps and sudden transitions from periodic to aperiodic vibration dynamics, can arise spontaneously from the intrinsic properties of the vocal production system (but see Elemans, Laje, Mindlin, & Goller, 2010). This is due to the fact that the syringeal voice source consists of vibrating labia or membranes with nonlinear properties. It has been observed in diverse scientific fields that coupled nonlinear oscillators in general can exhibit complex dynamics without any external, complex control ("chaos theory"; see Ott, 2002, for a textbook). The key point about nonlinear dynamics in vocalization is that a relatively simple system such as two coupled vibrating membranes can give rise to a series of different states of oscillatory behavior that appear as qualitatively different, even though the changes in control parameters are simple and continuous.

Nonlinear phenomena do not require the involvement of the two sides of the syrinx (Zollinger, Riede, & Suthers, 2008; Fee et al., 1998), and are not even restricted to birds with a bipartite syrinx: bird species that are single-voice also produce them (Figure 20.6). It has been shown in two species of cockatoo that their calls have a chaotic structure (Fletcher, 2000). Turtledoves, genus *Streptopelia*, also produce vocalizations that show signs of nonlinear dynamics, such as frequency jumps, subharmonics, and chaos (Beckers & ten Cate, 2006; Beckers et al., 2003a). Beckers and ten Cate (2006) hypothesize that such large, seemingly qualitative differences between species-specific song in this genus of doves may correspond to different attractor states of the same type of dynamic system. Large differences between species sounds are not necessarily due to correspondingly large differences in sound-production mechanisms or evolutionary differentiation.

Most bird vocalizations that are described as "noisy" may also be due to nonlinear dynamics and reflect a chaotic oscillatory state of the syrinx. It is interesting to note that this is in contrast with noisy sound sources in human speech, such as in whispered consonants (e.g., English /h/), which do not originate from vocal fold oscillations but from air turbulence in the vocal tract, a process that indeed is truly random or noisy. Chaos, however, is completely deterministic and only superficially resembles noise in spectrographic representation. Nevertheless, some bird vocalizations may actually originate from a true noise source, at least as judged by the human ear—for example, the already-mentioned hissing in swans.

Nonlinear dynamics also occur in human voice production. In speech they are normally associated with pathologies, but they can play an important role in other vocalizations, such as crying or shouting (e.g., Neubauer, Edgerton, & Herzel, 2004). The same seems true for many vocalizations of other mammals (Fitch, Neubauer, & Herzel, 2002; Wilden et al., 1998).

Vocal-Tract Formants

As outlined above, birds in general, and songbirds in particular, have evolved a specialized voice organ that in multiple ways is more versatile than the mammalian laryngeal voice organ. This may explain why a possible role for the vocal tract as a filter in bird vocal production has traditionally received relatively little consideration. Exceptions to this are studies testing the involvement of the vocal tract as a resonance space that influences syringeal voice frequency (so-called coupling between vocal tract and syrinx, as would for example be the case with a whistle mechanism). However, so far tests for the whistle hypothesis have turned out negative, both in songbirds (Nowicki, 1987) and nonsongbirds (Ballintijn & ten Cate, 1998), although a coupling between tracheal resonances and membrane vibration frequency has been found in a mechanical model of the syrinx that is based on a pneumatic valve (Elemans, Muller, Larsen, & Van Leeuwen, 2009).

In human speech, in contrast, the vocal tract has long been known to play a major role in modulating acoustic features that are produced independently from the voice source (Fant, 1960). These features, called formants, arise from acoustic resonances from the vocal tract that filter the voice sound, and thereby shape its spectral characteristics. As humans speak, they continually change the geometry of the vocal tract by articulation of the tongue, lips, and soft palate, in turn changing its resonance properties and thus the spectral characteristics of the sound produced. These formant-dependent spectral characteristics code important information that has linguistic meaning. Indeed, in many languages (e.g., English) formant patterns in vocalizations are much more important with respect to linguistic meaning than any feature of the voice (Ladefoged, 2005). This is exemplified by the fact that one can communicate the same information both in voiced speech and in whispered speech. In the latter, there is no voice activity while vocal-tract articulation-induced formant patterns are the same.

Vocal-Tract Filtering to Produce Pure-Tonal Song
Although one could get the impression from some of the earlier birdsong literature that vocal-tract filtering in birds might not play a role in vocalization at all, this is in a strict sense impossible because it is physically inevitable that an enclosed system of air cavities possesses resonant properties (Fletcher, 1992). The more relevant

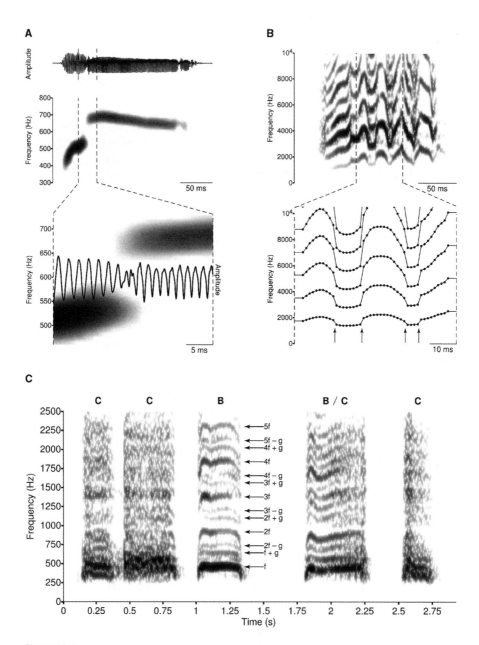

Figure 20.6
Nonlinear phenomena in bird vocalization. (a) A sudden frequency jump ("voice breaking")
in a coo note of the ringdove, *Streptopelia risoria*, that is not explained by clavicular air sac
pressure variation (see discontinuities in Figure 20.4c) and that may well be due to a bifurca-
tion in the dynamics of the underlying nonlinear voice system. (b) Similar jumps in a contact

question, then, is whether vocal-tract filtering is involved in shaping important features of time-frequency patterns, as in human speech.

The evidence that this is indeed the case is accumulating. Singing birds often modulate their beak gape during vocalization (Westneat, Long, Hoese, & Nowicki, 1993). Hoese, Podos, Boetticher, and Nowicki (2000) showed in canaries that these articulations are necessary to produce spectrally normal song: if they are disrupted, the normally pure-tonal song changes into song with harmonic overtones. This observation is consistent both with the idea that the vocal tract filters a multiharmonic sound into a pure-tonal sound, and with the alternative idea that vocal-tract properties cause the voice to produce a pure-tonal sound. The former idea corresponds to a linear interaction between voice source and vocal-tract filter, as is the case in human speech, while the latter mechanism depends on a nonlinear interaction between vocal-tract resonances and voice source, which is known to occur in human soprano singing (Rothenberg, 1987). Beckers et al. (2003b) showed that in turtledoves the vocal tract functions as a resonance filter, which converts a multiharmonic source sound into a pure-tonal one as it radiates from the animal (Figure 20.4). This provided the first direct evidence for a filtering function in birdsong, although the filter is relatively static and is not involved in causing complex time-frequency patterns. Its function probably is to amplify the fundamental frequency of the syringeal source sound, rather than remove harmonic overtones.

In songbirds a similar filtering mechanism may exist. Riede, Suthers, Fletcher, and Blevins (2006), using X-ray cinematography, found in Northern cardinals that the volume of the oropharyngeal-esophageal cavity is modulated during singing, and correlates with voice fundamental frequency. As in the disrupted beak-gape experiments, these results do not distinguish between the vocal tract as linear source-filter system as found in human speech and a nonlinear interaction between vocal tract and the syrinx as in human soprano song. Although it has been shown that vocal-tract articulations cause modulation of resonance filtering characteristics in zebra finches (Ohms, Snelderwaard, ten Cate, & Beckers, 2010), this species does not sing

---

call of a monk parakeet, *Myiopsitta monachus*; note that the frequency jumps (indicated with arrows) follow a particular pattern—for example, in the first jump, the 5th harmonic before the jump is continuous with the 6th harmonic after the jump. This is typical for such dynamic systems (Fee et al., 1998). (c) The perch-coo of the oriental turtledove, *Streptopelia orientalis*, shows other transitions in oscillatory dynamics that are typical of nonlinear systems: C = chaotic regime, B = biphonation regime. In the 3rd element, a harmonic signal (f, first five harmonics shown) is modulated by a lower-frequency component (g), which causes a sideband pattern around each harmonic of f. Note that in the 4th element, the regime changes from biphonation (or perhaps subharmonic) to chaotic. (a from Beckers et al., 2003a; c from Beckers & ten Cate 2006.)

pure-tonal songs, which complicates direct interpretation of X-ray film with respect to the relationship between vocal-tract articulation and acoustic modulation. It thus remains possible that a source-filter separation hypothesis does not hold in songbirds (Laje & Mindlin, 2005). This notwithstanding, the findings so far suggest that in addition to the beak gape, other parts of the vocal tract are modulated during phonation in order to track the changing source frequency. Beak movements alone seem to act as a variable low-pass filter (Nelson, Beckers, & Suthers, 2005; Ohms et al., 2010), and not as a variable bandpass filter.

Vocal-Tract Filtering to Produce Independent Formant Patterns

By themselves, the types of vocal-tract involvement described so far, both in songbirds and nonsongbirds, do not cause an additional level of vocal complexity; their resonance patterns are adjusted to the frequency of the sound produced by the syrinx, which at least in songbirds requires fast and precise coordination of neuromuscular control of craniomandibular, syringeal, and respiratory systems. Thus, in these cases the function of vocal-tract filtering or articulation does not resemble the situation in human speech, where vocal-tract articulations cause complex formant patterns independent of the voice source.

Recent work, however, has shown that this hallmark of human speech production also occurs in birds. Morphological (Homberger, 1986) and behavioral observations (Nottebohm, 1976; Patterson & Pepperberg, 1994, 1998; Warren, Patterson, & Pepperberg, 1996; Ohms, Beckers, ten Cate, & Suthers, 2012) suggest that in parrots, tongue movements modulate formant patterns independently from the voice source, analogously to speech. Beckers, Nelson, and Suthers (2004) have shown that such lingual articulation indeed causes significant changes in formant patterns of monk parakeets, *Myiopsitta monachus*. They replaced the syrinx with a small speaker that emitted broadband sounds, and determined the effect of vocal-tract filtering by analyzing the frequency patterns of the sound that radiated from the bird under different tongue placements. Both high-low and front-back placements in this species modulate the four formants that are present between 0.5 and 10 kHz (Figure 20.7b). These findings may not only explain the ability of parrots to mimic human speech; the natural calls of monk parakeets also have a broadband frequency spectrum in which formant patterns are indicated (see Figure 20.7a). However, whether these formant patterns have meaning to these birds, or are even learned by imitation, remains to be investigated.

In conclusion, recent research has shown convincingly that the vocal tract constitutes an important part of the avian vocal production system in different bird taxa, and can shape vocalizations in different ways (Figure 20.8). In doves and parrots the vocal tract can act as a linear filter, as is the case in the source-filter system in human speech production. In songbirds this may or may not be the case. However,

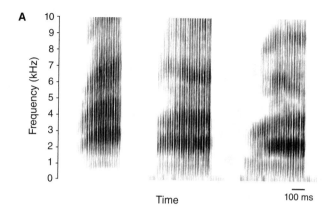

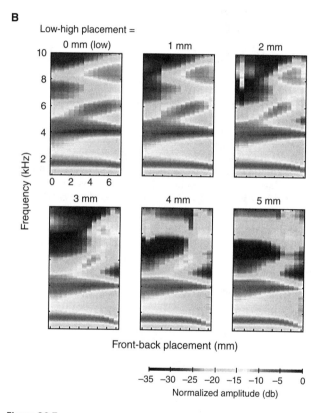

Figure 20.7
Formant modulation in the monk parakeet, *Myiopsitta monachus*. (a) Formant modulations that resemble those in human speech seem to occur in the greeting calls of this species (examples of three different individuals; compare with human voice in Figure so.1b). Spectrograms were calculated with a short-time Fourier transform, with a Gaussian 3 ms window and a 30 dB dynamic range. (b) Tongue placement and corresponding other articulation causes formant modulation: resonance characteristics of the vocal tract change with both low-high and front-back articulations. (Adapted from Beckers et al., 2004.)

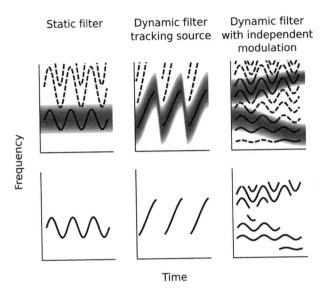

Static filter | Dynamic filter tracking source | Dynamic filter with independent modulation

Frequency

Time

Figure 20.8
Schematic representation of three models of formant use, as found in bird vocalization so far. In the simplest one (static filter), found in the ringdove, a strong, single formant shows little or no modulation and amplifies the fundamental frequency of the voice source, which is not modulated beyond the formant bandwidth. In many songbirds, however, the voice frequency is modulated over a wide range, which is tracked by modulating vocal-tract formants accordingly (dynamic filter tracking source). Voice and formant modulation may also occur independently, as is the case in monk parakeets, resembling the system used in human speech (dynamic filter with independent modulation). Solid lines: voice source frequencies that are radiated after formant filtering; dashed lines: voice source frequencies that are not supported by formants; gray bands: vocal-tract formants.

irrespective of the actual underlying acoustic mechanism, it has been shown in this group that vocal-tract articulations are necessary to produce spectrally normal vocalizations.

Future Directions

Although much progress has been made in understanding how birds vocalize, it is probably fair to say that, by far, most of the work remains to be done. The phrase "how birds vocalize" may even be unhelpful, if we consider the enormous diversity in birds (some 9,000 species), the sometimes fundamental differences in morphology of the vocal tract and other organs, and the extremely different vocalizations they can produce. Indeed, as outlined in this chapter, in some of the few species studied so far vocal mechanisms appear to be substantially different. We can only hope that future studies will include more species.

Traditionally, the focus in the field of bird vocal behavior has been on songbirds, preferably ones that sing beautiful songs. It is interesting to note that the two recent findings on peripheral production mechanisms that are of interest from a comparative perspective with human speech—vocal-tract filtering and articulation—both concerned nonsongbirds. Likewise, the existence of superfast muscles in bird vocalization has first been established in a nonsongbird, which subsequently led to similar studies in a songbird. Vocal production mechanisms in nonsongbirds may thus deserve more attention that they have received so far.

Perhaps most notable is that vocal production in parrots has so far received only sporadic attention, while from a phonetic and behavioral point of view their vocalizations may be more comparable to human speech than those of any other group of birds. Performing physiological studies in parrots is even more challenging than in songbirds because they often outsmart measures designed to prevent them from sabotaging sensitive—or even rugged—measurement devices. However, current fast developments in the miniaturization of transducers and telemetry will likely provide new opportunities to improve the situation within the foreseeable future, and stimulate more studies in parrots as well as other bird taxa.

From a comparative point of view it will be particularly interesting to test whether birds with broadband vocalizations use, or even learn, patterns of formant modulation for communication. Parrots may be a promising group in which to look for this, but some songbirds, such as zebra finches and crows, may also prove interesting in this respect.

References

Allan, S. E., & Suthers, R. A. (1994). Lateralization and motor stereotype of song production in the brown-headed cowbird. *Journal of Neurobiology, 25*, 1154–1166.

Amador, A., Goller, F., & Mindlin, G. B. (2008). Frequency modulation during song in a suboscine does not require vocal muscles. *Journal of Neurophysiology, 99*, 2383–2389.

Ames, P. (1971). The morphology of the syrinx in passerine birds. *Bulletin of the Peabody Museum of Natural History, 37*, 1–194.

Ballintijn, M., & ten Cate, C. (1998). Sound production in the collared dove: A test of the "whistle" hypothesis. *Journal of Experimental Biology, 201*, 1637–1649.

Ballintijn, M., ten Cate, C., Nuijens, F., & Berkhoudt, H. (1994). The syrinx of the collared dove (Streptopelia decaocto): Structure, inter-individual variation and development. *Netherlands Journal of Zoology, 45*, 455–479.

Beckers, G. J. L., & ten Cate, C. (2006). Nonlinear phenomena and song evolution in Streptopelia doves. *Acta Zoologica Sinica, 52*, 482–485.

Beckers, G. J. L., Nelson, B. S., & Suthers, R. A. (2004). Vocal-tract filtering by lingual articulation in a parrot. *Current Biology*, *14*, 1592–1597.

Beckers, G. J. L., Suthers, R. A., & ten Cate, C. (2003a). Mechanisms of frequency and amplitude modulation in ring dove song. *Journal of Experimental Biology*, *206*, 1833–1843.

Beckers, G. J. L., Suthers, R. A., & ten Cate, C. (2003b). Pure-tone birdsong by resonance filtering of harmonic overtones. *Proceedings of the National Academy of Sciences of the United States of America*, *100*, 7372–7376.

Beddard, F. (1898). *The structure and classification of birds*. New York: Longmans, Green, and Co.

Bolhuis, J. J., Okanoya, K., & Scharff, C. (2010). Twitter evolution: Converging mechanisms in birdsong and human speech. *Nature Reviews: Neuroscience*, *11*, 747–759.

Bostwick, K. S., & Prum, R. O. (2003). High-speed video analysis of wing-snapping in two manakin clades (Pipridae: Aves). *Journal of Experimental Biology*, *206*, 3693–3706.

Brackenbury, J. H. (1982). The structural basis of voice production and its relationship to sound characteristics. In D. E. Kroodsma, E. H. Miller, & H. Ouellet (Eds.), *Acoustic communication in birds* (pp. 53–73). New York: Academic Press.

Calder, W. A. (1970). Respiration during song in the canary (Serinus canaria). *Comparative Biochemistry and Physiology*, *32*, 251–258.

Casey, R. M., & Gaunt, A. S. (1985). Theoretical models of the avian syrinx. *Journal of Theoretical Biology*, *116*, 45–64.

Chamberlain, D., Gross, W., Cornwell, G., & Mosby, H. (1968). Syringeal anatomy in the common crow. *Auk*, *89*, 244–252.

Clark, C. J., Elias, D. O., & Prum, R. O. (2011). Aeroelastic flutter produces hummingbird feather songs. *Science*, *333*, 1430–1433.

Crystal, D. (2007). *The Cambridge encyclopedia of language*. Cambridge: Cambridge University Press.

Cuvier, G., & Duvernoy, G. L. (1846). *Leçons d'anatomie comparée*. Paris: Fortin, Masson et Cie.

Darwin, C. (1871). *The Descent of Man*. London: John Murray.

Doupe, A. J., & Kuhl, P. K. (1999). Birdsong and human speech: Common themes and mechanisms. *Annual Review of Neuroscience*, *22*, 567–631.

Elemans, C. P. H., Laje, R., Mindlin, G., & Goller, F. (2010). Smooth operator: Avoidance of subharmonic bifurcations through mechanical mechanisms simplifies song motor control in adult zebra finches. *Journal of Neuroscience*, *30*, 13246–13253.

Elemans, C. P. H., Mead, A. F., Rome, L. C., & Goller, F. (2008). Superfast vocal muscles control song production in songbirds. *PLoS ONE*, *3*, e2581.

Elemans, C. P. H., Muller, M., Larsen, O. N., & van Leeuwen, J. L. (2009). Amplitude and frequency modulation control of sound production in a mechanical model of the avian syrinx. *Journal of Experimental Biology*, *212*, 1212–1224.

Elemans, C. P. H., Spierts, I. L. Y., Muller, U. K., van Leeuwen, J. L., & Goller, F. (2004). Bird song: Superfast muscles control dove's trill. *Nature*, *431*, 146.

Elemans, C., Zaccarelli, R., & Herzel, H. (2008). Biomechanics and control of vocalization in a non-songbird. *Journal of the Royal Society Interface, 5*, 691–703. .

Fant, G. (1960). *Acoustic theory of speech production.* The Hague: Mouton.

Fedde, M. R. (1998). Relationship of structure and function of the avian respiratory system to disease susceptibility. *Poultry Science, 77*, 1130–1138.

Fee, M. S. (2002). Measurement of the linear and nonlinear mechanical properties of the oscine syrinx: Implications for function. *Journal of Comparative Physiology. A, Neuroethology, Sensory, Neural, and Behavioral Physiology, 188*, 829–839.

Fee, M. S., Shraiman, B., Pesaran, B., & Mitra, P. P. (1998). The role of nonlinear dynamics of the syrinx in the vocalizations of a songbird. *Nature, 395*, 67–71.

Finnegan, E. M., & Alipour, F. (2009). Phonatory effects of supraglottic structures in excised canine larynges. *Journal of Voice, 23*, 51–61.

Fitch, W. T., Neubauer, J., & Herzel, H. (2002). Calls out of chaos: The adaptive significance of nonlinear phenomena in mammalian vocal production. *Animal Behaviour, 63*, 407–418.

Fletcher, N. H. (1989). Acoustics of bird song—Some unresolved problems. *Comments on Theoretical Biology, 1*, 237–251.

Fletcher, N. H. (1992). *Acoustic systems in biology.* Oxford: Oxford University Press.

Fletcher, N. H. (2000). A class of chaotic bird calls? *Journal of the Acoustical Society of America, 108*, 821–826.

Fletcher, N. H., Riede, T., Beckers, G. J. L., & Suthers, R. A. (2004). Vocal tract filtering and the "coo" of doves. *Journal of the Acoustical Society of America, 116*, 3750–3756.

Gardner, T., Cecchi, G., Magnasco, M., Laje, R., & Mindlin, G. B. (2001). Simple motor gestures for birdsongs. *Physical Review Letters, 87*, 208101.

Gaunt, A. S. (1983). A hypothesis concerning the relationship of syringeal structure to vocal abilities. *Auk, 100*, 853–862.

Gaunt, A. S., Gaunt, S. L. L., & Casey, R. M. (1982). Syringeal mechanisms reassessed: Evidence from Streptopelia. *Auk, 99*, 474–494.

Gaunt, A. S., & Nowicki, S. (1998). Sound production in birds: Acoustics and physiology revisited. In S. L. Hopp, M. J. Owren, & C. S. Evans (Eds.), *Animal acoustic communication* (pp. 291–321). Berlin: Springer.

Goller, F., & Daley, M. A. (2001). Novel motor gestures for phonation during inspiration enhance the acoustic complexity of birdsong. *Proceedings of the Royal Society B: Biological Sciences, 268*, 2301–2305.

Goller, F., & Larsen, O. N. (1997a). A new mechanism of sound generation in songbirds. *Proceedings of the National Academy of Sciences of the United States of America, 94*, 14787–14791.

Goller, F., & Larsen, O. N. (1997b). In situ biomechanics of the syrinx and sound generation in pigeons. *Journal of Experimental Biology, 200*, 2165–2176.

Goller, F., & Suthers, R. A. (1996a). Role of syringeal muscles in controlling the phonology of bird song. *Journal of Neurophysiology, 76*, 287–300.

Goller, F., & Suthers, R. A. (1996b). Role of syringeal muscles in gating airflow and sound production in singing brown thrashers. *Journal of Neurophysiology, 75*, 867–876.

Greenwalt, C. H. (1968). *Bird song: Acoustics and physiology*. Washington, DC: Smithsonian Institution Press.

Häcker, V. (1900). *Der Gesang der Vögel*. Jena: Gustav Fisher.

Hartley, R. S., & Suthers, R. A. (1989). Airflow and pressure during canary song: Direct evidence for mini-breaths. *Journal of Comparative Physiology A: Neuroethology, Sensory, Neural, and Behavioral Physiology, 165*, 15–26.

Hérisant, M. (1753). Recherches sur les organes de la voix des quadrupèdes et de celle des oiseaux. *Mémoires de l'Academie Royale des Sciences*, 279–295.

Hoese, W. J., Podos, J., Boetticher, N. C., & Nowicki, S. (2000). Vocal tract function in birdsong production: Experimental manipulation of beak movements. *Journal of Experimental Biology, 203*, 1845–1855.

Hoh, J. (2005). Laryngeal muscle fibre types. *Acta Physiologica Scandinavica, 183*, 133–149.

Homberger, D. G. (1986). *The lingual apparatus of the African grey parrot, Psittacus erithacus Linné (Aves: Psittacidae): Description and theoretical mechanical analysis*. Washington, DC: American Ornithologists' Union.

Jensen, K. K., Cooper, B. G., Larsen, O. N., & Goller, F. (2007). Songbirds use pulse tone register in two voices to generate low-frequency sound. *Proceedings of the Royal Society B: Biological Sciences, 274*, 2703–2710.

King, A. S. (1989). Functional anatomy of the syrinx. In A. S. King & J. McLelland (Eds.), *Form and function in birds* (pp. 105–192). London: Academic Press.

Ladefoged, V. P. (2005). *Vowels and consonants*. Malden, MA: Wiley-Blackwell.

Laje, R., & Mindlin, G. B. (2005). Modeling source-source and source-filter acoustic interaction in birdsong. *Physical Review E: Statistical, Nonlinear, and Soft Matter Physics, 72*, 036218.

Laje, R., Sciamarella, D., Zanella, J., & Mindlin, G. B. (2008). Bilateral source acoustic interaction in a syrinx model of an oscine bird. *Physical Review E: Statistical, Nonlinear, and Soft Matter Physics, 77*, 011912.

Larsen, O. N., & Goller, F. (2002). Direct observation of syringeal muscle function in songbirds and a parrot. *Journal of Experimental Biology, 205*, 25–35.

McLelland, J. (1989). Anatomy of the lungs and air sacs. In A. King & J. McLelland (Eds.), *Form and function in birds* (pp. 221–279). New York: Academic Press.

Miller, A. H. (1934). The vocal apparatus of some North American owls. *Condor, 36*, 204–213.

Mindlin, G. B., & Laje, R. (2005). *The physics of birdsong*. Berlin: Springer.

Myers, J. A. (1917). Studies of the syrinx of Gallus domesticus. *Journal of Morphology, 29*, 165–215.

Nelson, B. S., Beckers, G. J. L., & Suthers, R. A. (2005). Vocal tract filtering and sound radiation in a songbird. *Journal of Experimental Biology, 208*, 297–308.

Neubauer, J., Edgerton, M., & Herzel, H. (2004). Nonlinear phenomena in contemporary vocal music. *Journal of Voice: Official Journal of the Voice Foundation, 18*, 1–12.

Nottebohm, F. (1976). Phonation in the orange-winged Amazon parrot, Amazona amazonica. *Journal of Comparative Physiology, 108A*, 157–170.

Nowicki, S. (1987). Vocal tract resonances in oscine bird sound production: Evidence from birdsongs in a helium atmosphere. *Nature, 325*, 53–55.

Nowicki, S., & Capranica, R. (1986). Bilateral syringeal interaction in vocal production of an oscine bird sound. *Science, 231*, 1297–1299.

Nowicki, S., & Marler, P. (1988). How do birds sing? *Music Perception, 5*, 391–426.

Ohms, V. R., Beckers, G. J. L., ten Cate, C., & Suthers, R. A. (2012). Vocal tract articulation revisited: The case of the monk parakeet. *Journal of Experimental Biology, 215*, 85–92.

Ohms, V. R., Snelderwaard, P. C., ten Cate, C., & Beckers, G. J. L. (2010). Vocal tract articulation in zebra finches. *PLoS ONE, 5*, e11923.

Ott, E. (2002). *Chaos in dynamical systems*. Cambridge: Cambridge University Press.

Patterson, D. K., & Pepperberg, I. M. (1994). A comparative study of human and parrot phonation: Acoustic and articulatory correlates of vowels. *Journal of the Acoustical Society of America, 96*, 634–648.

Patterson, D. K., & Pepperberg, I. M. (1998). Acoustic and articulatory correlates of stop consonants in a parrot and a human subject. *Journal of the Acoustical Society of America, 103*, 2197–2215.

Riede, T., Beckers, G. J. L., Blevins, W., & Suthers, R. A. (2004). Inflation of the esophagus and vocal tract filtering in ring doves. *Journal of Experimental Biology, 207*, 4025–4036.

Riede, T., Suthers, R. A., Fletcher, N. H., & Blevins, W. E. (2006). Songbirds tune their vocal tract to the fundamental frequency of their song. *Proceedings of the National Academy of Sciences of the United States of America, 103*, 5543–5548.

Rome, L. C. (2006). Design and function of superfast muscles: New insights into the physiology of skeletal muscle. *Annual Review of Physiology, 68*, 193–221.

Rothenberg, M. (1987). Cosi Fan Tuti and what it means, or, nonlinear source-tract interaction in the soprano voice and some implications for the definition of vocal effiencieny. In T. Baer, C. Sasaki, & K. S. Harris (Eds.), *Vocal fold physiology: Laryngeal function in phonation and respiration* (pp. 254–263). Boston: College Hill Press.

Rüppell, W. (1933). Physiologie und Akustik der Vogelstimme. *Journal fur Ornithologie, 81*, 433–542.

Stevens, K. N. (1999). *Acoustic phonetics*. Cambridge, MA: MIT Press.

Suthers, R. A. (1990). Contributions to birdsong from the left and right sides of the intact syrinx. *Nature, 347*, 473–477.

Suthers, R. A. (1997). Peripheral control and lateralization of birdsong. *Journal of Neurobiology, 33*, 632–652.

Suthers, R. A., & Goller, F. (1997). Motor correlates of vocal diversity in songbirds. In V. Nolan Jr., E. Ketterson, & S. F. Thompson (Eds.), *Current Ornithology* (pp. 235–288). New York: Plenum Press.

Suthers, R. A., Goller, F., & Pytte, C. (1999). The neuromuscular control of birdsong. *Philosophical Transactions of the Royal Society of London B: Biological Sciences, 354*, 927–939.

Suthers, R. A., Goller, F., & Wild, J. M. (2002). Somatosensory feedback modulates the respiratory motor program of crystallized birdsong. *Proceedings of the National Academy of Sciences of the United States of America, 99*, 5680–5685.

Suthers, R. A., & Zollinger, S. A. (2008). From brain to song: The vocal organ and vocal tract. In H. P. Zeigler & P. Marler (Eds.), *Neuroscience of Birdsong* (pp. 78–98). Cambridge: Cambridge University Press.

Titze, I. R. (1994). *Principles of voice production*. Upper Saddle River, NJ: Prentice Hall.

Warren, D. K., Patterson, D. K., & Pepperberg, I. M. (1996). Mechanisms of American English Vowel Production in a Grey Parrot (Psittacus erithacus). *Auk, 113*, 41–58.

Westneat, M., Long, J., Hoese, W., & Nowicki, S. (1993). Kinematics of birdsong: Functional correlation of cranial movements and acoustic features in sparrows. *Journal of Experimental Biology, 182*, 147–171.

Wild, J. M., Goller, F., & Suthers, R. A. (1998). Inspiratory muscle activity during bird song. *Journal of Neurobiology, 36*, 441–453.

Wilden, I., Herzel, H., Peters, G., & Tembrock, G. (1998). Subharmonics, biphonation, and deterministic chaos in mammal vocalization. *Bioacoustics, 9*, 171–196.

Zeigler, H. P., & Marler, P. (2004). *Behavioral neurobiology of bird song*. New York: New York Academy of Sciences.

Zollinger, S. A., Riede, T., & Suthers, R. A. (2008). Two-voice complexity from a single side of the syrinx in northern mockingbird Mimus polyglottos vocalizations. *Journal of Experimental Biology, 211*, 1978–1991.

V   GENES, SONG, SPEECH, AND LANGUAGE

Simon E. Fisher

It has been suspected for several decades that genetic factors contribute to speech and language development/function, based primarily on indirect data from twin studies and observations of familial clustering in relevant neurodevelopmental disorders (Fisher, Lai, & Monaco, 2003). In recent years, advances in modern genetics and genomics have led to the isolation of specific molecular variants that cause human speech and language deficits, which has spearheaded a paradigm shift in the field. This work is epitomized by studies of the *FOXP2* gene (Fisher & Scharff, 2009), rare mutations of which lead to problems mastering sequences of mouth movements during speech, accompanied by expressive and receptive language deficits. In what follows, I use the example of *FOXP2* to dispel some common myths about the relationship between genes and speech/language, and to show how molecular discoveries allow us to establish solid biological foundations for theories about the basis and origin of human traits.

An Unusual Family

The initial discovery of the *FOXP2* gene and its potential importance for speech and language stemmed from intensive studies of an unusual British multigenerational family, the "KE" family (Figure 21.1). In the early 1990s it was reported that around half the family—fifteen of the members—had unexplained problems with speech and language development in the absence of any obvious neurological, anatomical, or physiological cause; the other members of the family were unaffected (Hurst, Baraitser, Auger, Graham, & Norell, 1990). Developmental syndromes that disturb language often show some evidence of familial clustering (i.e., several cases within the same pedigree) (Fisher et al., 2003) but this particular family stood out, both in terms of the large numbers of members affected and the striking pattern of inheritance (see Figure 21.1). The KE family disorder seemed to be passed on from one generation to the next in a simple manner, one that appeared consistent with damage to one copy of just a single gene (the identity of which was unknown at

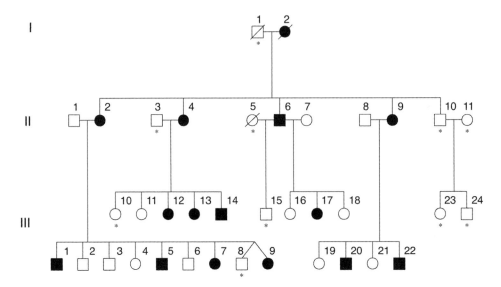

Figure 21.1
Pedigree of the KE family. About half the members of this three-generation (I–III) family
were found to be affected by a severe speech and language disorder. The pattern of inheritance suggested the disorder in this family may be caused by a dominant mutation in a single
autosomal gene, motivating a genomewide search to pinpoint the locus responsible. The
affected individuals are indicated by filled symbols. Squares represent males and circles represent females. A line through a symbol indicates that individual was deceased at the time
the genetic study was carried out. Asterisks indicate individuals for whom DNA samples were
not available. Reproduced with permission from Lai et al. (2001).

that point). Indeed the inheritance of the disorder was reminiscent of classic textbook examples of dominant single-gene (or Mendelian) traits, like Huntington's
disease or neurofibromatosis.

At this early stage, even though investigations of DNA samples from the KE
family had yet to be carried out, there was already excited discussion regarding the
existence of a possible "gene for language" or "gene for grammar." This kind of
speculation was fueled primarily by a report in which it was suggested that the KE
family disorder involved disruption of one selective part of the grammatical faculty:
"the accurate usage of syntactical-semantic features of language, such as the significance of number, gender, animacy, proper names, tense and aspect," a phenomenon
referred to as "feature-blindness" (Gopnik, 1990). For example, it was claimed that
the affected family members "lack a general rule for producing plurals" (Gopnik,
1990). Other researchers countered this perspective, noting that the most prominent
characteristics of the KE family disorder are difficulties mastering sequences of
movements of the mouth and face, yielding profound impairments in speech devel-

opment (Vargha-Khadem & Passingham, 1990). Further in-depth studies of the phenotype indicated that the affected members have at least some knowledge of morphosyntactic rules, evident from overregularization errors made during testing (Vargha-Khadem, Watkins, Alcock, Fletcher, & Passingham, 1995). Moreover, rather than being confined to specific aspects of grammar, their problems span a broad range of skills involved in language production and comprehension (Vargha-Khadem et al., 1995).

The Hunt for *FOXP2*

The apparently simple inheritance of the KE family disorder was crucial because it increased the ease of tracking down the underlying cause using contemporary gene-mapping techniques (Fisher, Vargha-Khadem, Watkins, Monaco, & Pembrey, 1998). From the mid-1980s onward, the field of human genetics has been revolutionized by successive advances in molecular methods, enabling researchers to identify genes implicated in a trait of interest without requiring any prior knowledge of its biological basis. Such efforts typically begin with a genomewide scan—DNA samples are collected from individuals or families affected with the trait and genetic markers with known locations on many different chromosomal regions are analyzed. Statistical testing is used to evaluate whether the status of any of the markers correlates with the status of the trait; a significant correlation suggests that the marker in question could be next to a gene that influences the trait. Due to their simple inheritance patterns and high genotype-phenotype correlations, Mendelian disorders were the first to yield to these kinds of gene-mapping strategies, although as technologies and methods continue to evolve, geneticists now have their sights set on common traits of considerable genetic complexity (McCarthy et al., 2008).

Fisher and colleagues began a genomewide search in the KE family, and it soon became clear that the speech and language impairment in this pedigree was strongly linked to a series of neighboring DNA markers located on one particular stretch of chromosome 7, in a chromosomal band called 7q31 (Fisher et al., 1998). These data empirically confirmed the suspected Mendelian inheritance of the disorder and narrowed the search for the damaged gene to a tiny proportion of the genome. Nevertheless, the region of 7q31 that was implicated still contained several million basepairs of DNA and harbored many different genes, some known, others completely uncharacterized. Since this work was carried out prior to completion of the full human genome sequence, the researchers had to assemble detailed maps of the region and started analyzing the most interesting candidate genes (based on available functional data) in affected KE members, looking for possible mutations (Lai et al., 2000). A potentially laborious search was circumvented by the team's fortunate discovery of another case of speech and language deficits involving the same region of chromosome 7. The child, CS, was unrelated to the KE family, and he carried a gross chromosomal rearrangement, visible under the microscope, in which part of his chromosome 7 had broken off and become reciprocally exchanged with part of his chromosome 5 (Lai et al., 2000). In this type of so-called balanced translocation, no significant genetic material is lost; thus, such mutations may often be benign, without phenotypic consequence. However, if a site of chromosomal breakage lies within an important gene its function can be disrupted, yielding a disorder. In the case of CS, the chromosome 7 breakpoint was located in 7q31, mapping within a candidate gene that had been poorly characterized in previous genomic studies; only a partial section of the gene transcript had ever been isolated, and its function was undetermined (Lai et al., 2000).

Lai and coworkers went on to identify the remaining parts of the candidate gene, and discovered that it encoded a special type of regulatory protein, known as a forkhead-box (FOX) protein (Lai, Fisher, Hurst, Vargha-Khadem, & Monaco, 2001). FOX proteins are defined by presence of a characteristic stretch of 80–100 amino acids (Cirillo & Barton, 2008), which folds up into a motif that can bind directly to DNA (the functional significance of which will be explained later in this chapter). When Lai et al. (2001) sequenced the gene—now given the official name *FOXP2*—in the KE family, they discovered that all fifteen affected individuals carried the same point mutation, a change affecting only one nucleotide of DNA. (Note: In the recently standardized nomenclature for forkhead-box genes, *FOXP2* is used for the human gene, *Foxp2* for the mouse version, and *FoxP2* for corresponding versions found in other species. Protein symbols are the same as the gene names, but are nonitalicized.) The KE mutation involved replacement of a G with an A in a key part of the coding region of the *FOXP2* gene (Figure 21.2), leading to alteration of a single residue in the amino acid sequence of the encoded protein. Geneticists refer to this type of mutation as a "missense mutation." Intriguingly, the resulting amino acid substitution lay at a crucial point within the DNA-binding domain of the FOXP2 protein, and was predicted to prevent it from functioning properly (Lai et al., 2001). This mutation was absent from the unaffected members of the KE family, and also could not be detected in a large number of control individuals from the general population, confirming that it is not a natural polymorphism (Lai et al., 2001). The missense mutation found in the KE family and the translocation observed in case CS (Figure 21.2) were each present in a heterozygous state, meaning that the affected people carried one damaged version of *FOXP2*, while their other copy of the gene was normal. (For genes on non–sex chromosomes, every human being typically carries two copies, one inherited from the father, the other inherited from the mother.) Based on the findings, Lai et al. (2001) proposed that damage to a single copy of *FOXP2* was sufficient to significantly disturb speech and language development. This was the first time that etio-

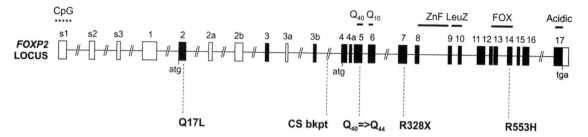

Figure 21.2

Schematic of the human *FOXP2* locus, which spans over 600 kilobases in chromosomal band 7q31. Black shading indicates exons that are translated into protein; "atg" and "tga" denote positions of initiation and termination codons. The main FOXP2 protein isoform, encoded by exons 2–17, contains 715 amino acids, with polyglutamine tracts of 40 (Q40) and 10 residues (Q10), a zinc-finger motif (ZnF), a leucine zipper (LeuZ), a forkhead domain (FOX), and an acidic C-terminus. Exon s1 overlaps with a type of regulatory region known as a CpG island. Additional information on features of this locus can be found in Fisher et al. (2003). Sites of coding variants reported in children with severe speech and language impairment are indicated below the locus schematic, including the R553H mutation initially identified in the KE family (Lai, Fisher, Hurst, Vargha-Khadem, & Monaco, 2001), and the three additional changes uncovered in a subsequent screening study of 49 other probands (MacDermot et al., 2005). The figure also shows the site of the translocation breakpoint found in case CS, mapping between exons 3b and 4 (Lai, Fisher, Hurst, Vargha-Khadem, & Monaco, 2001). Multiple additional translocation cases involving *FOXP2* disruption have since been reported. Adapted with permission from MacDermot et al. (2005).

logical changes at the molecular level could be clearly implicated in a human trait of this nature.

Nevertheless, it is important to recognize that identification of a gene does not represent an endpoint. Far from it, the discovery of *FOXP2* was instead a starting point for a new wave of research, raising a plethora of novel questions to be addressed through diverse methodologies spanning a range of distinct disciplines.

First Clues to *FOXP2* Function

Almost nothing was known about the function of *FOXP2* when it was first implicated in speech and language disorders, but initial clues were gleaned by examining the amino acid sequence of the encoded protein and comparing it to that of other proteins that had already been studied (Lai et al., 2001). As noted above, such analyses clearly placed it within the FOX group of regulatory proteins. The general role of these proteins is well defined in molecular terms—they are transcription factors, binding to regulatory regions associated with genes and influencing the quantities of messenger RNA (mRNA) that are transcribed from each target locus

(Carlsson & Mahlapuu, 2002; Cirillo & Barton, 2008). Since mRNA transcripts act as templates for translation of proteins, increased transcription of a gene can lead to higher amounts of its encoded protein. In other words, transcription factors act to modulate the levels of expression of other genes (its downstream targets), helping to govern how much protein product is ultimately generated from each. A complex multicellular organism contains a vast array of different cell types, all carrying a virtually identical set of genomic instructions; it is the unique transcriptional profile (i.e., which genes are switched on or off, and how much gene product is made from each) that determines the distinctive identity and functions of each cell. Moreover, many cells respond dynamically to incoming signals by up-regulating and/or down-regulating the expression of specific sets of genes. Transcription factors like the FOX proteins are often central elements (hubs) of the networks that regulate these molecular processes (Carlsson & Mahlapuu, 2002).

The human genome encodes at least 43 different FOX transcription factors, divided into subgroups denoted by the letters A–R (Cirillo & Barton, 2008). The unity of their general role at the molecular level broadens out dramatically when one considers cellular and developmental functions. FOX proteins are known to be important for a wide variety of biological processes: cellular proliferation, differentiation, pattern formation, signal transduction, and so on (Carlsson & Mahlapuu, 2002). Moreover, a single FOX protein may have multiple functions, and can play distinct roles depending on which cofactors (other interacting proteins) are present in its cellular environment. Developmental processes are highly sensitive to changes in the dosage of these genes—that is, the numbers of functional copies carried (Carlsson & Mahlapuu, 2002). Indeed, altered dosage of different *FOX* genes has been implicated in an array of major disease states, including developmental eye disorders (*FOXC1*), ovarian failure (*FOXL2*), immune deficiency (*FOXP3*), and diabetes (*FOXO1*) (Hannenhalli & Kaestner, 2009; Lehmann, Sowden, Carlsson, Jordan, & Bhattacharya, 2003). A number of FOX transcription factors make fundamental contributions to CNS development—for example, studies in mice show that Foxb1 loss yields severe abnormalities in diencephalon and midbrain (Wehr, Mansouri, De Maeyer, & Gruss, 1997), while Foxg1 is important for normal proliferation and differentiation of progenitor cells of the telencephalon (Hanashima, Li, Shen, Lai, & Fishell, 2004).

The functional diversity of different FOX transcription factors is explained by the fact that, outside of their characteristic DNA-binding domain, most of them show considerable divergence in protein sequence (Carlsson & Mahlapuu, 2002). The major form of human *FOXP2* encodes a sequence of 715 amino acids (Figure 21.3), in which the 80 residues comprising the DNA-binding domain are located toward the far end of the protein (Lai et al., 2001). As described earlier, the etiological missense mutation in the KE family yields a substitution within this essential domain;

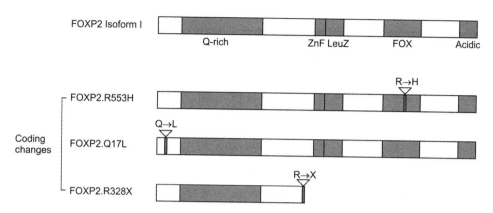

Figure 21.3

Representation of the full-length FOXP2 protein (main isoform), containing the glutamine-rich region (Q-rich), zinc finger (ZnF), leucine zipper (LeuZ) and forkhead-box (FOX) domains, and the C terminal acidic tail region (Acidic). The predicted protein products yielded by coding changes identified in cases of speech and language impairment (R553H, Q17L, R328X) are aligned beneath. The figure is adapted with permission from Vernes et al. (2006), which describes functional genetic analyses of each of these mutant products.

the substitution is located at residue 553 of the FOXP2 protein, changing an arginine residue (symbol R) to a histidine (H), and is thus referred to as R553H. A region near the beginning of the protein contains contiguous stretches of glutamine (Q) residues referred to as polyglutamine (or polyQ) tracts. PolyQ tracts are found in a number of important neural proteins and have been implicated in etiology of neurodegenerative disorders (Williams & Paulson, 2008). Another functional motif, close to the middle of the protein, enables two separate molecules (monomers) of FOXP2 to combine with each other and form a dimer, a unit containing two molecules bound together (Wang, Lin, Li, & Tucker, 2003). The key element of this dimerization domain is a structure called a leucine zipper, acting to "zip up" the two molecules. It is thought that FOXP2 needs to be in the dimeric form in order to efficiently bind to its target sites in the genome (Li, Weidenfeld, & Morrisey, 2004). Other defined parts of the protein allow it to interact with cofactors that mediate its role as a modulator of gene expression (Li et al., 2004). However, the structural and functional properties of much of the FOXP2 protein sequence still remain to be determined.

FOXP2 in Human Speech and Language Disorders

One of the most obvious questions to ask in light of the discovery of *FOXP2* is whether point mutations of this gene have a broader impact beyond the unusual example of the KE family (Fisher, 2005). It is essential to stress that *FOXP2* is a large gene at the genomic level (see Figure 21.2), with 25 known exons spanning >600kb of the 7q31 chromosomal band (Bruce & Margolis, 2002; Fisher et al., 2003; Lai et al., 2001). Mutations altering the amino acid sequence of the encoded product could feasibly occur anywhere within the *FOXP2* coding sequence, affecting any one or more positions of the 715-residue protein. Therefore, to properly evaluate potential contributions of *FOXP2* mutations to a phenotype of interest, screening efforts must search through the *entire* coding sequence (Fisher, 2005). Investigations in which cohorts are screened only for presence/absence of the specific R553H mutation (as carried out by Meaburn, Dale, Craig, & Plomin, 2002) are of limited value, especially since this particular mutation appears isolated to the KE family. Such a restricted screen would clearly miss etiological mutations located elsewhere in the DNA-binding domain, or in other crucial parts of the protein, including the polyQ tracts, leucine zipper, and other interacting structures. In effect, searching just for an R553H change is equivalent to a proofreader checking for one particular error in a single word of a >700-word text, and assuming that the entire remaining text must be correct (for further commentary see Fisher, 2005).

The first comprehensive *FOXP2* mutation searches focused on two typical forms of neurodevelopmental disorder: SLI (specific language impairment) and autism (Newbury et al., 2002; Wassink et al., 2002). SLI is formally defined as an unexpected failure to acquire normal expressive and/or receptive language skills, in the absence of explanatory factors such as deafness, cerebral palsy, mental retardation, or cleft lip/palette; it is highly heritable and has been estimated to affect up to 7% of pre-school children (Bishop, 2001; Tomblin et al., 1997). Screening of exons covering the entire coding sequence of *FOXP2* in 43 cases of SLI did not identify any mutations contributing to the disorder (Newbury et al., 2002). Autism represents another highly heritable trait involving disrupted language development (Abrahams & Geschwind, 2008). Classic definitions of autism depend on a core triad of co-occurring features: communication deficits, impaired social interaction, and rigid/repetitive behaviors (Tager-Flusberg, Joseph, & Folstein, 2001). In practice, a diagnosis of autism can encompass a considerable degree of variability in linguistic profile (Happe, Ronald, & Plomin, 2006). For example, some children may be completely nonverbal, while others can attain language competence that appears normal in terms of phonology and structure but remains abnormal with respect to pragmatics — use of language in a social context (Tager-Flusberg et al., 2001). Two independent studies searched through all coding exons of *FOXP2* in probands from autism families (48 families in Newbury et al., 2002; 135 families in Wassink et al., 2002) and both failed to find etiological mutations.

In considering the above findings, it is worth noting that the phenotypic profiles associated with typical forms of SLI and autism differ somewhat from those observed in the case of CS and the KE family. For the latter, although a broad range of lin-guistic skills is affected, the most robust diagnostic marker is impaired coordination of the rapid movement sequences that are important for speech articulation (Vargha-Khadem et al., 1998). This involves inconsistent speech errors that become more frequent with increased complexity of utterances, a phenomenon usually referred to as "developmental verbal dyspraxia" (or "childhood apraxia of speech") (Watkins, Dronkers, & Vargha-Khadem, 2002). Indeed, it has been hypothesized that the dif-ficulties reflect underlying problems with learning, planning, and execution of oro-facial motor sequences, since tests indicate reduced performance for nonspeech as well as speech-related movements (Vargha-Khadem, Gadian, Copp, & Mishkin, 2005). Perhaps then it is unsurprising that studies of SLI and autism did not uncover new *FOXP2* mutations; language problems in typical forms of SLI most frequently occur without overt deficits in control of speech articulation (Shriberg, Tomblin, & McSweeny, 1999), and while such deficits can sometimes be observed in autistic cases, they are not generally thought of as a central diagnostic feature (Tager-Flusberg et al., 2001).

These concerns led MacDermot and colleagues to focus their *FOXP2* screening efforts on children who had been given a formal clinical diagnosis of developmental verbal dyspraxia, and so might better match the phenotypic profile seen in the KE family and case CS (MacDermot et al., 2005). By searching through all known *FOXP2* exons in 49 unrelated probands, they were able to identify a novel point mutation disrupting the gene in one of the children. This was of a type known as a nonsense mutation, creating a premature "stop" signal around halfway through the coding sequence, at the point corresponding to amino acid residue number 328, which normally encodes an arginine (R) (Figure 21.2). The mutation, referred to as R328X, yields a severely truncated FOXP2 protein product (327 amino acids long instead of 715) that has lost many of the key functional elements, most importantly the DNA-binding domain (Figure 21.3). As in the KE family and case CS, the R328X mutation was found in a heterozygous state (one damaged copy of *FOXP2*, one normal copy). Moreover, it cosegregated with disorders in the proband—his sister, similarly diagnosed with verbal dyspraxia, and his mother, who had a history of speech and language difficulties—and it was not detected in a large number of control individuals from the general population (MacDermot et al., 2005). Two additional FOXP2 coding changes were identified in other probands; one carried a missense mutation yielding a glutamine-to-leucine substitution near the start of the protein (Q17L), and another had a slightly expanded polyglutamine tract (44 glu-tamines instead of 40). Neither change was present in panels of normal controls, but in each case the proband had a sibling who was affected with developmental verbal dyspraxia but did not carry a mutation. As such the etiological relevance of the Q17L and Q40→44 variants remains unclear (MacDermot et al., 2005). Through analyses of human neuronlike cells, grown in the laboratory, it has been possible to assess the functional impact of the different coding variants identified in cases of developmental verbal dyspraxia (Vernes et al., 2006). Such experiments demonstrated that the R553H and R328X mutations each clearly disturb the function of the resulting FOXP2 protein. These mutations were found to affect the localization of the protein within the cell (normally it is confined to the nucleus), disrupt its ability to bind to DNA, and impair its regulation of downstream target genes. By contrast the Q17L coding variant did not affect FOXP2 protein function in any of the above assays (Vernes et al., 2006).

Crucially, the MacDermot et al. findings indicate that coding variants of FOXP2 are not restricted to the KE family and are likely to account for a small yet significant proportion (~>2%) of children affected by developmental verbal dyspraxia. Furthermore, the CS translocation is far from being the only example of a genomic rearrangement disturbing the *FOXP2* locus; several cases of chromosomal abnormalities disrupting the 7q31 region have since been reported, including other translocations (Feuk et al., 2006; Kosho et al., 2008; Shriberg et al., 2006), as well as deletions, in which potentially large genomic sections are lost from the chromosome (Feuk et al., 2006; Lennon et al., 2007; Zeesman et al., 2006). The consensus from such reports is that when one copy of *FOXP2* is damaged or lost the affected person is highly likely to have problems with speech and language development. Note that these kinds of large-scale chromosomal rearrangements can simultaneously disturb other genes beyond *FOXP2* and so may show a more complex disorder involving additional problems (dysmorphologies, autism, etc.). Alternatively, given how large this gene is at the genomic level, some rearrangements (such as small deletions) could potentially disturb parts of the *FOXP2* locus (such as the noncoding regions) while still allowing expression of a normal protein. Therefore, it can be difficult to dissect links between genotype and phenotype in some cases.

Not a "Gene for Speech"

A superficial view of human evolution might require that any genes implicated in speech- and language-related phenotypes must be unique to *Homo sapiens*. However, it can be argued that our species' unique capacity for acquiring complex speech and language depends on the integrated functions of multiple neural systems supporting a range of different processes, and that these systems and processes will have precedents in ancestral species (Fisher & Marcus, 2006). Moreover, a gene does not specify behavioral or cognitive outputs, or even directly code for a particular brain "module" with a unique function. Instead, gene products (typically proteins) interact with each other in complicated networks to help build neural circuits and maintain a language-ready brain (Fisher, 2006). They do so by affecting an array of cellular properties and functions, such as neuronal proliferation and migration, neurite outgrowth, axon pathfinding, neurotransmitter production and reception, strengthening and weakening of synapses (the connections between neurons), and so on. All in all, as noted by Fisher & Scharff (2009), it is wishful thinking to expect that the emergence of spoken language will be explicable in terms of a single human-specific molecular agent.

FOXP2 provides a case in point. Far from being exclusively human, versions of this gene (orthologs) are found in remarkably similar form in many distantly related vertebrate species, including rodents, birds, reptiles, and fish (Bonkowsky & Chien, 2005; Haesler et al., 2004; Lai, Gerrelli, Monaco, Fisher, & Copp, 2003; Teramitsu, Kudo, London, Geschwind, & White, 2004). In genomewide comparisons of similarity between coding sequences of rodents and humans, *FOXP2* belongs among the top 5% of genes; only 3 substitutions and a 1-residue difference in polyglutamine tract length distinguish the major FOXP2 protein in *Homo sapiens* from the equivalent Foxp2 protein of *Mus musculus* (Enard et al., 2002). Of course, the functional impact of any given gene does not depend only on the amino acid sequence of the encoded protein, but also on the way its expression is regulated—that is, when and where the gene is switched on in the tissues of the body (Fisher, 2006). Studies of expression patterns of versions of the *FOXP2* gene found in diverse vertebrates again indicate striking concordances with the human situation. As will be elaborated on elsewhere in this chapter, expression is observed in corresponding brain structures in different species ranging from humans to fish, with enrichment in similar neuronal subpopulations (Bonkowsky & Chien, 2005; Campbell, Reep, Stoll, Ophir, & Phelps, 2009; Ferland, Cherry, Preware, Morrisey, & Walsh, 2003; Haesler et al., 2004; Shah, Medina-Martinez, Chu, Samaco, & Jamrich, 2006; Takahashi, Liu, Hirokawa, & Takahashi, 2003; Takahashi et al., 2008; Teramitsu et al., 2004). Taken together, the available sequence and expression data indicate that this gene is evolutionarily ancient and likely played important roles in brain development and function in a common vertebrate ancestor (Fisher & Marcus, 2006; Fisher & Scharff, 2009).

Broader analyses of expression have shown that the gene is also switched on in other tissues, including lung, cardiovascular, and intestinal cells, suggesting it has additional roles outside of the brain (Lai et al., 2001; Shu, Yang, Zhang, Lu, & Morrisey, 2001). For example, the distal airway epithelium of the lung and the outflow tract of the heart are notable nonneural sites where the *Foxp2* gene is expressed during embryogenesis (Shu et al., 2001). This observation of multiple functions is typical of transcription factors; routinely, the same gene will be recruited in different contexts to support diverse processes, depending on precisely which interacting cofactors are present in any given cell or situation (Carlsson & Mahlapuu, 2002; Cirillo & Barton, 2008; Lehmann et al., 2003). Indeed, the combinatorial action of regulatory factors represents a central principle of developmental biology, allowing exceptional levels of multicellular complexity to emerge from the activities of a surprisingly small number of genes (Fisher, 2006).

Given that expression of this gene is not confined to the central nervous system (CNS), it may seem paradoxical that people with *FOXP2* mutations show a disorder that appears to be largely restricted to the brain. Why are there no reports of problems with lung, heart, or intestinal development in the affected human individuals? Perhaps there are subtle nonneural correlates of the *FOXP2*-related speech and language disorder that have escaped detection? In addressing this issue, it should be remembered that although these people carry one damaged copy of *FOXP2*, they still retain one normal copy of the gene. Thus far, no human has ever been found to be homozygous for *FOXP2* mutation. As such, it can be hypothesized that a half-dosage of the *FOXP2* gene yields deficits in certain neural circuits where it is expressed, because these cells are most vulnerable to reductions in levels of functional FOXP2 protein, while expression sites in other tissues accommodate this partial decrease and can still develop normally. This explanation gains support from studies of other FOX transcription factors, where reduced functional dosage represents a common mechanism mediating developmental disorders, with certain cell types being more tolerant of decreased levels than others. For example, heterozygous mutations of *FOXC1* typically yield eye-related disorders (e.g., glaucoma), but in fact *FOXC1* (like *FOXP2*) is not specific to the affected tissue, and has a range of different expression sites; the developing eye appears particularly sensitive to partial reductions in FOXC1 protein levels (Lehmann et al., 2003).

Clearly, then, although there is little doubt that mutations of *FOXP2* yield speech and language deficits, this does not mean that it can be accurately referred to as a "gene for speech" or "gene for language" (Fisher, 2006). In other words, *FOXP2* does not exist in order to endow us with these unique human gifts. Instead, it is an evolutionarily ancient regulatory gene with multiple functions in the brain and elsewhere in the body (Fisher & Marcus, 2006; Fisher & Scharff, 2009). I have argued extensively elsewhere that, while this may serve as a useful shorthand for emphasizing the importance of inherited influences, terms like "the speech gene" or "the language gene" lead to damaging misconceptions about the actions of genetic factors and the ways that they are able to affect behavior and cognition (Fisher, 2005, 2006). I will not dwell on the issues here, except to note that these arguments do not diminish the value of *FOXP2* as an entry point into pathways that are relevant for speech and language. But they do require that the findings are incorporated into a sophisticated framework, one that is rooted in our wider knowledge of constraints on biological systems (Fisher & Scharff, 2009).

Lessons from Gene Expression

FOXP2 is not transcribed ubiquitously all over the brain, nor is this isolated to one selected spot. (It is rare to find that a gene is exclusively switched on in just one single part of the CNS.) Instead, it is expressed in a number of different neural structures, most notably the cortex, basal ganglia, thalamus, and cerebellum (Lai et al., 2003). Expression begins during embryogenesis and continues through fetal development; in some brain structures it persists postnatally and even into adulthood (Ferland et al., 2003).

Detailed examination of neural expression patterns of human *FOXP2* and nonhuman orthologs has revealed several intriguing features. Within the brain, expression of the gene appears to be primarily found in neurons (excitable cells that transmit information via electrochemical signaling), rather than glia (the support cells of the CNS) (Ferland et al., 2003). In the structures in which it is expressed, it usually shows enrichment in particular subpopulations of neurons, which display distinct patterns of connectivity. For example, in the mammalian cortex, which forms a characteristic six-layered structure, Foxp2 protein is typically confined to neurons in the deepest layers—cells generated early in cortical development (Ferland et al., 2003). In the basal ganglia, there is strong expression in the medium spiny neurons of the striatum (Lai et al., 2003; Takahashi et al., 2003). Within the cerebellum—a structure comprising a highly heterogeneous mixture of many different cell types—the gene is switched on only in the Purkinje cells (large neurons that integrate signals from other cells to provide the major output from this structure) and deep nuclei (Ferland et al., 2003; Lai et al., 2003). As far as it is possible to compare brain tissue in different species, it has been found that humans, primates, rodents, birds, reptiles, and fish display generally similar patterns of neural expression for this gene, which in many cases includes localization to corresponding neuronal subpopulations (Bonkowsky & Chien, 2005; Ferland et al., 2003; Haesler et al., 2004; Takahashi et al., 2008; Teramitsu et al., 2004). In-depth analyses of the distribution of Foxp2 protein in the brains of four species of mice found high conservation, but also documented subtle interspecific differences in some structures (Campbell et al., 2009), although the functional significance of this diversity has not yet been established.

Characterization of expression patterns is a useful step toward understanding the role(s) of a brain-related gene, because such patterns can give hints as to the neural structures and circuits that it is likely to affect. In the case of the *FOXP2* gene, it is notable that the main areas of CNS expression belong to networks implicated in multimodal sensory processing, sensorimotor integration, and modulation of motor output (Campbell et al., 2009). In particular, the cortex, basal ganglia, and cerebellum (key expression sites) form distributed circuits that are implicated in various motor-related functions, including the procedural learning of patterns of movement

(De Zeeuw & Yeo, 2005; Graybiel, 2005). The basal ganglia modulate activity of the premotor and prefrontal cortex via complex connections projecting through the globus pallidus, substantia nigra, and thalamus (Middleton & Strick, 2000). Purkinje cells in the cerebellum are also fundamental in regulating motor coordination, and receive strong synaptic excitation from climbing fibers that originate in the inferior olives (De Zeeuw & Yeo, 2005), which is also a conserved site where the *FOXP2/ Foxp2* gene is expressed (Lai et al., 2003).

One might ask if these conserved vertebrate gene expression patterns are really relevant for understanding etiological pathways in human cases of speech and language disorder. The question has been addressed by comparing expression findings to completely independent data obtained from neuroimaging studies of the KE family (Lai et al., 2003), some of which were completed well in advance of identifying the rogue gene itself (Vargha-Khadem et al., 1998). On magnetic resonance imaging (MRI) scanning, the brains of affected KE individuals appear overtly normal. However, detailed statistical analyses of the different voxels (3-D pixels) of the scans have revealed several sites showing subtle but significant structural differences, as compared to unaffected people (Belton, Salmond, Watkins, Vargha-Khadem, & Gadian, 2003; Vargha-Khadem et al., 1998; Watkins, Vargha-Khadem, et al., 2002). These include bilateral reductions in gray-matter density in the inferior frontal gyrus (which contains Broca's area in the left hemisphere), the caudate nucleus, the precentral gyrus, the temporal pole, and the cerebellum, as well as increases in the posterior superior temporal gyrus, angular gyrus, and putamen (Belton et al., 2003; Vargha-Khadem et al., 1998; Watkins, Vargha-Khadem, et al., 2002). Moreover, functional MRI studies using verb-generation tasks uncovered abnormal activation patterns in affected KE members, with significant underactivation of Broca's area and the putamen, even when the tasks are "covert"—that is, carried out silently without any vocal output (Liegeois et al., 2003). Thus, key sites of pathology in people carrying FOXP2 mutations include parts of the cortex, the striatum (caudate/putamen), and the cerebellum, overlapping with structures that typically display high expression levels of the gene (Lai et al., 2003).

Modeling Mutations in the Mouse

Rather than detracting from interest in *FOXP2*, the high levels of similarity for orthologs of this gene in diverse vertebrate species make it feasible to use animal models to study its in vivo functions, in ways that are not currently possible in humans. This enables researchers to move forward from initial speculations based on neural expression patterns to formal empirical testing of potential roles in the living brain (Fisher & Scharff, 2009). Using genetic manipulation in animal systems, it is possible to dissect the causal effects of gene variants not only at molecular and cellular levels, but also on the development and patterning of the CNS, testing for roles in the formation and function of particular neural structures and circuits. Findings from animal models provide the solid bedrock to help build more accurate models of how a gene functions in the circuitry of the human brain, especially for a case like *FOXP2* that is so well conserved across distant species.

The mouse is a very commonly used model for studying mammalian gene function, due to a combination of short generation time, small physical size, and ease of genetic manipulation, facilitated by increasingly ingenious experimental tools. The traditional method is to simply "knock out" the gene of interest (completely disabling its function in all cells) and assess the consequences for the resulting animals, as carried out in one of the early *Foxp2* mouse studies (Shu et al., 2005). An alternative is to engineer mice that carry the same specific mutations as those found in cases of human disorder; this strategy was employed in subsequent investigations that recapitulated the missense mutation found in the KE family (Groszer et al., 2008) (see Figure 21.4). Even more sophisticated approaches enable the experimenter to control when and where (in which tissues or cells of the living organism) the gene is disrupted. For example, mice carrying a "conditional" allele of *Foxp2* have been generated (French et al., 2007); here the gene will function normally unless it encounters an unusual bacteriophage protein known as Cre recombinase, which inactivates it. These conditional mice can be bred with specially made transgenic lines in which Cre is expressed only in particular cell types and/or at specific developmental stages yielding selective *Foxp2* disruption (French et al., 2007).

Studies of mouse models rapidly established that a complete lack of any functional Foxp2 protein is lethal (French et al., 2007; Groszer et al., 2008; Shu et al., 2005). Homozygous mice (those in which both gene copies of *Foxp2* are damaged in all cells) show slower rates of postnatal weight gain than their littermates, despite adequate feeding (Figure 21.4a), and also have multiple additional symptoms of general developmental delay. They have obvious dysfunctions of the motor system (Figure 21.4b), and typically die only 3 to 4 weeks after being born. Detailed examination of the lungs of homozygous knockouts uncovered some evidence of increased postnatal dilation of distal airspaces (Shu et al., 2007) but it is not clear that this subtle alteration can explain the lethality, and it remains to be established what exactly causes these mice to die. Examination of brains of homozygous mice revealed that a total absence of functional Foxp2 protein leads to a disproportionately small cerebellum with reduced foliation (French et al., 2007; Groszer et al., 2008). Interestingly, in normal mice there are chemical signals from the Purkinje cells (a key site where Foxp2 protein is found) that play important roles in stimulating cerebellar growth during the first few weeks of life. In addition, some researchers have reported misplaced Purkinje cells and reduced dendritic aborization in homozygous mice (Fujita et al., 2008; Shu et al., 2005). No obvious morphological abnormalities have

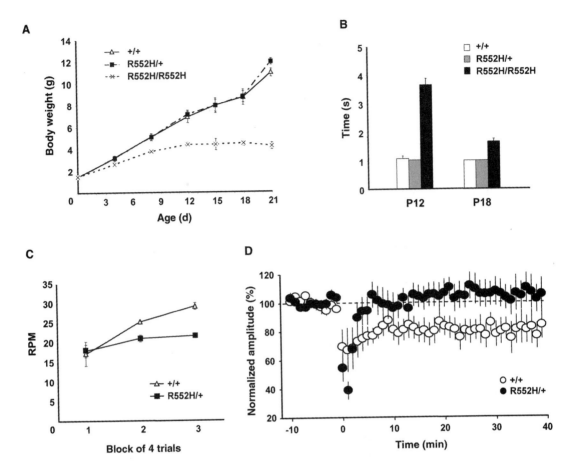

Figure 21.4

Investigations of mice with an etiological *Foxp2* mutation. Groszer et al. (2008) generated mutant mice carrying an R552H missense mutation in mouse Foxp2; this matches the FOXP2-R553H mutation that causes speech and language disorder in the KE family. (a) Time course of postnatal bodyweight development. Homozygotes (mice that carry two mutant copies of R552H mice; R552H/R552H) show reduced weight gain, while heterozygotes (with only one mutant copy; R552H/+) are indistinguishable from wildtype littermates (+/+). (b) Postnatal righting-reflex development. Homozygotes display severely impaired righting reflexes, here assessed at postnatal day 12 and 18, reflecting a substantial general motor impairment. Heterozygotes are again indistinguishable from wildtype littermates. (c) Motor-skill learning. Despite their normal general motor development, heterozygous R552H mice display deficits compared to wildtypes in motor-skill learning on the accelerating rotarod (ANOVA of trials × genotype, p < 0.0001) (RPM = revolutions per minute). Consistent findings were also observed on voluntary tilted running wheels placed in the home cage (data not shown). (d) Synaptic plasticity. Electrophysiological recordings of brain slices demonstrated that heterozygous R552H mice have impaired synaptic plasticity. The panel shows summary LTD data in the dorsolateral striatum, including mean amplitudes for each minute. Following high-frequency stimulation, wildtype mice show significant striatal LTD while heterozygous mutants do not. All panels show (mean+/-SEM). Adapted with permission from Groszer et al. (2008).

been observed in other CNS structures when mice completely lack functional Foxp2 protein, with the caveat that systematic volumetric analyses have not yet been carried out.

The above data from homozygous knockout/mutant mice point to fundamental functions of the gene in early development. However, no humans with homozygous mutations of *FOXP2* have ever been reported in the literature, which may not be surprising given the severe and lethal phenotype found in the homozygous mouse models. Since all identified cases of *FOXP2*-related speech/language disorder involve damage to only one copy of the gene, investigations of heterozygous mouse models are likely to prove most pertinent for illuminating the relevant neurogenetic mechanisms that go awry (Fisher & Scharff, 2009). The issue is illustrated by the work of Groszer and colleagues, who used a gene-driven mutagenesis screening strategy to generate mice that carried an R552H substitution in the mouse Foxp2 protein (Groszer et al., 2008). This is identical to the R553H substitution in the human FOXP2 protein that causes speech and language impairment in the KE family—the slightly different numbering reflects a minor difference in polyglutamine tract length in the two species. Homozygous R552H mutants showed the severe lethal phenotype typical of a complete loss of Foxp2 protein function (Figure 21.4a–b). In contrast, heterozygous mice, carrying one damaged (*Foxp2-R552H*) copy and one intact copy of the gene, were indistinguishable from wildtype littermates (Groszer et al., 2008). They were entirely healthy, with normal rates of postnatal growth and baseline motor skills (Figure 21.4a–b), and no detectable anomalies in gross neuroanatomy. The researchers looked closely at motor behaviors of these heterozygous mice, by automatically recording voluntary activity on tilted running-wheel systems introduced into the home cage, and assessing how movement patterns changed over a period of several days (Groszer et al., 2008). The data suggested that the R552H heterozygotes had impairments in motor-skill learning, which were also apparent on more traditional testing paradigms, like the accelerating rotarod (Figure 21.4c).

Since they were studying mice rather than humans, Groszer and colleagues were able to go further and directly examine neurophysiological functioning of relevant circuits in the R552H heterozygotes (Groszer et al., 2008). They focused attention on Foxp2-expressing networks that are known to be important for motor-skill learning, carrying out electrophysiological recordings in brain slices from the striatum and cerebellum. Both regions showed obvious signs of aberrant synaptic plasticity in the R552H heterozygotes. For example, there was a striking lack of long-term depression (LTD) at glutamatergic synapses of the dorsolateral striatum (Figure 21.4d), while parallel fiber–Purkinje cell synapses in the cerebellum showed increased paired-pulse facilitation at short interstimulus intervals (Groszer et al., 2008). In a follow-up study, in vivo electrophysiological recordings in awake behaving mice demonstrated that R552H heterozygotes have aberrant striatal activity and altered plasticity during the process of motor-skill learning (French et al., 2011). Further work is now underway to define the molecular mechanisms in the different brain structures that underlie alterations in plasticity, and to demonstrate that a causal relationship links the electrophysiological anomalies with the behavioral impairments. It is likely that the latter will only be resolved through use of the conditional mouse lines to selectively inactivate *Foxp2* (French et al., 2007); moreover, such lines will enable scientists to separate out contributions of *Foxp2* to corticostriatal and cerebellar circuitry and relate these to motor-skill phenotypes.

Foxp2 in Animal Vocalization

What is the impact of *Foxp2* gene disruption on mouse vocalization? This has proved to be a controversial question, in part because very little is known at present about rodent vocal behaviors and the underlying neural mechanisms. (The dearth of understanding in this area stands in stark contrast to the admirably thorough descriptions of the neurobiological bases of avian vocalization, particularly for vocal-learning birds, as discussed in detail in several other chapters in this book.) The vocal repertoire of mice includes three main types of vocalization: (1) sonic broadband calls in a range that is audible to humans, generated by vibration of the larynx; (2) pure ultrasonic whistles, above 20 KHz, produced by expiration of air through tightly opposed nonvibrating vocal cords; and (3) broadband click sounds, emitted via unknown mechanisms (Ehret, 2005) (Figure 21.5). In general, the neural structures and circuits that underpin vocalization behaviors of mice have not yet been clearly defined (although it is hoped that this situation will begin to be remedied in the coming years).

At the time of writing, several published reports have assessed the consequences of *Foxp2* damage for mouse vocalization (Fujita et al., 2008; Gaub et al., 2010; Groszer et al., 2008; Shu et al., 2005), all focusing on innately specified pup calls, which are automatically produced on change of arousal state, without requiring auditory feedback (Ehret, 2005). These investigations have shown that a total lack of functional Foxp2 protein leads to a virtual absence of ultrasonic isolation calls, those calls that are normally produced by pups on isolation from their mother and littermates (Fujita et al., 2008; Gaub et al., 2010; Groszer et al., 2008; Shu et al., 2005). Some have argued that the findings demonstrate a specific effect of the *Foxp2* gene on mouse vocalization to directly parallel its involvement in human speech, and that the gene is essential for ultrasound production (Fujita et al., 2008). But there is cause for skepticism, on a number of grounds. First, despite the absence of functional Foxp2 protein, homozygous mice *do* produce ultrasonic calls if they are placed in situations of greater arousal (i.e., more stressful than the simple isolation

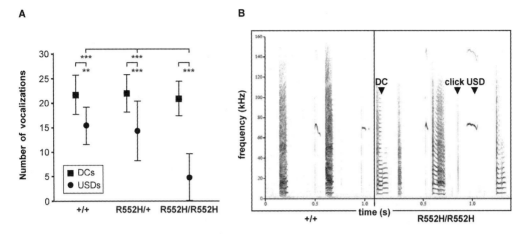

Figure 21.5
Vocalizations from mutant *Foxp2* mouse pups. Although it has been suggested that homozygous *Foxp2* mouse mutants do not produce any ultrasonic vocalizations (Shu et al., 2005), analyses of pups under appropriate conditions indicate that this is not the case (Groszer et al. 2008). (a) When mouse pups are lifted gently by the tail, they emit audible distress calls (DCs) accompanied by ultrasounds (USDs) and clicks. The panel shows the number of DCs and USDs (recorded over a period of 10 seconds) from heterozygous and homozygous mice carrying the R552H mutation (mean+/-SD). The mutation has no impact on the number of DCs produced. Heterozygotes emit similar numbers of USDs to their wildtype littermates. Homozygotes produce significantly fewer USDs, but they also have substantial developmental delays and severe general motor impairment (**$p<0.01$, ***$p<0.001$). (b) Sample sonograms of DCs, USDs, and clicks for wildtype and R552H homozygous pups, demonstrating that homozygotes can still produce all call types.

condition) (Gaub et al., 2010; Groszer et al., 2008) (Figure 21.5). Second, as noted above, the reduced vocalization of these homozygotes occurs in the context of a severe and lethal phenotype involving substantial delays in general development and gross impairments of basic motor functions (cf. Figure 21.4a–b); there is insufficient evidence here of a primary effect on ultrasound production. Third, it remains contentious whether or not heterozygous pups (matching the status of humans with *FOXP2* gene mutations) show consistent reductions in ultrasonic vocalization, and structural properties of those ultrasounds that are emitted appear to be well preserved (Gaub et al. 2010; Groszer et al., 2008; Shu et al., 2005). Finally, unlike innately specified pup calls, speech is a learned activity that brings the articulatory system under voluntary control, so it is worth being cautious when trying to draw parallels between the two (Fisher & Scharff, 2009).

Perhaps more definitive insights will emerge by studying the sequences of ultrasonic vocalizations produced by young adolescent male mice on exposure to females (or their pheromones). Holy and Guo (2005) demonstrated that the streams of ultrasounds elicited (between 30 and 110 Khz) have many features that are typical of song, including a repertoire of distinct syllable types, which are put together in structured temporal sequences involving repeated motifs. Efforts are underway to establish whether any aspects of these ultrasonic "songs" might be acquired through auditory-guided vocal learning; initial data suggest that such learning is very limited in mice (Kikusui et al., 2011). Regardless of the answer to this important question, by recording and analyzing the songs of adult mice carrying *Foxp2* disruptions, we might be in a better position to discuss the effects of this gene on sensorimotor functions and motor sequencing in general, and vocalization pathways in particular.

Compelling support for involvement of this gene in animal vocalization has come not from mice, but instead from an elegant series of studies in songbirds (Haesler et al., 2007; Haesler et al., 2004; Miller et al., 2008; Rochefort, He, Scotto-Lomassese, & Scharff, 2007; Teramitsu et al., 2004; Teramitsu & White, 2006). These investigations will not be described here, since they are covered in detail in Scharff and Thompson (chapter 22, this volume). One exciting conclusion from this avenue of research is that levels of FoxP2 in Area X (a striatal nucleus) of the zebra finch brain are functionally linked to accurate imitation of tutor song (Haesler et al., 2007), and could be implicated in vocal plasticity in other song-learning birds, such as canaries (Haesler et al., 2004).

FOXP2 and the Evolution of Spoken Language

Given that damage to human *FOXP2* primarily disrupts speech and language, key characteristics of our species, it is of interest to ask if and how the gene has changed during hominid evolution. When evolutionary anthropologists began to address this question they uncovered signs of accelerated change in FOXP2 protein sequence on the lineage that led to modern humans (Enard et al., 2002; Zhang, Webb, & Podlaha, 2002). Out of the three amino acid substitutions that distinguish the mouse protein from that in modern humans, two occurred on the human lineage after splitting from the chimpanzee (i.e., at some point within the last 5–6 million years). Although just two amino acid substitutions in a protein of over 700 residues might be considered only a small degree of change, the findings stand out against the general background of high conservation of FoxP2 in most vertebrates. On examining variability in *FOXP2* genomic sequences among individuals from different parts of the world, researchers uncovered further evidence of Darwinian selection at the locus, and estimated that the modern human version had arisen and spread through the population within the past 200,000 years (Enard et al., 2002; Zhang et al., 2002).

With dramatic developments in DNA sequencing technologies, Svante Pääbo's group was more recently able to revisit the question of hominid *FOXP2* evolution by directly querying the status of the gene in fossilized bone tissue from two Neandertals (Krause et al., 2007). The human lineage split off from Neandertals a minimum of 300,000 years ago (Weaver, Roseman, & Stringer, 2008). So it was expected that the protein encoded by the Neandertal gene would not carry the two amino acid substitutions that are fixed in modern human populations, but should instead look like the chimpanzee version, thus confirming the recent origin of these key evolutionary changes. Surprisingly, the prediction turned out to be wrong; samples from both Neandertals clearly matched *Homo sapiens* at the two crucial sites of the FOXP2 protein, suggesting that the substitutions predated the human-Neandertal split (Krause et al., 2007). Although it remains possible that such data represent artifacts, produced by contamination of the ancient samples with modern human DNA (Coop, Bullaughey, Luca, & Przeworski, 2008), this explanation is unlikely given the careful measures and controls that were incorporated into the analyses. Indeed subsequent work as part of the Neandertal genome sequencing project has confirmed the findings.

In trying to resolve the conflicting dating estimates (Coop et al., 2008; Krause et al., 2007), Pääbo's team has suggested that the *FOXP2* locus might have undergone multiple selective events on the lineage that led to modern humans (Ptak et al., 2009). The older event (before the human-Neandertal split) would have involved the two amino acid substitutions, while the more recent one (within the past 200,000 years) may have involved neighboring noncoding sequences, affecting the way the gene is regulated. Under this model, the signature of recent selection found in extant DNA samples does not relate to the amino acid changes themselves but instead to some as yet undetected human-specific variant(s) at the locus that might affect splicing or expression of the gene (Ptak et al., 2009).

A major caveat when considering these types of molecular evolution data is that inferences must often be drawn from DNA sequence alone, without the means to assess whether the relevant changes really had any functional significance in our ancestors. For example, the amino acid substitutions of FOXP2 that occurred in hominid evolution are located in poorly understood parts of the protein, outside the known functional domains. It is essential to understand that these evolutionary changes (while they might be considered "mutations" in the evolutionary sense of the word) are entirely distinct from the etiological mutations found in the KE family and other cases of language disorder (Figure 21.6). Although there were some initial speculations about how these substitutions might affect properties of the protein (Enard et al., 2002), several years of intensive cell-based analyses failed to uncover obvious differences in the functions of human and chimpanzee versions of FOXP2.

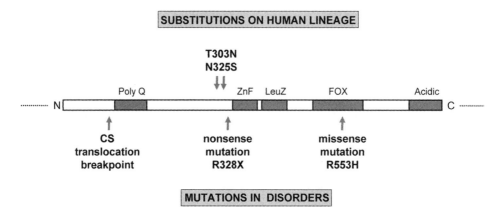

Figure 21.6
The crucial distinction between evolutionary substitutions and etiological mutations. A schematic of the human FOXP2 protein (main isoform) is shown, with the key domains shaded (see Figure 21.3 for labeling). Below this, examples are given of the most well-studied etiological mutations of FOXP2 that cause speech and language disorder. These include the R553H missense mutation, the R328X nonsense mutation, and the CS translocation—each yields a major disruption of FOXP2 function (Lai et al., 2001; MacDermot et al., 2005). The etiological mutations are found in a heterozygous state in the affected people, such that these people carry half the normal functional dosage of FOXP2 protein. Contrast the etiological mutations with the two amino acid substitutions that occurred during human evolution, indicated above the schematic. The evolutionary changes, T303N and N325S, are located at different positions from the etiological mutations, lying outside the known functional domains, and they would be expected to have only very minor effects on the structure and properties of the FOXP2 protein. In fact, until recently, there was no empirical evidence of any functional impact for T303N and/or N325S. But papers published in 2009 suggested that introduction of the evolutionary changes into model systems does indeed affect FOXP2 function, albeit in a rather subtle manner (Enard et al., 2009; Konopka et al., 2009).

However, in 2009, Enard and colleagues reported the consequences of inserting the two amino acid changes into the endogenous *Foxp2* locus of a mouse, using a "knock-in" strategy (Enard et al., 2009). The resulting mice showed a number of intriguing differences from their wildtype littermates in aspects of brain function, including reduced exploratory behavior and decreased dopamine levels. Examination of nonneural tissues, including organs where Foxp2 protein is normally found, like the heart and lung, failed to find any significant changes in the knock-in mice. When the researchers focused their attention on neurons in the striatum (already established as a site of particular interest in previous *FOXP2/Foxp2* studies, as described above) they observed that the knock-in animals had longer dendrites than controls, as well as increased LTD at glutamatergic synapses (Enard et al., 2009). These findings are notably distinct from those observed for mice carrying loss-of-function alleles; contrast the enhanced striatal synaptic plasticity in these animals, for example, with the impaired synaptic plasticity previously shown for R552H heterozygous mice (Groszer et al., 2008). The mice carrying the human amino acids also displayed some qualitative differences in the properties of pup isolation calls, primarily a slight alteration in frequency when compared to littermate controls. As discussed earlier in this chapter, very little is known about the neurobiology underlying mouse vocalizations, and innate pup calls are far from being a model of human speech. Thus, while these vocalization changes are intriguing, further study is required to establish their significance, and how they may relate to the other phenotypic findings.

Overall, Enard and colleagues provided the first experimental evidence to support the idea that the amino acid changes that occurred during human evolution had a functional impact, and it seems that they may be especially relevant for cortico–basal ganglia circuits (for a review, see Enard, 2011). It is particularly noteworthy that these substitutions affected dendritic growth and synaptic plasticity, CNS processes that are known to be influenced by ancestral versions of the gene (Groszer et al., 2008; Vernes et al., 2011), but in an opposite direction to loss-of-function alleles. More recently, Konopka et al. (2009) engineered human neuronal cell models to express either human or chimpanzee versions of FOXP2 protein and reported quantitative differences in the resulting profiles of gene expression (see below), albeit without describing any consequent changes in neurobiological properties of the cells.

The investigations of *FOXP2*'s potential role in human evolution have led to something of a revival of the "speech gene"/"language gene" tag, particularly in the media. It is worth reiterating here that it is unlikely that any single gene is responsible for the emergence of the complex suite of skills that allows members of our species to acquire spoken language. So how should we view the evolutionary findings, and assimilate them with our broader knowledge of how *FOXP2* functions? One parsimonious model is as follows. *FOXP2* is an ancient gene and is found in similar form in nonspeaking vertebrates, where we suspect it affects plasticity of circuits involved in sensorimotor integration and motor-skill learning. Perhaps the alterations of *FOXP2* in the human lineage were important in enhancing these processes, at time points when spoken language was emerging and evolving (driven in part by other genetic and nongenetic factors). Such modifications may have had wider ramifications, beyond facilitating sequencing of articulatory movements, if *FOXP2* also plays roles in neural plasticity during procedural learning, for example. This fits in with the idea that our speech and language skills did not appear fully formed and out of the blue, instead involving recruitment and refinement of existing anatomical, physiological, and neurological systems (Fisher & Marcus, 2006). It is fascinating that our first glimpse into the evolutionary history of the genetic architecture underlying human speech and language emphasizes not only the unique nature of *Homo sapiens*, but also our close biological connections to other non-speaking species.

A Functional Genomic Perspective on Speech and Language

FOXP2 mutations are only a rare cause of speech and language impairments in the human population (MacDermot et al., 2005). Given the high heritability of these neurodevelopmental disorders, it is likely that additional contributing genes remain to be discovered, and that they will also shed light on the biological underpinnings of normal spoken language. Remarkably, *FOXP2* itself offers powerful routes toward uncovering the other key genes, given that it is a regulatory factor modulating the expression of downstream targets in relevant brain circuits. The value of this strategy was demonstrated in 2008 when Vernes and colleagues isolated the *CNTNAP2* gene as a neural target directly regulated by the FOXP2 protein; they found that FOXP2 binds to a genomic site near the start of this target gene and acts to repress its expression (Vernes et al., 2008). *CNTNAP2* encodes an important protein, found in the membranes of CNS neurons, which is implicated in several processes in brain development (including cell adhesion, neuronal recognition, and localization/maintenance of voltage-gated potassium channels) and shows enriched expression in frontal gray matter in the developing cerebral cortex (Alarcon et al., 2008). Armed with the knowledge that *CNTNAP2* is a FOXP2 target, the researchers tested it for relevance to language phenotypes, and found that genomic variants in this target were correlated with reduced skills in children showing typical forms of language impairment (Vernes et al., 2008). Similar *CNTNAP2* variants are associated with delayed language in autistic disorder (Alarcon et al., 2008) and with normal variation in early language performance in the general population (Whitehouse et al., 2011).

The *CNTNAP2* study is noteworthy in that it identifies functional genetic links between the rare severe monogenic syndrome caused by *FOXP2* mutation and common language-related disorders with complex genetic architecture. More fundamentally, it illustrates that the neurogenetic networks in which *FOXP2* is embedded have broader relevance for understanding the basis of human speech and language. The other elements of these networks are being teased apart through state-of-the-art functional genomic approaches (Spiteri et al., 2007; Vernes et al., 2007; Konopka et al., 2009; Vernes et al., 2011), and future work will determine whether additional targets and interacting partners of FOXP2 are involved in speech- and language-related phenotypes (see O'Roak et al., 2011, for a recent example). This heralds an exciting new perspective on age-old questions regarding the origins of one of the most fascinating aspects of the human condition.

In conclusion, although we are still at the early stages of deciphering *FOXP2-*dependent pathways in humans and animals, the story so far provides a template for how scientists can effectively begin building bridges between genes, neurons, brains, and language. It is clear that success in this endeavor requires a multidisciplinary and multispecies perspective, integrating findings from psychology, neuroscience, genetics, developmental biology, and evolutionary anthropology.

References

Abrahams, B. S., & Geschwind, D. H. (2008). Advances in autism genetics: On the threshold of a new neurobiology. *Nature Reviews: Genetics, 9*, 341–355.

Alarcon, M., Abrahams, B. S., Stone, J. L., Duvall, J. A., Perederiy, J. V., Bomar, J. M., et al. (2008). Linkage, association, and gene-expression analyses identify CNTNAP2 as an autism-susceptibility gene. *American Journal of Human Genetics, 82*, 150–159.

Belton, E., Salmond, C. H., Watkins, K. E., Vargha-Khadem, F., & Gadian, D. G. (2003). Bilateral brain abnormalities associated with dominantly inherited verbal and orofacial dyspraxia. *Human Brain Mapping, 18*, 194–200.

Bishop, D. V. (2001). Genetic and environmental risks for specific language impairment in children. *Philosophical Transactions of the Royal Society B: Biological Sciences, 356*, 369–380.

Bonkowsky, J. L., & Chien, C. B. (2005). Molecular cloning and developmental expression of foxP2 in zebrafish. *Developmental Dynamics, 234*, 740–746.

Bruce, H. A., & Margolis, R. L. (2002). FOXP2: Novel exons, splice variants, and CAG repeat length stability. *Human Genetics, 111*, 136–144.

Campbell, P., Reep, R. L., Stoll, M. L., Ophir, A. G., & Phelps, S. M. (2009). Conservation and diversity of Foxp2 expression in muroid rodents: Functional implications. *Journal of Comparative Neurology, 512*, 84–100.

Carlsson, P., & Mahlapuu, M. (2002). Forkhead transcription factors: Key players in development and metabolism. *Developmental Biology, 250*, 1–23.

Cirillo, L. A., & Barton, M. C. (2008). Many forkheads in the road to regulation: Symposium on Forkhead Transcription Factor Networks in Development, Signalling and Disease. *EMBO Reports, 9*, 721–724.

Coop, G., Bullaughey, K., Luca, F., & Przeworski, M. (2008). The timing of selection at the human FOXP2 gene. *Molecular Biology and Evolution, 25*, 1257–1259.

De Zeeuw, C. I., & Yeo, C. H. (2005). Time and tide in cerebellar memory formation. *Current Opinion in Neurobiology, 15*, 667–674.

Ehret, G. (2005). Infant rodent ultrasounds—a gate to the understanding of sound communication. *Behavior Genetics, 35*, 19–29.

Enard, W. (2011). FOXP2 and the role of cortico-basal ganglia circuits in speech and language evolution. *Current Opinion in Neurobiology, 21*, 415–424.

Enard, W., Gehre, S., Hammerschmidt, K., Holter, S. M., Blass, T., Somel, M., et al. (2009). A humanized version of Foxp2 affects cortico-basal ganglia circuits in mice. *Cell, 137*, 961–971.

Enard, W., Przeworski, M., Fisher, S. E., Lai, C. S., Wiebe, V., Kitano, T., et al. (2002). Molecular evolution of FOXP2, a gene involved in speech and language. *Nature*, *418*, 869–872.

Ferland, R. J., Cherry, T. J., Preware, P. O., Morrisey, E. E., & Walsh, C. A. (2003). Characterization of Foxp2 and Foxp1 mRNA and protein in the developing and mature brain. *Journal of Comparative Neurology*, *460*, 266–279.

Feuk, L., Kalervo, A., Lipsanen-Nyman, M., Skaug, J., Nakabayashi, K., Finucane, B., et al. (2006). Absence of a paternally inherited FOXP2 gene in developmental verbal dyspraxia. *American Journal of Human Genetics*, *79*, 965–972.

Fisher, S. E. (2005). Dissection of molecular mechanisms underlying speech and language disorders. *Applied Psycholinguistics*, *26*, 111–128.

Fisher, S. E. (2006). Tangled webs: Tracing the connections between genes and cognition. *Cognition*, *101*, 270–297.

Fisher, S. E., Lai, C. S., & Monaco, A. P. (2003). Deciphering the genetic basis of speech and language disorders. *Annual Review of Neuroscience*, *26*, 57–80.

Fisher, S. E., & Marcus, G. F. (2006). The eloquent ape: Genes, brains and the evolution of language. *Nature Reviews: Genetics*, *7*, 9–20.

Fisher, S. E., & Scharff, C. (2009). FOXP2 as a molecular window into speech and language. *Trends in Genetics*, *25*, 166–177.

Fisher, S. E., Vargha-Khadem, F., Watkins, K. E., Monaco, A. P., & Pembrey, M. E. (1998). Localisation of a gene implicated in a severe speech and language disorder. *Nature Genetics*, *18*, 168–170.

French, C. A., Groszer, M., Preece, C., Coupe, A. M., Rajewsky, K., & Fisher, S. E. (2007). Generation of mice with a conditional Foxp2 null allele. *Genesis (New York)*, *45*, 440–446.

French, C. A., Jin, X., Campbell, T. G., Gerfen, E., Groszer, M., Fisher, S. E., et al. (2011). An aetiological Foxp2 mutation causes aberrant striatal activity and alters plasticity during skill learning. *Molecular Psychiatry*. [Epub ahead of print]. doi:10.1038/mp.2011.105.

Fujita, E., Tanabe, Y., Shiota, A., Ueda, M., Suwa, K., Momoi, M. Y., et al. (2008). Ultrasonic vocalization impairment of Foxp2 (R552H) knockin mice related to speech-language disorder and abnormality of Purkinje cells. *Proceedings of the National Academy of Sciences of the United States of America*, *105*, 3117–3122.

Gaub, S., Groszer, M., Fisher, S. E., & Ehret, G. (2010). The structure of innate vocalizations in Foxp2-deficient mouse pups. *Genes, Brain & Behavior*, *9*, 390–401.

Gopnik, M. (1990). Feature-blind grammar and dysphasia. *Nature*, *344*, 715.

Graybiel, A. M. (2005). The basal ganglia: Learning new tricks and loving it. *Current Opinion in Neurobiology*, *15*, 638–644.

Groszer, M., Keays, D. A., Deacon, R. M., de Bono, J. P., Prasad-Mulcare, S., Gaub, S., et al. (2008). Impaired synaptic plasticity and motor learning in mice with a point mutation implicated in human speech deficits. *Current Biology*, *18*, 354–362.

Haesler, S., Rochefort, C., Georgi, B., Licznerski, P., Osten, P., & Scharff, C. (2007). Incomplete and inaccurate vocal imitation after knockdown of FoxP2 in songbird basal ganglia nucleus Area X. *PLoS Biology*, *5*, e321.

Haesler, S., Wada, K., Nshdejan, A., Morrisey, E. E., Lints, T., Jarvis, E. D., et al. (2004). FoxP2 expression in avian vocal learners and non-learners. *Journal of Neuroscience*, *24*, 3164–3175.

Hanashima, C., Li, S. C., Shen, L., Lai, E., & Fishell, G. (2004). Foxg1 suppresses early cortical cell fate. *Science*, *303*, 56–59.

Hannenhalli, S., & Kaestner, K. H. (2009). The evolution of Fox genes and their role in development and disease. *Nature Reviews: Genetics*, *10*, 233–240.

Happe, F., Ronald, A., & Plomin, R. (2006). Time to give up on a single explanation for autism. *Nature Neuroscience*, *9*, 1218–1220.

Holy, T. E., & Guo, Z. (2005). Ultrasonic songs of male mice. *PLoS Biology*, *3*, e386.

Hurst, J. A., Baraitser, M., Auger, E., Graham, F., & Norell, S. (1990). An extended family with a dominantly inherited speech disorder. *Developmental Medicine and Child Neurology*, *32*, 352–355.

Kikusui, T., Nakanishi, K., Nakagawa, R., Nagasawa, M., Mogi, K., & Okanoya, K. (2011). Cross fostering experiments suggest that mice songs are innate. *PLoS ONE*, *6*, e17721.

Konopka, G., Bomar, J. M., Winden, K., Coppola, G., Jonsson, Z. O., Gao, F., et al. (2009). Human-specific transcriptional regulation of CNS development genes by FOXP2. *Nature*, *462*, 213–217.

Kosho, T., Sakazume, S., Kawame, H., Wakui, K., Wada, T., Okoshi, Y., et al. (2008). De-novo balanced translocation between 7q31 and 10p14 in a girl with central precocious puberty, moderate mental retardation, and severe speech impairment. *Clinical Dysmorphology*, *17*, 31–34.

Krause, J., Lalueza-Fox, C., Orlando, L., Enard, W., Green, R. E., Burbano, H. A., et al. (2007). The derived FOXP2 variant of modern humans was shared with Neandertals. *Current Biology*, *17*, 1908–1912.

Lai, C. S., Fisher, S. E., Hurst, J. A., Levy, E. R., Hodgson, S., Fox, M., et al. (2000). The SPCH1 region on human 7q31: Genomic characterization of the critical interval and localization of translocations associated with speech and language disorder. *American Journal of Human Genetics*, *67*, 357–368.

Lai, C. S., Fisher, S. E., Hurst, J. A., Vargha-Khadem, F., & Monaco, A. P. (2001). A forkhead-domain gene is mutated in a severe speech and language disorder. *Nature*, *413*, 519–523.

Lai, C. S., Gerrelli, D., Monaco, A. P., Fisher, S. E., & Copp, A. J. (2003). FOXP2 expression during brain development coincides with adult sites of pathology in a severe speech and language disorder. *Brain*, *126*, 2455–2462.

Lehmann, O. J., Sowden, J. C., Carlsson, P., Jordan, T., & Bhattacharya, S. S. (2003). Fox's in development and disease. *Trends in Genetics*, *19*, 339–344.

Lennon, P. A., Cooper, M. L., Peiffer, D. A., Gunderson, K. L., Patel, A., Peters, S., et al. (2007). Deletion of 7q31.1 supports involvement of FOXP2 in language impairment: Clinical report and review. *American Journal of Medical Genetics, Part A*, *143A*, 791–798.

Li, S., Weidenfeld, J., & Morrisey, E. E. (2004). Transcriptional and DNA binding activity of the Foxp1/2/4 family is modulated by heterotypic and homotypic protein interactions. *Molecular and Cellular Biology*, *24*, 809–822.

Liegeois, F., Baldeweg, T., Connelly, A., Gadian, D. G., Mishkin, M., & Vargha-Khadem, F. (2003). Language fMRI abnormalities associated with FOXP2 gene mutation. *Nature Neuroscience*, *6*, 1230–1237.

MacDermot, K. D., Bonora, E., Sykes, N., Coupe, A. M., Lai, C. S., Vernes, S. C., et al. (2005). Identification of FOXP2 truncation as a novel cause of developmental speech and language deficits. *American Journal of Human Genetics*, *76*, 1074–1080.

McCarthy, M. I., Abecasis, G. R., Cardon, L. R., Goldstein, D. B., Little, J., Ioannidis, J. P., et al. (2008). Genome-wide association studies for complex traits: Consensus, uncertainty and challenges. *Nature Reviews: Genetics*, *9*, 356–369.

Meaburn, E., Dale, P. S., Craig, I. W., & Plomin, R. (2002). Language-impaired children: No sign of the FOXP2 mutation. *NeuroReport*, *13*, 1075–1077.

Middleton, F. A., & Strick, P. L. (2000). Basal ganglia and cerebellar loops: Motor and cognitive circuits. *Brain Research: Brain Research Reviews*, *31*, 236–250.

Miller, J. E., Spiteri, E., Condro, M. C., Dosumu-Johnson, R. T., Geschwind, D. H., & White, S. A. (2008). Birdsong decreases protein levels of FoxP2, a molecule required for human speech. *Journal of Neurophysiology*, *100*, 2015–2025.

Newbury, D. F., Bonora, E., Lamb, J. A., Fisher, S. E., Lai, C. S., Baird, G., et al. (2002). FOXP2 is not a major susceptibility gene for autism or specific language impairment. *American Journal of Human Genetics*, *70*, 1318–1327.

O'Roak, B. J., Deriziotis, P., Lee, C., Vives, L., Schwartz, J. J., Girirajan, S., et al. (2011). Exome sequencing in sporadic autism spectrum disorders identifies severe de novo mutations. *Nature Genetics*, *43*, 585–589.

Ptak, S. E., Enard, W., Wiebe, V., Hellmann, I., Krause, J., Lachmann, M., et al. (2009). Linkage disequilibrium extends across putative selected sites in FOXP2. *Molecular Biology and Evolution*, *26*, 2181–2184.

Rochefort, C., He, X., Scotto-Lomassese, S., & Scharff, C. (2007). Recruitment of FoxP2-expressing neurons to area X varies during song development. *Developmental Neurobiology*, *67*, 809–817.

Shah, R., Medina-Martinez, O., Chu, L. F., Samaco, R. C., & Jamrich, M. (2006). Expression of FoxP2 during zebrafish development and in the adult brain. *International Journal of Developmental Biology*, *50*, 435–438.

Shriberg, L. D., Ballard, K. J., Tomblin, J. B., Duffy, J. R., Odell, K. H., & Williams, C. A. (2006). Speech, prosody, and voice characteristics of a mother and daughter with a 7;13 translocation affecting FOXP2. *Journal of Speech, Language, and Hearing Research: JSLHR*, *49*, 500–525.

Shriberg, L. D., Tomblin, J. B., & McSweeny, J. L. (1999). Prevalence of speech delay in 6-year-old children and comorbidity with language impairment. *Journal of Speech, Language, and Hearing Research: JSLHR*, *42*, 1461–1481.

Shu, W., Cho, J. Y., Jiang, Y., Zhang, M., Weisz, D., Elder, G. A., et al. (2005). Altered ultrasonic vocalization in mice with a disruption in the Foxp2 gene. *Proceedings of the National Academy of Sciences of the United States of America*, *102*, 9643–9648.

Shu, W., Lu, M. M., Zhang, Y., Tucker, P. W., Zhou, D., & Morrisey, E. E. (2007). Foxp2 and Foxp1 cooperatively regulate lung and esophagus development. *Development*, *134*, 1991–2000.

Shu, W., Yang, H., Zhang, L., Lu, M. M., & Morrisey, E. E. (2001). Characterization of a new subfamily of winged-helix/forkhead (Fox) genes that are expressed in the lung and act as transcriptional repressors. *Journal of Biological Chemistry*, *276*, 27488–27497.

Spiteri, E., Konopka, G., Coppola, G., Bomar, J., Oldham, M., Ou, J., et al. (2007). Identification of the transcriptional targets of FOXP2, a gene linked to speech and language, in developing human brain. *American Journal of Human Genetics, 81*, 1144–1157.

Tager-Flusberg, H., Joseph, R., & Folstein, S. (2001). Current directions in research on autism. *Mental Retardation and Developmental Disabilities Research Reviews, 7*, 21–29.

Takahashi, K., Liu, F. C., Hirokawa, K., & Takahashi, H. (2003). Expression of Foxp2, a gene involved in speech and language, in the developing and adult striatum. *Journal of Neuroscience Research, 73*, 61–72.

Takahashi, K., Liu, F. C., Oishi, T., Mori, T., Higo, N., Hayashi, M., et al. (2008). Expression of FOXP2 in the developing monkey forebrain: Comparison with the expression of the genes FOXP1, PBX3, and MEIS2. *Journal of Comparative Neurology, 509*, 180–189.

Teramitsu, I., Kudo, L. C., London, S. E., Geschwind, D. H., & White, S. A. (2004). Parallel FoxP1 and FoxP2 expression in songbird and human brain predicts functional interaction. *Journal of Neuroscience, 24*, 3152–3163.

Teramitsu, I., & White, S. A. (2006). FoxP2 regulation during undirected singing in adult songbirds. *Journal of Neuroscience, 26*, 7390–7394.

Tomblin, J. B., Records, N. L., Buckwalter, P., Zhang, X., Smith, E., & O'Brien, M. (1997). Prevalence of specific language impairment in kindergarten children. *Journal of Speech, Language, and Hearing Research: JSLHR, 40*, 1245–1260.

Vargha-Khadem, F., Gadian, D. G., Copp, A., & Mishkin, M. (2005). FOXP2 and the neuroanatomy of speech and language. *Nature Reviews: Neuroscience, 6*, 131–138.

Vargha-Khadem, F., & Passingham, R. E. (1990). Speech and language defects. *Nature, 346*, 226.

Vargha-Khadem, F., Watkins, K., Alcock, K., Fletcher, P., & Passingham, R. (1995). Praxic and nonverbal cognitive deficits in a large family with a genetically transmitted speech and language disorder. *Proceedings of the National Academy of Sciences of the United States of America, 92*, 930–933.

Vargha-Khadem, F., Watkins, K. E., Price, C. J., Ashburner, J., Alcock, K. J., Connelly, A., et al. (1998). Neural basis of an inherited speech and language disorder. *Proceedings of the National Academy of Sciences of the United States of America, 95*, 12695–12700.

Vernes, S. C., Newbury, D. F., Abrahams, B. S., Winchester, L., Nicod, J., Groszer, M., et al. (2008). A functional genetic link between distinct developmental language disorders. *New England Journal of Medicine, 359*, 2337–2345.

Vernes, S. C., Nicod, J., Elahi, F. M., Coventry, J. A., Kenny, N., Coupe, A. M., et al. (2006). Functional genetic analysis of mutations implicated in a human speech and language disorder. *Human Molecular Genetics, 15*, 3154–3167.

Vernes, S. C., Oliver, P. L., Spiteri, E., Lockstone, H. E., Puliyadi, R., Taylor, J. M., et al. (2011). Foxp2 regulates gene networks implicated in neurite outgrowth in the developing brain. *PLOS Genetics, 7*, e1002145.

Vernes, S. C., Spiteri, E., Nicod, J., Groszer, M., Taylor, J. M., Davies, K. E., et al. (2007). High-throughput analysis of promoter occupancy reveals direct neural targets of FOXP2, a gene mutated in speech and language disorders. *American Journal of Human Genetics, 81*, 1232–1250.

Wang, B., Lin, D., Li, C., & Tucker, P. (2003). Multiple domains define the expression and regulatory properties of Foxp1 forkhead transcriptional repressors. *Journal of Biological Chemistry, 278,* 24259–24268.

Wassink, T. H., Piven, J., Vieland, V. J., Pietila, J., Goedken, R. J., Folstein, S. E., et al. (2002). Evaluation of FOXP2 as an autism susceptibility gene. *American Journal of Medical Genetics, 114,* 566–569.

Watkins, K. E., Dronkers, N. F., & Vargha-Khadem, F. (2002). Behavioural analysis of an inherited speech and language disorder: Comparison with acquired aphasia. *Brain, 125,* 452–464.

Watkins, K. E., Vargha-Khadem, F., Ashburner, J., Passingham, R. E., Connelly, A., Friston, K. J., et al. (2002). MRI analysis of an inherited speech and language disorder: Structural brain abnormalities. *Brain, 125,* 465–478.

Weaver, T. D., Roseman, C. C., & Stringer, C. B. (2008). Close correspondence between quantitative- and molecular-genetic divergence times for Neandertals and modern humans. *Proceedings of the National Academy of Sciences of the United States of America, 105,* 4645–4649.

Wehr, R., Mansouri, A., de Maeyer, T., & Gruss, P. (1997). Fkh5-deficient mice show dysgenesis in the caudal midbrain and hypothalamic mammillary body. *Development, 124,* 4447–4456.

Whitehouse, A. J., Bishop, D. V., Ang, Q. W., Pennell, C. E., & Fisher, S. E. (2011). CNTNAP2 variants affect early language development in the general population. *Genes, Brain & Behavior, 10,* 451–456.

Williams, A. J., & Paulson, H. L. (2008). Polyglutamine neurodegeneration: Protein misfolding revisited. *Trends in Neurosciences, 31,* 521–528.

Zeesman, S., Nowaczyk, M. J., Teshima, I., Roberts, W., Cardy, J. O., Brian, J., et al. (2006). Speech and language impairment and oromotor dyspraxia due to deletion of 7q31 that involves FOXP2. *American Journal of Medical Genetics, Part A, 140,* 509–514.

Zhang, J., Webb, D. M., & Podlaha, O. (2002). Accelerated protein evolution and origins of human-specific features: Foxp2 as an example. *Genetics, 162,* 1825–1835.

Constance Scharff and Christopher K. Thompson

Of the nearly 10,000 different species of birds, almost half fall into the suborder of the oscine passerines, more commonly known as songbirds. Songbirds learn their vocalizations, a process that shares a number of similarities with language development in humans (Bolhuis, Okanoya, & Scharff, 2010; Doupe & Kuhl, 1999; Moorman & Bolhuis, chapter 5, this volume). Like language in humans, song in songbirds is typically learned early in life (though many species can add additional elements to their repertoire after sexual maturation). Juvenile songbirds (usually only males in most species) learn their song from an adult male tutor. In many songbird species, such as the zebra finch, song learning has two phases: a sensory phase, during which the tutor song is committed to memory, and a sensorimotor phase, during which the bird's own vocal output is "matched" with the memorized tutor song. When a young songbird starts to sing during the sensorimotor phase, its song output is not yet stereotyped and does not resemble the tutor song very well. This kind of vocalization is known as "subsong" (Catchpole & Slater, 1995). The production of subsong is reminiscent of "babbling" in human infants (Doupe & Kuhl, 1999; Bolhuis et al., 2010). As the sensorimotor phase progresses, the bird more closely matches its own output with the tutor song and, after sexual maturation, eventually produces a highly stereotyped song known as crystallized song.

The similarities of birdsong and human language are not as well investigated when one considers the functional significance of these behaviors, however. Human speech can express vast amounts of information via its combinatorial use of a finite repertoire of phonemes, whereas the combinatorial use of individual elements of a bird's vocalization have by and large not been associated with specific meaning. However, little research has addressed this, and some evidence for specific information being conveyed by particular vocalizations exists (Templeton, Greene, & Davis, 2005). What is well established is that birdsong is used in a reproductive context, either to attract mates or to ward off rivals from a home territory (Catchpole & Slater, 1995).

The Song Control System

The production, learning, and perception of song are regulated by a series of discrete brain nuclei collectively known as the song control system (Figure 22.1). Many components of the song control system been characterized since it was first described (Nottebohm & Arnold, 1976; Nottebohm, Stokes, & Leonard, 1976), but new components and connections continue to be discovered. The nomenclature for avian neuroanatomy, including the song control system, was revised in 2004 to better reflect its evolutionary continuity with other vertebrates, including mammals (Jarvis et al., 2005; Reiner, Perkel, Bruce, et al., 2004). The integration of song perception and motor output is likely to involve the transfer of auditory information from various brain areas in the auditory telencephalon to HVC (formerly known as "high vocal center"; see Reiner, Perkel, Mello, & Jarvis, 2004). From HVC, the song control system diverges into two distinct pathways: the descending motor pathway and the anterior forebrain pathway (AFP). The descending motor pathway consists of a series of projections from HVC eventually leading to the muscles of the syrinx, the vocal organ acting as the sound source in songbirds. Lesions of nuclei in the descending motor pathway result in a severe disruption of song (Nottebohm et al., 1976). The AFP emanates from a subset of HVC neurons that project to Area X, a basal ganglia nucleus made up of striatal and pallidal elements, which projects to the thalamus, then back to another song control nucleus in the telencephalon, before connecting to the descending motor pathway. The AFP shares homology with the cortico–basal ganglia–thalamo–cortical loop in mammals (Farries, 2004; Gale & Perkel, 2010). In zebra finches, lesions of nuclei in the AFP disrupt the learning of song in juveniles (Nordeen & Nordeen, 1993; Scharff & Nottebohm, 1991; Sohrabji, Nordeen, & Nordeen, 1990). Once song is stably learned, Area X and other AFP nuclei continue to be relevant for online monitoring of song (Kao, Doupe, & Brainard, 2005; Olveczky, Andalman, & Fee, 2005; Scharff, Nottebohm, & Cynx, 1998). In adult Bengalese finches, partial lesions of Area X do not abolish song but instead induce a type of stuttering, which is consistent with the role Area X may play as a basal ganglia nucleus in regulating the timing of precise motor control (Kobayashi, Uno, & Okanoya, 2001). Most of this review will be focused on this song system nucleus, Area X.

Genetic Contributions to Song Learning

The capacity to learn vocalizations in songbirds is subject to a number of constraints: the nature of the auditory filters that decide what is relevant imitating, the physical and physiological properties that affect what sounds the syrinx can produce, and the interplay between breathing and singing. These factors, as well as the learning faculty

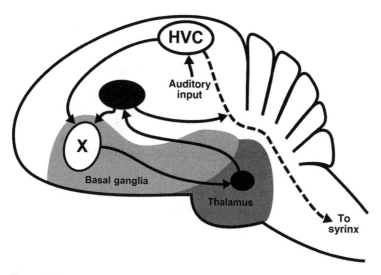

Figure 22.1
Simplified sagittal schematic of the song control system. Auditory input enters the song control system via HVC. The descending motor pathway (dashed line from HVC to the syrinx) controls song production. The anterior forebrain pathway (dark lines) contributes to the acquisition of a song template. Dashed lines indicate indirect connection.

itself, depend on the activity of genes. While there has been much progress regarding the neural mechanisms of song learning in birds, the role of genes has only recently come into focus (Hilliard, Miller, Fraley, Horvath, & White, 2012). Findings on other learning-related or song-system-specific genes are summarized in a comprehensive recent review (White, 2009). This chapter highlights studies on songbirds that address the role of one gene, FoxP2, which is relevant for human speech and language.

A Little about FoxP2

FOXP2 is a member of the winged helix transcription factor family, characterized by a highly conserved forkhead (Fox) domain that binds to distinct DNA sequences in the regulatory regions of its target genes. The Fox transcription factor family is highly conserved across taxa and implicated in many developmental processes and diseases (Hannenhalli & Kaestner, 2009). FOXP2 in particular is the first gene identified to be causally related to a fairly specific speech and language phenotype, developmental verbal dyspraxia (DVD, alternatively called childhood apraxia of speech, CAS; for definition, see the American Speech-Language-Hearing Association, www.asha.org) (Lai, Fisher, Hurst, Vargha-Khadem, & Monaco, 2001). DVD's core symptoms are inaccurate and incomplete pronunciation of words, difficulties with repeating multisyllable nonsense words, and impaired receptive speech (Simms, 2007). Structural and functional brain imaging of humans with FOXP2 mutations shows subtle volume differences and striking activation differences during language tasks, particularly in corticocerebellar and corticostriatal circuits (Vargha-Khadem, Gadian, Copp, & Mishkin, 2005). Since the discovery of FOXP2 mutations in humans, in vitro and in vivo studies have made considerable progress in addressing the molecular, neural, and evolutionary function of FoxP2 in different systems. The details about FoxP2's role in language and functional significance in the mammalian brain are reviewed elsewhere in this book (Fisher, chapter 21, this volume). Here we highlight recent findings in songbirds on the developmental, seasonal, and behavioral regulation of FoxP2 expression, its importance for song learning and the structural plasticity of Area X neurons, and the expression of target genes of FoxP2 in the song control system.

FoxP2 in Pre- and Postnatal Development

Consistent with a developmental role of other Fox proteins, FoxP2 is expressed in regions of the vertebrate embryo in which inductive signals organize adjacent proliferation of neural progenitors and subsequent migration (Scharff & Haesler, 2005), a feature that persists in adult avian but not mammalian neuroproliferative zones (Rochefort, He, Scotto-Lomassese, & Scharff, 2007). As noted above, structural and functional brain imaging of humans with FOXP2 mutations shows subtle volume differences and striking activation differences during language tasks—for example, in corticostriatal and other circuits (Belton, Salmond, Watkins, Vargha-Khadem, & Gadian, 2003; Liegeois et al., 2003). These findings strongly suggest that FoxP2 plays an important role in early brain development in mammals; what role that might be in songbirds is not yet clear, however, especially since a FoxP2 knockout bird is not yet available.

Though FoxP2 expression studies in avian and mammalian embryos are consistent with a role for FoxP2 in early brain development, this does not rule out the possibility that FoxP2 continues to play a role later in life. Indeed, FoxP2 expression persists in the striatum, dorsal thalamus, cerebellum, and other regions of adult birds and rodents (Fisher & Scharff, 2009). In the posthatch songbird striatum, including Area X neurons, FoxP2 is expressed by medium spiny neurons, colocalizing with DARPP32 (Haesler et al., 2004). In addition, results indicate that FoxP2 plays a role in neural circuits relevant for auditory-guided vocal motor learning, at the time learning takes place (see below). Because of the previous findings implicating the AFP circuit in song development, it is interesting that in juvenile zebra finches, *FoxP2* mRNA expression levels in Area X are 10%–20% higher than in the surrounding striatum during vocal sensorimotor learning (Haesler et al., 2004) (Figure 22.2). Other regions

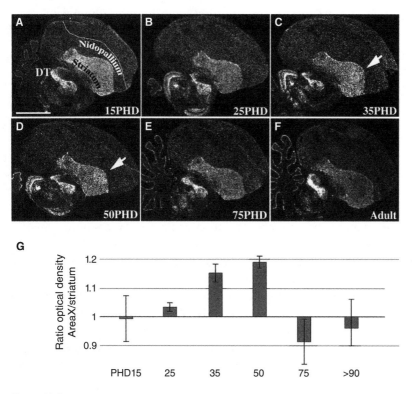

Figure 22.2
In situ hybridization in sagittal brain sections from male zebra finches shows that expression
of FoxP2 mRNA in Area X varies across development. In the telencephalon, FoxP2 is primarily expressed in the striatum (A). At posthatch day (PHD) 35 and 50 (C and D, respectively),
Area X neurons (arrows) up-regulate expression of FoxP2 relative to the rest of striatum.
Expression of FoxP2 then decreases as birds age (E and F). (G) Quantification of FoxP2
mRNA expression in Area X relative to expression in the rest of striatum. Figure reproduced
with permission from Haesler et al. (2004).

involved in controlling the learning and production of song show very low *FoxP2*
expression (Haesler, et al., 2004; Teramitsu, Kudo, London, Geschwind, & White,
2004). This change in *FoxP2* expression is not related to immediately prior singing
activity, because birds in this study had not sung prior to sacrifice. The up-regulation
of FoxP2 expression in juveniles may be related to fine-tuning synaptic connectivity,
because at this time, neurons in the AFP have established synapses and are functioning, but the topography typical for this pathway is still actively remodeled (Iyengar,
Viswanathan, & Bottjer, 1999).

Seasonal Change in Expression of FoxP2

A further correlation between song learning and levels of FoxP2 expression exists in another species of songbird, canaries. During the breeding season, male canaries sing highly regular and stereotyped song, and FoxP2 expression in Area X is low. After the breeding season, song stereotypy decreases and there is an increase in the rate of incorporation of new song elements in the song repertoire (Nottebohm, Nottebohm, Crane, & Wingfield, 1987). Concomitantly, FoxP2 in Area X is up-regulated (Haesler et al., 2004). As in the juvenile zebra finches described above, singing activity in these canaries did not seem to contribute to the amounts of FoxP2 expressed in Area X, though more thorough analysis of song-driven changes in FoxP2 expression in songbird species beyond zebra finches is needed.

Though the mechanism for seasonal regulation of FoxP2 expression in canaries is unknown, the pattern of change in expression suggests a potential role for sex steroid hormones. During the spring months, adult male canaries enter breeding condition, which is characterized by increased reproductive activity, increased singing rate, and high levels of circulating testosterone. When the reproductive season ends, circulating levels of testosterone decrease to undetectable levels. In canaries there is another increase in circulating testosterone starting in October, which is associated with feather molting and not related to reproduction. Figure 22.3 shows that FoxP2 expression in Area X appears to be inversely correlated with the

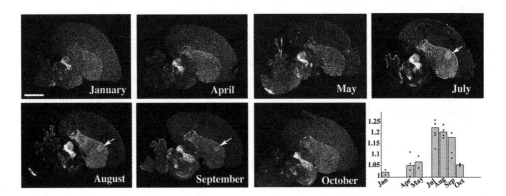

Figure 22.3
In situ hybridization in sagittal brain sections from male canaries shows that expression of FoxP2 mRNA in Area X varies seasonally. Expression of FoxP2 in Area X is up-regulated in months of the year when birds are in nonbreeding condition, have low levels of circulating testosterone, and are most likely to incorporate new syllables into their repertoires. Panel in lower right illustrates quantification of FoxP2 mRNA expression in Area X relative to expression in the rest of striatum. Figure reproduced with permission from Haesler et al. (2004).

circulating levels of testosterone, which suggests that testosterone, and/or it metabolites, may regulate expression of FoxP2 in Area X. This hypothesis still must be experimentally tested, but sex steroid regulation of expression of FoxP2 (and FoxP1) have been observed in pancreatic cancer cell lines, for example (Takayama et al., 2008).

Behavioral Regulation of FoxP2 Expression

In adult zebra finches, song rate and social context affect expression of FoxP2 mRNA and protein in Area X at a minute-to-hour time scale. When juvenile and adult male zebra finches sing in the absence of another bird or outside of courtship context (undirected singing), *FoxP2 mRNA* is significantly lower in Area X than in the surrounding striatum (Teramitsu, Poopatanapong, Torrisi, & White, 2010; Teramitsu & White, 2006). Conversely, FoxP2 expression is higher in adult males that did not sing and in males that sang toward females (i.e., directed singing) (Teramitsu & White, 2006). Singing appears to regulate expression of *FoxP2 mRNA* and protein differentially, however. Levels of FoxP2 protein extracted from tissue punches of Area X from birds that sang either directed or undirected song were very variable and did not differ with singing context (Miller et al., 2008). This discrepancy between mRNA and protein levels could be due to differential rates of mRNA and protein processing. Alternatively, protein levels might have yielded different results than mRNA levels because the authors normalized the amount of FoxP2 protein to the amount of glyceraldehyde-3-phosphate dehydrogenase (GAPDH) as a housekeeping gene. Since GAPDH is itself regulated by singing in HVC–to–Area X projecting neurons (Lombardino et al., 2006), levels in Area X might also be affected by singing. In addition, the most consistent and strongest difference in FoxP2 expression is an increase that occurs in nonsinging birds during the first two hours of daylight, before they sing (Miller et al., 2008). During the developmental phase of song learning, FoxP2 mRNA levels, similar to brains of adult birds, decrease after two hours of singing in Area X in hearing as well as in deafened birds. The acute decrease in FoxP2 levels might be important for modifying the song during learning in juveniles as well as for allowing subtle adjustments involved in song maintenance in adulthood (Teramitsu et al., 2010).

Auditory feedback is not necessary for the singing-induced regulation of FoxP2 in zebra finches, because *FoxP2* expression is also down-regulated in deafened birds that sing undirected song (Teramitsu et al., 2010). Nevertheless, auditory input appears to shape the change in *FoxP2* expression, because the expression of *FoxP2* is correlated with the amount of singing in hearing birds but not in deafened birds (Teramitsu et al., 2010). Interestingly, levels of FoxP2 protein in the medial, but not lateral, geniculate nucleus also change after auditory stimulation in mice (Horng et al., 2009), emphasizing that neural activity can regulate the expression of FoxP2 in specific subsets of neurons in different species. Future experiments should address the integration of auditory input and motor control and its relationship to FoxP2 expression in Area X.

Functional Role of FoxP2 in Songbirds

To address a causal relationship between FoxP2 expression and vocal learning, Haesler et al. (2007) experimentally reduced FoxP2 levels using lentivirus-mediated RNA interference in Area X of juvenile zebra finches throughout the sensorimotor song learning phase (Figure 22.4). The FoxP2 knockdown birds copied tutor songs only partially, imitating some elements but omitting others, imitating less accurately, and producing song elements more variably during each rendition (Haesler et al., 2007). Despite incomplete and inaccurate copying of tutor song, zebra finches with knocked-down FoxP2 in Area X remain able to generate a normal range of sounds. Interestingly, mice with reduced or absent FoxP2 are also able to produce the entire repertoire of ultrasonic distress and isolation calls (Gaub, Groszer, Fisher, & Ehret, 2010). Together, these data suggest that the sensorimotor integration necessary for the imitative learning of sounds is more likely to be affected by altered FoxP2 levels than the motor production itself. The song phenotype of FoxP2 knockdown zebra finches strikingly echoes the incomplete and inaccurate renditions of words and highly variable pronunciation in humans with a mutated FOXP2 gene (Hurst, Baraitser, Auger, Graham, & Norell, 1990). FoxP2 levels were not manipulated during embryonic development in these experiments but only when song control brain circuits were already largely assembled, suggesting that a reduction of FoxP2 affects postnatal function independently from effects on early nervous system development.

Within Area X, FoxP2 is expressed in spiny neurons that exhibit many features of mammalian striatal medium spiny neurons. Spiny neurons in Area X are innervated by glutamatergic HVC neurons (Farries, Ding, & Perkel, 2005), and this projection is modulated presynaptically by midbrain dopaminergic input (Ding, Perkel, & Farries, 2003). Because midbrain dopamine acts on many behavioral systems, including reward learning, the integration of pallial and dopaminergic signals in FoxP2-expressing spiny neurons may be essential for fine-tuning song motor output to match the tutor song model. Modulation of FoxP2 expression might up- or down-regulate neural plasticity-relevant genes that in turn could affect motor learning and motor performance via structural and functional changes of the spiny neurons. This hypothesis is supported by recent data showing that spiny neurons in adult Area X exhibit significantly fewer spines after receiving lentivirally-mediated FoxP2 knockdown than control knockdowns (Schulz, Haesler, Scharff, & Rochefort, 2010). To

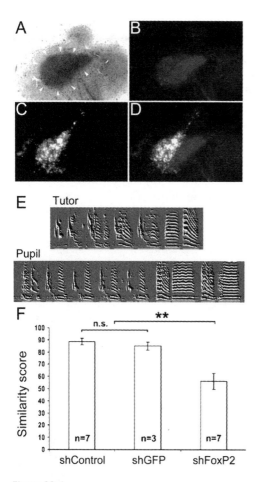

Figure 22.4
Lentiviral-mediated knockdown of FoxP2 mRNA in Area X leads to incomplete and inaccurate copying of song in zebra finches. Phase-contrast (A) and fluorescence (B) imaging elucidates the borders of Area X. The extent of the viral infection is made visible through virally mediated green fluorescent protein (GFP) expression (C), covering part of Area X (D). (E) Spectrograms of tutor (bottom) and of adult pupil (top) that received knockdown of FoxP2 in Area X as a juvenile. Note incomplete and inaccurate song imitation. (F) Quantification of similarity with tutor song of control pupils and pupils with FoxP2 knockdown. Control experiments were carried out with nontargeting shRNA sequences (shControl) and shRNA sequences targeting GFP (shGFP). (n.s., not significant) Images and figures are reproduced from Haesler et al. (2007) under CCAL.

investigate whether FoxP2 might also play a role in specifying the neural fate during neurogenesis, knockdown virus was also injected into the ventricular zone where striatal spiny neurons are born before migrating to Area X and the surrounding striatum. In spite of FoxP2 knockdown, neurons developed into spiny neurons that migrated and integrated into Area X, albeit carrying fewer spines than neurons infected with the control virus. These findings show that a reduction of FoxP2 protein levels in newly adult-generated neurons does not prevent them from differentiating into spiny Area X neurons, but knockdown does influence synaptic spine density and presumably synaptic plasticity. Also consistent with this interpretation are the findings of altered synaptic plasticity and impaired motor learning in the striatum of mice with experimentally reduced amounts of FoxP2 (Groszer et al., 2008).

The results from the FoxP2 knockdown experiment and those showing singing-related changes in FoxP2 expression suggest that FoxP2 may function as a "plasticity gate," though the exact direction gate swings open is not exactly clear. For instance, 75-day-old birds, whose song imitation is already quite good but not perfect, sing with lower stereotypy after 2 hours of singing than after 2 hours of silence (Miller, Hilliard, & White, 2010). This result suggests that the gate is open when birds need to increase song variability for matching the tutor's song which is accomplished by actively down-regulating FoxP2 levels. On the other hand, levels of *FoxP2* mRNA in zebra finch Area X are higher in juveniles than in adults (Haesler et al., 2004), and a hallmark of sexual maturation in zebra finches is song crystallization (i.e. decrease in song variability). Thus, it appears that down-regulation of FoxP2 as zebra finches reach sexual maturity closes the plasticity gate and song stereotypy increases. Nevertheless, it is clear that FoxP2 plays some role in gating plasticity, perhaps by translating synaptic activity into network adjustment via spine pruning. Yet changes in expression FoxP2 expression may have opposing effects depending upon the context.

Downstream and Upstream of FoxP2

FoxP2 can act as a transcriptional repressor as well as an activator of downstream genes. Recently, direct neural targets of FOXP2 were identified in human neuronal cell models in an unbiased genomic screen (Vernes et al., 2007). One of the downstream targets of FOXP2, CNTNAP2 (also known as CASPR2), recently received special attention since particular single nucleotide polymorphism signatures are independently associated with speech delays in autism and language deficits in common forms of language impairment (Vernes et al., 2008). In songbirds, CNTNAP2 is expressed in some song control nuclei, and its expression is sexually dimorphic in some respects (Panaitof, Abrahams, Dong, Geschwind, & White, 2010). Whether

FoxP2 specifically regulates CNTNAP2 in songbirds has not yet been addressed, though evidence from the human fetal cortex suggests that FOXP2 represses CNTNAP2 expression (Vernes et al., 2008). Regardless, these findings further point toward possibly shared molecular and neural substrates involved in speech and song.

To understand the role of FOXP2 for cellular and behavioral function and how this might have changed during the course of evolution, one also needs to identify which molecules regulate the transcription of FoxP2 itself. In juvenile zebra finches, administration of a cannabinoid agonist increases FoxP2 expression in the striatum, persisting into adulthood (Soderstrom & Luo, 2010). This suggests a potential inter-action of cannabinoid signaling and FoxP2 expression in brain regions relevant for learning and practicing song. Also, as noted above, the seasonal pattern of FoxP2 expression in canary Area X suggests that sex steroid hormones may play a role in regulating FoxP2 expression.

Partners of FoxP2

Though the evidence is overwhelming that FoxP2 is an important regulator of the development of human language and birdsong, no single gene can be solely respon-sible for vocal learning. In fact, FoxP1, another protein of the FoxP gene family, dimerizes with FoxP2 (Li, Weidenfeld, & Morrisey, 2004). FoxP1 is expressed in Area X and unlike FoxP2 is also substantially expressed in HVC and other song control system nuclei (Haesler et al., 2004; Teramitsu et al., 2004). In mammals, FoxP1 expression appears to be regulated by alpha-synuclein because expression of FoxP1 is significantly reduced in the brains of alpha-synuclein knockout mice (Kurz et al., 2010). Interestingly, the ultrasonic vocalizations of these mice occur more frequently and with higher peak amplitudes than with wildtype mice. These findings and others suggest that FOXP1 may also play a role in human speech. In fact, muta-tions of FOXP1 have been associated with a number of patients with significant speech and language maladies, confirming this prediction (Carr et al., 2010; Hamdan et al., 2010; Horn et al., 2010; O'Roak et al., 2011). FoxP4 is another member of the FoxP family, and like FoxP1, dimerizes with FoxP2. FoxP4 and FoxP2 were recently shown to be critical regulators of neural progenitor maintenance via suppression of key components of adherens junctions (Rousso, et al., 2012). The functional role of FoxP1 and FoxP4 and its relationship to FoxP2 in songbirds is not currently known but is an important gap in our understanding that must be filled.

Evolution of FoxP2

Comparative genetic studies can help elucidate how vocal learning evolved. The last common ancestor of humans and birds lived about 310 million years ago. The similarity of basal ganglia circuits and their functions in amniotes is consistent with the idea that vocal learning in the divergent lineages of synapsids (leading to mammals) and diapsids (leading to birds) may have exapted existing pallial-basal ganglia features, including FoxP2's role in the striatum. Comparing genes relevant for speech and learned birdsong may uncover shared key molecular networks relevant for vocal learning in distantly related species. While the amino acid sequence between mice, songbirds, chimps, and humans differs by less than 5%, comparison of FoxP2 sequences in bird species that differ with respect to the trait of vocal learning did not reveal FoxP2 sequence variants that segregated with the ability to imitate communication sounds (Scharff & Haesler, 2005). This emphasizes the fact that while FoxP2 plays a role in song, particular FoxP2 protein versions apparently do not correlate with vocal learning in birds. Similar comparisons have, however, not yet been done for the regulatory regions of FoxP2, because regulatory regions are much more elusive than coding regions and correspondingly less information is available about them. But these regions are thought to be particularly relevant for bringing about evolutionary changes in morphology and behavior (Carroll, 2003; Scharff & Petri, 2011). It is therefore conceivable that changes in these regions of FoxP2 may relate to differences in vocal learning in different species of songbirds as well as birds that do not learn to produce their vocalizations. Investigating the relative contribution of regulatory and coding regions of FoxP2 and other positively selected genes during the course of hominin evolution and their possible relationship to the evolution of language is already providing exciting insights (see the review by Fisher, chapter 21, this volume).

Conclusion

FoxP2 is an example of a gene that is relevant for human language and can be studied in the songbird model system. Investigating the role of genes operating at the different levels of organ systems and neural circuits underlying vocal learning and production in songbirds will advance efforts to understand the development and neural control of human language. During the last few years, songbird research has entered the age of molecular genetics. The first songbird genome, that of the zebra finch, has been sequenced (Warren et al., 2010), and transgenesis is feasible, if inefficient (Agate, Scott, Haripal, Lois, & Nottebohm, 2009). The next decade will increasingly harness gene manipulations, gene expression analyses, and genome and transcriptome sequencing techniques to gain insight into the molecular underpinnings of song learning and song production, identifying gene networks associated with developing vocal-learning circuits and their function, manipulating them, and comparing them in different species. This is increasingly feasible because genome sequencing is becoming faster and cheaper. Molecular studies in birds comparing non–vocal learners with vocal learners promise particularly exciting insights into the evolution of this trait and maybe also other analogies between humans and songbirds.

References

Agate, R. J., Scott, B. B., Haripal, B., Lois, C., & Nottebohm, F. (2009). Transgenic songbirds offer an opportunity to develop a genetic model for vocal learning. *Proceedings of the National Academy of Sciences of the United States of America, 106*, 17963–17967.

Belton, E., Salmond, C. H., Watkins, K. E., Vargha-Khadem, F., & Gadian, D. G. (2003). Bilateral brain abnormalities associated with dominantly inherited verbal and orofacial dyspraxia. *Human Brain Mapping, 18*, 194–200.

Bolhuis, J. J., Okanoya, K., & Scharff, C. (2010). Twitter evolution: Converging mechanisms in birdsong and human speech. *Nature Reviews: Neuroscience, 11*, 747–759.

Carr, C. W., Moreno-De-Luca, D., Parker, C., Zimmerman, H. H., Ledbetter, N., Martin, C. L., et al. (2010). Chiari I malformation, delayed gross motor skills, severe speech delay, and epileptiform discharges in a child with FOXP1 haploinsufficiency. *European Journal of Human Genetics, 18*, 1216–1220.

Carroll, S. B. (2003). Genetics and the making of Homo sapiens. *Nature, 422*, 849–857.

Catchpole, C., & Slater, P. J. B. (1995). *Bird song: Biological themes and variations.* Cambridge: Cambridge University Press.

Deregnaucourt, S., Mitra, P. P., Feher, O., Pytte, C., & Tchernichovski, O. (2005). How sleep affects the developmental learning of bird song. *Nature, 433*, 710–716.

Ding, L., Perkel, D. J., & Farries, M. A. (2003). Presynaptic depression of glutamatergic synaptic transmission by D1-like dopamine receptor activation in the avian basal ganglia. *Journal of Neuroscience, 23* (14), 6086–6095.

Doupe, A. J., & Kuhl, P. K. (1999). Birdsong and human speech: Common themes and mechanisms. *Annual Review of Neuroscience, 22*, 567–631.

Farries, M. A. (2004). The avian song system in comparative perspective. *Annals of the New York Academy of Sciences, 1016*, 61–76.

Farries, M. A., Ding, L., & Perkel, D. J. (2005). Evidence for "direct" and "indirect" pathways through the song system basal ganglia. *Journal of Comparative Neurology, 484*, 93–104.

Fisher, S. E., & Scharff, C. (2009). FOXP2 as a molecular window into speech and language. *Trends in Genetics, 25*, 166–177.

Gale, S. D., & Perkel, D. J. (2010). Anatomy of a songbird basal ganglia circuit essential for vocal learning and plasticity. *Journal of Chemical Neuroanatomy, 39*, 124–131.

Gaub, S., Groszer, M., Fisher, S. E., & Ehret, G. (2010). The structure of innate vocalizations in Foxp2-deficient mouse pups. *Genes, Brain & Behavior, 9*, 390–401.

Groszer, M., Keays, D. A., Deacon, R. M., de Bono, J. P., Prasad-Mulcare, S., Gaub, S., et al. (2008). Impaired synaptic plasticity and motor learning in mice with a point mutation implicated in human speech deficits. *Current Biology, 18*, 354–362.

Haesler, S., Rochefort, C., Georgi, B., Licznerski, P., Osten, P., & Scharff, C. (2007). Incomplete and inaccurate vocal imitation after knockdown of FoxP2 in songbird basal ganglia nucleus Area X. *PLoS Biology*, *5*, e321.

Haesler, S., Wada, K., Nshdejan, A., Morrisey, E. E., Lints, T., Jarvis, E. D., et al. (2004). FoxP2 expression in avian vocal learners and non-learners. *Journal of Neuroscience*, *24*, 3164–3175.

Hamdan, F. F., Daoud, H., Rochefort, D., Piton, A., Gauthier, J., Langlois, M., et al. (2010). De novo mutations in FOXP1 in cases with intellectual disability, autism, and language impairment. *American Journal of Human Genetics*, *87*, 671–678.

Hannenhalli, S., & Kaestner, K. H. (2009). The evolution of Fox genes and their role in development and disease. *Nature Reviews: Genetics*, *10*, 233–240.

Hilliard, A. T., Miller, J. E., Fraley, E. R., Horvath, S., & White, S. A. (2012). Molecular microcircuitry underlies functional specification in a basal ganglia circuit dedicated to vocal learning. *Neuron*, *73*(3), 537–552.

Horn, D., Kapeller, J., Rivera-Brugues, N., Moog, U., Lorenz-Depiereux, B., Eck, S., et al. (2010). Identification of FOXP1 deletions in three unrelated patients with mental retardation and significant speech and language deficits. *Human Mutation*, *31*, E1851–E1860.

Horng, S., Kreiman, G., Ellsworth, C., Page, D., Blank, M., Millen, K., et al. (2009). Differential gene expression in the developing lateral geniculate nucleus and medial geniculate nucleus reveals novel roles for Zic4 and Foxp2 in visual and auditory pathway development. *Journal of Neuroscience*, *29*, 13672–13683.

Hurst, J. A., Baraitser, M., Auger, E., Graham, F., & Norell, S. (1990). An extended family with a dominantly inherited speech disorder. *Developmental Medicine and Child Neurology*, *32*, 352–355.

Iyengar, S., Viswanathan, S. S., & Bottjer, S. W. (1999). Development of topography within song control circuitry of zebra finches during the sensitive period for song learning. *Journal of Neuroscience*, *19*, 6037–6057.

Jarvis, E. D., Gunturkun, O., Bruce, L., Csillag, A., Karten, H., Kuenzel, W., et al. (2005). Avian brains and a new understanding of vertebrate brain evolution. *Nature Reviews: Neuroscience*, *6*, 151–159.

Kao, M. H., Doupe, A. J., & Brainard, M. S. (2005). Contributions of an avian basal ganglia-forebrain circuit to real-time modulation of song. *Nature*, *433*, 638–643.

Kobayashi, K., Uno, H., & Okanoya, K. (2001). Partial lesions in the anterior forebrain pathway affect song production in adult Bengalese finches. *NeuroReport*, *12*, 353–358.

Kurz, A., Wohr, M., Walter, M., Bonin, M., Auburger, G., Gispert, S., et al. (2010). Alpha-synuclein deficiency affects brain Foxp1 expression and ultrasonic vocalization. *Neuroscience*, *166*, 785–795.

Lai, C. S., Fisher, S. E., Hurst, J. A., Vargha-Khadem, F., & Monaco, A. P. (2001). A forkhead-domain gene is mutated in a severe speech and language disorder. *Nature*, *413*, 519–523.

Li, S., Weidenfeld, J., & Morrisey, E. E. (2004). Transcriptional and DNA binding activity of the Foxp1/2/4 family is modulated by heterotypic and homotypic protein interactions. *Molecular and Cellular Biology*, *24*, 809–822.

Liegeois, F., Baldeweg, T., Connelly, A., Gadian, D. G., Mishkin, M., & Vargha-Khadem, F. (2003). Language fMRI abnormalities associated with FOXP2 gene mutation. *Nature Neuroscience*, *6*, 1230–1237.

Lombardino, A. J., Hertel, M., Li, X. C., Haripal, B., Martin-Harris, L., Pariser, E., et al. (2006). Expression profiling of intermingled long-range projection neurons harvested by laser capture microdissection. *Journal of Neuroscience Methods*, *157*, 195–207.

Miller, J. E., Hilliard, A. T., & White, S. A. (2010). Song practice promotes acute vocal variability at a key stage of sensorimotor learning. *PLoS ONE*, *5*, e8592.

Miller, J. E., Spiteri, E., Condro, M. C., Dosumu-Johnson, R. T., Geschwind, D. H., & White, S. A. (2008). Birdsong decreases protein levels of FoxP2, a molecule required for human speech. *Journal of Neurophysiology*, *100*, 2015–2025.

Nordeen, K. W., & Nordeen, E. J. (1993). Long-term maintenance of song in adult zebra finches is not affected by lesions of a forebrain region involved in song learning. *Behavioral and Neural Biology*, *59*, 79–82.

Nottebohm, F., & Arnold, A. P. (1976). Sexual dimorphism in vocal control areas of the songbird brain. *Science*, *194*, 211–213.

Nottebohm, F., Nottebohm, M. E., Crane, L. A., & Wingfield, J. C. (1987). Seasonal changes in gonadal hormone levels of adult male canaries and their relation to song. *Behavioral and Neural Biology*, *47*, 197–211.

Nottebohm, F., Stokes, T. M., & Leonard, C. M. (1976). Central control of song in the canary, Serinus canarius. *Journal of Comparative Neurology*, *165*, 457–486.

Olveczky, B. P., Andalman, A. S., & Fee, M. S. (2005). Vocal experimentation in the juvenile songbird requires a basal ganglia circuit. *PLoS Biology*, *3*, e153.

O'Roak, B. J., Deriziotis, P., Lee, C., Vives, L., Schwartz, J. J., Girirajan, S., et al. (2011). Exome sequencing in sporadic autism spectrum disord4ers identifies severe de novo mutations. *Nature Genetics*, *43*, 585–589.

Panaitof, S. C., Abrahams, B. S., Dong, H., Geschwind, D. H., & White, S. A. (2010). Language-related Cntnap2 gene is differentially expressed in sexually dimorphic song nuclei essential for vocal learning in songbirds. *Journal of Comparative Neurology*, *518*, 1995–2018.

Reiner, A., Perkel, D. J., Bruce, L. L., Butler, A. B., Csillag, A., Kuenzel, W., et al. (2004). Revised nomenclature for avian telencephalon and some related brainstem nuclei. *Journal of Comparative Neurology*, *473*, 377–414.

Reiner, A., Perkel, D. J., Mello, C. V., & Jarvis, E. D. (2004). Songbirds and the revised avian brain nomenclature. *Annals of the New York Academy of Sciences*, *1016*, 77–108.

Rochefort, C., He, X., Scotto-Lomassese, S., & Scharff, C. (2007). Recruitment of FoxP2-expressing neurons to area X varies during song development. *Developmental Neurobiology*, *67*, 809–817.

Rousso, D. L., Pearson, C. A., Gaber, Z. B., Miquelajauregui, A., Li, S., Portera-Cailliau, C., Morrisey, E. E., & Novitch, B. G. (2012). Foxp-mediated suppression of N-cadherin regulates neuroepithelial character and progenitor maintenance in the CNS. *Neuron*, *74*(2), 314–330.

Scharff, C., & Haesler, S. (2005). An evolutionary perspective on FoxP2: strictly for the birds? *Current Opinion in Neurobiology*, *15*, 694–703.

Scharff, C., & Nottebohm, F. (1991). A comparative study of the behavioral deficits following lesions of various parts of the zebra finch song system: Implications for vocal learning. *Journal of Neuroscience, 11*, 2896–2913.

Scharff, C., Nottebohm, F., & Cynx, J. (1998). Conspecific and heterospecific song discrimination in male zebra finches with lesions in the anterior forebrain pathway. *Journal of Neurobiology, 36*, 81–90.

Scharff, C., & Petri, J. (2011). Evo-devo, deep homology and FoxP2: Implications for the evolution of speech and language. *Philosophical Transactions of the Royal Society of London B: Biological Sciences, 366*, 2124–2140.

Schulz, S. B., Haesler, S., Scharff, C., & Rochefort, C. (2010). Knockdown of FoxP2 alters spine density in Area X of the zebra finch. *Genes, Brain & Behavior, 9*, 732–740.

Simms, M. D. (2007). Language disorders in children: Classification and clinical syndromes. *Pediatric Clinics of North America, 54*, 437–467.

Soderstrom, K., & Luo, B. (2010). Late-postnatal cannabinoid exposure persistently increases FoxP2 expression within zebra finch striatum. *Developmental Neurobiology, 70*, 195–203.

Sohrabji, F., Nordeen, E. J., & Nordeen, K. W. (1990). Selective impairment of song learning following lesions of a forebrain nucleus in the juvenile zebra finch. *Behavioral and Neural Biology, 53*, 51–63.

Takayama, K., Horie-Inoue, K., Ikeda, K., Urano, T., Murakami, K., Hayashizaki, Y., et al. (2008). FOXP1 is an androgen-responsive transcription factor that negatively regulates androgen receptor signaling in prostate cancer cells. *Biochemical and Biophysical Research Communications, 374*, 388–393.

Templeton, C. N., Greene, E., & Davis, K. (2005). Allometry of alarm calls: black-capped chickadees encode information about predator size. *Science, 308*, 1934–1937.

Teramitsu, I., Kudo, L. C., London, S. E., Geschwind, D. H., & White, S. A. (2004). Parallel FoxP1 and FoxP2 expression in songbird and human brain predicts functional interaction. *Journal of Neuroscience, 24*, 3152–3163.

Teramitsu, I., Poopatanapong, A., Torrisi, S., & White, S. A. (2010). Striatal FoxP2 is actively regulated during songbird sensorimotor learning. *PLoS ONE, 5*, e8548.

Teramitsu, I., & White, S. A. (2006). FoxP2 regulation during undirected singing in adult songbirds. *Journal of Neuroscience, 26*, 7390–7394.

Vargha-Khadem, F., Gadian, D. G., Copp, A., & Mishkin, M. (2005). FOXP2 and the neuroanatomy of speech and language. *Nature Reviews: Neuroscience, 6*, 131–138.

Vernes, S. C., Newbury, D. F., Abrahams, B. S., Winchester, L., Nicod, J., Groszer, M., et al. (2008). A functional genetic link between distinct developmental language disorders. *New England Journal of Medicine, 359*, 2337–2345.

Vernes, S. C., Spiteri, E., Nicod, J., Groszer, M., Taylor, J. M., Davies, K. E., et al. (2007). High-throughput analysis of promoter occupancy reveals direct neural targets of FOXP2, a gene mutated in speech and language disorders. *American Journal of Human Genetics, 81*, 1232–1250.

Warren, W. C., Clayton, D. F., Ellegren, H., Arnold, A. P., Hillier, L. W., Kunstner, A., et al. (2010). The genome of a songbird. *Nature, 464*, 757–762.

White, S. A. (2009). Genes and vocal learning. *Brain and Language, 115*, 21–28.

23 Genetic Basis of Language: Insights from Developmental Dyslexia

Franck Ramus

Since the beginning of the cognitive revolution, it has been hypothesized that the human capacity to acquire a language is "innate"—that is, part of our species' biological makeup, and, therefore, encoded in some way in our genetic program (Chomsky, 1959). Over the years, a wide variety of arguments have been advanced in support of this view: the universality of some properties of human languages (Chomsky, 1957), the "poverty of the stimulus" available for language acquisition (Chomsky, 1965), the spontaneous emergence of languages (Bickerton, 1984; Goldin-Meadow & Mylander, 1998), biological adaptations such as that of the vocal tract (Lenneberg, 1967), the existence of genetic disorders specifically affecting language (Gopnik & Crago, 1991), the heritability of language abilities and disorders (Stromswold, 2001), the adaptiveness of language as a communication system (Pinker & Bloom, 1990), and the plausibility of a gradual evolution of the language faculty (Jackendoff, 1999; on the special topic of language evolution, see Fitch, chapter 24, this volume).

Although the evidence gathered in the last few decades in favor of a biological basis for language looks convincing to many scientists, genetic evidence has remained until recently relatively indirect, in the sense that it has not addressed the fundamental questions: If there is a genetic basis for language, then what exactly is there in the human genome that is different from other species, and that gives us language? How does it build a brain that can learn a human language?

There is no easy way to obtain direct answers to these fascinating questions. Genetic differences between species are only beginning to be systematically searched, and the many differences that are found are not straightforwardly identifiable as associated with language. However, part of the answer will likely come from addressing a related but different question: What human genetic variations are associated with variations in the ability to learn a language? Indeed, most genetic methods rely on detecting correlations between variations in the genotype and variations in the phenotype. As with many other traits, language abilities vary along a normal distribution. Cases in the lower end of the distribution ("disorders") are typically the most informative, because they highlight specific causal relationships between genes, brain, and cognition that are often not readily apparent in normal development. Indeed, disorders of language acquisition have so far provided almost all the available data on language genetics. Furthermore, language disorders are diverse, affecting different aspects of language, therefore promising to illuminate genetic influences on more specific components of language (phonology, morphology, syntax, articulation . . .). This chapter reviews the genetic data gathered on developmental dyslexia and reflects on what they teach us about the genetic basis of language.

Definition and Cognitive Phenotype

Developmental dyslexia is by definition a disorder of reading acquisition. However, it has been well established over the last three decades that most cases of dyslexia can be attributed to a subtle disorder of oral language (the "phonological deficit"),[1] whose symptoms happen to surface most prominently in reading acquisition (Snowling, 2000; Lyon, Shaywitz, & Shaywitz, 2003; Ramus, 2003). Therefore dyslexia is expected to ultimately reveal something about genetic factors implicated in language, and in particular in phonology. However, exactly what aspect of phonology is not entirely clear.

Indeed, the main symptoms of the "phonological deficit in dyslexia" are poor phonological awareness (the ability to pay attention to and explicitly manipulate speech sounds), poor verbal short-term memory, and slow lexical retrieval (evidenced in rapid naming tasks where subjects must name series of objects, colors, or digits in quick succession). This diversity of impairments has led many researchers to hypothesize that dyslexics' phonological representations are somewhat degraded, fuzzy, or noisy, lacking either in temporal or spectral resolution, or insufficiently attuned to the categories of the native language. This degradation is assumed either to be specific to the speech-processing system (Snowling, 2000; Adlard & Hazan, 1998; Serniclaes et al., 2004), or to follow from a lower-level auditory deficit (Tallal, 1980; Goswami et al., 2002). The latter view has been much challenged in recent years (Ramus, 2003; Rosen, 2003; White, Milne, et al., 2006; White, Frith, et al., 2006). As will become apparent below, the neurobiological and genetic data are consistent with the view that an auditory disorder is not necessary to engender dyslexics' phonological deficit (Ramus, 2004), although a possible compromise would be a disruption in the fine-tuning of auditory cortical analysis for the specific needs of the processing of speech sounds, as suggested by recent data (Lehongre, Ramus, Villiermet, Schwartz, & Giraud, 2011). An alternative view is that dyslexics' phonological representations are intrinsically normal, and that their difficulties in certain (but not all) phonological tasks arise from a deficit in the access to these representations—that is, particularly recruited for short-term memory and conscious manipulations (Marshall, Ramus, & Van der Lely, 2011; Ramus & Szenkovits, 2008). The elucidation of the precise nature of the phonological deficit will therefore determine whether dyslexia can inform us on the links between genes and phonological representations per se, or rather between genes and some cognitive processes operating on phonological representations (which might nevertheless be to some extent specific to language).

Neurological Phenotype

In the late 1970s, Galaburda and colleagues began to dissect human brains whose medical records indicated a diagnosis of developmental dyslexia[2] (Galaburda & Kemper, 1979). After dissecting four consecutive brains, and finding evidence for abnormalities of neuronal migration in all four, they hypothesized that this was unlikely to occur by chance, and that such brain development aberrations (ectopias, microgyria) might provide an explanation for dyslexia (Galaburda, Sherman, Rosen, Aboitiz, & Geschwind, 1985).[3] Most interestingly, neuronal migration disruptions were found predominantly in left perisylvian areas traditionally associated with language.[4] Galaburda and colleagues subsequently confirmed these findings in three more (female) brains (Humphreys, Kaufmann, & Galaburda, 1990), as well as the rarity of such abnormalities in control brains (Kaufmann & Galaburda, 1989). Unfortunately, no attempt at an independent replication was ever published. Nevertheless, brain imaging studies have largely confirmed structural and functional abnormalities in dyslexics' left perisylvian areas, although at a different level of description. Findings from MRI studies typically consist of reduced gray matter density, reduced anisotropy of the underlying white matter, and hypo- or hyperactivations (Démonet, Taylor, & Chaix, 2004; Eckert, 2004; Temple, 2002). At the moment it is impossible to establish their relationship with putative perturbations of neuronal migration, which are not visible in MRI scans. Thus, the dyslexia research community came to consider these findings as intriguing, but inconclusive. However, new results from genetic studies now suggest a reappraisal of the old neuronal migration hypothesis.

Genetic Findings

Historically, the first hint of a genetic influence on language abilities came from the observation that language disorders, including developmental dyslexia, tend to run in families (Stephenson, 1907; Hallgren, 1950): when one person has a language disorder, the risk in 1st-degree relatives is around 50%, far above the prevalence of these disorders. Although the affection pattern in many families suggests an autosomal dominant transmission,[5] this is not sufficient to prove genetic transmission, because members of a family share not only genes but also a linguistic environment. It is conceivable that parents with a language disorder would constitute a less favorable environment for the acquisition of language or of reading by their children, so family studies inevitably confound genetic and nongenetic factors.

Twin and adoption studies are the usual method to try and disentangle genetic and environmental factors. In the most classic twin studies, one compares the concordance of a given disorder[6] between monozygotic (MZ) and dizygotic (DZ) twins.[7] For instance, in a meta-analysis of twin studies by Stromswold (2001), the concordance of written language disorders is around 75% for MZ twins and 43% for DZ twins. Both figures are far above the typical prevalence of written language disorder (5%), and the substantial difference between MZ and DZ twins can largely be attributed to differences in their genetic similarity. Such concordance measures thus lead to the estimation of heritability—that is, the proportion of phenotypic variance than can be attributed to genetic variance. In Stromswold's review, the heritability estimate for written language disorders was 64%. This number has not been significantly challenged, either by more recent studies, or by adoption studies that rely on slightly different assumptions. However, it is important to emphasize that heritability estimates should not be taken at face value, because they are dependent on the population sampled and on the range of environmental influences it is exposed to. Beyond the elusive search for "true" heritability values, the point of heritability studies is to highlight those phenotypes, such as written language ability, that seem genetically influenced, and that therefore justify further research at the molecular level.

Until recently, linkage studies had provided six reliable chromosomal loci suspected to harbor genes associated with dyslexia, on chromosomes 1, 2, 3, 6, 15, and 18 (Grigorenko, 2003). Now six genes associated with dyslexia have been identified in some of these loci: DYX1C1 on 15q21 (Taipale et al., 2003), KIAA0319 on 6p22 (Paracchini et al., 2006; Cope, Harold, et al., 2005), DCDC2 just a few markers away on 6p22 (Meng, Smith, et al., 2005), ROBO1 on 3p12 (Hannula-Jouppi et al., 2005), and MRPL19 and C2ORF3 on 2p12 (Anthoni et al., 2007). The association of KIAA0319 and DCDC2 with dyslexia has been replicated in at least some independent studies (Harold et al., 2006; Schumacher et al., 2005; Luciano et al., 2007; Ludwig et al., 2008).

In two of these genes (DYX1C1, ROBO1), mutations or at least rare patterns (haplotypes) have been found in the dyslexic members of some isolated families, but these mutations are too rare to play a significant role in explaining dyslexia in general. It is unclear yet whether more common variants of these genes might modulate the susceptibility to dyslexia in the general population (Meng, Hager, et al., 2005; Brkanac et al., 2007; Scerri et al., 2004; Wigg et al., 2004; Bellini et al.,

2005; Marino et al., 2005; Cope, Hill, et al., 2005). As far as the other genes are concerned, the associated variants are alleles or haplotypes that are relatively frequent in the population. Thus, the mere possession of such a susceptibility allele is not a necessary and sufficient condition to cause dyslexia. Rather, it increases the probability of developing the disorder (typically multiplying it by 1.5 to 2). Therefore, it seems that the most common cases of dyslexia belong to the family of "complex genetic diseases" (like diabetes or certain cancers), where a multitude of genetic factors intervene, interacting with each other and with environmental factors, thereby modulating the susceptibility to the disorder. Rather than altering the amino acid sequence of the protein, such susceptibility alleles typically produce more gradual effects, altering quantitatively the expression of the protein (Velayos-Baeza, Toma, Da Roza, Paracchini, & Monaco, 2007; Velayos-Baeza, Toma, Paracchini, & Monaco, 2008; Tapia-Paez, Tammimies, Massinen, Roy, & Kere, 2008; Dennis et al., 2009). Follow-up investigations focus on understanding the precise functional role of these alleles by studying more directly the structure of the protein and its subdomains, as well as its expression patterns across the cortex and at different stages of brain development (Meng, Smith, et al., 2005; Paracchini, Scerri, & Monaco, 2007). It turns out that genes associated with dyslexia are highly expressed in the brain, in the cerebral cortex, and particularly so during fetal development (Bai et al., 2003; Dennis et al., 2006; Taipale et al., 2003).

On top of these relatively classic functional studies, LoTurco and colleagues have used a particularly innovative technique to study the role of three of these genes in brain development. They have produced "functional knockout" rats using in vivo RNA interference. This technique has allowed them to specifically block the translation of the gene of interest, in vivo, locally, and at a chosen stage of development (indeed, in utero during neuronal migration). Using this technique, they have shown that DYX1C1 is involved in radial neuronal migration, and that the part of the protein that is truncated in a Finnish dyslexic family is necessary and sufficient for normal neuronal migration (Wang et al., 2006). They have further shown that cortical ectopias (like the ones observed in dyslexic brains) sometimes occur as a result of the DYX1C1-induced disruption of neuronal migration, and that more generally the laminar organization is locally disrupted, with a distribution of neurons skewed in favor of layers I and II as well as toward the white matter (Rosen et al., 2007). The same team has been able to conduct similar studies on both DCDC2 (Meng, Smith, et al., 2005; Burbridge et al., 2008) and KIAA0319 (Paracchini et al., 2006; Peschansky et al., 2009), again concluding that these genes are crucially implicated in neuronal migration and in the laminar organization of the cortex. Finally, ROBO1 is a homolog of a well-known drosophila gene that is involved in interhemispheric axon guidance and in the migration of cortical interneurons (Andrews et al., 2006, 2008; Lopez-Bendito et al., 2007).

It would seem a priori highly unlikely that the first four genes associated with developmental dyslexia should all be implicated in neuronal migration. The fact that they are suggests that there is a real link between disturbances of neuronal migration and dyslexia. Thus, 20 years after the first postmortem studies, genetic findings finally seem to confirm Galaburda et al.'s original hypothesis (Galaburda, LoTurco, Ramus, Fitch, & Rosen, 2006; Ramus, 2004, 2006) and suggest a relatively coherent scenario of the etiology of dyslexia that can be summarized as follows. Certain variants (alleles or mutations) of certain genes increase the susceptibility to disruptions of neuronal migration, sometimes engendering ectopias or microgyri, but most importantly locally disrupting the laminar organization of the cortex. Through mechanisms that are not understood yet, these disruptions may, in certain individuals, accumulate in left perisylvian areas that are involved in speech processing and phonology and that are later recruited for reading acquisition. The disruption of these areas also surfaces more macroscopically in the MRI in the form of reduced gray matter density and reduced anisotropy of the underlying white matter. It engenders subtle deficits of phonological abilities that may have little impact on the acquisition of oral language, but manifest most remarkably during the acquisition of written language, which recruits those abilities particularly intensively.

Perspectives for Language Genetics

Until now I have described developmental dyslexia as a distinct entity from other language disorders; however, this is an oversimplification. Many children with specific language impairment (SLI), although not all of them, grow up to become dyslexic (Bishop & Snowling, 2004; McArthur, Hogben, Edwards, Heath, & Mengler, 2000; Flax et al., 2003; Marshall et al., 2011). Some children with dyslexia or SLI also present some form of speech sound disorder (SSD), if only in early development (Bishop & Adams, 1990; Shriberg, Tomblin, & McSweeny, 1999). This pattern of multiple comorbidities is hardly surprising if one considers that the different components of language, albeit functionally independent, may partly depend on each other in the course of development. But beyond this observation, it is likely that the comorbidity can be largely ascribed to common underlying biological factors. This is indeed suggested by several lines of converging evidence:

• The neural bases of dyslexia and SLI partly overlap (in left perisylvian regions traditionally associated with phonology) (Démonet, Thierry, & Cardebat, 2005).
• Familial aggregation studies have found that in families having one member with SLI or SSD, the likelihood of other members showing another form of language impairment (whether dyslexia, SLI, or SSD) increases (Flax et al., 2003; Lewis, 1992).
• Genetic linkage sites seem to overlap between dyslexia and SSD (Smith et al., 2005; Stein et al., 2004, 2006; Miscimarra et al., 2007). However, the fact that linkage

sites overlap does not guarantee that a single gene is associated with both disorders: linkage sites may contain many genes, including two affecting different disorders. On the other hand, there is no hint of any overlap between dyslexia and SLI linkage sites, which may not be all that surprising, given the statistical power of most linkage analyses (Marlow et al., 2003). None of the genes associated with dyslexia has been associated with SLI or SSD so far, but this may change sooner or later (Newbury et al., 2009, 2011).

The possibility that some gene variants might increase the susceptibility to several disorders makes sense in functional terms. For instance, there is no reason to expect that dyslexia is the only disorder arising from slight disturbances in neuronal migration. Therefore, neuronal migration genes associated with dyslexia should plausibly be expected to be associated with other disorders such as SLI (Ramus, 2004).

Furthermore, genes typically have more than one function, and therefore can have effects on multiple phenotypes: this is known as pleiotropy. For instance, all the genes discussed in this chapter are expressed not only in the developing brain, but also in other organs at various stages of life, showing that they have multiple functions, some as remote from cognition as digestion or reproduction.

These considerations have led Kovas and Plomin (2006) to hypothesize that genes affecting cognition are "generalist genes" affecting most cognitive functions and disorders, and indeed that they produce their effects relatively uniformly on a "generalist brain." It is certainly true that many genes affect many brain areas and many cognitive functions, yet the "generalist genes" hypothesis may be an overgeneralization. Some twin studies show that not all cognitive functions share genetic variance (e.g., Ronald, Happé, & Plomin, 2005), including phonological and morphosyntactic abilities (Bishop, Adams, & Norbury, 2006). And although many genes seem to be expressed more or less uniformly across the cortex, few studies have actually compared the expression of the genes of interest across different cortical areas. FOXP2 is a good case in point. It may have multiple effects on development, but it clearly does not have uniform effects on the brain: it is expressed in specific brain areas that turn out to bear a direct relationship with the neurological and cognitive phenotype associated with a FOXP2 mutation (see Fisher, chapter 21, this volume). Such neuroanatomical specificity is not uncommon among transcription factors. Performing a systematic search over more than 1,000 known transcription factors, Gray and colleagues (2004) have found 349 whose expression pattern is restricted to specific areas of the mouse brain and that are together sufficient to explain its architecture. Far from being generalist genes, their expression is specific and has equally specific functional consequences.

In the case of genes associated with dyslexia, expression patterns are available, but only from adult human brains, and with a relatively rough cortical parcelation (lobe by lobe, without distinguishing left from right hemisphere). Yet they do not turn out be particularly uniform (Paracchini et al., 2007; Meng, Smith, et al., 2005). Most importantly, the sites of brain disturbance themselves are clearly not uniform, whether one looks at histological studies, brain morphometry, or diffusion tensor imaging. The relationship between genes and neuropathological sites remains to be fully understood. More detailed studies might reveal that genes associated with dyslexia are expressed more in left perisylvian areas, but this can be considered unlikely for genes generally involved in neuronal migration. Then why do the disruptions occur precisely there? One reason could be just chance: in many individuals with the same gene variants, they may by chance occur elsewhere, and produce other effects (SLI, SSD, or any other cognitive deficit for that matter). We would see them in left perisylvian areas because we look only at dyslexic individuals. Yet, if chance was the only factor at play, one would predict complete cross-transmission between disorders: dyslexic parents would be as likely to beget SLI as dyslexic children. However, this is not the case (Flax et al., 2003; Lewis, 1992). An alternative would be that neuronal migration genes interact with other genes, which do have more specific expression patterns. The combination of certain alleles in neuronal migration genes and in restricted expression genes could result in disruptions of neuronal migration confined to certain cortical areas (Ramus, 2004). Yet another possibility would be that left perisylvian areas are, for unrelated (say, vascular) reasons, be more vulnerable to all forms of insult, including disturbances of neuronal migration (McBride & Kemper, 1982; Geschwind & Galaburda, 1985). One way or another, neuroanatomical location matters, more than anything else, for determining the precise nature of a cognitive phenotype.

In light of the above discussion on comorbidity and pleiotropy, one does expect to find genes associated with dyslexia as well as SSD and/or SLI, and perhaps even with other developmental disorders. However, this does not imply that all disorders are the same or that genes are "genes for everything." Not all dyslexic children have SSD or SLI, not all brain areas are involved in all language functions, not all genes affect all brain areas and functions, and therefore it is also to be expected that some genes will be uniquely associated with one disorder, alongside other genes that will be more general susceptibility factors for a certain class of neurodevelopmental disorders.

One final area where entirely novel results should be expected in the coming years is that of gene × environment interactions. All genetic studies of language disorders have until now focused on detecting the main effects of gene variants. This is of course the first step necessary in the identification of candidate genes. However, the effects of genes sometimes differ as a function of other factors, some genetic, some environmental. Evidence for nonadditive effects between genetic and environmental factors have begun to be investigated in the case of other disorders, such as conduct disorder (Caspi et al., 2002) or depression (Caspi et al., 2003). Does a susceptibility allele for developmental dyslexia produce a different effect depending on the presence of other risk factors (such as mild hearing impairment)? Or on the familial linguistic environment (Kremen et al., 2005)? Or on the language itself? Or on schooling practices? Or symmetrically, does a given environmental factor produce a different effect depending on the genotype of the child? The answers to these fascinating questions are now within reach.

Acknowledgments

This chapter was adapted from Ramus and Fisher (2009). The work is supported by Agence Nationale de la Recherche (Genedys) and the European Commission (Neurodys).

Notes

1. A minority of cases of dyslexia are likely due to disorders in the visual modality. They are not further discussed here, because they are less well understood and are of course not relevant to language genetics.

2. Some of these cases also presented with a number of comorbidities, some quite usual (speech and language delay in early childhood), some less (migraine, epilepsy). They nevertheless all had IQ scores in the normal range (85 or greater).

3. They also reported other differences between dyslexic and control individuals, including disruptions in the thalamus and abnormal asymmetry patterns in the planum temporale. A more complete and integrative view of those differences is provided in Ramus (2004).

4. More specifically, these areas are the left inferior frontal, posterior superior temporal, supramarginal, and angular gyri.

5. That is, the transmission of a dominant gene variant carried by a non–sex chromosome.

6. The probability that the disorder, when present in one twin, is present in the other.

7. Monozygotic twins share 100% of their genome, while dizygotic twins share only 50% of their gene variants (like ordinary siblings).

References

Adlard, A., & Hazan, V. (1998). Speech perception in children with specific reading difficulties (dyslexia). *Quarterly Journal of Experimental Psychology, 51A,* 153–177.

Andrews, W., Barber, M., Hernadez-Miranda, L. R., Xian, J., Rakic, S., Sundaresan, V., et al. (2008). The role of Slit-Robo signaling in the generation, migration and morphological differentiation of cortical interneurons. *Developmental Biology, 313,* 648–658.

Andrews, W., Liapi, A., Plachez, C., Camurri, L., Zhang, J., Mori, S., et al. (2006). Robo1 regulates the development of major axon tracts and interneuron migration in the forebrain. *Development, 133,* 2243–2252.

Anthoni, H., Zucchelli, M., Matsson, H., Muller-Myhsok, B., Fransson, I., Schumacher, J., et al. (2007). A locus on 2p12 containing the co-regulated MRPL19 and C2ORF3 genes is associated to dyslexia. *Human Molecular Genetics*, *16*, 667–677.

Bai, J. L., Ramos, R. L., Ackman, J. B., Thomas, A. M., Lee, R. V., & LoTurco, J. J. (2003). RNAi reveals doublecortin is required for radial migration in rat neocortex. *Nature Neuroscience*, *6*, 1277–1283.

Bellini, G., Bravaccio, C., Calamoneri, F., Donatella Cocuzza, M., Fiorillo, P., Gagliano, A., et al. (2005). No evidence for association between dyslexia and DYX1C1 functional variants in a group of children and adolescents from Southern Italy. *Journal of Molecular Neuroscience*, *27*, 311–314.

Bickerton, D. (1984). The language bioprogram hypothesis. *Behavioral and Brain Sciences*, *7*, 173–221.

Bishop, D. V. M., & Adams, C. (1990). A prospective study of the relationship between specific language impairment, phonological disorders and reading retardation. *Journal of Child Psychology and Psychiatry, and Allied Disciplines*, *31*, 1027–1050.

Bishop, D. V. M., Adams, C. V., & Norbury, C. F. (2006). Distinct genetic influences on grammar and phonological short-term memory deficits: Evidence from 6-year-old twins. *Genes, Brain & Behavior*, *5*, 158–169.

Bishop, D. V. M., & Snowling, M. J. (2004). Developmental dyslexia and specific language impairment: Same or different? *Psychological Bulletin*, *130*, 858–886.

Brkanac, Z., Chapman, N. H., Matsushita, M. M., Chun, L., Nielsen, K., Cochrane, E., et al. (2007). Evaluation of candidate genes for DYX1 and DYX2 in families with dyslexia. *American Journal of Medical Genetics, Part B: Neuropsychiatric Genetics*, *144B*, 556–560.

Burbridge, T. J., Wang, Y., Volz, A. J., Peschansky, V. J., Lisann, L., Galaburda, A. M., et al. (2008). Postnatal analysis of the effect of embryonic knockdown and overexpression of candidate dyslexia susceptibility gene homolog Dcdc2 in the rat. *Neuroscience*, *152*, 723–733.

Caspi, A., McClay, J., Moffitt, T. E., Mill, J., Martin, J., Craig, I. W., et al. (2002). Role of genotype in the cycle of violence in maltreated children. *Science*, *297*, 851–854.

Caspi, A., Sugden, K., Moffitt, T. E., Taylor, A., Craig, I. W., Harrington, H., et al. (2003). Influence of life stress on depression: Moderation by a polymorphism in the 5-HTT gene. *Science*, *301*, 386–389.

Chomsky, N. (1957). *Syntactic structures*. The Hague: Mouton.

Chomsky, N. (1959). A review of B. F. Skinner's *Verbal Behavior*. *Language*, *35*, 26–58.

Chomsky, N. (1965). *Aspects of the theory of syntax*. Cambridge, MA: MIT Press.

Cope, N., Harold, D., Hill, G., Moskvina, V., Stevenson, J., Holmans, P., et al. (2005). Strong evidence that KIAA0319 on chromosome 6p is a susceptibility gene for developmental dyslexia. *American Journal of Human Genetics*, *76*, 581–591.

Cope, N., Hill, G., van den Bree, M., Harold, D., Moskvina, V., Green, E. K., et al. (2005). No support for association between dyslexia susceptibility 1 candidate 1 and developmental dyslexia. *Molecular Psychiatry*, *10*, 237–238.

Démonet, J.-F., Taylor, M. J., & Chaix, Y. (2004). Developmental dyslexia. *Lancet*, *363*, 1451–1460.

Démonet, J. F., Thierry, G., & Cardebat, D. (2005). Renewal of the neurophysiology of language: Functional neuroimaging. *Physiological Reviews, 85*, 49–95.

Dennis, M., Paracchini, S., Scerri, T. S., Prokunina-Olsson, L., Knight, J. C., Wade-Martins, R., et al. (2009). A common variant associated with dyslexia reduces expression of the KIAA0319 gene. *PLOS Genetics, 5*, e1000436.

Dennis, M., Paracchini, S., Wade-Martins, R., Velayos-Baeza, A., Green, E. D., & Monaco, A. P. (2006). Expression of the KIAA0319 gene from a haplotype associated with developmental dyslexia. *American Journal of Medical Genetics, Part B: Neuropsychiatric Genetics, 141B*, 710–710.

Eckert, M. (2004). Neuroanatomical markers for dyslexia: A review of dyslexia structural imaging studies. *Neuroscientist, 10*, 362–371.

Flax, J. F., Realpe-Bonilla, T., Hirsch, L. S., Brzustowicz, L. M., Bartlett, C. W., & Tallal, P. (2003). Specific language impairment in families: Evidence for co-occurrence with reading impairments. *Journal of Speech, Language, and Hearing Research: JSLHR, 46*, 530–543.

Galaburda, A. M., & Kemper, T. L. (1979). Cytoarchitectonic abnormalities in developmental dyslexia: a case study. *Annals of Neurology, 6*, 94–100.

Galaburda, A. M., LoTurco, J., Ramus, F., Fitch, R. H., & Rosen, G. D. (2006). From genes to behavior in developmental dyslexia. *Nature Neuroscience, 9*, 1213–1217.

Galaburda, A. M., Sherman, G. F., Rosen, G. D., Aboitiz, F., & Geschwind, N. (1985). Developmental dyslexia: four consecutive patients with cortical anomalies. *Annals of Neurology, 18*, 222–233.

Geschwind, N., & Galaburda, A. M. (1985). Cerebral lateralization. Biological mechanisms, associations, and pathology: I. A hypothesis and a program for research. *Archives of Neurology, 42*, 428–459.

Goldin-Meadow, S., & Mylander, C. (1998). Spontaneous sign systems created by deaf children in two cultures. *Nature, 391*, 279–281.

Gopnik, M., & Crago, M. B. (1991). Familial aggregation of a developmental language disorder. *Cognition, 39*, 1–50.

Goswami, U., Thomson, J., Richardson, U., Stainthorp, R., Hughes, D., Rosen, S., et al. (2002). Amplitude envelope onsets and developmental dyslexia: A new hypothesis. *Proceedings of the National Academy of Sciences of the United States of America, 99*, 10911–10916.

Gray, P. A., Fu, H., Luo, P., Zhao, Q., Yu, J., Ferrari, A., et al. (2004). Mouse brain organization revealed through direct genome-scale TF expression analysis. *Science, 306*, 2255–2257.

Grigorenko, E. L. (2003). The first candidate gene for dyslexia: Turning the page of a new chapter of research. *Proceedings of the National Academy of Sciences of the United States of America, 100*, 11190–11192.

Hallgren, B. (1950). Specific dyslexia (congenital word-blindness): A clinical and genetic study. *Acta Psychiatrica et Neurologica, Supplementum, 65*, 1–287.

Hannula-Jouppi, K., Kaminen-Ahola, N., Taipale, M., Eklund, R., Nopola-Hemmi, J., Helena Kääriäinen, H., et al. (2005). The axon guidance receptor gene ROBO1 is a candidate gene for developmental dyslexia. *PLOS Genetics, 1*, e50.

Harold, D., Paracchini, S., Scerri, T., Dennis, M., Cope, N., Hill, G., et al. (2006). Further evidence that the KIAA0319 gene confers susceptibility to developmental dyslexia. *Molecular Psychiatry, 11*, 1085–1091.

Humphreys, P., Kaufmann, W. E., & Galaburda, A. M. (1990). Developmental dyslexia in women: Neuropathological findings in three patients. *Annals of Neurology, 28*, 727–738.

Jackendoff, R. (1999). Possible stages in the evolution of the language capacity. *Trends in Cognitive Sciences, 3*, 272–279.

Kaufmann, W. E., & Galaburda, A. M. (1989). Cerebrocortical microdysgenesis in neurologically normal subjects: A histopathologic study. *Neurology, 39*, 238–244.

Kovas, Y., & Plomin, R. (2006). Generalist genes: Implications for the cognitive sciences. *Trends in Cognitive Sciences, 10*, 198–203.

Kremen, W. S., Jacobson, K. C., Xian, H., Eisen, S. A., Waterman, B., Toomey, R., et al. (2005). Heritability of word recognition in middle-aged men varies as a function of parental education. *Behavior Genetics, 35*, 417–433.

Lehongre, K., Ramus, F., Villiermet, N., Schwartz, D., & Giraud, A. L. (in press). Altered low-gamma sampling in auditory cortex accounts for the three main facets of dyslexia. *Neuron*.

Lenneberg, E. (1967). *Biological foundations of language*. New York: Wiley.

Lewis, B. A. (1992). Pedigree analysis of children with phonology disorders. *Journal of Learning Disabilities, 25*, 586–597.

Lopez-Bendito, G., Flames, N., Ma, L., Fouquet, C., Di Meglio, T., Chedotal, A., et al. (2007). Robo1 and Robo2 cooperate to control the guidance of major axonal tracts in the mammalian forebrain. *Journal of Neuroscience, 27*, 3395–3407.

Luciano, M., Lind, P. A., Duffy, D. L., Castles, A., Wright, M. J., Montgomery, G. W., et al. (2007). A haplotype spanning KIAA0319 and TTRAP is associated with normal variation in reading and spelling ability. *Biological Psychiatry, 62*, 811–817.

Ludwig, K. U., Roeske, D., Schumacher, J., Schulte-Korne, G., Konig, I. R., Warnke, A., et al. (2008). Investigation of interaction between DCDC2 and KIAA0319 in a large German dyslexia sample. *Journal of Neural Transmission, 115*, 1587–1589.

Lyon, G. R., Shaywitz, S. E., & Shaywitz, B. A. (2003). A definition of dyslexia. *Annals of Dyslexia, 53*, 1–14.

Marino, C., Giorda, R., Luisa Lorusso, M., Vanzin, L., Salandi, N., Nobile, M., et al. (2005). A family-based association study does not support DYX1C1 on 15q21.3 as a candidate gene in developmental dyslexia. *European Journal of Human Genetics, 13*, 491–499.

Marlow, A. J., Fisher, S. E., Francks, C., MacPhie, I. L., Cherny, S. S., Richardson, A. J., et al. (2003). Use of multivariate linkage analysis for dissection of a complex cognitive trait. *American Journal of Human Genetics, 72*, 561–570.

Marshall, C. R., Ramus, F., & van der Lely, H. (2011). Do children with dyslexia and/or specific language impairment compensate for place assimilation? Insight into phonological grammar and representations. *Cognitive Neuropsychology, 27*, 563–586.

McArthur, G. M., Hogben, J. H., Edwards, V. T., Heath, S. M., & Mengler, E. D. (2000). On the "specifics" of specific reading disability and specific language impairment. *Journal of Child Psychology and Psychiatry, and Allied Disciplines, 41*, 869–874.

McBride, M. C., & Kemper, T. L. (1982). Pathogenesis of four-layered microgyric cortex in man. *Acta Neuropathologica, 57,* 93–98.

Meng, H., Hager, K., Held, M., Page, G. P., Olson, R. K., Pennington, B. F., et al. (2005). TDT-association analysis of EKN1 and dyslexia in a Colorado twin cohort. *Human Genetics, 118,* 87–90.

Meng, H., Smith, S. D., Hager, K., Held, M., Liu, J., Olson, R. K., et al. (2005). DCDC2 is associated with reading disability and modulates neuronal development in the brain. *Proceedings of the National Academy of Sciences of the United States of America, 102,* 17053–17058.

Miscimarra, L., Stein, C., Millard, C., Kluge, A., Cartier, K., Freebairn, L., et al. (2007). Further evidence of pleiotropy influencing speech and language: Analysis of the DYX8 region. *Human Heredity, 63,* 47–58.

Newbury, D. F., Paracchini, S., Scerri, T. S., Winchester, L., Addis, L., Richardson, A. J., et al. (2011). Investigation of dyslexia and SLI risk variants in reading- and language-impaired subjects. *Behavior Genetics, 41,* 90–104.

Newbury, D. F., Winchester, L., Addis, L., Paracchini, S., Buckingham, L. L., Clark, A., et al. (2009). CMIP and ATP2C2 modulate phonological short-term memory in language impairment. *American Journal of Human Genetics, 85,* 264–272.

Paracchini, S., Scerri, T., & Monaco, A. P. (2007). The genetic lexicon of dyslexia. *Annual Review of Genomics and Human Genetics, 8,* 57–79.

Paracchini, S., Thomas, A., Castro, S., Lai, C., Paramasivam, M., Wang, Y., et al. (2006). The chromosome 6p22 haplotype associated with dyslexia reduces the expression of KIAA0319, a novel gene involved in neuronal migration. *Human Molecular Genetics, 15,* 1659–1666.

Peschansky, V. J., Burbridge, T. J., Volz, A. J., Fiondella, C., Wissner-Gross, Z., Galaburda, A. M., et al. (2009). The effect of variation in expression of the candidate dyslexia susceptibility gene homolog Kiaa0319 on neuronal migration and dendritic morphology in the rat. *Cerebral Cortex, 20,* 884–897.

Pinker, S., & Bloom, P. (1990). Natural language and natural selection. *Behavioral and Brain Sciences, 13,* 707–784.

Ramus, F. (2003). Developmental dyslexia: Specific phonological deficit or general sensorimotor dysfunction? *Current Opinion in Neurobiology, 13,* 212–218.

Ramus, F. (2004). Neurobiology of dyslexia: A reinterpretation of the data. *Trends in Neurosciences, 27,* 720–726.

Ramus, F. (2006). Genes, brain, and cognition: A roadmap for the cognitive scientist. *Cognition, 101,* 247–269.

Ramus, F., & Fisher, S. E. (2009). Genetics of language. In M. S. Gazzaniga (Ed.), *The cognitive neurosciences* (pp. 855–871). Cambridge, MA: MIT Press.

Ramus, F., & Szenkovits, G. (2008). What phonological deficit? *Quarterly Journal of Experimental Psychology, 61,* 129–141.

Ronald, A., Happé, F., & Plomin, R. (2005). The genetic relationship between individual differences in social and nonsocial behaviours characteristic of autism. *Developmental Science, 8,* 444–458.

Rosen, G. D., Bai, J., Wang, Y., Fiondella, C. G., Threlkeld, S. W., Joseph, J., et al. (2007). Disruption of neuronal migration by RNAi of Dyx1c1 results in neocortical and hippocampal malformations. *Cerebral Cortex*, *17*, 2562–2572.

Rosen, S. (2003). Auditory processing in dyslexia and specific language impairment: Is there a deficit? What is its nature? Does it explain anything? *Journal of Phonetics*, *31*, 509–527.

Scerri, T. S., Fisher, S. E., Francks, C., MacPhie, I. L., Paracchini, S., Richardson, A. J., et al. (2004). Putative functional alleles of DYX1C1 are not associated with dyslexia susceptibility in a large sample of sibling pairs from the UK. *Journal of Medical Genetics*, *41*, 853–857.

Schumacher, J., Anthoni, H., Dahdouh, F., König, I. R., Hillmer, A. M., Kluck, N., et al. (2005). Strong genetic evidence for DCDC2 as a susceptibility gene for dyslexia. *American Journal of Human Genetics*, *78*, 52–62.

Serniclaes, W., van Heghe, S., Mousty, P., Carré, R., & Sprenger-Charolles, L. (2004). Allophonic mode of speech perception in dyslexia. *Journal of Experimental Child Psychology*, *87*, 336–361.

Shriberg, L. D., Tomblin, J. B., & McSweeny, J. L. (1999). Prevalence of speech delay in 6-year-old children and comorbidity with language impairment. *Journal of Speech, Language, and Hearing Research: JSLHR*, *42*, 1461–1481.

Smith, S. D., Pennington, B. F., Boada, R., & Shriberg, L. D. (2005). Linkage of speech sound disorder to reading disability loci. *Journal of Child Psychology and Psychiatry, and Allied Disciplines*, *46*, 1057–1066.

Snowling, M. J. (2000). *Dyslexia* (2nd ed.). Oxford: Blackwell.

Stein, C. M., Millard, C., Kluge, A., Miscimarra, L. E., Cartier, K. C., Freebairn, L. A., et al. (2006). Speech sound disorder influenced by a locus in 15q14 region. *Behavior Genetics*, *36*, 858–868.

Stein, C. M., Schick, J. H., Gerry Taylor, H., Shriberg, L. D., Millard, C., Kundtz-Kluge, A., et al. (2004). Pleiotropic effects of a chromosome 3 locus on speech-sound disorder and reading. *American Journal of Human Genetics*, *74*, 283–297.

Stephenson, S. (1907). Six cases of congenital word-blindness affecting three generations of one family. *Ophthalmoscope*, *5*, 482–484.

Stromswold, K. (2001). The heritability of language: A review and metaanalysis of twin, adoption, and linkage studies. *Language*, *77*, 647–723.

Taipale, M., Kaminen, N., Nopola-Hemmi, J., Haltia, T., Myllyluoma, B., Lyytinen, H., et al. (2003). A candidate gene for developmental dyslexia encodes a nuclear tetratricopeptide repeat domain protein dynamically regulated in brain. *Proceedings of the National Academy of Sciences of the United States of America*, *100*, 11553–11558.

Tallal, P. (1980). Auditory temporal perception, phonics, and reading disabilities in children. *Brain and Language*, *9*, 182–198.

Tapia-Paez, I., Tammimies, K., Massinen, S., Roy, A. L., & Kere, J. (2008). The complex of TFII-I, PARP1, and SFPQ proteins regulates the DYX1C1 gene implicated in neuronal migration and dyslexia. *FASEB Journal*, *22*, 3001–3009.

Temple, E. (2002). Brain mechanisms in normal and dyslexic readers. *Current Opinion in Neurobiology*, *12*, 178–183.

Velayos-Baeza, A., Toma, C., da Roza, S., Paracchini, S., & Monaco, A. P. (2007). Alternative splicing in the dyslexia-associated gene KIAA0319. *Mammalian Genome, 18,* 627–634.

Velayos-Baeza, A., Toma, C., Paracchini, S., & Monaco, A. P. (2008). The dyslexia-associated gene KIAA0319 encodes highly N- and O-glycosylated plasma membrane and secreted iso-forms. *Human Molecular Genetics, 17,* 859–871.

Wang, Y., Paramasivam, M., Thomas, A., Bai, J., Kaminen-Ahola, N., Kere, J., et al. (2006). DYX1C1 functions in neuronal migration in developing neocortex. *Neuroscience, 143,* 515–522.

White, S., Frith, U., Milne, E., Rosen, S., Swettenham, J., & Ramus, F. (2006). A double dis-sociation between sensorimotor impairments and reading disability: A comparison of autistic and dyslexic children. *Cognitive Neuropsychology, 23,* 748–761.

White, S., Milne, E., Rosen, S., Hansen, P. C., Swettenham, J., Frith, U., et al. (2006). The role of sensorimotor impairments in dyslexia: A multiple case study of dyslexic children. *Devel-opmental Science, 9,* 237–255.

Wigg, K. G., Couto, J. M., Feng, Y., Barr, C. L., Anderson, B., Cate-Carter, T. D., et al. (2004). Support for EKN1 as the susceptibility locus for dyslexia on 15q21. *Molecular Psychiatry, 9,* 1111–1121.

VI EVOLUTION OF SONG, SPEECH, AND LANGUAGE

24 Musical Protolanguage: Darwin's Theory of Language Evolution Revisited

W. Tecumseh Fitch

Darwin's *Origin of Species* (1859) made little mention of human evolution. This initial avoidance of human evolution was no oversight, but rather a carefully calculated move: Darwin was well aware of the widespread resistance his theory would meet from scientists, clergymen, and the lay public, and mention of human evolution might have generated insuperable opposition. But Darwin's many opponents quickly seized on the human mind, and language in particular, as a potent weapon in the battle against his new way of thinking. Alfred Wallace, whose independent discovery of the principle of natural selection spurred Darwin to finally publish his long-developing "outline" of the theory in 1859, did not help by arguing that natural selection was unable to explain the origins of the human mind. Although Wallace had reservations about all evolutionary approaches to the mind, human language provided the most powerful argument, due to the respectable position of linguistics and philology in Victorian science.

Darwin's most formidable foe on the linguistic front was Friederich Max Müller, professor of linguistics at Oxford University, a very well-known and well-respected scholar (Stam, 1976). In his "Lectures on the Science of Language," delivered at the Royal Institution of Great Britain in 1861 and rapidly published thereafter (Müller, 1861), Müller launched a full frontal attack on Darwin and Darwinism, using his credentials in the "science of language" as a powerful bludgeon. Müller's position was uncomplicated: "Language is the Rubicon which divides man from beast, and no animal will ever cross it. . . . The science of language will yet enable us to withstand the extreme theories of the Darwinians, and to draw a hard and fast line between man and brute" (Quoted in Noiré, 1917, p. 73–74). For Müller, language was the key feature distinguishing humans from all animals. Müller's arguments were seen by many as convincing: his student Noiré dubbed him the "Darwin of the mind" and considered Müller "the only equal, not to say superior, antagonist, who has entered the arena against Darwin" (Noiré, 1917, p. 73). Müller's argument about the unbridgeable, qualitative difference between human language and all forms of animal communication, combined with Wallace's opinions, provided arguments that Darwin by necessity took very seriously.

Thus, when Darwin finally broached the subject of human evolution in his second great book, *The Descent of Man and Selection in Relation to Sex* (1871), the need to provide a credible explanation for language evolution was a central concern. He rose to the challenge: his "musical protolanguage" model represents a powerful marriage of comparative data, evolutionary insight, and a biological perspective on language. Darwin's view of language was ahead of its time, and his model and arguments remain surprisingly relevant to contemporary debates. He clearly adopted a "multicomponent" view of language, one that recognized the necessity of several distinct mechanisms to produce the complex product that we now call language, rather than privileging any one factor as the single key to language in a monolithic sense. Among these several components, he presciently recognized the necessity for complex vocal learning, and recognized that this biological capacity, while unusual among mammals, is shared with many birds. The importance of vocal learning has often been forgotten, but also frequently reaffirmed by later scholars (Egnor & Hauser, 2004; Fitch, 2000; Janik & Slater, 1997; Marler, 1976; Nottebohm, 1976).

Darwin also adopted an empirical, data-driven approach to the problem at hand, exploiting what Botha (2009) has termed "windows" into language evolution. In particular, Darwin exploited a wide comparative database, drawing on not just his knowledge of nonhuman primate behavior, but also insights from many other vertebrates. Finally, and most characteristically, he resisted any special pleading about human evolution. He intended his model of human evolution to fit within, and remain consistent with, a broader theory of evolution that applies to beetles, flowers, and birds. Unlike Wallace (1905), who remained a human exceptionalist to his death, Darwin aimed to uncover general principles, like sexual selection and shifts of function, to provide explanations of unusual or unique human traits. While gradualistic, his model does not assume any simple continuity of function between nonhuman primate calls and language, and he clearly recognized the uniqueness of language in our species. In many ways, then, Darwin's model of language evolution finds a natural place in the landscape of the contemporary debate on language evolution, and it is surprising that his model has received relatively little detailed consideration in the modern literature (for exceptions see Donald, 1991; Fitch, 2006).

In this chapter, I aim to redress this neglect by considering Darwin's model of language evolution in detail. After discussing Darwin's main points and arguments, I briefly review additional data supporting Darwin's model that has appeared since his death. I also discuss the issue of meaning, about which Darwin had too little to say, but that can be resolved by the addition of a hypothesis due to Jespersen (1922). My conclusion is that, suitably modified in light of contemporary understanding, Darwin's model of language evolution, based on a "protolanguage" more musical than linguistic, provides one of the most convincing frameworks available for understanding language evolution. The present book provides an appropriate place to discuss Darwin's model, given the heavy reliance of both on comparative data concerning birdsong. The timing of my writing, on the 150th anniversary of the *Origin* and the 200th of Darwin's birth, is also appropriate for a revival of interest in Darwin's compelling and well-supported hypothesis.

Language as an "Instinct to Learn"

Chapter 2 of the *Descent of Man* (Darwin, 1871), titled "Comparison of the Mental Powers of Man and the Lower Animals," is one of the most remarkable portions of the entire Darwinian corpus, noteworthy for its conciseness and its breadth of argument, in considering the evolution of the human mind. The first half of the chapter lays the groundwork for modern research in comparative cognition, arguing that animals have emotions, attention, and memory as well as many other mental traits in common with humans. However, Darwin's opponents, notably Müller, had already ceded the point that animals have memory, experience emotions, and so on. Language was the key issue, and one can imagine considerable anticipation of both pro- and anti-Darwinian readers as they turned to the section simply titled "Language."

In ten densely argued pages, Darwin considers some theoretical preliminaries, then lays out his theory of language evolution. The first stage involved a general increase in intelligence and complex mental abilities, and the second involved a sexually selected attainment of the specific capacity for complex vocal control: singing. The third stage was the addition of meaning to the "songs" of the second stage, which was both driven by, and in turn fueled, further increases in intelligence.

Theoretically, Darwin makes a number of important observations. First, he recognizes the crucial distinction between the language *faculty* (the biological capacity that enables humans to acquire language) and particular languages (like Latin or English). The former capacity, which Darwin refers to as "an instinctive tendency to acquire an art" (p. 56), is shared by all members of the human species. Darwin neatly bypasses the unproductive nature-nurture debate that has consumed so much scholarly energy by observing that language "is not a true instinct, as every language has to be learnt. It differs, however, from all ordinary arts, for man has an instinctive tendency to speak, as we see in the babble of our young children" (p. 55). As ethologist Peter Marler has put it, language is not an instinct, but an "instinct to learn" whose expression entails that both biological and environmental preconditions be fulfilled. It is this "instinct to learn" for which a biological, evolutionary explanation must be sought: a thoroughly modern perspective.

Second, although Darwin was well aware of the peculiarities of the human vocal tract, he argues that the human capacity for language must be sought in the brain, rather than the peripheral vocal tract. He acknowledges that "articulate speech" (by which he means vocalization augmented by controlled movement of the lips and tongue, p. 59) is "peculiar to man," but he denies that this mere power of articulation suffices to distinguish human language, "for as every one knows, parrots can talk." Darwin states that it is not speech, but humans' "large power of connecting definite sounds with definite ideas" that is definitive of language, and that this capacity "obviously depends on the development of the mental faculties" (p. 54). By locating the language capacity in the human brain, Darwin's viewpoint is again thoroughly modern.

Finally, Darwin recognized the relevance to language evolution of birdsong, which he considered the "nearest analogy to language." Like humans, birds have fully instinctive calls, and an instinct to sing. But the songs themselves are learned. He recognized the parallel between infant babbling and songbird "subsong," and recognized the key fact that *cultural* transmission ensures the formation of regional dialects in both birdsong and speech. Finally, he recognized that physiology is not enough for learned song: crows have a syrinx as complex as a nightingale's but use it only in unmusical croaking. All of these parallels have been amply confirmed, and further explored, by modern researchers (Doupe & Kuhl, 1999; Marler, 1970; Nottebohm, 1972, 1975).

Darwin's "Musical Protolanguage" Hypothesis

Darwin's model of the phylogenesis of the language faculty, like most models today, posits that different aspects of language were acquired sequentially, in a particular order, and under the influence of distinguishable selection pressures. The hypothetical systems characterized by each addition can be termed, following Bickerton (1990) and Hewes (1973), "protolanguages." Darwin's first hypothetical stage in the progression from an apelike ancestor to modern humans was a greater development of protohuman cognition: "The mental powers in some early progenitor of man must have been more highly developed than in any existing ape, before even the most imperfect form of speech could have come into use" (p. 57). He elsewhere suggests that both social and technological factors may have driven this increase in cognitive power.

Next, Darwin outlines the crucial second step: what I have dubbed "musical protolanguage" (Fitch, 2006). Having noted multiple similarities with birdsong, he argues that the evolution of a key aspect of spoken language, vocal imitation, was driven by sexual selection, and used largely "in producing true musical cadences, that is in singing" (p. 56). He suggests that this musical protolanguage would have been used in both courtship and territoriality (as a "challenge to rivals"), as well as in the expression of emotions like love, jealousy, and triumph. Darwin concludes "from a widely-spread analogy" (amply documented with comparative data later in the book) that sexual selection played a crucial role in driving this stage of language evolution, in particular suggesting that the capacity to imitate vocally evolved analogously in humans and songbirds.

The crucial remaining question is how emotionally expressive musical protolanguage made the transition to true meaningful language—how, in Humboldt's words, humans became "a singing creature, only associating thoughts with the tones" (Von Humboldt, 1836, p. 76). This leap, from nonpropositional song to propositionally meaningful speech, remains the greatest explanatory challenge for all musical protolanguage theories (see Mithen, 2005). Darwin (1871, p. 56), citing the previous writings of Müller and Farrar (1870), suggests that articulate language "owes its origins to the imitation and modification, aided by signs and gestures, of various natural sounds, the voices of other animals, and man's own instinctive cries." Darwin thus embraces all three of the major leading theories of word origins of his contemporaries (see Fitch, 2010). Once protohumans had the capacity to imitate vocally, and to combine such signals with meanings, virtually any source of word forms and meanings would suffice, including onomatopoeia (an imitated roar for *lion*, or "whoosh" for *wind*) and controlled imitation of human emotional vocalizations (mock laughter for *play* or *happiness*). The attachment of specific and flexible meanings to vocalizations required only that "some unusually wise ape-like animal should have thought of imitating the growl of a beast of prey. . . . And this would have been a first step in the formation of a language" (p. 57).

Darwin does not suggest that the evolutionary process would stop with the initial acquisition of meaning. For "as the voice was used more and more, the vocal organs would have been strengthened and perfected" (p. 57). Additionally, language would have "reacted on the mind by enabling and encouraging it to carry on long trains of thought," which "can no more be carried on without the aid of words, whether spoken or silent, than a long calculation without the use of figures or algebra" (p. 57). Thus began the interactive evolutionary spiral that led to modern human language, and human intelligence, today.

Signaling Modality: Vocalization or Gesture?

Darwin also explicitly acknowledges the role of gesture in conveying meaning, echoing Condillac's earlier arguments (Condillac, 1747/1971) and presaging contemporary discussions (Arbib, 2005; Corballis, 2003; Hewes, 1973; Stokoe, 1974; Tomasello & Call, 2007). Darwin is aware of the power of signed language: he reminds us that using his fingers "a person with practice can report to a deaf man every word of a speech rapidly delivered at a public meeting" (p. 58). He also acknowledges the value of gesture in conveying meaning, and allows that vocal communication would have been "aided by signs and gestures" (p. 56). Nevertheless, he argues against gestural theorists, because the preexistence in all mammals of "vocal organs, constructed on the same general plan as ours" would lead any further development of communication to target the vocal organs rather than the fingers.

Darwin clearly believes that the power of speech is neural, not peripheral, citing the early aphasia literature as a demonstration of "the intimate connection between the brain, as it is now developed in us, and the faculty of speech" (p. 58) Comparing the vocal organs and brain, he concludes that "the development of the brain has no doubt been far more important" (p. 57). And although he uses a continuity argument to support the early and sustained role of speech, he firmly acknowledges the abrupt modern *discontinuity* in the linguistic system that has evolved. Thus, like many other insightful commentators (e.g., Donald, 1991; Hockett & Ascher, 1964), Darwin recognizes that posing phylogenetic continuity and modern discontinuity as in any way opposed is to create a false dichotomy. The treelike nature of phylogeny guarantees that both are core parts of the evolutionary process.

Darwin Redux: Modern Comparative Data

Summarizing, Darwin suggests that the first step on the road to human language was a general increase in intelligence in the hominid lineage. In a typically pluralistic fashion, he recognizes both "social intelligence" ("Machiavellian intelligence" in the modern trope (Byrne & Whiten, 1988)) and technological/ecological intelligence (e.g., for tool use) as playing important selective roles. Given our modern understanding of hominid evolution, this first stage might be provisionally linked to the genus *Australopithecus* or perhaps early *Homo* (e.g., *Homo habilis*).

The second stage is the least intuitive: that before vocalizations were used meaningfully they were used, so to speak, aesthetically, to fulfill many of the same functions for which modern humans use music today (courtship, bonding, territorial advertisement and defense, competitive displays, etc.). This idea that complex vocalizations (and thus some aspects of phonology and syntax) might have preceded the ability of speech to convey propositions and distinct meanings is the most challenging aspect of Darwin's model. But Darwin uses the comparative database, and particularly detailed analogy between learned birdsong and human song and speech, to show that this step is not just plausible but well documented: it has occurred in many other species. Indeed, modern data shows that vocal learning, without propositional meaning, has evolved independently in at least three other clades of mammals (cetaceans, pinnipeds, and bats) and three clades of birds (parrots, hummingbirds, and oscine songbirds) (Janik & Slater, 1997; Jarvis, 2004). Such convergent evolution, or repeated independent evolutionary development of a comparable ability, provides our strongest empirical basis for estimating the likelihood of a particular type of evolutionary event (Harvey & Pagel, 1991). Many of the chapters in this book affirm, and extend, the observations of parallels between language learning and birdsong that Darwin offered in 1871. Thus, whether intuitive or not, Darwin's focus on, and hypothesis for, the evolution of vocal learning is consistent with a wealth of evolutionary and comparative data.

Difficulties with Darwin's Model: Evolving Phrasal Semantics

How did man become, as Humboldt somewhere defined him, "a singing creature, only associating thoughts with the tones"?
—Otto Jespersen (1922, p. 437)

Despite its many virtues, some important problems remain with Darwin's model that have impeded its acceptance today. The first and most important is his explanation of the addition of meaning. Darwin's explanation, typical for his day, was concerned only with *word meanings* (what today would be termed "lexical semantics"). But from the viewpoint of modern linguistics, his model seems wholly inadequate to deal with large swaths of semantics, particularly those aspects tied in with the interpretation of whole phrases and sentences ("phrasal semantics"). Modern formal semantics has developed rigorous models of this aspect of linguistic meaning (Dowty, Wall, & Peters, 1981; Guttenplan, 1986; Montague, 1974; Portner, 2005), and it is far more complex and difficult to explain than lexical semantics. Although one can hardly blame Darwin for not foreseeing these relatively recent developments in linguistics, they nonetheless raise substantial difficulties for his model. For much of the syntactic "glue" that binds sentences together into large, meaningful wholes (function words, inflection, bound morphemes, word order, and a host of other elements) cannot be understood as resulting from onomatopoeia or imitation of emotional expressions. Nor can they be readily understood as "inventions" of some uniquely intelligent individual: all evidence suggests that these indispensable linguistic tools develop reliably in individuals of normal intelligence (Bickerton, 1981; Kegl, 2002; Mufwene, 2001; Mühlhäusler, 1997; Senghas, Kita, & Özyürek, 2005). This key aspect of language thus seems to have a biological basis. Darwin does recognize the phenomenon today called "grammaticalization": he states that "conjugations, declensions &c., originally existed as distinct words, since joined together" (p. 61). But he offers no model for the origin of these distinct words, and it is hard to see how onomatopoeia or similar processes could have generated this original syntactic and semantic "glue." Thus, complex phrasal semantics remains unexplained by Darwin's model.

However, this oversight was remedied long ago by the linguist Otto Jespersen (1922). Jespersen's basic insight involves recognizing the link, in humans, between musical and linguistic phrases, and working conceptually backward from there. Jespersen suggested a form of protolanguage in which, initially, whole propositional meanings attached to entire sung phrases, but where there was no consistent link between the individual *conceptual* components of the meaning, and component parts of the musical phrases (syllables and notes). Thus, there were no "words" as we now understand them. From this "holistic" starting point, Jespersen argued that a cognitive process of analysis began, which slowly isolated individual chunks of the musical phrase (syllables, or multisyllabic "phraselets"—what today we call "words") and associated them with individual components of the meaning (e.g., nouns, verbs, and adjectives, whose precursors were already present in the conceptual systems of our prelinguistic ancestors).

Jespersen's hypothesis of a "holistic protolanguage" has recently been rediscovered and championed by linguist Alison Wray (1998, 2000) and neuroscientist Michael Arbib (2005). Both cite considerable additional evidence supporting this "analytic" model, including data from modern adult language, child language acquisition, and cognitive neuroscience. Supporters of the more intuitive "synthetic" model of protolanguage, in which words evolved first followed by syntactic operations for combining them (e.g., Bickerton, 1990), have subjected holistic models to extensive criticisms (Bickerton, 2007; Tallerman, 2007, 2008). However, I argue that most of these critiques miss their mark if the notion of a musical protolanguage is accepted as a starting point (see Fitch, 2010). Jespersen/Wray's model of holistic protolanguage thus dovetails nicely with the musical protolanguage hypothesis, in ways that I believe resolve many, if not all, of these criticisms (Fitch, 2006; Mithen, 2005).

Sexual Selection

A second problem with Darwin's model remains unresolved at present: his focus on sexual selection as the force driving the evolution of musical protolanguage. Appearing as it did as a few pages in a lengthy tome introducing and then extensively documenting the very idea of sexual selection, this aspect of Darwin's theory has the virtue of explaining a core aspect of human evolution using a broad principle abundantly demonstrated in the evolution of other species. As throughout his work, Darwin eschewed "special pleading" for our own species. The central difficulty for this beautiful hypothesis is posed by two ugly facts about modern human language: it is equally developed in males and females, and it is expressed very early in ontogeny, essentially at birth (Fitch, 2005a). These aspects of language differentiate it sharply from most sexually selected traits, which are strongly biased to develop in the more competitive sex (typically males), and only at sexual maturity. If anything, human females have superior language skills to those of males (Henton, 1992; Kimura, 1983; Maccoby & Jacklin, 1974), and language is remarkable in its very early development, with at least some early tuning to phonology already occurring in utero before birth (DeCasper & Fifer, 1980; Mehler et al., 1988; Spence & Freeman, 1996).

There are several potential responses to the difficulty that these facts pose: one is to argue that during the musical protolanguage stage, sexual selection was the driving force, and song was (as in most bird species) expressed mainly in males at sexual maturity. Then, at a later stage (presumably during the evolution of meaningful language) some other selective force kicked in, so that language became equally (or better) expressed in females, and was pushed to develop early. A candidate selective force is kin communication: that selection for information transmission between parents and their offspring, or more generally between adults and their younger kin. I have suggested that kin selection drove this second stage of the evolution of propositional semantic content (Fitch, 2004, 2007; for an exploration and critique of this idea, see Zawidzki, 2006). This kin-selection scenario neatly explains the early ontogenetic appearance of language in infants (the earlier offspring begin absorbing their elders' knowledge, the better), and its bias toward females (who are the primary caregivers in all hominoids). The continued presence of meaningful speech in males is easily explained by the dual facts that immature males must also learn, and that, unusually in humans, adult males play an important role in childrearing (whether the father, or male siblings of the mother, is irrelevant to this fact). Finally, this kin-selection model has the virtue of explaining why language evolved in humans and *not* in other "musical" lineages. Humans combine an extended childhood, with ample time to acquire knowledge, with very small reproductive output. The fact that ape babies are born singly, and rarely, makes the survival of each individual hominid infant a crucial component of reproductive success in the great ape lineage (Fitch, 2007; Hrdy, 1999, 2004).

An alternative possibility is that sexual selection was, and remains, an important driving force in human cognitive evolution, including language (Miller, 2001), but that human pair bonding has "changed the rules" in significant ways, so that both sexes are choosy, and both compete for high-quality mates. Some comparative data can be cited in support of this second option. Recent data shows that female birdsong is not so uncommon as thought by Darwin, who considered female song to be a simple aberration (Langmore, 2000; Riebel, 2003; Ritchison, 1986). There is some evidence suggesting that sexual selection can indeed drive female birdsong, though it seems clear that female song is a secondary derivation of male song in most lineages (Langmore, 1996). While these observations provide some support for the idea that the dual-sex expression of human language could result from sexual selection, it is important to recognize that female song still appears to be numerically speaking exceptional and that *any* model based on sexual selection will have difficulty explaining the extremely early development, and productive use, of language in human infants.

A final possibility is that sexual selection *never* played a role in the evolution of music or of language. The popular notion that music evolved for courtship (Miller, 2000, 2001) stands on a surprisingly weak empirical footing compared to a less obvious but better documented function of music: mother-infant communication (Trainor, 1996; Trehub, 2003a, 2003b). Mothers sing to their infants all over the world, even those who claim to be unable to sing (Street, Young, Tafuri, & Ilari, 2003), and infants both prefer song to speech, and respond to song in manifestly adaptive ways (e.g., engaging with and getting excited by play songs, and being lulled to sleep by lullabies (Trehub & Trainor, 1998)). These observations suggest that music originally functioned in a childcare context, as it continues to do today. By this model, the use of music in bonding among adults is simply a side effect of this central function, and its occasional use in courtship is a red herring (Dissanayake, 2000; Falk, 2004; Trehub & Trainor, 1998). This final possibility is clearly compatible with the kin-selection arguments advanced above, but here there would be no intervening stage of language evolution in which sexual selection ever played a dominating role. Even Darwin was occasionally wrong.

Terminological Niceties: Musical or Prosodic Protolanguage?

A final, less crucial difficulty with Darwin's model is terminological. Darwin himself seemed to conceive of his presemantic protolangage in terms directly comparable to modern-day music (or at least he provides no indication that this is *not* the case). He concludes that "musical notes and rhythm" were present in this protolanguage, and that they were deployed "in producing true musical cadences, that is in singing" (1871, p. 56). This is why I term his model "musical protolanguage." However, modern human music consists not just of song, but also instrumental music, so this appellation might immediately have connotations of drumming, whistling, or flutes that are not, strictly speaking, relevant to language evolution. More pertinently, if we take the musical protolanguage model seriously, we must acknowledge that modern music may not necessarily preserve the state of this protolanguage precisely, and that both music and language have changed in the interim (Brown, 2000). That is, Darwin's hypothetical communication system was protomusic, not music per se. Adopting the logic of comparative reconstruction, we can then ask which aspects of modern speech and song are shared, and thereby reconstruct this earlier system (Fitch, 2005b). The central shared aspects are prosodic and phonological: the use of a set of primitives (syllables) to produce larger, hierarchically structured units (phrases) that are discretely distinctive. But two key "musical" aspects are not shared between speech and song: namely discrete-pitched notes, and temporal iso-chrony (a steady beat). I have used this comparison of modern speech and song to argue for a subtly different model from that of Darwin, which I termed "prosodic" rather than "musical" protolanguage, in which protolanguage consisted of sung syllables, but *not* of notes that could be arranged in a scale, nor produced with a steady rhythm (Fitch, 2006). This prosodic protolanguage model thus includes the "sung cadence" aspect of Darwin's model, while rejecting both his "notes" and "rhythm" (at least as normally construed). Both of these aspects of (most) modern song are, by hypothesis, more recent developments in music, not present in protolanguage. I see this as an adjustment of Darwin's hypothesis, fully in keeping with its spirit. Furthermore, it is unclear from his writings whether Darwin would have disagreed with this adjustment.

A different reconstruction of the common ancestor of music and language, involving both discrete pitches and isochronic rhythm (as well as tone-based meaning) is given by Brown (2000). Brown also argues that his hypothetical protolanguage, which he dubs "musilanguage," could not have evolved by normal neo-Darwinian selection and thus demands a group selection explanation. This remains its clearest, and most dubious, distinction from what is otherwise just a rediscovery of Darwin's basic hypothesis (for critiques see Botha, 2008; Fitch, 2010).

Conclusions

I have argued that Darwin's model for language evolution, "musical protolanguage," suitably updated, provides a compelling fit to both the phenomenology of modern music and language, and to a wealth of comparative data. By placing vocal control at the center of his model, Darwin availed himself of the rich comparative database of other species who have independently evolved complex vocal imitation, and he thus explains two of the features of human language that set it off most sharply from nonhuman primate communication systems: vocal learning and cultural transmission. The biggest missing piece in Darwin's model, as I see it, is a reasonable explanation of phrasal semantics (and the aspects of syntax that go with it), but this gap was filled by Jespersen by 1922. Together, these hypotheses provide one of the leading models of language evolution available today (for an enthusiastic book-length exploration see Mithen, 2005), and one that has been repeatedly rediscovered by later scholars (e.g., Brown, 2000; Livingstone, 1973; Richman, 1993). While many aspects of what has now become a family of models remain to be explored empirically (the issues surrounding sexual, kin, and group selection remain particularly unclear), this is a model worthy of detailed consideration and elaboration today. Most importantly, Darwin's model makes numerous testable empirical predictions (for example, about the partially overlapping nature of the brain mechanisms underlying music and spoken language, and their genetic basis) that can be addressed in the coming decades. The fact that it was born of, and supported by, the similarities between birdsong and human speech and song makes it particularly relevant to the current book, and the 200th anniversary of Darwin's birth seems an opportune time for his model of language evolution to regain the prominence it deserves.

References

Arbib, M. A. (2005). From monkey-like action recognition to human language: An evolutionary framework for neurolinguistics. *Behavioral and Brain Sciences*, *28*, 105–167.

Bickerton, D. (1981). *Roots of language*. Ann Arbor, MI: Karoma Press.

Bickerton, D. (1990). *Language and species*. Chicago: University of Chicago Press.

Bickerton, D. (2007). Language evolution: A brief guide for linguists. *Lingua*, *117*, 510–526.

Botha, R. (2008). On musilanguage/"Hmmmmm" as an evolutionary precursor to language. *Language & Communication*, *29*, 61–76.

Botha, R. (2009). Theoretical underpinnings of inferences about language evolution. In R. Botha & C. Knight (Eds.), *The cradle of language* (pp. 93–111). Oxford: Oxford University Press.

Brown, S. (2000). The "Musilanguage" model of music evolution. In N. L. Wallin, B. Merker, & S. Brown (Eds.), *The origins of music* (pp. 271–300). Cambridge, MA: MIT Press.

Byrne, R. W., & Whiten, A. (1988). *Machiavellian intelligence: Social expertise and the evolution of intellect in monkeys, apes and humans*. Oxford: Clarendon Press.

Condillac, É. B. de. (1747/1971). *Essai sur l'origine des connaissances humaines* (Nugent, T., Trans.). Gainesville, FL: Scholar's Facsimiles and Reprints.

Corballis, M. C. (2003). From mouth to hand: Gesture, speech and the evolution of right-handedness. *Behavioral and Brain Sciences*, *26*, 199–260.

Darwin, C. (1859). *On the origin of species*. London: John Murray.

Darwin, C. (1871). *The descent of man and selection in relation to sex*. London: John Murray.

DeCasper, A. J., & Fifer, W. P. (1980). Of human bonding: Newborns prefer their mothers' voices. *Science*, *208*, 1174–1176.

Dissanayake, E. (2000). Antecedents of the temporal arts in early mother-infant interaction. In N. L. Wallin, B. Merker, & S. Brown (Eds.), *The origins of music* (pp. 389–410). Cambridge, MA: MIT Press.

Donald, M. (1991). *Origins of the modern mind*. Cambridge, MA: Harvard University Press.

Doupe, A. J., & Kuhl, P. K. (1999). Birdsong and human speech: Common themes and mechanisms. *Annual Review of Neuroscience*, *22*, 567–631.

Dowty, D. R., Wall, R. E., & Peters, S. (1981). *Introduction to Montague semantics*. Dordrecht: Reidel.

Egnor, S. E. R., & Hauser, M. D. (2004). A paradox in the evolution of primate vocal learning. *Trends in Neurosciences*, *27*, 649–654.

Falk, D. (2004). Prelinguistic evolution in early hominins: Whence motherese? *Behavioral and Brain Sciences, 27,* 491–503.

Farrar, F. W. (1870). Philology & Darwinism. *Nature, 1,* 527–529.

Fitch, W. T. (2000). The evolution of speech: A comparative review. *Trends in Cognitive Sciences, 4,* 258–267.

Fitch, W. T. (2004). Kin selection and "Mother Tongues": A neglected component in language evolution. In D. K. Oller & U. Griebel (Eds.), *Evolution of communication systems: A comparative approach* (pp. 275–296). Cambridge, MA: MIT Press.

Fitch, W. T. (2005a). The evolution of language: A comparative review. *Biology and Philosophy, 20,* 193–230.

Fitch, W. T. (2005b). The evolution of music in comparative perspective. In G. Avanzini, L. Lopez, S. Koelsch, & M. Majno (Eds.), *The neurosciences and music II: From perception to performance* (Vol. 1060, pp. 29–49). New York: New York Academy of Sciences.

Fitch, W. T. (2006). The biology and evolution of music: A comparative perspective. *Cognition, 100,* 173–215.

Fitch, W. T. (2007). Evolving meaning: The roles of kin selection, allomothering and paternal care in language evolution. In C. Lyon, C. Nehaniv, & A. Cangelosi (Eds.), *Emergence of communication and language* (pp. 29–51). New York: Springer.

Fitch, W. T. (2010). *The evolution of language.* Cambridge: Cambridge University Press.

Guttenplan, S. (1986). *The languages of logic.* Oxford: Blackwell.

Harvey, P. H., & Pagel, M. D. (1991). *The comparative method in evolutionary biology.* Oxford: Oxford University Press.

Henton, C. (1992). The abnormality of male speech. In G. Wolf (Ed.), *New departures in linguistics* (pp. 27–59). New York: Garland.

Hewes, G. W. (1973). Primate communication and the gestural origin of language. *Current Anthropology, 14,* 5–24.

Hockett, C. F., & Ascher, R. (1964). The human revolution. *Current Anthropology, 5,* 135–147.

Hrdy, S. B. (1999). *Mother Nature.* New York: Pantheon Books.

Hrdy, S. B. (2004). Comes the child before man: How cooperative breeding and prolonged postweaning dependence shaped human potentials. In B. Hewlett & M. Lamb (Eds.), *Hunter gatherer childhoods* (pp. 65–91). Hawthorne, NY: Aldine/de Gruyter.

von Humboldt, W. (1836). *Über die Kawi-Sprache auf der Insel Java.* Berlin: Druckerei der Königlichen Akademie der Wissenschaften.

Janik, V. M., & Slater, P. B. (1997). Vocal learning in mammals. *Advances in the Study of Behavior, 26,* 59–99.

Jarvis, E. D. (2004). Learned birdsong and the neurobiology of human language. *Annals of the New York Academy of Sciences, 1016,* 749–777.

Jespersen, O. (1922). *Language: Its nature, development and origin.* New York: Norton.

Kegl, J. (2002). Language emergence in a language-ready brain: Acquisition issues. In G. Morgan & B. Woll (Eds.), *Language acquisition in signed languages* (pp. 207–254). Cambridge: Cambridge University Press.

Kimura, D. (1983). Sex differences in cerebral organization for speech and praxic functions. *Canadian Journal of Psychology, 37,* 19–35.

Langmore, N. E. (1996). Female song attracts males in the alpine accentor *Prunella collaris. Proceedings of the Royal Society B: Biological Sciences, 263,* 141–146.

Langmore, N. E. (2000). Why female birds sing. In Y. Espmark, T. Amundsen, & G. Rosenqvist (Eds.), *Signalling and signal design in animal communication* (pp. 317–327). Trondheim, Norway: Tapir Academic Press.

Livingstone, F. B. (1973). Did the Australopithecines sing? *Current Anthropology, 14,* 25–29.

Maccoby, E. E., & Jacklin, C. N. (1974). *The psychology of sex differences* (Vol. 1). Stanford, CA: Stanford University Press.

Marler, P. (1970). Birdsong and speech development: could there be parallels? *American Scientist, 58,* 669–673.

Marler, P. (1976). An ethological theory of the origin of vocal learning. *Annals of the New York Academy of Sciences, 280,* 386–395.

Mehler, J., Jusczyk, P., Lambertz, G., Halsted, N., Bertoncini, J., & Amiel-Tison, C. (1988). A precursor of language acquisition in young infants. *Cognition, 29,* 143–178.

Miller, G. F. (2000). Evolution of music through sexual selection. In N. L. Wallin, B. Merker, & S. Brown (Eds.), *The origins of music* (pp. 329–360). Cambridge, MA: MIT Press.

Miller, G. F. (2001). *The mating mind: How sexual choice shaped the evolution of human nature.* New York: Doubleday.

Mithen, S. (2005). *The singing Neanderthals: The origins of music, language, mind, and body.* London: Weidenfeld & Nicolson.

Montague, R. (1974). Universal Grammar. In R. H. Thomason (Ed.), *Formal philosophy: Selected papers of Richard Montague* (pp. 222–246). New Haven, CT: Yale University Press.

Mufwene, S. S. (2001). *The ecology of language evolution.* New York: Cambridge University Press.

Mühlhäusler, P. (1997). *Pidgin and creole linguistics* (Rev. ed.). London: University of Westminster Press.

Müller, F. M. (1861). The theoretical stage, and the origin of language. In F. M. Müller (Ed.), *Lectures on the science of language.* London: Longman, Green, Longman, and Roberts.

Noiré, L. (1917). *The origin and philosophy of language.* Chicago: Open Court.

Nottebohm, F. (1972). The origins of vocal learning. *American Naturalist, 106,* 116–140.

Nottebohm, F. (1975). A zoologist's view of some language phenomena, with particular emphasis on vocal learning. In E. H. Lenneberg & E. Lenneberg (Eds.), *Foundations of language development* (pp. 61–103). New York: Academic Press.

Nottebohm, F. (1976). Vocal tract and brain: A search for evolutionary bottlenecks. *Annals of the New York Academy of Sciences, 280,* 643–649.

Portner, P. H. (2005). *What is meaning: Fundamentals of formal semantics.* Oxford: Blackwell.

Richman, B. (1993). On the evolution of speech: Singing as the middle term. *Current Anthropology, 34,* 721–722.

Riebel, K. (2003). The "mute" sex revisited: Vocal production and perception learning in female songbirds. *Advances in the Study of Behavior, 33*, 49–86.

Ritchison, G. (1986). The singing behavior of female Northern Cardinals. *Condor, 88*, 156–159.

Senghas, A., Kita, S., & Özyürek, A. (2005). Children creating core properties of language: Evidence from an emerging sign language in Nicaragua. *Science, 305*, 1779–1782.

Spence, M. J., & Freeman, M. (1996). Newborn infants prefer the maternal low-pass filtered voice, but not the maternal whispered voice. *Infant Behavior and Development, 19*, 199–212.

Stam, J. H. (1976). *Inquiries into the origin of language*. New York: Harper & Row.

Stokoe, W. C. (1974). Motor signs as the first form of language. In R. W. Wescott (Ed.), *Language origins* (pp. 35–49). Silver Spring, MD: Linstock Press.

Street, A., Young, S., Tafuri, J., & Ilari, B. (2003). Mothers' attitudes towards singing to their infants. *Proceedings of the 5th Triennial ESCOM Conference, 5*, 628–631.

Tallerman, M. (2007). Did our ancestors speak a holistic protolanguage? *Lingua, 117*, 579–604.

Tallerman, M. (2008). Holophrastic protolanguage: Planning, processing, storage, and retrieval. *Interaction Studies: Social Behaviour and Communication in Biological and Artificial Systems, 9* (1), 84–99.

Tomasello, M., & Call, J. (2007). Ape gestures and the origins of language. In J. Call & M. Tomasello (Eds.), *The gestural communication of apes and monkeys* (pp. 221–239). London: Erlbaum.

Trainor, L. J. (1996). Infant preferences for infant-directed versus noninfant-directed playsongs and lullabies. *Infant Behavior and Development, 19*, 83–92.

Trehub, S. E. (2003a). The developmental origins of musicality. *Nature Neuroscience, 6*, 669–673.

Trehub, S. E. (2003b). Musical predispositions in infancy: An update. In I. Peretz & R. J. Zatorre (Eds.), *The cognitive neuroscience of music* (pp. 3–20). Oxford: Oxford University Press.

Trehub, S. E., & Trainor, L. J. (1998). Singing to infants: Lullabies and play songs. *Advances in Infant Research, 12*, 43–77.

Wallace, A. R. (1905). *Darwinism: An exposition of the theory of natural selection with some of its applications*. New York: Macmillan.

Wray, A. (1998). Protolanguage as a holistic system for social interaction. *Language & Communication, 18* (1), 47–67.

Wray, A. (2000). Holistic utterances in protolanguage: The link from primates to humans. In C. Knight, M. Studdert-Kennedy, & J. R. Hurford (Eds.), *The evolutionary emergence of language: Social function and the origins of linguistic form* (pp. 285–302). Cambridge: Cambridge University Press.

Zawidzki, T. W. (2006). Sexual selection for syntax and kin selection for semantics: Problems and prospects. *Biology and Philosophy, 21*, 453–470.

25 Birdsong as a Model for Studying Factors and Mechanisms Affecting Signal Evolution

Kazuo Okanoya

Birdsong is a learned behavior that is culturally transmitted within a set of biological constraints, and it can serve as an important biological model for human language with respect to interactions between culture and heredity (Bolhuis, Okanoya, & Scharff, 2010). The study reported on here examined the differences between wild and domesticated strains of white-rumped (or backed) munias (*Lonchura striata*) in terms of their songs. The comparison between the two strains revealed evolutionary factors affecting the acoustic and syntactic morphology of species-specific songs; these factors might also be relevant to the emergence of language in humans.

Wild white-rumped munias were originally imported from the Sappo port in Sekkou-shou, China, to Nagasaki, Japan, by a federal king of the Kyu-syu prefecture in 1763 (Washio, 1996). Since that time, they have frequently been imported from China to Japan, particularly during 1804–1829, when aviculture flourished in Japan. The white-rumped munia is generally brown with a white patch on the rump, as its name implies (Restall, 1996). However, in 1856, birds with totally white plumage were distinguished from white-rumped munias and called *Juushimatsu*, society finches. Although these birds were actually imported from China, European aviculturists believed that they came from India, and domesticated white-rumped munias imported from Japan to Europe were called Bengalese finches (Buchan, 1976). In what follows, the Japanese strain of domesticated wild white-rumped munias will be referred to as Bengalese finches. Bengalese finches were domesticated for their reproductive efficiency and their ability to foster other bird species, as well as for their plumage (Taka-Tsukasa, 1917). During the approximately 250 years of their domestication, however, song characteristics have changed substantially from those observed in the wild strain, and the purpose of this chapter is to discuss the possible behavioral and evolutionary reasons behind these differences. Figure 25.1 presents a photograph of a wild white-rumped munia and a domesticated Bengalese finch.

Figure 25.1
A Bengalese finch (left) and a white-rumped munia (right). Photo by Maki Ikebuchi.

Song Differences in Wild and Domesticated Strains

Representative sonograms from a Bengalese finch and a white-rumped munia are shown in Figure 25.2. Brief inspection of the sonograms suggested that these two songs were very different in acoustic morphology and the order of elements. In general, the songs of the wild strain were noiselike and the notes were ordered simply and in stereotyped fashion, whereas the songs of the domesticated strain were more narrow-banded and had complex note-to-note transition patterns. We initially confirmed these impressions with acoustic analyses of song notes and then by transition analysis of note sequences (Honda & Okanoya, 1999).

Acoustic analyses revealed that the frequency of the maximum amplitude was higher in Bengalese finches than in white-rumped munias, and bandwidths 15 dB below the maximum amplitude were wider in white-rumped munias than in Bengalese finches. Furthermore, the sound density (root mean square value of 5 sec of continuous singing) was, on average, 14 dB higher in Bengalese finches than in white-rumped munias when recordings were made with identical settings. However, no differences in the number of types of song elements were found between Bengalese finches (average 9.3) and white-rumped munias (average 8.4). Thus, Bengalese finch songs were higher pitched, more narrow-banded, and louder than were white-rumped munia songs, but the strains did not differ with regard to repertoire size.

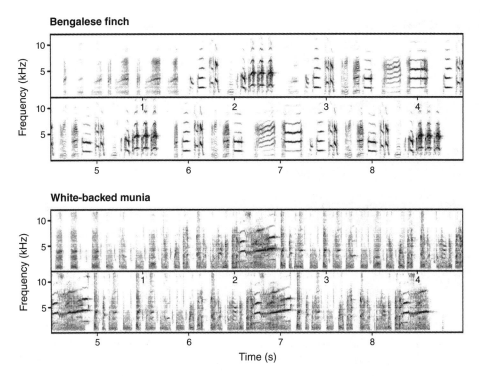

Figure 25.2
Sonogram of a Bengalese finch song (upper) and a white-rumped munia song (lower).

The sequential complexity of the songs was evaluated with the linearity index (Scharff & Nottebohm, 1991), which is the number obtained by dividing the number of unique types of song notes by the number of observed transition patterns from one note type to another. This index is 1.0 (when N is the number of note types, then this will be $N/N = 1$) when the element sequence in the song is always identical, and it will approach 0 ($N/N^2 = 1/N$) when the element sequence is completely random. Results of this analysis showed that the average linearity index was significantly lower, signifying greater complexity, in Bengalese finches (0.33) than in white-rumped munias (0.61). Representative transition diagrams from both strains are shown in Figure 25.3.

Female Reactions to Song Complexity

What are the functions of song complexity in the Bengalese finch? Although this species is domesticated, we hypothesized that function evolved in part as a result of sexual selection by females (Anderson, 1994; Catchpole, 2000; Okanoya, 2002). Because the Japanese avicultural literature does not contain evidence that songs

Bengalese finch

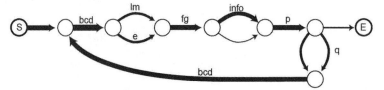

White-backed munia

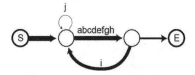

Figure 25.3
Transition diagram of a Bengalese finch song (upper) and a white-rumped munia song (lower).

were artificially selected by breeders (Washio, 1996), we assumed that breeders selected only successful pairs and that this indirectly resulted in the selection of good singers. Therefore, we further hypothesized that males and females differed with regard to song perception and that song complexity efficiently stimulated sexual behavior in females. We evaluated the former hypothesis using heart-rate measurements and the latter using several assays that supplemented one another (Searcy, 1992; Searcy & Yasukawa, 1996). We first measured the reinforcing properties of a complex song using an operant task involving perch selection. Next, we measured the degree of nest-building behavior by female Bengalese finches as a function of stimulus songs. In addition, we measured the serum estradiol levels in females stimulated with complex versus simple songs.

Heart Rate
Birdsong might be assessed and processed differently by each sex because its production and functional use are often sexually dimorphic. However, straightforward examination of this hypothesis has been difficult because different behavioral measures have been used to describe the process of song assessment in the two sexes. We analyzed changes in heart rate as an index of song assessment in the Bengalese finch (Ikebuchi, Futamatsu, & Okanoya, 2003). In this species, only males sing and song is used exclusively for mate attraction. Bengalese finches are not territorial, and the songs are not used in aggressive contexts. When a song stimulus was presented for the first time, the heart rate of the study participants increased. The

duration of this increase in heart rate was defined as the period in which the heart rate increased by two standard deviations above that measured in the baseline interval, which was 10 sec before song presentation. In both sexes, the repeated presentation of one song resulted in a reduction in the heart-rate response. The presentation of heterospecific (zebra finch) songs did not increase the heart rate of Bengalese finches. When a novel conspecific song was presented, the heart rate increased only in female and not in male birds with each presentation of the stimulus. These findings confirmed the differential responses to songs by each sex in this species: males ignored the songs of other birds, whereas females were attentive. These patterns were not due to sex differences in memory capacity; operant conditioning studies have demonstrated that males and females do not differ in their memory capacity for songs (Ikebuchi & Okanoya, 2000). The results suggested that syntactically complex songs might be more potent than simple songs in maintaining arousal in females.

Reinforcing Properties of Song Complexity

To examine the preferences of female Bengalese finches with regard to song complexity, we employed an operant conditioning technique using the song as a reinforcer (Morisaka, Katahira, & Okanoya, 2008). The protocol and apparatus used by Gentner and Hulse (1998) to test song preference in female European starlings were modified for Bengalese finches. We prepared a large metal cage and placed pot-shaped nests in two upper corners. We also placed small speakers for song playback inside the nests and fastened a perch in front of each of the nest pots. A natural song sung by a male Bengalese finch was used to prepare a simple (order of song notes fixed) and a complex (order of song notes varied according to a finite-state rule) song, both of which were played back from the relevant speaker when the bird sat on the perch. A female bird was placed inside this apparatus. Four of the eight birds tested chose the complex song, one chose the simple song, and the remaining three birds chose both songs at random. These results suggested that the song preferences of female Bengalese finches varied depending on the individual, although more tended to prefer complex to simple songs. Because only one type of song was used in the experiment, the experiment is pseudoreplication and the results should be interpreted with caution. Nevertheless, such female preferences could potentially contribute to sexual selection that facilitates the evolution of complex songs in male Bengalese finches (Morisaka et al., 2008).

Nest-Building Behavior

To further demonstrate the function of song complexity, we examined the nest-building behaviors of females (Eisner, 1961, 1963) in response to songs with complex or simple syntax (Okanoya & Takashima, 1997) using an approach first developed by Hinde and Steel (1976) and Kroodsma (1976). Hinde and Steel (1976) demonstrated that female domesticated canaries engaged in more transportation of nest material when stimulated with conspecific songs than with songs of other species. Kroodsma (1976) found that female canaries performed more nest building and laid more eggs when stimulated with a large rather than a small repertoire of songs.

We analyzed the song recordings of a male Bengalese finch and identified four distinctive song phrases (Okanoya & Takashima, 1997). The four phrases in this bird's song were organized such that phrase A or B was repeated several times and phrase C or D followed this repetition, but phrases C and D were never repeated. After phrase C or D was sung once, phrase A or B was repeated. We designed a computer program to produce this sequence of song phrases (complex syntax song) or one that repeated only phrase B (simple syntax song). Phrase B contained most of the song notes that occurred in phrases A, C, and D.

We examined three groups of four female Bengalese finches; each finch was kept in a separate cage and they were kept together in a sound isolation box. The first group was stimulated with the song characterized by the complex syntax, the second group with the song characterized by the simple syntax, and the third group was not stimulated with any song. The number of nesting items carried each day was counted and compared among the groups. Females stimulated with complex songs carried more nesting material (Figure 25.4). We further examined whether randomly generated note sequences were more effective than were syntactically synthesized ones. Females who were stimulated with random note sequences were less responsive and

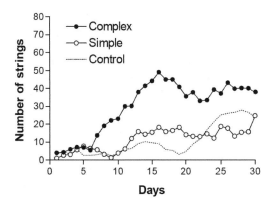

Figure 25.4
Results of the nest-building assay. When stimulated with the complex song, female Bengalese fiches carried significantly more nest material than did the birds stimulated with simple song or no song. Each line indicates the median of four birds. Data are smoothed by taking running averages for each four days.

carried comparable numbers of nest items compared to females stimulated with the simple sequence. Although random sequences resulted in complex orderings of song notes, randomness did not generate the same female response as did complexity produced by syntax (Okanoya, 2004).

Estradiol Levels

Three groups of female Bengalese finches were used in this experiment; each group consisted of four separately caged females kept together in a sound isolation box (Okanoya, 2004). The first group was stimulated with the song characterized by complex syntax, the second group with the song characterized by simple syntax, and the third group received no song stimulation. The levels of serum estradiol were compared among the groups before and after the experiment in order to consider baseline differences. Serum estradiol levels before and after the experiment were, on average, 0.37 and 0.76 ng mg^{-1}, respectively, in females stimulated with the complex song; 0.55 and 0.67 ng mg^{-1}, respectively, in females stimulated with the simple song; and 0.46 and 0.52 ng mg^{-1}, respectively, in females who heard a blank tape. Therefore, the complex song was more effective in stimulating female Bengalese finches into the reproductive condition ($t = 2.858$, $p < 0.05$ by post hoc tests after two-way ANOVA comparing stimulus condition and experimental periods).

Cross-Fostering Studies between the Wild and Domesticated Strains

Bengalese finch songs are sequentially and phonologically complex, whereas white-rumped munia songs are simpler. To elucidate the degree to which environmental and genetic factors contributed to these differences in song structure, we cross-fostered white-rumped munia and Bengalese finch chicks (i.e., we used seven pairs of Bengalese finches and four pairs of white-rumped munias and exchanged some of the eggs during incubation) (Okanoya & Takahasi, 2008; Takahasi & Okanoya, 2010). As a result, we obtained 14 Bengalese finch–reared male white-rumped munias and seven white-rumped munia–reared male Bengalese finches. For comparison, we also examined 12 normally reared male Bengalese finches and 7 normally reared male white-rumped munias. When the chicks had fully matured, their songs were recorded, and phonological and syntactic comparisons were performed. Inspection of sonograms revealed that munia-fostered Bengalese finches were able to learn most of the songs sung by fostering fathers but Bengalese-fostered munias had some difficulty learning the songs sung by fostering fathers (Figure 25.5).

Constraints in Phonological Learning

The accuracy of song-note learning was measured as the percentage of song elements shared between the chick and the father. Detailed phonological analyses

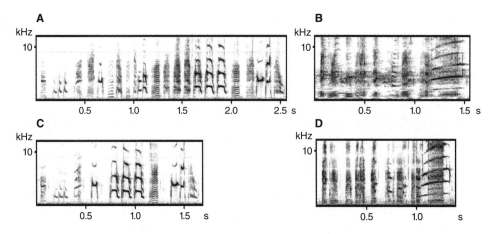

Figure 25.5
Results of cross-fostering experiment. A munia chick (C) fostered by a Bengalese finch father
(A) did not learn part of the song, but a Bengalese finch chick (D) fostered by a munia father
(B) had no apparent difficulty learning the song.

revealed that the accuracy of song-note learning was highest in white-rumped
munia chicks reared by white-rumped munias (98%) and lowest in white-rumped
munia chicks cross-fostered by Bengalese finches (82%). In contrast, Bengalese
finch chicks exhibited an intermediate degree of learning accuracy, irrespective of
whether they were reared by white-rumped munias (92%) or conspecifics (94%).
A two-way ANOVA detected a significant interaction between genetic background
and rearing environment, indicating that white-rumped munias were sensitive to
their rearing environments, whereas Bengalese finches were not ($p < 0.04$). These
results suggest that white-rumped munias are highly specialized for learning the
phonology of their own songs but are less adaptable to learning the phonology of
Bengalese finch songs. In contrast, Bengalese finches are less specialized for learning
the phonology of their own strain and more able to generalize their capacities to
learn the songs sung by white-rumped munias. These findings suggested an innate
bias toward species-specific phonology in white-rumped munias that might have
been lost in Bengalese finches during domestication (Okanoya & Takahasi, 2008;
Takahasi & Okanoya, 2010).

Constraints in Syntax Learning

We used the same data set to test for a learning bias for song syntax. Similarities
between the two types of song syntax were evaluated by first expressing the two
songs under study as Markovian transition matrices (Okanoya & Takahasi, 2008;
Takahasi & Okanoya, 2010). In constructing these matrices, we considered the song

notes shared by the songs of both tutor and pupil, as well as the song notes in songs sung only by tutors or only by pupils. The correlation coefficient calculated from the nonzero elements of the two matrices was used as an index for syntactic similarity. Using this method, we calculated average similarities between the songs of tutor and pupil in the four cross-fostered groups (Bengalese finches tutored by Bengalese finches, white-rumped munias tutored by white-rumped munias, Bengalese finches tutored by white-rumped munias, and white-rumped munias tutored by Bengalese finches).

Consistent with the results of the phonological learning experiment, the similarity between the songs of tutors and pupils was highest for white-rumped munias tutored by white-rumped munias (0.91) and lowest for white-rumped munias tutored by Bengalese finches (0.70). The similarities of Bengalese finches tutored by Bengalese finches (0.82) or by white-rumped munias (0.75) were intermediate in comparison with the two more extreme cases. Thus, when learning to sequence song elements, white-rumped munias were biased toward learning the linear syntax associated with their own strain and were far less adept at learning the complex syntax associated with Bengalese finches. These results supported our previous conclusion that white-rumped munias might have an innate bias toward learning species-specific syntax and that this bias might have disappeared in Bengalese finches during domestication.

Field Studies of Wild White-Rumped Munia Populations

We assumed that songs were kept simple in wild white-rumped munias because of specific pressures in the wild. Singing loud, complex songs in the wild is costly for at least three reasons. First, such songs attract predators. Second, they draw on cognitive resources necessary for reacting to dangers, including predation. Third, the evolution and maintenance of the brain mechanisms underlying complex songs are costly. We began our fieldwork in Taiwan (Republic of China) to examine these hypotheses. However, the observation of predation in the wild is very rare, and this strategy did not allow for quantitative assessment of the hypotheses.

We examined the other factors that might have accounted for the loss of the innate bias in Bengalese finches with regard to learning songs. One reason for the loss of this bias might involve an important function served by songs in the wild. Before a song can function as a mating signal to attract conspecific females, the singer must be identified as conspecific by the female. Toward this end, the song should possess species-specific characteristics. This function as an identifying mechanism might degenerate in a domestic environment because birds are paired by humans in these settings at least with birds of the same species and thus do not need to seek mates on their own.

Several field studies might support this hypothesis. In the wild, white-rumped munias coexist with various sympatric species, including spotted munia. A strong innate bias toward conspecific phonology should be adaptive for species of munia in avoiding futile attempts at hybridization. In contrast, Bengalese finches are domesticated and have been subject to controlled breeding. In such an environment, a species-specific bias would be neutral and might degenerate rapidly, perhaps allowing Bengalese finches to gain a more general ability to learn a wide-ranging phonology.

We have preliminary data on the relationship between the degree of colony mixing (with spotted munia) and song linearity in wild populations of white-rumped munia (Kagawa, Yamada, Lin, Mizuta, Hasegawa, & Okanoya, 2012). When the level of heterogeneity in the colony was higher, songs of white-rumped munia were more linear and exhibited less variable phonology. This might indicate that when more sympatric birds are present, species of munia must exaggerate their species-specific characteristics through their songs.

Discussion

I have described acoustic and syntactic differences between wild white-rumped munias and Bengalese finches, female responses to song complexity, effects of cross-fostering, and fieldwork to identify geographic variations in songs. Integrating these findings, I can now suggest a testable scenario for song evolution in Bengalese finches.

Domestication and Sexual Selection

The cross-fostering study revealed that white-rumped munias had a fairly narrow tuned learning mechanism for strain-specific phonology, whereas Bengalese finches had a more broadly tuned but less accurate learning mechanism. This finding should be considered in light of the results of fieldwork that showed that higher sympatric ratios were associated with lower levels of song complexity.

Birdsong must initially operate as a species identifier, and then it can function for sexual selection. In this regard, songs do not need to function as species markers in the absence of sympatric, closely related species. In environments characterized by the latter, however, songs should serve a sexual purpose. Domestication represents a special case in which no sympatric species exists. Because Bengalese finches no longer need to identify their species, they might have gradually lost the bias toward learning and producing species-specific characteristics in their songs. As a result of broadening the species-specific filter, Bengalese finches might have developed the ability to produce phonologically and syntactically complex songs. In this sense, song complexity might have arisen from a loss of species-specific bias rather than representing a gain in general learnability. Once constraints are weakened, female preferences might reinforce this tendency toward more complex songs. Indeed, male songs can be complex in syntactic and phonological domains to satisfy females' preference for variations.

A Scenario for the Evolution of Song Complexity

Based on the experimental results reviewed in this chapter, we suggest several steps that might underlie the evolution of complex song syntax in the Bengalese finch. In most estrildid finches, songs are used solely for the purpose of mating and not in male-male interactions. Thus, sexual selection is likely to have enhanced those song properties on which females base their choices, resulting in traits that are handicaps in the wild environment (Darwin, 1871; Zahavi & Zahavi, 1996). The following is one possible scenario that might explain the emergence of finite-state syntax in the Bengalese finch.

Complexity in song-note transitions became a sexually selected trait in white-rumped munias and was subject to individual variations due to genetic differences in neural capabilities and cultural differences in song traditions. However, the wild environment restricted the degree of possible song complexity in white-rumped munias due to the various costs associated with the maintenance of such traits, possibly including predation costs, foraging time, immunological costs associated with the production of testosterone, and a metabolic cost associated with maintaining sufficient brain capacity to underpin the song system. Furthermore, songs are needed to identify species in the wild, requiring that songs avoid phonological and syntactic complexity. Thus, mutations leading to greater song complexity would not have become fixed in a population of wild white-rumped munias, especially when sympatric species were living near them.

However, domestication eliminated many of these potential costs, especially those associated with predation and foraging time. Thus, domestication relaxed the restrictions imposed on the evolution of song complexity (Okanoya, 2002; Ritchie & Kirby, 2007). Furthermore, it reduced the necessity for identifying members of the species via song. Therefore, mutations leading to song complexity through the loss of a rigid song structure were not fixed in the natural environment and were not eliminated in the domesticated environment. Changes in brain structure then allowed more elaborate songs to be learned and gave rise to the improvisation of song syntax. Genes that allowed for the learning of complex songs were selected because of the preferences of females. Female preference for complexity, in turn, may be maintained as a leftover property from the wild that ensured selection of well-fed individuals (Soma et al., 2006).

Conclusion

Additional evidence is necessary to reinforce the above scenario. We need fieldwork data from a wider variety of geographic locations with different sympatric ratios. Data supporting female preferences for complexity should be examined at different levels of female receptivity, including approach, preference, acceptance, breeding behavior, breeding effort, and maternal effects (Soma, Hiraiwa-Hasegawa, & Okanoya, 2009). In the context of such reservations, I propose that explanations of the emergence of human language might benefit from observations of distantly related species such as Bengalese finches (Deacon, 2003; Okanoya, 2007; Merker & Okanoya, 2007; Okanoya & Merker, 2007).

References

Anderson, M. (1994). *Sexual selection*. Princeton, NJ: Princeton University Press.

Bolhuis, J. J., Okanoya, K., & Scharff, C. (2010). Twitter evolution: Converging mechanisms in birdsong and human speech. *Nature Reviews. Neuroscience, 11*, 747–759.

Buchan, J. (1976). *Bengalese finches*. Bristol: Isles d'Avon.

Catchpole, C. K. (2000). Sexual selection and the evolution of song and brain structure in the *Acrocephalus* warblers. *Advances in the Study of Behavior, 29*, 45–97.

Darwin, C. (1871). *The descent of man, and selection in relation to sex*. New York: Appleton.

Deacon, T. (2003). Multilevel selection in a complex adaptive system: The problem of language origins. In B. Weber & D. Depew (Eds.), *Evolution and learning: The Baldwin effect reconsidered* (pp. 81–106). Cambridge, MA: MIT Press.

Eisner, E. (1961). The behaviour of the Bengalese finch in the nest. *Ardea, 49*, 51–69.

Eisner, E. (1963). A quantitative study of parental behaviour in the Bengalese finch. *Behaviour, 20*, 134–206.

Gentner, T. Q., & Hulse, S. H. (1998). Perceptual mechanisms for individual vocal recognition in European starlings, *Sturnus vulgaris. Animal Behaviour, 56*, 579–594.

Hinde, R. A., & Steel, E. (1976). The effect of male song on an oestrogen-dependent behaviour pattern in the female canary (*Serinus canariensis*). *Hormones and Behavior, 7*, 293–304.

Honda, E., & Okanoya, K. (1999). Acoustical and syntactical comparisons between songs of the white-backed munia (*Lonchura striata*) and its domesticated strain, the Bengalese finch (*Lonchura striata* var. *domestica*). *Zoological Science, 16*, 319–326.

Ikebuchi, M., Futamatsu, M., & Okanoya, K. (2003). Sex differences in song perception by Bengalese finches as measured by cardiac responses. *Animal Behaviour, 65*, 123–130.

Ikebuchi, M., & Okanoya, K. (2000). Limited auditory memory for conspecific songs in a non-territorial songbird. *Neuroreport, 11*, 3915–3939.

Kagawa, H., Yamada, H., Lin, R.-S., Mizuta, T., Hasegawa, T., & Okanoya, K. (2012). Ecological correlates of song complexity in white-rumped munias: The implication of relaxation of selection as a cause for signal variation in birdsong. *Interaction Studies: Social Behaviour and Communication in Biological and Artificial Systems, 13*, 263–284.

Kroodsma, D. E. (1976). Reproductive development in a female songbird: Differential stimulation by quality of male song. *Science, 192*, 574–575.

Merker, B., & Okanoya, K. (2007). The natural history of human language: Bridging the gap without magic. In C. Lyon, C. L. Nehaniv, & A. Cangelosi (Eds.), *Emergence of communication and language* (pp. 403–420). London: Springer-Verlag.

Morisaka, T., Katahira, K., & Okanoya, K. (2008). Variability in preference for conspecific songs with syntactical complexity in female Bengalese finches: Towards an understanding of song evolution. *Ornithological Science, 7*, 75–84.

Okanoya, K. (2002). Sexual selection as a syntactical vehicle: The evolution of syntax in birdsong and human language through sexual selection. In A. Wray (Ed.), *Transitions to language* (pp. 46–63). Oxford: Oxford University Press.

Okanoya, K. (2004). Song syntax in Bengalese finches: Proximate and ultimate analyses. *Advances in the Study of Behavior, 34*, 297–346.

Okanoya, K. (2007). Language evolution and an emergent property. *Current Opinion in Neurobiology, 17*, 271–276.

Okanoya, K., & Merker, B. (2007). Neural substrates for string-context mutual segmentation: A path to human language. In C. Lyon, C. L. Nehaniv, & A. Cangelosi (Eds.), *Emergence of communication and language* (pp. 421–434). London: Springer-Verlag.

Okanoya, K., & Takahasi, M. (2008). Evolution of phonological complexity: Loss of species-specific bias leads to more generalized learnability in a species of songbirds. In A. D. M. Smith, K. Smith, & R. F. Cancho (Eds.), *The evolution of language: Proceedings of the 7th International Conference* (pp. 476–477). Barcelona: World Scientific.

Okanoya, K., & Takashima, A. (1997). Auditory preference of the female as a factor directing the evolution of Bengalese finch songs. *Transactions of the Technical Committee on Psychological and Physiological Acoustics, 1–6*, 27–81.

Restall, R. L. (1996). *Munias and manikins*. London: Christopher Helm.

Ritchie, G., & Kirby, S. (2007). A possible role for selective masking in the evolution of complex, learned communication systems. In C. Lyon, C. L. Nehaniv, & A. Cangelosi (Eds.), *Emergence of communication and language* (pp. 387–402). London: Springer-Verlag.

Scharff, C., & Nottebohm, F. (1991). A comparative study of the behavioral deficits following lesions of various parts of the zebra finch song system: Implications for vocal learning. *Journal of Neuroscience, 11*, 2896–2913.

Searcy, W. A. (1992). Measuring responses of female birds to male song. In P. K. McGregor (Ed.), *Playback and studies of animal communication* (pp. 175–189). New York: Plenum Press.

Searcy, W. A., & Yasukawa, K. (1996). Song and female choice. In D. E. Kroodsma & E. H. Miller (Eds.), *Ecology and evolution of acoustic communication in birds* (pp. 454–473). Ithaca, NY: Cornell University Press.

Soma, M., Hiraiwa-Hasegawa, M., & Okanoya, K. (2009). Early ontogenetic effects on song quality in the Bengalese finch (*Lonchura striata* var. *domestica*): Laying order, sibling competition and song syntax. *Behavioral Ecology and Sociobiology, 63*, 363–370.

Soma, M., Takahashi, M., Ikebuchi, M., Yamada, H., Suzuki, M., Hasegawa, T., et al. (2006). Early rearing conditions affect the development of body size and song in Bengalese fiches. *Ethology, 112*, 1071–1078.

Takahasi, M., & Okanoya, K. (2010). Song learning in the wild and domesticated strains of white-rumped munia, *Lonchura striata*, compared by cross-fostering procedures: Domestication increases song variability by decreasing strain-specific bias. *Ethology, 116*, 396–405.

Taka-Tsukasa, N. (1917). *Kahidori*. Tokyo: Shokabo. (In Japanese)

Washio, K. (1996). *Enigma of Bengalese finches*. Tokyo: Kindai Bungei-sha. (In Japanese)

Zahavi, A., & Zahavi, A. (1996). *The handicap principle: A missing piece of Darwin's puzzle*. Oxford: Oxford University Press.

26 Evolution of Vocal Communication: An Avian Model

Irene M. Pepperberg

And, indeed, if we now look at the birds together with all the mammals other than man, we have little hesitation in saying that the birds are by far the most advanced both in their control of their vocalisations and by the way in which they can adapt them collectively and individually to function as a most powerful communication system. . . .

I sometimes amuse myself by imagining an intelligent visitor from another planet arriving on this earth just before the differentiation of the human stock—say somewhere about one million years ago. If such a visitor had been asked by an all-seeing Creator which group of animals he supposed would the most easily be able to achieve a true language, I feel little doubt that he would have said unhesitatingly, "Why, of course, the birds."
—W. H. Thorpe, *Animal and Human Nature* (1974)

Studies on the evolution of communication often concentrate on the primate lineage, focusing on close phylogenetic relationships between present-day human and non-human primates: a common primate ancestor and the many neurological, anatomical, and resultant behavioral parallels (e.g., Deacon, 1997), including recent discoveries concerning mirror neurons (MNs; see Arbib, 2005; Fogassi & Gallese, 2002). An exclusively primate-centric model, however, overlooks parallel or convergent evolution: the likelihood that, through evolutionary pressures and exploitation of ecological niches, similar communicative abilities evolved in somewhat different ways in different species, and that birds—with their advanced cognitive and communicative abilities (e.g., Emery & Clayton, 2004; Pepperberg, 2007, 2012)—can provide models for the evolution of communication, particularly for *vocal learning*, and possibly even language. Even what once seemed to be different neuroanatomical structures subserving vocal behavior in birds and humans are now evaluated for similarity (see many chapters, this volume; cf. Person, Gale, Farries, & Perkel, 2008). Thus, I suggest that examining avian subjects, particularly their learning and use of various vocal systems, will shed light on the evolution of vocal communication. I first discuss such evolution within the primate line and reasons for emphasis therein, then suggest why and how birds may be more appropriate models, particularly with respect to a possible "missing-link" model.

Much has been made of what appear to be precursors of human vocal communication in nonhuman primates. Considerable data have been collected on alarm calls in vervets (Struhsaker, 1967; Seyfarth, Cheney, & Marler, 1980) and Diana, Campbell, and putty-nosed monkeys (e.g., Arnold & Zuberbühler, 2006; Ouattara, Lemassona, & Zuberbüler, 2009; Zuberbühler, Noë, & Seyfarth, 1997), differential food calling in tamarins (Roush & Snowdon, 2001), and variation in social calls in Campbell monkeys (Lemasson & Hausberger, 2011). The researchers involved suggest the existence of some levels of reference (i.e., statistically significant correlations between the calls and specific types of predators/foods or at least specific states of arousal to which the calls might refer) and even possible combinatorial ability (wherein animals hearing two sequential calls react somewhat differently than they would after hearing the calls individually). Evans, Evans, and Marler (1993) coined the term *functional reference* for such correlations because the level is not isomorphic with what is generally meant by reference in human language (e.g., the intentional use of a learned symbol to denote a specific object; see Tallerman & Gibson, 2011). Although the actual degree to which nonhuman utterances might be referential in the human sense is still a matter of debate, researchers do agree that, to any significant extent, vocal *learning* is all but absent in nonhuman primates,[1] engendering gestural (or motor) theories of language evolution to explain how vocal communication arose in hominids (Hewes, 1973); interestingly, a similar proposal exists for birds (Williams & Nottebohm, 1985).

What are the tenets of such theories? Hewes (1973) suggested that gestural communication—initially, voluntary use of various manual signals—arose fairly early in the hominid line and only later was subsumed by a vocal system, as manual gestures became associated with nonspeech movements of body parts used for cries and calls, sucking and feeding; these nonspeech movements were precursors of what were, when adapted for communicative intent, then termed *articulatory gestures* (e.g., see Fogassi & Ferrari, 2004; Studdert-Kennedy, 2005)—the *hidden* constrictions and releases of different parts of the vocal tract. But how did the brain transfer voluntary control from manual to vocal gestures? One idea (Corballis, 1989, 1991, 2003) was that the left hemisphere took control of voluntary manual communicative gestures (e.g., pointing actions), and that this laterality and the voluntary nature of the behavior were preserved when facial motions connected with manual gestures. That hypothesis, however, did not explain another part of the motor theory—that an individual understands vocal communication by representing others' speech as motor articulatory behavior (Liberman & Mattingly, 1985; Vihman, 1993). Such representation would be assisted if, for example, your articulatory system responded to my voice as if you were talking. Interestingly, some MNs are involved in exactly that kind of parity (e.g., Arbib, 2005; see discussion in Corballis, 2010). Moreover, because MNs have an inhibitory component (Baldissera, Cavallari, Craighero, & Fadiga, 2001), allowing you not to repeat my utterance if you so choose, they do not preclude voluntary choice. The specific connection to language is that MNs are found in Broca's area in humans—one of the "language centers"—and in Broca's homolog in monkeys, an area designated F5; the monkeys' system reacts to grasping, mouthing, and related actions (e.g., Fadiga, Fogassi, Pavesi, & Rizzolatti, 1995; Fogassi & Gallese, 2002). In squirrel monkeys, for example, F5 is activated during both production and perception of manual gestures but not vocalizations (Jürgens, 1998). Rizzolatti and Arbib (1998) hypothesized that development of vocal from manual communication involves evolution of Broca's area from F5: an evolved mirror system likely being a neural "missing link" between communication abilities of our nonhuman ancestors and modern human language. Instead of requiring a major evolutionary brain reorganization to go from the proposed voluntary control of manual to articulatory gestures in the hominid line—that is, for adding voluntary control to vocalizations—all that might be needed was a shift in (or evolution of) the monkey-like MN system (Arbib, 2005), likely an expansion of the projection from F5 that controls vocal folds to control the tongue and lips (Arbib, 2008).

A similar avian evolutionary scenario could be proposed, including lateralization (note Corballis, 2008), concerning all but manual gestures (Williams & Nottebohm, 1985). The beak, however, is an avian equivalent of forelimbs; moreover, motor control of the beak resides in areas separate from, but near to, the neural song system (Wild, Arends, & Zeigler, 1985) and the responsible neural areas likely relate to those controlling human jaw movements (e.g., articulatory gestures; Wild, 1997). Also, the tongue/beak system could easily be adapted from feeding to singing (Homberger, 1986), and many songbirds use their beaks and tongues in specific patterns for building nests, suggesting that ordering of what also could be considered gestures is part of their biology (see Pepperberg, 2007). I will come back to these avian systems shortly; first I look at learning.

Specifically, how do we get from voluntary control of vocalizations to vocal learning (Pepperberg, 2007, 2011)? Even if the MN system is involved in voluntary control of communication, vocal behavior that is under voluntary control is not necessarily learned; context for use—but not the sounds used—may be learned, as in vervet alarm calls (Cheney & Seyfarth, 1990; see Wich & Sterck, 2003, for data on voluntary use of alarm calls in langurs). Understanding connections among learning, voluntary control, and MNs takes us to another aspect of the MN system: involvement in imitation. Imitation is a form of learning—and one implicated in certain aspects of vocal communication. Initially, researchers proposed that an MN system enabled its owner to recognize an action through resonance and, because such recognition is one of the first steps in being able to imitate the action, that MNs were the basis for imitation (see Fogassi & Ferrari, 2004; Vauclair, 2004). That is, on seeing a novel action (whether manual or vocal), individuals somehow configure their own body parts so as to replicate the action (even if initially only roughly), and this ability resides in the MN system. But monkeys, where MNs were first discovered, do not imitate (Visalberghi & Fragaszy, 1990, 2002).[2] In fact, monkeys' MNs respond only when an observed action is in their repertoire, not to a novel action (Chaminade et al., 2001; Rizzolatti, Fogassi, & Gallese, 2001). Human MNs, in contrast, seem to recognize a novel behavior as a combination of novel actions that can be *approximated* by variants of actions already in the repertoire (see Arbib, 2005, but note Dinstein, Hasson, Rubin, & Heeger, 2007; Dinstein, Thomas, Behmann, & Heeger, 2008), thus assisting in imitation of the behavior. But what does this mean for the evolution of vocal learning?

Various levels of imitation exist and, likely, various types of MNs relate to these levels of imitation and learning (Fogassi & Ferrari, 2004; Pepperberg, 2005a, 2005b, 2007), both for different species and along evolutionary and developmental pathways. Details about these proposed intermediate forms are presented elsewhere (Pepperberg, 2005a, 2005b, 2007); the implication (e.g., Arbib, 2005, 2008) is that some intermediate form of MN system existed in our human ancestors, somewhat between that of present humans and nonhuman primates, that enabled imitative learning of a simple vocal system. A complication, however, is the impossibility of finding fossil evidence for appropriate language-ready or protolinguistic brain structures in this so-called missing link.

And here is where birds can become our model. Avian brain structures are now thought to be derived from the same pallial structures as mammalian brains and many birds are assumed to have large cortical-like structures (Jarvis et al., 2005; Jarvis, chapter 4, this volume), likely containing some form of MN system for vocal learners. Recently, studies (Prather & Mooney, chapter 19, this volume) have suggested the existence of an oscine MN system that responds in ways similar to that of primates. Why not also assume that birdsong evolution involved an intermediate MN system but that, unlike the primate line, species remain that provide the "missing link"? Let's look at the many parallels between avian and primate species, from those that have little in the way of learned vocal communication (e.g., the nonoscine and suboscine birds, nonhuman primates), to those that have many traits in common with humans (e.g., oscine songbirds), to those possible missing links (birds that seem to bridge the suboscine/oscine or nonlearning/vocal learning divide).

Let's begin with birds that do not learn their vocalizations (not counting maturation effects), whose communicative behavior is primarily (though not exclusively) genetically determined (e.g., de Kort & ten Cate, 2001). Particularly good parallels between birds and primates in this case involve unlearned alarm calls in chickens (Evans et al., 1993) and vervets (Cheney & Seyfarth, 1990). These species

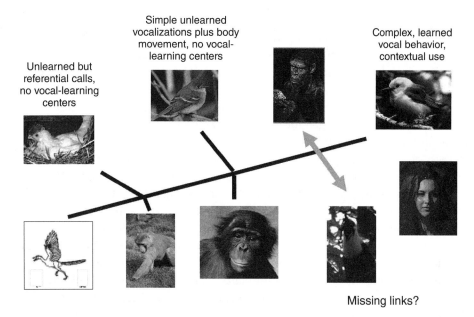

Simple unlearned vocalizations plus body movement, no vocal-learning centers

Complex, learned vocal behavior, contextual use

Unlearned but referential calls, no vocal-learning centers

Missing links?

Figure 26.1
Diagram of possible evolutionary correlations in vocal learning.

do, however, learn appropriate contexts for call use. Parallels also exist between other avian and primate species, both having somewhat more complex vocal communicative behavior (e.g., more flexibility in the context of use and interpreting meaning): suboscines such as flycatchers, and apes. Flycatchers' relatively simple songs are unlearned, but these birds learn from interactions how the context in which they use these vocalizations alters meaning. Louisana waterthrushes (*Seiurus motacilla*), for example, have two main song forms that are used for different levels of territorial defense (i.e., stating their presence versus active defense); they switch between forms, combine approach/withdrawal flights and perching with these forms, and alter the numbers of repetitions of these forms to demonstrate various levels of aggressive response to interactive playback (Smith & Smith, 1996). Eastern kingbirds (*Tyrannus tyrannus*) also alter the proportion of different song types as they alter their behavior toward intruders (Smith & Smith, 1992)[3]. Male vermillion flycatchers (*Pyrocephalus rubinus*) respond to interactive playback of longer song bouts as more threatening than shorter bouts, and may use song bout length to signal their own quality/fighting ability (Rivera-Cáceres, Garcia, Quirós-Guerrero, & Ríos-Chelén, 2011). Different birds do not respond identically to similar intrusions, which strongly suggests (although cannot prove) that the responses are not simply indications of emotional arousal, but likely have some voluntary components. Such behavior appears related to that of bonobos, whose combinations of actions and vocalizations more effectively elicit responses than do gestures alone (Pollick & de Waal, 2007). And an argument can be made that flycatchers who alter the number of repetitions of their single song or alter flight patterns or body postures while singing in order to signal different levels of aggression (i.e., engage in a very simple combinatory syntax; Smith & Smith, 1992, 1996) are living models of our ancestors whose mixtures of grunts and gestures may have served a similar purpose (Pepperberg, 2007; cf. Bickerton, 2003).[4]

All these cited existent species have brain nuclei that control the physical production of vocalizations but lack significant brain centers for vocal *learning* (Kroodsma & Konishi, 1991; Jürgens, personal communication cited in Arbib, 2008). If, as Smith (1997) argues, communication involves parity for both sender and receiver, and an MN system is what allows a brain to process this parity (e.g., Arbib, 2005), their underlying neurobiology likely involves a simple form of MN system that codes relationships among another agent's action (e.g., adults' calling), context of the action (e.g., presence of a particular predator), and the ability to replicate the action—allowing for choice in *whether* to execute the action (i.e., some control over inhibitory neurons) but with strong limitations as to *what* vocal action can be expressed.

Parallels between vocal learning in oscines and humans are well known (e.g., Baptista, 1983, 1988; Jarvis et al., 2005; Kroodsma, 1988; Marler, 1970). For both birds and humans there exist (1) a sensitive period during which exposure to the adult system allows development to proceed most rapidly, although acquisition is indeed possible beyond this period, particularly if social interaction is involved; (2) a babbling or practice stage wherein juveniles experiment with sounds that will ultimately become part of their repertoire; (3) a need to learn not just what to produce but to understand the appropriate context in which to produce specific vocalizations; and (4) the existence of possibly homologous brain structures devoted to the acquisition, storage, and production of vocalizations. Recent discovery of what appear to be MNs in the avian song system (Prather & Mooney, chapter 19, this volume) suggest even stronger correlations (cf. Person, Gale, Farries, & Perkel, 2008).

Not only are the different forms of avian communication described above reflected in different neuroanatomical systems (Jarvis & Mello, 2000; Jarvis et al., 2005; Kroodsma & Konishi, 1991; Nottebohm, 1980), but studies also suggest how the fully developed song system evolved as a specialization from preexisting motor pathways (e.g., Farries, 2001; Perkel, 2004), via paths such as addition and subtraction of certain projections between brain nuclei (e.g., Farries, 2004). Quite possibly mammalian structures that were eventually co-opted for the evolution of language in the hominid line have parallels in avian brain structures that were co-opted for the evolution of song learning and song decoding. Notably, Lieberman (2000) has argued that communication structures in humans evolved from reptilian basal ganglian circuits (i.e., brain areas responsible for learning particular patterns of motor activity—specifically, sequences he compares to syntax—especially those yielding reward, in ways that are sensitive to contextual inputs). If Lieberman is correct, certainly the same case can be made for birds (Medina & Reiner, 2000; Fee & Goldberg, 2011).

At the very least, in both songbirds and humans we see the evolution of brain and other anatomical features (e.g., tongue, syrinx/larynx, basilar papilla/cochlea) that enable both the production and processing of rapid sound sequences (Carr & Soares, 2002; Lieberman, 1991; Mann & Kelly, 2011; Stevens, 1998; Williams, 1989) that, even in birds, require some level of rule-governed behavior (maybe even a simplistic form of syntax, Gentner, Fenn, Margoliash, & Nusbaum, 2006; Abe & Watanabe, 2011; note Berwick, Okanoya, Beckers, & Bolhuis, 2011). Margoliash (2003) argues that the organization of auditory information (in terms of internal representations and issues of timing) may also reflect similar physical constraints that are expressed in related biological solutions. Birds, for example, have neurons that uniquely respond to their own individual vocalizations (Dave & Margoliash, 2000) that seem to assist in vocal learning. Parrots, quail, humans, and nonhuman mammals parse phonological space similarly (Kluender et al., 1987; Kuhl, 1981; Patterson & Pepperberg, 1994, 1998; Pepperberg, chapter 13, this volume), suggesting that speech phonology evolved so as to make use of existing auditory sensitivities basic not just to humans or even mammals, but at least to vertebrates (e.g., Dent, Brittan-Powell, Dooling, & Pierce, 1997; cf. Locke, 1997). Maybe other valid evolutionary comparisons exist (Pepperberg, 2007).

But crucial to birds as evolutionary models are two species that do not quite fit into the current picture of oscine versus suboscine classification—that is, species that somehow seem to bridge the vocal learning/nonlearning divide and thus might be considered as models for a hominid "missing link." Supposedly, one of these species, the three-wattled bellbird (*Procnias tricarunculata*), a close relative of flycatchers, *is* a suboscine, but early on Snow (1973) suggested it learned its songs. More recent reports (Kroodsma, 2005) provide evidence that males have dialects, that they can be bilingual with respect to these dialects (at least for several years), and that a close relative, the bare-throated bellbird (*P. nudicollis*), learns heterospecific song. Collection of DNA samples in the three-wattled bellbird have indeed shown that the different dialects come from the same species (Saranathan, Hamilton, Powell, Kroodsma, & Prum, 2007). These facts, along with the knowledge that some bellbirds do not begin to sound like (or even look like) adults until they are four or five years old (Kroodsma, 2005), suggest something radically different from what is expected of a suboscine. Even oscines that are open-ended learners usually have a recognizable song in their first year as an adult. And, although bellbirds supposedly do not change their overall dialects in adulthood, they seem to shift frequency over the years;[5] the suggestion is that older males shift, forcing younger ones to shift as well or lose status (and possibly mating chances) within the group (Kroodsma, 2005). Their learning seems rather oscinelike, except for the extraordinarily long juvenile stage and the fact that they are classified as suboscines. Interestingly, bellbirds also have more K-selected species traits than most suboscines and oscines, strengthening parallels with hominids: longer lives (possibly over 20 years), longer maturation periods, larger body size, fewer young, and intense male-male competition in which older, stronger males get the most matings (Powell & Bjork, 2004; Snow, 1977; Snow, 1982). Do they have specific areas in the brain devoted to song learning like oscines? No one knows, but the likelihood is strong. Is their prolonged babbling stage a consequence of a brain that is "differently" equipped for learning? If so, might they have a primitive MN system that is slow to mature, slow to take it beyond the babbling stage (Pepperberg, 2007)?

Despite the lack of answers to these questions, I suggest using the behavior and brain of bellbirds as a "missing-link" model for early hominid communication—that is, as a model for the MN system in the species or multiple species that likely bridged the gap between *Homo sapiens* and our nonhuman primate ancestors.[6] Given the bellbirds' seemingly intermediate status in terms of vocal learning, use of these birds as models could help us determine what is innate and what is learned—and the likelihood of there being a continuum, rather than a sharp break, between innate and learned. Might those evolutionary pressures that led from the innate, relatively simple song of true suboscines to the fairly simple but apparently slowly learned song of the bellbird to the amazing complexity of, for example, the brown thrasher's hundreds of songs similarly have been exerted on the nonhuman-to-hominid line? Could these evolutionary pressures have been exerted on an MN system, such that the complexity of the MN system and the complexity of the relevant behavior evolved in parallel, synergistically supporting the next evolutionary stage? That hypothesis would predict an intermediate bellbird MN system unlike any avian vocal-learning system observed to date. The existence of articulatory gestures grounded in feeding behavior and contact calls/cries that can be co-opted for other uses is not likely limited to primates; possibly an avian MN system shifted in the same manner, and the bellbird MN system might reflect such a shift.

In sum, whether avian and human abilities evolved convergently—whether similar adaptive responses independently evolved in association with similar environmental pressures—is unclear, but a common core of skills likely underlies complex cognitive and communicative behavior across species, even if specific skills manifest differently. Because many birds, like humans (but few other mammals) *learn* their vocal communication systems, the study of birds allows us to focus on both learning and vocal behavior. Few theses concerning the origins of language focus on the evolution of *learning* as the basis for communicative skills. Vocal learning is important not only because humans communicate primarily in the vocal mode, but also because it is one of the most transparent of modes for study (Pepperberg, 1999, 2011).

We thus should examine many species for information on evolutionary pressures that helped shape existing systems (Pepperberg, 1999, 2004, 2007, 2011). Such pressures were exerted not only on primates; hence the existence of analogous avian complex communication systems and their bases in what likely are homologous neural architectures. Although we no longer have access to the precursor neuroanatomy that gave rise to current human language abilities, the parallels between the acquisition, development, and use of current human communication and some avian systems (see Pepperberg, 2011; Pepperberg & Schinke-Llano, 1991) suggest that parallels likely existed in their evolutionary history. Species such as the bellbird could be a model for the missing human precursor.

Arguably, much of the above is speculation. I do not claim to answer the very difficult questions about the origins of communicative abilities but rather suggest lines of research. I do not posit how language developed from what was likely the simple communication system of our hominid ancestors—a concatenation of cultural, social, and neuroanatomical changes likely were involved (e.g., Corballis, 2010). And many important references and intermediate steps have been omitted so that only the basics could be presented. My hope is that readers will investigate the area for themselves, in far greater detail, and devise ways to answer these questions through rigorous scientific experimentation.

Acknowledgments

Many of the ideas for this chapter were initially developed during a Radcliffe Institute Fellowship (2004–2005). The writing of the chapter was supported in part by the Marc Haas Foundation and by donors to *The Alex Foundation*.

Notes

1. Note that Crockford, Herbinger, Vigilant, and Boesch (2004) suggest some form of vocal dialect learning in apes, but critical rearing experiments have not been done.

2. Note that Voelkl and Huber (2000) claim true imitation in marmosets (*Callithrix jacchus*), but the actions—opening a container with hands versus mouths—are not novel actions, and thus do not qualify as true imitation (Thorpe, 1963). Arbib (personal communication, November 2005) also suggests that the described behavior involves priming rather than action recognition. Huber et al. (2009) might disagree.

3. Leger (2005) has shown that the Flammulated Attila (*Attila flammulatus*) actually varies the order of its notes in its two songs, but does not describe any functional reasons for this behavior.

4. If such parallels seem a bit far-fetched, note that earlier studies of parallels between human bilingualism and a related form of avian song acquisition (e.g., Pepperberg & Schinke-Llano, 1991) have demonstrated how birds can indeed be used as models for human behavior (see also Pepperberg, 2011).

5. Other features of the song have also changed over the years (Kroodsma, personal communication, September 2005), but the change in frequency has been emphasized because it is the most obvious (Kroodsma, 2005).

6. The bellbird is endangered, but conceivably data might be obtained in the future from captive birds in a noninvasive manner.

References

Abe, K., & Watanabe, D. (2011). Songbirds possess the spontaneous ability to discriminate syntactic rules. *Nature Neuroscience*, *14*, 1067–1074.

Arbib, M. A. (2005). From monkey-like action recognition to human language: An evolutionary framework for neurolinguistics. *Behavioral and Brain Sciences*, *28*, 105–167.

Arbib, M. A. (2008). From grasp to language: Embodied concepts and the challenge of abstraction. *Journal of Physiology, Paris*, *102*, 4–20.

Arnold, K., & Zuberbühler, K. (2006). Semantic combinations of primate calls. *Nature*, *441*, 303.

Baldissera, F., Cavallari, P., Craighero, L., & Fadiga, L. (2001). Modulation of spinal excitability during observation of hand actions in humans. *European Journal of Neuroscience*, *13*, 190–194.

Baptista, L. F. (1983). Song learning. In A. H. Brush & G. A. Clark, Jr., (Eds.), *Perspectives in ornithology* (pp. 500–506). Cambridge: Cambridge University Press.

Baptista, L. F. (1988). Song learning in white-crowned sparrows (*Zonotrichia leucophrys*): Sensitive phases and stimulus filtering revisited. In R. van den Elzen, K.-L. Schuchmann, & K. Schmidt-Koenig (Eds.), *Proceedings of the 100th Deutsche Ornithologische Gesellschaft: Current Topics in Avian Biology* (pp. 143–152). Bonn.

Berwick, R. C., Okanoya, K., Beckers, G. J. L., & Bolhuis, J. J. (2011). Songs to syntax: The linguistics of birdsong. *Trends in Cognitive Sciences*, *15*, 113–121.

Bickerton, D. (2003). Symbol and structure: A comprehensive framework for language evolution. In M. H. Christiansen & S. Kirby (Eds.), *Language evolution* (pp. 77–93). Oxford: Oxford University Press.

Carr, C. E., & Soares, D. (2002). Evolutionary convergence and shared computational principles in the auditory system. *Brain, Behavior and Evolution*, *59*, 294–311.

Chaminade, T., Meary, D., Orliaguet, J.-P., & Decety, J. (2001). Is perceptual anticipation a motor simulation? A PET study. *Brain Imaging*, *12*, 3669–3674.

Cheney, D. L., & Seyfarth, R. M. (1990). *How monkeys see the world*. Chicago: University of Chicago Press.

Corballis, M. C. (1989). Laterality and human evolution. *Psychological Review*, *96*, 492–505.

Corballis, M. C. (1991). *The lopsided ape: evolution of the generative mind*. Oxford: Oxford University Press.

Corballis, M. C. (2003). From mouth to hand: Gesture, speech, and the evolution of right-handedness. *Behavioral and Brain Sciences*, *26*, 199–260.

Corballis, M. C. (2008). Of mice and men—and lopsided birds. *Cortex*, *44*, 3–7.

Corballis, M. C. (2010). Mirror neurons and the evolution of language. *Brain and Language*, *112*, 25–35.

Crockford, C., Herbinger, I., Vigilant, L., & Boesch, C. (2004). Wild chimpanzees produce group-specific calls: A case for vocal learning? *Ethology*, *110*, 221–243.

Dave, A. S., & Margoliash, D. (2000). Song replay during sleep and computational rules for sensorimotor vocal learning. *Science*, *290*, 812–816.

Deacon, T. W. (1997). *The symbolic species: The co-evolution of language and the brain*. New York: Norton.

de Kort, S. R., & ten Cate, C. (2001). Response to interspecific vocalizations is affected by degree of phylogenetic relatedness in Streptopelia doves. *Animal Behaviour*, *61*, 239–247.

Dent, M. L., Brittan-Powell, E. F., Dooling, R. J., & Pierce, A. (1997). Perception of synthetic /ba/-/wa/ speech continuum by budgerigars (*Melopsittacus undulatus*). *Journal of the Acoustical Society of America*, *102*, 1891–1897.

Dinstein, I., Hasson, U., Rubin, N., & Heeger, D. J. (2007). Brain areas selective for both observed and executed movements. *Journal of Neurophysiology*, *98*, 1415–1427.

Dinstein, I., Thomas, M., Behmann, M., & Heeger, D. J. (2008). A mirror up to nature. *Current Biology*, *18*, R13–R18.

Emery, N. J., & Clayton, N. S. (2004). The mentality of crows: Convergent evolution of intelligence in corvids and apes. *Science*, *306*, 1903–1907.

Evans, C. S., Evans, L., & Marler, P. (1993). On the meaning of alarm calls: Functional reference in an avian vocal system. *Animal Behaviour*, *46*, 23–38.

Fadiga, L., Fogassi, L., Pavesi, G., & Rizzolatti, G. (1995). Motor facilitation during action observation: A magnetic simulation study. *Journal of Neurophysiology*, *73*, 2608–2611.

Farries, M. A. (2001). The oscine song system considered in the context of the avian brain: Lessons learned from comparative neurobiology. *Brain, Behavior and Evolution*, *58*, 80–100.

Farries, M. A. (2004). The avian song system in comparative perspective. *Annals of the New York Academy of Sciences*, *1016*, 61–76.

Fee, M. S., & Goldberg, J. H. (2011). A hypothesis for basal ganglia–dependent reinforcement learning in the songbird. *Neuroscience*, *198*, 152–170.

Fogassi, L., & Ferrari, P. F. (2004). Mirror neurons, gestures, and language evolution. *Interaction Studies: Social Behaviour and Communication in Biological and Artificial Systems*, *5*, 345–363.

Fogassi, L., & Gallese, V. (2002). The neural correlates of action understanding in non-human primates. In M. I. Stamenov (Ed.), *Mirror neurons and the evolution of brain and language* (pp. 21–43). Philadelphia: John Benjamins.

Gentner, T. Q., Fenn, K. J., Margoliash, D., & Nusbaum, H. C. (2006). Recursive syntactic pattern learning by songbirds. *Nature*, *440*, 1204–1207.

Hewes, G. W. (1973). Primate communication and the gestural origin of language. *Current Anthropology*, *33*, 65–84.

Homberger, D. G. (1986). *The lingual apparatus of the African grey parrot, Psittacus erithacus Linne (Aves: Psittacidae): Description and theoretical mechanical analysis. Ornithological Monographs, No. 39*. Washington, DC: American Ornithologists' Union.

Huber, L., Range, F., Voelkl, B., Szucsich, A., Virányi, Z., & Miklosi, A. (2009). The evolution of imitation: What do the capacities of non-human animals tell us about the mechanisms of imitation? *Philosophical Transactions of the Royal Society B: Biological Sciences*, *364*, 2299–2309.

Jarvis, J. D., Güntürkün, O., Bruce, L., Csillag, A., Karten, H., Kuenzel, W., et al. (2005). Avian brains and a new understanding of vertebrate evolution. *Nature Reviews Neuroscience*, *6*, 151–159.

Jarvis, J. D., & Mello, C. V. (2000). Molecular mapping of brain areas involved in parrot vocal communication. *Journal of Comparative Neurology*, *419*, 1–31.

Jürgens, U. (1998). Neuronal control of mammalian vocalization, with special reference to the squirrel monkey. *Naturwissenschaften*, *85*, 376–388.

Kluender, K. R., Diehl, R. L., & Killeen, P. R. (1987). Japanese quail can learn phonetic categories. *Science*, *237*, 1195–1197.

Kroodsma, D. E. (1988). Song types and their use: Developmental flexibility of the male blue-winged warbler. *Ethology*, *79*, 235–247.

Kroodsma, D. E. (2005). *The singing life of birds* (pp. 96–101). New York: Houghton Mifflin.

Kroodsma, D. E., & Konishi, M. (1991). A suboscine bird (Eastern phoebe, *Sayornis phoebe*) develops normal song without auditory feedback. *Animal Behaviour*, *42*, 477–487.

Kuhl, P. K. (1981). Discrimination of speech by nonhuman animals: Basic auditory sensitivities conducive to the perception of speech-sound categories. *Journal of the Acoustical Society of America*, *70*, 340–349.

Leger, D. W. (2005). First documentation of combinatorial song syntax in a suboscine passerine species. *Condor*, *107*, 765–774.

Lemasson, A., & Hausberger, M. (2011). Acoustic variability and social significance of calls in female Campbell's monkeys (*Cercopithecus campbelli campbelli*). *Journal of the Acoustical Society of America*, *129*, 3341–3352.

Liberman, A. M., & Mattingly, I. G. (1985). The motor theory of speech perception revised. *Cognition*, *21*, 1–36.

Lieberman, P. (1991). Preadaptation, natural selection, and function. *Language & Communication*, *11*, 63–65.

Lieberman, P. (2000). *Human language and our reptilian brain*. Cambridge, MA: Harvard University Press.

Locke, J. L. (1997). A theory of neurolinguistic development. *Brain and Language*, *58*, 265–326.

Mann, Z. F., & Kelly, M. W. (2011). Development of tonotopy in the auditory periphery. *Hearing Research*, *276* (1–2), 2–15. doi:10.1016/j.heares.2011.01.011.

Margoliash, D. (2003). Offline learning and the role of autogenous speech: New suggestions from birdsong research. *Speech Communication*, *41*, 165–178.

Marler, P. (1970). A comparative approach to vocal learning: Song development in white-crowned sparrows. *Journal of Comparative and Physiological Psychology*, *71*, 1–25.

Medina, L., & Reiner, A. (2000). Do birds possess homologues of mammalian primary visual, somatosensory and motor cortices? *Trends in Neurosciences*, *23*, 1–12.

Nottebohm, F. (1980). Brain pathways for vocal learning in birds: A review of the first ten years. *Progress in Psychobiology and Physiological Psychology*, *9*, 85–124.

Ouattara, K., Lemasson, A., & Zuberbühler, K. (2009). Campbell's monkeys concatenate vocalizations into context-specific call sequences. *Proceedings of the National Academy of Sciences of the United States of America*, *106*, 22026–22031.

Patterson, D. K., & Pepperberg, I. M. (1994). A comparative study of human and parrot phonation: Acoustic and articulatory correlates of vowels. *Journal of the Acoustical Society of America*, *96*, 634–648.

Patterson, D. K., & Pepperberg, I. M. (1998). A comparative study of human and grey parrot phonation: Acoustic and articulatory correlates of stop consonants. *Journal of the Acoustical Society of America*, *103*, 2197–2213.

Pepperberg, I. M. (1999). *The Alex studies*. Cambridge, MA: Harvard University Press.

Pepperberg, I. M. (2004). The evolution of communication from an avian perspective. In D. K. Oller & U. Griebel (Eds.), *Evolution of communication systems: A comparative approach* (pp. 171–192). Cambridge, MA: MIT Press.

Pepperberg, I. M. (2005a). Evolution of language from an avian perspective. In M. Tallerman (Ed.), *Language origins: Perspectives on evolution* (pp. 239–261). Oxford: Oxford University Press.

Pepperberg, I. M. (2005b). Insights into vocal imitation in grey parrots (*Psittacus erithacus*). In S. Hurley & N. Chader (Eds.), *Perspectives on imitation: From mirror neurons to memes* (Vol. 1, pp. 243–262). Cambridge, MA: MIT Press.

Pepperberg, I. M. (2007). Emergence of linguistic communication: Studies on grey parrots. In C. Lyon, C. L. Nehaniv, & A. Cangelosi (Eds.), *Emergence of communication and language* (pp. 355–386). London: Springer.

Pepperberg, I. M. (2011). Evolution of communication and language: Insights from parrots and songbirds. In M. Tallerman & K. Gibson (Eds.), *Oxford handbook of language evolution* (pp. 109–119). Oxford: Oxford University Press.

Pepperberg, I. M. (2012). Further evidence for addition and numerical competence by a Grey parrot (*Psittacus erithacus*). *Animal Cognition*, *15*, 711–717.

Pepperberg, I. M., & Schinke-Llano, L. (1991). Language acquisition and use in a bilingual environment: A framework for studying birdsong in zones of sympatry. A peer-reviewed essay on contemporary issues. *Ethology*, *89*, 1–28.

Perkel, D. J. (2004). Origin of the anterior forebrain pathway. *Annals of the New York Academy of Sciences*, *1016*, 736–748.

Person, A. L., Gale, S. D., Farries, M. A., & Perkel, D. J. (2008). Organization of the songbird basal ganglia, including Area X. *Journal of Comparative Neurology*, *508*, 840–866.

Pollick, A. S., & de Waal, F. B. M. (2007). Ape gestures and language evolution. *Proceedings of the National Academy of Sciences of the United States of America, 104*, 8184–8189.

Powell, G. V. N., & Bjork, R. D. (2004). Habitat linkages and the conservation of tropical biodiversity as indicated by seasonal migrations of three-wattled bellbirds. *Conservation Biology, 18*, 500–509.

Rivera-Cáceres, K., Garcia, C. M., Quirós-Guerrero, E., & Ríos-Chelén, A. A. (2011). An interactive playback experiment shows song bout size discrimination in the suboscine vermilion flycatcher (*Pyrocephalus rubinus*). *Ethology, 117*, 1120–1127.

Rizzolatti, G., & Arbib, M. (1998). Language within our grasp. *Trends in Neurosciences, 21*, 188–194.

Rizzolatti, G., Fogassi, L., & Gallese, V. (2001). Neurophysiological mechanisms underlying the understanding and imitation of actions. *Nature Reviews Neurology, 2*, 661–670.

Roush, R. S., & Snowdon, C. T. (2001). Food transfer and development of feeding behavior and food-associated vocalizations in cotton-top tamarins. *Ethology, 107*, 415–429.

Saranathan, V., Hamilton, D., Powell, G. V. N., Kroodsma, D. E., & Prum, R. O. (2007). Genetic evidence supports song-learning in the three-wattled bellbird *Procnias trucarunculata* (Cotingidae). *Molecular Ecology, 16*, 3689–3702.

Seyfarth, R., Cheney, D., & Marler, P. (1980). Monkey responses to three different alarm calls: Evidence for predator classification and semantic communication. *Science, 210*, 801–803.

Smith, W. J. (1997). The behavior of communicating, after twenty years. In D. H. Owings, M. D. Beecher, & N. S. Thompson (Eds.), *Perspectives in ethology* (Vol. 12, pp. 7–53). New York: Plenum Press.

Smith, W. J., & Smith, A. M. (1992). Behavioral information provided by two song forms of the Eastern kingbird, *T. tyrannus. Behaviour, 120*, 90–102.

Smith, W. J., & Smith, A. M. (1996). Information about behavior provided by Louisiana waterthrush, *Seurus motacilla* (Parulinae), songs. *Animal Behaviour, 51*, 785–799.

Snow, B. K. (1977). Territorial behaviour and courtship of the male three-wattled bellbird. *Auk, 94*, 623–645.

Snow, D. W. (1973). Distribution, ecology, and evolution of the bellbirds (*Procnias*, Cotingidae). *Bulletin of the British Museum of Natural History, 25*, 369–391.

Snow, D. W. (1982). *The cotingas*. Ithaca, NY: Cornell University Press.

Stevens, K. N. (1998). *Acoustic phonetics*. Cambridge, MA: MIT Press.

Struhsaker, T. (1967). Auditory communication among vervet monkeys (*Ceropithecus aethiops*). In S. Altmann & K. Gibson (Eds.), *Social communication among primates* (pp. 281–324). Chicago: University of Chicago Press.

Studdert-Kennedy, M. (2005). How did language go discrete? In M. Tallerman (Ed.), *Language origins: Perspectives on evolution* (pp. 48–67). Oxford: Oxford University Press.

Tallerman, M., & Gibson, K. (Eds.). (2011). *Oxford handbook of language evolution*. Oxford: Oxford University Press.

Thorpe, W. H. (1963). *Learning and instinct in animals*. London: Methuen.

Thorpe, W. H. (1974). *Animal and human nature*. New York: Doubleday.

Vauclair, J. (2004). Lateralization of communicative signals in nonhuman primates and the hypothesis of the gestural origin of language. *Interaction Studies: Social Behaviour and Communication in Biological and Artificial Systems, 5*, 365–386.

Vihman, M. H. (1993). Variable paths to early word production. *Journal of Phonetics, 21*, 61–82.

Visalberghi, E., & Fragaszy, D. M. (1990). Do monkeys ape? In S. T. Parker & K. R. Gibson (Eds.), *"Language" and intelligence in monkeys and apes* (pp. 247–273). Cambridge: Cambridge University Press.

Visalberghi, E., & Fragaszy, D. M. (2002). "Do monkeys ape?" Ten years after. In K. Dautenhahn & C. L. Nehaniv (Eds.), *Imitation in animals and artifacts* (pp. 471–499). Cambridge, MA: MIT Press.

Voelkl, B., & Huber, L. (2000). True imitation in marmosets. *Animal Behaviour, 60*, 195–202.

Wich, S. A., & Sterck, E. A. (2003). Possible audience effect in Thomas langurs (*Presbytis thomasi*): An experimental study on male loud calls in response to a tiger model. *American Journal of Primatology, 60*, 155–159.

Wild, M. (1997). Neural pathways for the control of birdsong production. *Journal of Neurobiology, 33*, 653–670.

Wild, M., Arends, J. J. A., & Zeigler, H. P. (1985). Telencephalic connections of the trigeminal system in the pigeon (*Columba livia*): A trigeminal sensorimotor circuit. *Journal of Comparative Neurology, 234*, 441–464.

Williams, H. (1989). Multiple representations and auditory-motor interactions in the avian song system. *Annals of the New York Academy of Sciences, 563*, 148–164.

Williams, H., & Nottebohm, F. (1985). Auditory responses in avian vocal motor neurons: A motor theory for song perception in birds. *Science, 229*, 279–282.

Zuberbühler, K., Noë, R., & Seyfarth, R. M. (1997). Diana monkey long-distance calls: Messages for conspecifics and predators. *Animal Behaviour, 53*, 589–604.

List of Contributors

Hermann Ackermann Department of General Neurology, Hertie Institute for Clinical Brain Research, University of Tübingen

Gabriël J. L. Beckers Department of Behavioural Neurobiology, Max Planck Institute for Ornithology

Robert C. Berwick Department of Electrical Engineering and Computer Science and Brain and Cognitive Sciences, Massachusetts Institute of Technology

Johan J. Bolhuis Cognitive Neurobiology and Helmholtz Institute, Departments of Psychology and Biology, Utrecht University

Noam Chomsky Department of Linguistics and Philosophy, Massachusetts Institute of Technology

Frank Eisner Institute of Cognitive Neuroscience, University College London

Martin Everaert Utrecht Institute of Linguistics OTS, Utrecht University

Michale S. Fee McGovern Institute for Brain Research, Department of Brain and Cognitive Sciences, Massachusetts Institute of Technology

Olga Fehér Department of Psychology, Hunter College

Simon E. Fisher Language and Genetics Department, Max Planck Institute for Psycholinguistics, and Donders Institute for Brain, Cognition and Behaviour, Radboud University Nijmegen

W. Tecumseh Fitch Department of Cognitive Biology, University of Vienna

Jonathan B. Fritz Center for Auditory and Acoustic Research, Institute for Systems Research, University of Maryland College Park

Sharon M. H. Gobes Neuroscience Program, Wellesley College

Riny Huybregts Department of Linguistics, Utrecht University

Erich D. Jarvis Howard Hughes Medical Institute, Department of Neurobiology

Robert Lachlan Department of Biology, Duke University

Ann Law   School of Law, Queen Mary, University of London

Michael A. Long   Department of Psychology & Neuroscience, NYU School of Medicine

Gary F. Marcus   Department of Psychology, New York University

Carolyn McGettigan   Institute of Cognitive Neuroscience

Daniel Mietchen   EvoMRI Communications

Richard Mooney   Department of Neurobiology, Duke University

Sanne Moorman   Cognitive Neurobiology and Helmholtz Institute, Departments of Psychology and Biology, Utrecht University

Kazuo Okanoya   Department of Cognitive and Behavioral Sciences, University of Tokyo

Christophe Pallier   Unité de Neuroimagerie Cognitive INSERM-CEA, Neurospin Center

Irene M. Pepperberg   Department of Psychology, Harvard University

Jonathan F. Prather   Department of Zoology and Physiology, Program in Neuroscience, University of Wyoming

Franck Ramus   Laboratoire de Sciences Cognitives et Psycholinguistique, Ecole Normale Supérieure, EHESS, CNRS, DEC-ENS

Eric Reuland   Utrecht Institute of Linguistics OTS, Utrecht University

Constance Scharff   Animal Behavior, Freie Universität Berlin

Sophie K. Scott   Institute of Cognitive Neuroscience, University College London

Neil Smith   Research Department of Linguistics, University College London

Ofer Tchernichovski   Department of Psychology, Hunter College,

Carel ten Cate   Behavioural Biology, Leiden University

Christopher K. Thompson   Department of Cell Biology, Scripps Research Institute

Frank Wijnen   Utrecht Institute of Linguistics OTS, Utrecht University

Moira Yip   Department of Linguistics, Division of Psychology and Language Sciences, University College London

Wolfram Ziegler   EKN—Clinical Neuropsychology Research Group, Clinic for Neuropsychology, City Hospital

Willem Zuidema   Institute for Logic, Language and Computation, University of Amsterdam

Index